ATMOSPHERIC REMOTE SENSING

EARTH OBSERVATION

ATMOSPHERIC REMOTE SENSING

Principles and Applications

Edited by

ABHAY KUMAR SINGH

Department of Physics, Banaras Hindu University, India

SHANI TIWARI

CSIR - National Institute of Oceanography, Goa, India

ELSEVIER

Elsevier
Radarweg 29, PO Box 211, 1000 AE Amsterdam, Netherlands
The Boulevard, Langford Lane, Kidlington, Oxford OX5 1GB, United Kingdom
50 Hampshire Street, 5th Floor, Cambridge, MA 02139, United States

ISBN: 978-0-323-99262-6

For Information on all Elsevier publications visit our website at
https://www.elsevier.com/books-and-journals

Publisher: Candice G Janco
Acquisitions Editor: Peter Llewellyn
Editorial Project Manager: Rupinder K Heron
Production Project Manager: R. Vijay Bharath
Cover Designer: Mark Rogers

Typeset by Aptara, New Delhi, India

CONTENTS

Contributors

Md. Arfan Ali
School of Marine Sciences, Nanjing University of Information Science and Technology (NUIST), Nanjing, China

N. Arun
Department of Civil Engineering, Indian Institute of Technology Madras, Chennai, India

Muhammad Bilal
School of Surveying and Land Information Engineering, Henan Polytechnic University, Jiaozuo, China; School of Marine Sciences, Nanjing University of Information Science and Technology (NUIST), Nanjing, China

H.C. Chandola
Department of Physics (UGC-Centre of Advanced Study), Kumaun University, Nainital, India

Shivam Kumar Chaubey
Atmospheric Research Laboratory, Department of Physics, Institute of Science, Banaras Hindu University, Varanasi, Uttar Pradesh, India

Prashant Kumar Chauhan
Atmospheric Research Laboratory, Department of Physics, Institute of Science, Banaras Hindu University, Varanasi, Uttar Pradesh, India

Sagnik Dey
Centre for Atmospheric Sciences, Indian Institute of Technology, Delhi, India

S.K. Dhaka
Radio and Atmospheric Physics Lab., Rajdhani College, University of Delhi, India

Soumi Dutta
Centre for Atmospheric Sciences, Indian Institute of Technology, Delhi, India

Larry Di Girolamo
Department of Atmospheric Sciences, University of Illinois at Urbana-Champaign, United States

Nishita Jaiswal
Remote Sensing Laboratory, Institute of Environment and Sustainable Development, Banaras Hindu University, Varanasi, Uttar Pradesh, India

Manish Jangid
School of Environmental Sciences, Jawaharlal Nehru University, New Delhi, India

Raja Obul Reddy Kalluri
Aerosol & Atmospheric Research Laboratory, Department of Physics, Sri Krishnadevaraya University, Anantapur, Andhra Pradesh, India

Vijay P. Kanawade
Centre for Earth, Ocean and Atmospheric Sciences, University of Hyderabad, Hyderabad, India

Yogesh Kant
Marine & Atmospheric Sciences Department, Indian Institute of Remote Sensing, ISRO, Dehradun, India

S. Kishore
Calaver Technologies Pvt. Ltd., Bengaluru, India

Rama Gopal Kotalo
Aerosol & Atmospheric Research Laboratory, Department of Physics, Sri Krishnadevaraya University, Anantapur, Andhra Pradesh, India

Vinay Kumar
Radio and Atmospheric Physics Lab., Rajdhani College, University of Delhi, India

Manish Kumar
Department of Environmental Science and Analytical Chemistry (ACES), Stockholm University, Stockholm, Sweden

Sushil Kumar
School of Information Technology, Engineering, Mathematics and Physics, The University of the South Pacific, Suva, Fiji

Harshbardhan Kumar
Academy of Scientific and Innovative Research (AcSIR), Ghaziabad, India; CSIR – National Institute of Oceanography, DonaPaula, Goa, India

Sarvan Kumar
Department of Earth and Planetary Sciences, Veer Bahadur Singh Purvanchal University, Jaunpur, India

Sanjay Kumar
Atmospheric Research Laboratory, Department of Physics, Institute of Science, Banaras Hindu University, Varanasi, Uttar Pradesh, India

Subodh Kumar
Department of Earth and Atmospheric Sciences, National Institute of Technology Rourkela, Rourkela, Odisha, India

Akhilesh Kumar
Atmospheric Research Laboratory, Department of Physics, Institute of Science, Banaras Hindu University, Varanasi, Uttar Pradesh, India; Kashi Naresh Government Post Graduate College Gyanpur, Bhadohi, India

Anil Kumar
Photogrammetry and Remote Sensing Department Indian Institute of Remote Sensing ISRO, Govt. of India, New Delhi, India

Anuska Kumari
GIS Engineer, National Informatics Centre, New Delhi, India

Ajeet Kumar Maurya
Department of Physics, Babasaheb Bhimrao Ambedkar University, Lucknow, Uttar Pradesh, India

Akansha Mehra
Independent Researcher, New Delhi, India

Alaa Mhawish
School of Marine Sciences, Nanjing University of Information Science and Technology (NUIST), Nanjing, China

Varun Narayan Mishra
Amity Institute of Geoinformatics & Remote Sensing (AIGIRS), Amity University, Noida, Gautam Buddha Nagar, India

Amit Kumar Mishra
School of Environmental Sciences, Jawaharlal Nehru University, New Delhi, India

Smrutisikha Mohanty
Remote Sensing Laboratory, Institute of Environment and Sustainable Development, Banaras Hindu University, Varanasi, Uttar Pradesh, India; Department of Environmental Studies, School of Interdisciplinary and Applied Sciences, Central University of Haryana, Haryana, India

Mona Sharma
Remote Sensing Laboratory, Institute of Environment and Sustainable Development, Banaras Hindu University, Varanasi, Uttar Pradesh; Department of Environmental Studies, School of Interdisciplinary and Applied Sciences, Central University of Haryana, Haryana, India

Mukulika Mondal
Atmospheric Research Laboratory, Department of Physics, Institute of Science, Banaras Hindu University, Varanasi, Uttar Pradesh, India

Jagabandhu Panda
Department of Earth and Atmospheric Sciences, National Institute of Technology Rourkela, Rourkela, Odisha, India

Kalpana Patel
SRM Institute of Science and Technology, Delhi-NCR Campus, Modinagar, Ghaziabad, Uttar Pradesh, India

Debashis Paul
Department of Earth and Atmospheric Sciences, National Institute of Technology Rourkela, Rourkela, Odisha, India

Swagata Payra
Department of Remote Sensing, Birla Institute of Technology Mesra, Ranchi, Jharkhand, India

Subhendu Pradhan
Department of Geography, Institute of Science, Banaras Hindu University, Varanasi, India

Vineet Pratap
Atmospheric Research Laboratory, Department of Physics, Institute of Science, Banaras Hindu University, Varanasi, Uttar Pradesh, India

Zhongfeng Qiu
School of Marine Sciences, Nanjing University of Information Science and Technology (NUIST), Nanjing, China

Rabi-ul-Islam
Haryana Space Applications Centre, Hisar, India

Praveen Kumar Rai
Department of Geography, K.M.C. Language University, Lucknow, India

S.S. Rao
Atmospheric Research Laboratory, Department of Physics, Institute of Science, Banaras Hindu University, Varanasi, Uttar Pradesh, India

Deependra Singh Rawat
Department of Allied Sciences (Physics), Graphic Era Hill University, Bhimtal Campus, Nainital, India

Rajeeb Samanta
Department of Geography, Prabhat Kumar College, Purba, Medinipur, West-Bengal, India

Chandan Sarangi
Department of Civil Engineering, Indian Institute of Technology Madras, Chennai, India

M. Sateesh
Climate Change Centre, King Abdullah University of Science and Technology, Thuwal, Saudi Arabia

Gopi Krishna Seemala
Indian Institute of Geomagnetism, Navi Mumbai, India

Arjun Pratap Shahi
Independent Researcher, Varanasi, India

Darga Saheb Shaik
Marine & Atmospheric Sciences Department, Indian Institute of Remote Sensing, ISRO, Dehradun, India; National Atmospheric Research Laboratory, Department of Space, Gadanki, India

Vikram Sharma
Department of Geography, Banaras Hindu University, Varanasi, India

Ajay Sharma
Department of Environment and Sustainable Development, Banaras Hindu University, Varanasi, Uttar Pradesh, India

Shubhra Sharma
Geosciences Division, Physical Research Laboratory, Ahmedabad, India

Megha Sharma
Department of Computer Science and Engineering, Amrita School of Computing, Amrita Vishwa Vidyapeetham, Bengaluru, India

A.K. Singh
Atmospheric Research Laboratory, Department of Physics, Institute of Science, Banaras Hindu University, Varanasi, Uttar Pradesh, India

S.B. Singh
Atmospheric Research Laboratory, Department of Physics, Institute of Science, Banaras Hindu University, Varanasi, Uttar Pradesh, India

Rajesh Singh
KSK Geomagnetic Research Laboratory, IIG, Prayagraj (Allahabad), India

Vivek Singh
Indian Institute of Tropical Meteorology, Ministry of Earth Sciences, Government of India, New Delhi, India

Abhay Kumar Singh
Atmospheric Research Laboratory, Department of Physics, Institute of Science, Banaras Hindu University, Varanasi, Uttar Pradesh, India; DST - Mahamana Centre of Excellence in Climate Change Research, Banaras Hindu University, Varanasi, India

Prafull Singh
Department of Geology, School of Earth, Biological and Environmental Science, Central University of South Bihar, Gaya, India

Prashant K. Srivastava
Remote Sensing Laboratory, Institute of Environment and Sustainable Development, Banaras Hindu University, Varanasi, Uttar Pradesh, India

Atul Kumar Srivastava
Indian Institute of Tropical Meteorology, Ministry of Earth Sciences, Government of India, New Delhi, India

Swati Suman
Remote Sensing Laboratory, Institute of Environment and Sustainable Development, Banaras Hindu University, Varanasi, Uttar Pradesh, India

M. Supriya
Department of Computer Science and Engineering, Amrita School of Computing, Amrita Vishwa Vidyapeetham, Bengaluru, India

Abin Thomas
Centre for Earth, Ocean and Atmospheric Sciences, University of Hyderabad, Hyderabad, India

Shani Tiwari
Academy of Scientific and Innovative Research (AcSIR), Ghaziabad, India; CSIR – National Institute of Oceanography, DonaPaula, Goa, India

Gaurish Tripathi
Atmospheric Research Laboratory, Department of Physics, Institute of Science, Banaras Hindu University, Varanasi, Uttar Pradesh, India

Sunita Verma
Department of Environment and Sustainable Development, Banaras Hindu University, Varanasi, Uttar Pradesh, India

Gerrit de Leeuw
Royal Netherlands Meteorological Institute (KNMI), R & D Satellite Observations, De Bilt, The Netherlands; Aerospace Information Research Institute, Chinese Academy of Sciences (AirCAS), Beijing, China; School of Environment Science and Spatial Informatics, University of Mining and Technology, Xuzhou, Jiangsu, China

CHAPTER 1

Composition and thermal structure of the earth's atmosphere

S.K. Dhaka, Vinay Kumar
Radio and Atmospheric Physics Lab., Rajdhani College, University of Delhi, India

1.1 Atmosphere

The earth's atmosphere is surrounded by a thin layer of gaseous environment, which is retained by the gravity. The atmospheric air is commonly known as mixture of different gases, containing on a dry molar basis ~78% of molecular nitrogen (N_2), ~21% of molecular oxygen (O_2), ~1% of argon (Ar), ~.04 % of carbon dioxide, and other multitude of trace species. Water vapor (H_2O) is present at a highly variable concentration that contribute up to a few percent over tropical oceans. In addition to these gases some fine liquid and solid particles are present in the air, named aerosol particles. Nearly 50% mass of the air is present below 5.5 km altitude and 90% below 16 km [1].

Essentially, it is the absorption of the solar radiation, which determines temperature of the atmosphere as a function of altitude due to selective absorption of certain wavelength in distinct regions. For instance, near about 100 km, solar radiation of very short wavelength is absorbed by the molecular oxygen and resulting photo chemical process, which is responsible for high temperature. In the tropical region, between the altitudes 35–70 km, the ultraviolet radiation is absorbed by ozone (O_3) giving rise to another temperature which is maximized near 55 km. In the lower atmosphere (less than 15 km) re-radiated IR part of the spectrum is absorbed by water and CO_2, and greenhouse gases. Infrared (IR) part lies in the complex vibrational-rotational spectrum of these trace gases [2]. For the processes cited above, following description is important to understand.

Solar radiation from the sun is the main source of energy for the earth's system. The sun emits radiation as a black body at effective temperature T_S = 5800 K. The energy flux corresponding to this temperature is

$$F_S = \sigma T_{s^4} \qquad (1.1a)$$

where σ = Stefan-Boltzmannconstant = 5.67 × 10^{-8} $Wm^{-2}K^{-4}$

The incoming radiation extends over all wavelengths but peaks in the visible (0.5 μm). The incident solar energy flux at the earth's mean distance from the sun is nearly 1368 W m^{-2}. This intercepted quantity is called the solar constant and denoted by F_S. Thus, the total radiation per unit time on the area cross section πa^2 is therefore $F_S \pi a^2$. Nearly 30–35%

Atmospheric Remote Sensing
DOI: https://doi.org/10.1016/B978-0-323-99262-6.00023-7

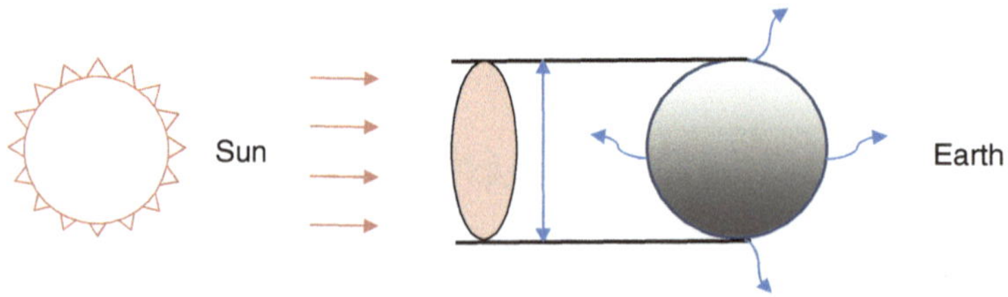

Fig. 1.1 ***Illustration for the incoming and outgoing radiation from earth's surface.*** **The parallel light arrows indicate incoming short wave (SW) solar radiation confined within cross section πa^2. The wavy arrows, on the right, indicate the thermal outgoing long wave radiation (OLR) from the total surface area $4\pi a^2$.**

of this energy is reflected to space by atmospheric system; this is called the planetary albedo (a). So, the earth's reflected back a $F_S \pi a^2$ of the incoming radiation and rest (1–a) $F_S \pi a^2$ is absorbed by its surface. This absorbed energy is emitted by the earth's as IR radiation of low energy (i.e., high wavelength) in all the direction from the cross-section $4\pi a^2$. The emitted energy from the earth's surface is called outgoing long wave radiation (OLR). OLR is a critical component of the earth's radiation budget and represents the total radiation going to space emitted by the atmosphere. According to the Stefan-Boltzmann law, the flux emitted by the earth's surface at temperature T_E amounts to be σT_{E^4}, therefore total emitted energy per unit time will be $4\pi a^2 \sigma T_{E^4}$. According to simple radiative model, in the thermal equilibrium, the total incoming and outgoing radiations must balance, that is, $(1-a) F_S \pi a^2 = 4\pi a^2 \sigma T_{E^4}$. By making use of the values of a and F_S, the mean temperature of earth's surface should be around $T_E \approx 255$ K. This value is found lower than the actual value of 288 K due to the exclusion of trapping effect from the greenhouse gases in this simple radiative model. Therefore, we must include the heating due to the presence of greenhouse gases. The greenhouse gases such as methane (CH_4), nitrogen dioxide (N_2O), water vapor (H_2O), and carbon dioxide (CO_2) absorb certain wavelengths of OLR adding heat to the atmosphere. The OLR is dependent on the temperature of the radiating body. It is affected by the earth's skin temperature, skin surface emissivity, atmospheric temperature, water vapor profile, cloud cover, and moisture [20]. The atmosphere's temperature is the most important property controlling its structure, and it is modulated by the different factor at different altitude level. A study on the thermal structure is significant for exploring the atmospheric behavior. The vertical temperature structure of earth's atmosphere is explored in the following section (Fig. 1.1). Fig 1.1 shows how short wave solar radiation converts into long wave radiation after reflection from the earth's surface.

1.2 Vertical temperature structure and nomenclature of the earth's atmosphere

The vertical temperature profile of the atmosphere in a steady state depends upon a balance at each level of the divergence of heat fluxes due to radiative transfer and

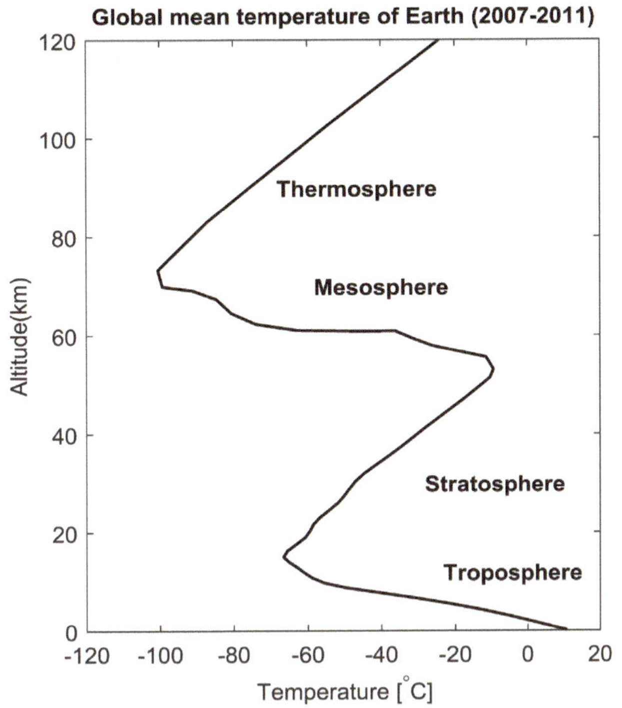

Fig. 1.2 ***Global mean temperature profile of the earth's atmosphere using data from 2007 to 2011.*** COSMIC and MLS satellite data are used to show mean temperature profile.

heat transfer by atmospheric motions. Atmospheric scientists made a partition the atmosphere vertically on the basis of this thermal structure. On the basis of temperature structure, the atmosphere can be divided mainly in four distinct layers namely as "troposphere" (~0–18 km), "stratosphere" (~18–50 km), "mesosphere" (~50–90 km) and "thermosphere" (above 90 km). The vertical structure of atmosphere is shown in Fig. 1.2, confined up to 120 km, based on the Microwave Limb Sounder (hereinafter MLS) satellite data [3]. The subsequent subsections provide a briefly discussion on each layer of the earth's atmosphere, later the main focus will be on the details of temperature variability in the troposphere and stratosphere.

1.2.1 Troposphere

Troposphere layer starts from the surface of earth and extends up to ~8–18 km that depends upon the latitude, longitude, and seasons. In equatorial region, it has maximum height nearly ~18 km and in polar region it shrinks up to minimum height ~7 km. In this layer, the temperature decreases as altitude increases with average laps rate 7°C/km in the wet condition. The troposphere contains near about 75–80% mass of the atmosphere's and all essentially such as water vapor and primary greenhouse gases. The tropospheric temperature is governed by H_2O, and greenhouse gases due to radiative and convective exchange. The vertical variation in the troposphere is controlled by convection carrying heat up from the lower surface to upward direction. Almost all weather phenomena and clouds formation take place in this region. From surface to top, the air density in this region decreases from ρ_0(~1.29 kg m^{-3}) to $\frac{\rho_0}{10}$. The layer that separates troposphere to next layer the stratosphere is called tropopause.

1.2.2 Stratosphere

The layer next to the tropopause is called the stratosphere which is extended up to ~50 km. The stratosphere is less dense, and dry in compare to the troposphere. This layer contains 20–25% mass of total atmosphere. The formation of the O_3 layer occurs in this region due to stability and absence of water vapor. O_3 is a primary absorber for solar UV radiations and thus shields our life at the surface from the damaging effects of this radiation. In reverse to the troposphere, the temperature in this region increases due to the heating from the absorption of UV radiation by O_3 and get maximize at the top because of less density. In this region, the density decreases from $\frac{\rho_0}{10}$ to $\frac{\rho_0}{10^3}$. The layer that separates stratosphere to next layer the mesosphere is called stratopause.

1.2.3 Mesosphere

The layer next to the stratopause is called the mesosphere which is extended up to maximum ~80 km. The mesospheric layer contains nearly 1% mass of total atmosphere. In this layer, very small amount of the ozone is available to absorb solar UV radiation but the radiative cooling by CO_2 is still effective, and due to this cooling effect, the temperature decreases again with altitude. In contrast to stratosphere, the mesospheric layer is unstable due to convective currents. The turbulence and wavy structure occur frequently in this region. The lower frequency gravity waves (GWs) from lower atmosphere break here in spatially and temporally variable patches [4–8]. The breaking of GWs makes this region thermally unstable. In this region, density decreases from $\frac{\rho_0}{10^3}$ to $\frac{\rho_0}{10^5}$. The layer that separates mesosphere to next layer the thermosphere is called the mesopause.

1.2.4 Thermosphere

The region above mesopause (90 km) is called thermosphere. The temperature in this region is very high and variable. In this region, very short wavelength UV radiations are absorbed by molecular oxygen O_2, molecular nitrogen N_2, and atomic oxygen O, and collectively it produces the heating. Molecules like CO_2 and O_2 are dissociated by high energy UV radiations ($\lambda < 0.1$ μm). The air of this region is so tenuous that assumptions of local thermodynamic equilibrium, as in black-body radiation, are not applicable. Collisions are so rare, that a stable population of ions can be sustained resulting plasma (ionized gas). In the region of the atmosphere, light atoms (such as hydrogen) can overcome the gravitational force and escape into outer space, if their velocity exceeds to the threshold value (escape velocity), and at that point the atmosphere effectively merges with outer space.

However, as mentioned above, in this chapter the main focus would remain on the troposphere and stratosphere; and how the thermal structure in this important part of the atmosphere modulated by the large-scale dynamical motions is shown. The temperature profile of earth's lower atmosphere (troposphere and stratosphere) reflects a balance between the convective, radiative, and dynamical heating/cooling of the surface-atmosphere system. The tropospheric and stratospheric regions are dynamically coupled to each other both on the latitudinal as well as longitudinal scale, hence, this coupling influences the temperature jointly. The troposphere is a layer where all weather phenomena occur, but this is also influenced by the stratospheric variation on seasonal and interannual scale.

1.3 Basics climatology of tropospheric and stratospheric region and inter-hemispheric coupling

Fig. 1.3 presents the basic climatology (covering years from 2007 to 2011) of temperature structure in the troposphere and stratosphere. The vertical temperature from the COSMIC data [9] is shown with altitude from southern hemisphere (SH) to northern hemisphere (NH) at every 5° latitude bands for two different seasons, winter (January) and summer (July).

The vertical temperature profiles (Fig. 1.3) show different seasonal characteristics in SH and NH, and in mid and higher latitudes bands. The curvature in the profiles depicts the latitudinal and geographical characteristic of the region. It is not necessary that tropopause is the coldest temperature; in fact, beyond the extra-tropical region toward polar region the coldest temperature can be seen in the stratospheric region, especially during winter. In general, in the higher latitudinal zone (30° and 45°) of both the hemispheres, two inversion temperatures are noticed with the coldest temperature occurs at the upper side; these minima are separated vertically by 3–6 km altitude. During winter season, in the latitude regions of 45–90°, a thick layer of coldest temperature in the stratospheric region is seen, which is located around 25 km [22]. The stronger seasonal variations in the solar radiation, its zenith angle, are the main causes for the cooler stratosphere in compare to tropopause. The occurrence of the coldest temperature in stratospheric region indicates that stratosphere radiative balancing is crucial for thermal balancing within this region.

Further, due to the uneven heating in the tropical, mid-latitudes, and polar region, the air mass circulations exist from tropical to polar side in both the hemispheres that is known as Brewer-Dobson circulation (BDC). The uneven heating creates the temperature gradient between topical and polar region. During the winter season of each

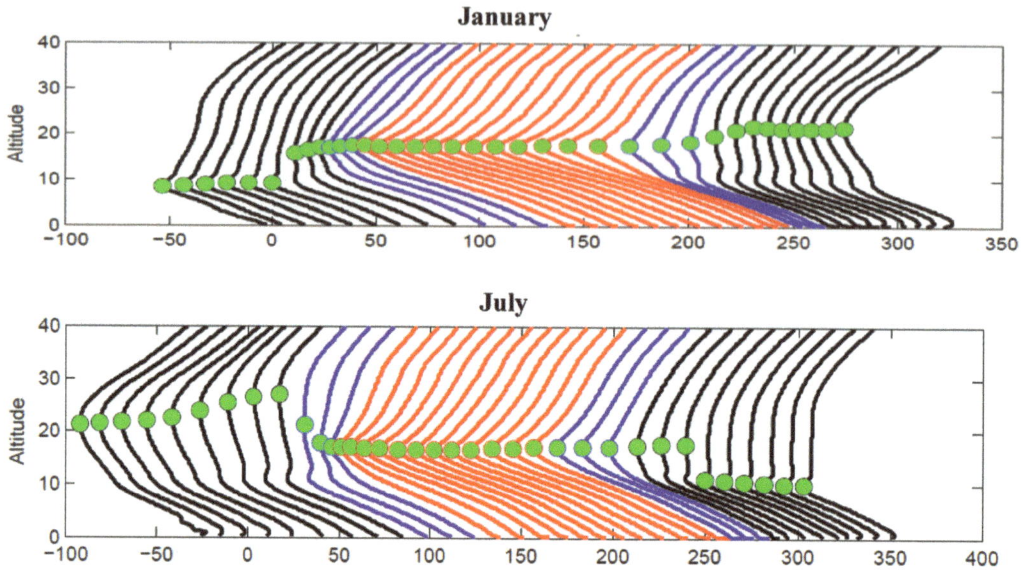

Fig. 1.3 ***The climatology (2007–2011) of vertical temperature structure of earth's atmosphere with altitude from SH to NH at every 5° band for two different seasons.*** Large number of profiles over land and oceans were used to create this structure; COSMIC satellite data are used. The coldest temperature point, in each profile identified, is shown with solid green dots. Red color used for tropical and extratropical 30°S and 30°N region. Blue and black color for 30°–45° and 45°–90° in each hemisphere, respectively. Each profile is shifted by 10°C from its preceding profile, and 85°S–90°S latitude band is a starting profile on the left side. [Adapted from Fig. 2 of [22] with permission].

hemisphere, the maximum temperature gradient exists between tropical and polar region, and due to that, the warm tropical air rises from troposphere to stratosphere. Further these air masses get transported towards high latitudes stratospheric region. While traveling, this air mass in the mid-latitude's stratospheric region partly descends back to the troposphere; way further ahead in the polar region the air mass accumulates in the lower stratosphere. In the absence of BDC, the coldest stratospheric temperature in the polar region should have been nearly ~ −110°C, but due to the BDC it is much higher [10].

It can also be noted from Fig. 1.3 that during winter, the coldest stratospheric temperature in SH polar region drops down to <−92°C (around 30 km), on the other hand in the NH, there is a little warmer coldest temperature of the order of ~ −80°C at slightly lower height (around 25 km). This contrast behavior indicates that NH lower stratospheric region is warmer than SH. Due to different topography of SH and NH, and different heat capacity of ocean and land, the BDC is stronger in NH in comparison to SH. Thus, NH polar lower stratosphere is warmer than SH because of the asymmetric magnitudes of BDC. The extreme low temperature (<−90°C) in the lower stratosphere of SH is suitable for formation of polar stratospheric cloud (PSC) which

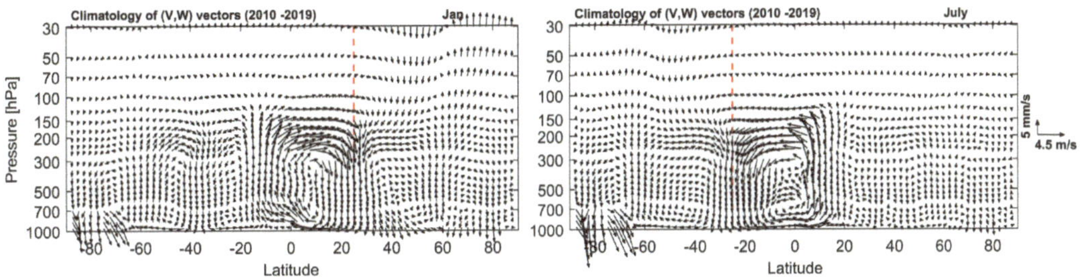

Fig. 1.4 ***The climatology (2010–2019) of meridional and vertical (V, W) wind in the latitude domain of 90°S to 90°N created using ERA-5 data for the two different seasons, winter (January) and summer (July).*** The vertical red dashed lines indicate the 25°N and 25°S during January, and July, respectively.

provides the surface for chemical reactions between chlorofluorocarbons (CFCs) and O_3 that depletes the O_3. This is the reason for the O_3 hole occurrence in the SH lower stratospheric polar region. Note that PSC does not exist in polar stratospheric region of NH because of warmer temperature (~ −80°C) due to the stronger BDC.

The source of CFCs also exists in the tropical region; BDC supports transportation of these species to the polar region. Hence, the BDC is playing a crucial role to modulate the temperature, dynamics, and chemistry from the tropical region to the polar region. Different latitude bands from tropical to polar regions are not only coupled via stratospheric BDC but also a remarkable coupling exists within the troposphere. In fact, the uneven heating in the tropical and polar region is responsible for developing such tropospheric circulations. Fig. 1.4 shows the basic climatology (covering period from 2010 to 2019) of meridional and vertical (V, W) wind vectors plotted in the latitude domain 90°S to 90°N for the two different seasons, winter (January) and summer (July). ERA-5 data [11] for the two different seasons are used to create the distinction.

On the basis of coupling in the different latitudes zone the tropospheric circulations can be divided into three cells, for example, Hadley cell, Ferrel cell, and the Polar cell, which are described briefly in the following.

1.3.1 Hadley cell

The Hadley cell exists from the equator to extra-topical (30°) region in both the hemispheres. The earth's surface is intensely heated by the sun near equator/tropical belt. Thus, warm air rises from the equator and creating the low-pressure region. As the air goes up it cools in the upper troposphere (300 hPa to 100 hPa) and spreads in both the hemispheres. Interestingly, the stronger air circulation takes place in the winter hemisphere (Fig. 1.4), that is, in NH during boreal winter, and SH during austral winter. Gradually, when air mass arrives around ~30 °N and ~30 °S, the cool air mass sinks toward earth's surface and thus develop the subtropical high-pressure zone. This part of the air mass flow completes the cycle from subtropical high to the equator and termed as a trade wind. Hadley cell depicts the asymmetrical variation in both the hemispheres happening due to different topography, and different heat capacity of ocean and land. In the NH, Hadley cell is stronger with wider coverage than in the SH (Fig. 1.4).

1.3.2 Ferrel cell

The Ferrel cell exists in the mid and high latitudes band prominently between 30° and 60° in both the hemispheres. Air mass in the subtropical high-pressure zone also moves toward the polar region, where ever this air mass wind travels over the oceans, it picks up the moisture, and on the way toward pole meets the cold air coming from the polar region around 60°N/S. As the air from the tropics bearing less density with warmth merge with the cold polar air mass, it creates the uprising air mass. As cited above, this newly formed uprising air mass completes the cycle as it gets cool while reaching in the upper troposphere and starts flowing back to the tropical latitude region Fig 1.4.

1.3.3 Polar cell

The Polar cell is the weakest among three cells and exists in the high latitudes band between 60° and 90° in both the hemispheres. The polar air mass is colder and dense thus forming the high-pressure belt, which is known as polar high. Air mass flow from the polar high to mid-latitudes joining the warm air mass near 60° and this scenario is making it rising upward. As it rises in the upper troposphere, this air mass is sufficiently cool and hence it moves over a path toward polar as well as the tropical region. Note the air mass completes the cycle in the polar region. The surface boundary around 60° N/S where the tropical warm and polar cold air meets is known as polar front.

Next section describes the mathematical formulation for the variation of temperature and pressure with altitudes in the tropospheric region.

1.4 Vertical variation of atmospheric temperature in the troposphere

Earth's acquires heat from the sun rays, thus air immediately above the surface gets the maximum heat and attain the maximum temperature; this process termed as conduction. Transfer of the heat from the lower to the upper layers is called convection. Thus, the atmosphere is heated from below and its temperature within the troposphere is determined by convection. When air mass rises in the vertical direction in the troposphere, it expands as pressure decreases with altitude. In the process of expansion, it uses its own internal energy (i.e., adiabatic expansion) resulting decrease in the temperature in vertical direction.

Let us consider a small air parcel thickness dz and surface area A at altitude z as shown in Fig. 1.5. If ρ is the density of air parcel. Fig 1.5 describes the situation of the air parcel in the atmosphere.

$$\text{mass of air parcel} = \rho A dz$$

$$\text{weight of air parcel} = \rho g A dz$$

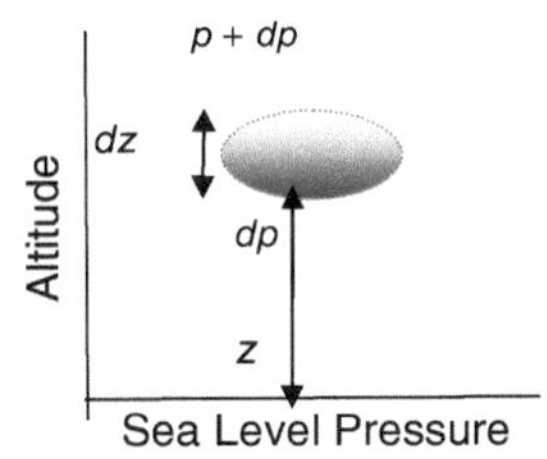

Fig. 1.5 ***A sketch depicting movement of air parcel of thickness dz and surface area A at located at an altitude z.***

where "g" is gravity of earth
"p" is upward pressure on the lower face.

So upward force on the air parcel due to pressure on the parcel = pA.

This would constitute the upward force.

$$\text{Total downward force} = pA \tag{1.1}$$

$p + dp$ is downward pressure on the upper surface. Fig. 1.6 is quite helpful to understand the pressure variation with height, one can visualise the real situation in the troposphere and stratosphere.

So downward force on the air parcel due to pressure on the parcel = $(p + dp)A$.

$$\begin{aligned}\text{Total downward force} &= \text{Force due to downward pressure} + \text{Weight} \\ &= (p + dp)A + \rho g A dz\end{aligned} \tag{1.2}$$

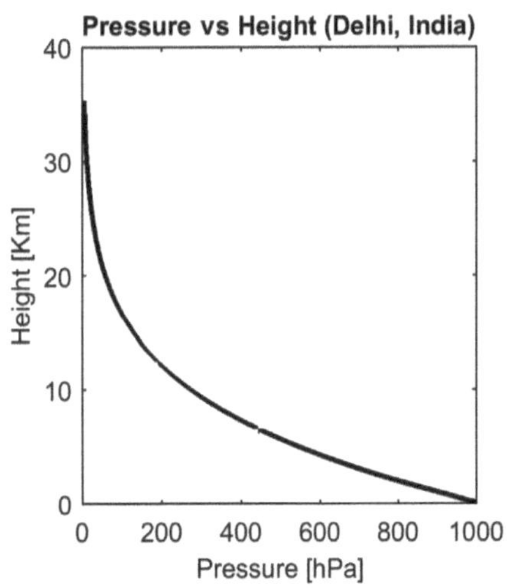

Fig. 1.6 ***Vertical variation of pressure with height is shown (IMD radiosonde data used of January 2022 at Delhi).***

In the equilibrium, the resultant force must be zero, that is,

Total upward force = total downward force

From Eqs. (1.1) and (1.2)

$$\begin{aligned} pA &= (p + dp)A + \rho g A dz \\ dp &= -\rho g d \end{aligned} \tag{1.3}$$

The above (Eq. 1.3) is called the hydrostatic equilibrium equation.

For 1 mole of gas $\rho = \frac{M}{V}$, where M is molecular weight and V is volume occupied by 1 mole of a gas. Thus

$$dp = -\frac{M}{V} g dz$$

From Ideal gas $pV = RT$, $\frac{1}{V} = \frac{p}{RT}$ so

$$\frac{dp}{p} = -\frac{Mg}{RT} dz \tag{1.4}$$

The adiabatic expansion follows the Poisson's Law,

$$PV^{\gamma} = \mathrm{k}\left(\mathrm{a\ constant}\right), \mathrm{or}\ \frac{T^{\gamma}}{p^{r-1}} = k \tag{1.5}$$

On differentiation we will get

$$\gamma T^{\gamma-1} dT = k(\gamma - 1) p^{\gamma-2} dp \tag{1.6}$$

By eliminating "k" from (5) and (6)

$$\frac{dp}{p} = \frac{\gamma}{\gamma - 1} \frac{dT}{T} \tag{1.7}$$

From Eqs. (1.4) and (1.7)

$$\frac{\gamma}{\gamma - 1} \frac{dT}{T} = -\frac{Mg}{RT} dz$$

$$dT = -\left(\frac{\gamma - 1}{\gamma}\right) \frac{Mg}{R} dz$$

Integrating

$$T = -\left(\frac{\gamma - 1}{\gamma}\right) \frac{Mg}{R} z + C$$

where C is integration constant at surface $z = 0$ let $T = T_0$, so $C = T_0$

Hence

$$T = T_0 - \left(\frac{\gamma - 1}{\gamma}\right)\frac{Mg}{R}z$$

Lapse rate

$$\frac{dT}{dz} = \left(\frac{\gamma - 1}{\gamma}\right)\frac{Mg}{R}$$

For the dry air, adiabatic lapse rate = ~9.8°C/Km and in the moist condition it drops to ~6.5°C/Km.

From Eq. (1.4)

$$\frac{dp}{p} = -\frac{Mg}{RT}dz$$

If m is the mass of gas molecules and N is the Avogadro number than $M = mN$ then

$$\frac{dp}{p} = -\frac{mNg}{RT}dz$$

Let p_0 and p is the atmospheric pressure at the bottom $z = 0$, and height z, respectively. So

$$\int\frac{dp}{p} = -\frac{mNg}{RT}\int_0^z dz$$

$$\log_e\frac{p}{p_0} = -\frac{mNg}{RT}z$$

$$p = p_0 e^{-\frac{mNg}{RT}z}$$

Above expression relating the atmospheric pressure with altitude shows an exponential decrease. As the atmospheric pressure is directly proportional to the density of a gas, therefore above expression indicates that density of a gas also decreases exponentially with altitude.

A diagram between atmospheric pressure and height is shown below:

Next section describes seasonal variation of tropospheric and stratospheric temperature up to altitude of 40 km in the equatorial regions, this is the region where almost uniform radiation reaches throughout the year.

1.5 Seasonal variation of tropospheric and stratospheric temperature in the equatorial regions (5° N–5° S)

Climatology for the seasonal anomalies of vertical temperature is shows in the Fig. 1.7. COSMIC data were used for the equatorial regions (5° N–5° S) over a period of

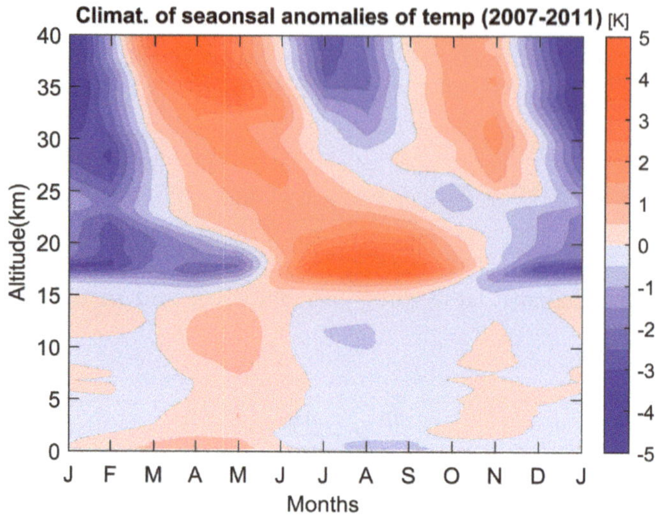

Fig. 1.7 ***Climatological view of seasonal anomalies of atmospheric temperature confined at the equator (5°N–5°S).*** Time axis denotes derived monthly anomalies, while altitude is shown in y-axis. COSMIC data were used to show the seasonal distinct features.

2007–2011. Anomalies computed in a month represent its deviation from the 5-year mean annual cycle. Hence, these anomalies represent the annual seasonal cycle.

It is seen that seasonal cycle is very weak (<1°C) within the troposphere (<15 km). This finding is in synchronization with the fact that almost uniform solar radiation reaching on the earth's surface. Such weak seasonal cycle suggests that solar radiation controls the temperature of the entire troposphere via the earth's surface. As tropospheric seasonal cycle is very weak, such low value may be associated with the observational error, as in the troposphere, COSMIC data are affected by the topography and moisture presents in the lower atmosphere. However, around the equinoxes there are some patterns with larger values suggesting their appearance due to seasonal variation in the incoming solar radiation. A significant seasonal cycle with about 8°C temperature difference exists in a 5 km thin layer (from 17 km to 22 km) of upper tropospheric and lower stratospheric (UTLS) region. The UTLS region is the coldest during January and warmest during August. Annual cycle in the UTLS have same phase relationship with the annual cycle of diabetic heating by ozone [21]. Such results confirm y the radiative heating taking place due to absorption of the UV radiation by O_3. This is a major source of heat in the UTLS.

Half yearly periodical motions can be seen in the upper stratosphere (>35 km) that is referred as semi-annual oscillation (SAO). The momentum advection owing to the meridional circulation generate SAO as sun crosses the equator twice a year. SAO dominates in the altitude range from 35 km to 50 km, and regulated by the seasonal cycle. The amplitude for the first phase (December, January, and February) is larger than the second phase (June, July, and August). These features are linked with the asymmetrical BDC and extra tropical planetary wave forcing [12]. Both the BDC and the amplitudes of SAO show larger values during NH winter (i.e., December, January, and February). Such coupling between BDC, planetary waves, and SAO amplitudes points that there is a crucial role of the atmospheric circulation in settling the seasonal cycle of temperature in the upper troposphere.

Further, in addition to seasonal cycle, the tropospheric and stratospheric regions also exhibit the interannual variation, which can be explained by various dynamical motions. Tropospheric El Niño Southern Oscillation (ENSO), and stratospheric quasi-biennial oscillation (QBO) are the major equatorial dynamical motions that controls the temperature of these regions up to 2–3°C. Next section provides a brief discussion on these two dynamical motions.

1.6 Interannual variation of tropospheric and stratospheric temperature and its linkage with QBO and ENSO

De-seasonalized temperature anomalies of the tropospheric and stratospheric regions are shown in the Fig. 1.8. Anomalies represent its deviation from the 8 years (2007–2014) climatological annual cycle. Stratospheric temperature shows the strong interannual variation which is associated with stratospheric quasi-biennial oscillation (QBO). QBO is a dominant oscillation in the equatorial stratosphere with a "quasi-biennial" period approximately 28 months that varies from 24 to 34 months (e.g., [13]).

QBO have two phases, the westerly and easterly phase (herein after W-QBO, E-QBO). The alternating westerly and easterly wind regimes develop in the upper stratospheric region and propagate downwards with speed 1 km/month until they are dissipated in the tropopause region. QBO is driven by tropospheric waves from intense

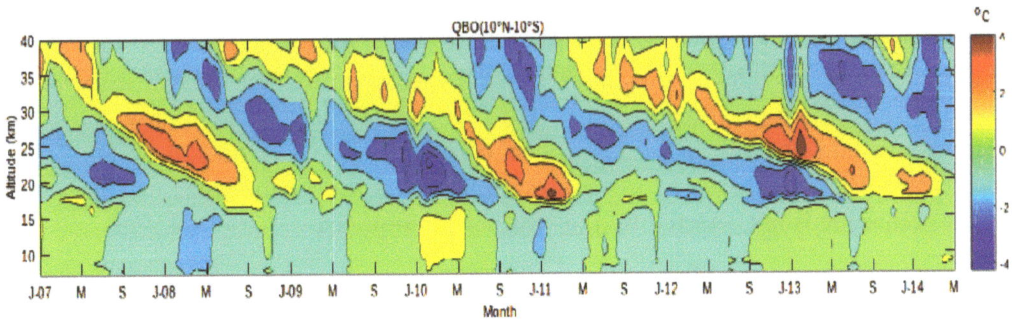

Fig. 1.8 ***Time-altitude sections of the monthly interannual anomalies of zonal-mean temperature, for 2007–2014.*** Red color represents positive (westerly) winds and blue represents negative (easterly) winds. Downward phase propagation is clearly seen, which usually terminate near 17 km altitude. Tropospheric region shows a very clean and significant cold and warm years in a time section of 2007–2014. Relatively, warn years coincide with ENSO and vice versa. COSMIC data were used with large number of profiles over land and ocean.

weather system. These waves break in the upper troposphere and force to push the wind descend with time. Downward propagation of W-QBO descends more regular and rapid than E-QBO. However, amplitude of the E-QBO show near about twice that of W-QBO. The W-QBO and E-QBO generates anomalies in the stratospheric temperature on interannual scale according to summer wind relationship. These alternate opposite anomalies are clearly seen in Fig. 1.8. However, the impact of QBO is not limited to equatorial stratosphere rather it shows a well-developed remote influence up to polar vortex. Such features are studied by Holton and Tan [14,15]. The mean meridional circulation (MMC) from equator to extra tropical region is also associated with QBO. Interestingly, QBO MMC generates a checkerboard pattern in temperature, and out of phase relationship exists in temperature anomalies of tropical and extratopical regions.

QBO show larger influence, for instance, it can modulate subtropical jet steam and tropical convective system via stratospheric and tropospheric pathway that depends on the season [16]. Equatorial QBO modulates the temperature of stratospheric region right from equator to polar regions. This also explains how the dynamics controls the stratospheric temperature on interannual scale.

In the tropical troposphere, there exists an important dynamical motion known as El Niño Southern Oscillation (ENSO), or one can say alternate occurrence of El Niño and La Niña over the Pacific sector with a time scale of several years. The El Niño generates warming near about ~2°C in the entire tropical troposphere, on the other hand, La Niña generates cooling. These warming and cooling effects associated with ENSO can be seen in Fig. 1.8. The clear positive anomalies patterns exist during El Niño (2007) and negative during La Niña (2008). Impact of ENSO is not limited in the tropical region rather it is a major source of global tele-connections. van Loon et al. [17] observed features of warm and weak wintertime stratospheric polar vortex during El Niño. Hence both, the stratospheric QBO and the tropospheric ENSO in the tropics have a wider global remote influence. Therefore, dynamics plays a very crucial role to determine the temperature variation on the interannual time scale.

Next section provides a brief discussion on the methodology for the space-based measurement of temperature structure as in the present era more focuses on these observations due to uniform measurements over land as well as on the ocean. Radio occultation (RO) technique is one of powerful tool as the temperature measurements of this technique is not affected by the weather phenomena.

1.7 Global positioning system (GPS) and radio occultation (RO) technique for the temperature observation

There are several remote sensing techniques to take observations in the earth's atmosphere. In the past, radiosonde and rocket observation were extensively used to present the thermal structure of the troposphere and up to 30 km altitude in the stratosphere [7,19]. However, they are limited to land regions. Later, Lidar, radar, and other in situ observations were in practice with their limitations. With the advent of the satellite observations, it has opened a vast capacity and possibility to cover land as well as oceans in all weather conditions. We mention here briefly a radio occultation technique. Radio occultation is a powerful technique for the global atmospheric observation with high vertical resolution, precision, and accuracy, in all weather conditions with uniform measurement over the both land as well as the ocean. This technique unable us for study ocean atmosphere also as the past ship sonde observation was limited. RO technology offers great potential for atmospheric research, for the improvement of numerical weather forecasts, space, weather monitoring, and climate change detection. This technique was applied for investigation of planetary atmospheres in the sixties [18]. The University Corporation for Atmospheric Research (UCAR), jointly with the University of Arizona and Jet Propulsion Laboratory (JPL), established the GPS/MET (GPS/Meteorology) program in 1993, to demonstrate active limb sounding of the ionosphere and earth's neutral atmosphere using the radio occultation technique. During the radio occultation GPS receivers in low earth orbit set or rise behind the earth's limb, while the signal from GPS slices through the atmosphere which produces the bending in the signal. The receivers onboard in Low-Earth Orbit (LEO) track the bending of in the signals emitted by GPS satellites while traversing the atmosphere. Due to the earths atmospheric, there is a delay in the signal between the GPS and the LEO. This delay allows for reconstruction of the vertical profile of bending angle α with the refractive index (n). In the earth's atmosphere the largest horizontal change occurs in refractive index (n) near the surface where signal can traverse from very dry to very wet condition specially whenever its crosses a coastline. The wet to dry index might change from 1.0004 to 1.0003, that is, 0.01 %. To see the changing in the appropriate unit the refractive index coverts in to atmospheric refractivity (N) according to following formula:

$$N = (n-1) \times 10^6 \tag{1.8}$$

where N is called atmospheric refractivity.

In the neutral atmosphere, the atmospheric refractivity N is related to air pressure (P), air temperature (T), and water vapor pressure (P_w)

$$N = 77.6\frac{P}{T} + 3.73 \times 10^5 \frac{P_w}{T^2} \tag{1.9}$$

The equation is also known as *Smith and Weintraub* equation. Using this formula, the vertical atmospheric temperature profile within tropospheric and stratospheric can be obtained.

1.8 Conclusion

General features of the atmosphere, which includes temperature structure in the lower and middle atmosphere, are shown based on observations using satellite data of latest technology.

Presentations of unique features of thermal structure covering from equator to polar regions, temperature anomalies, time-height sections over a long period have given an opportunity to the readers to understand the interaction of ocean and atmosphere. Differences in the northern and southern hemispheres is also noteworthy, which requires to absorb by the undergraduate and research scholars. Basic equations, which connects different atmospheric parameters, are presented in a simplest way. This chapter would be useful for the general readers as well for the researchers, who are willing to contribute to the climate change understanding and new salient features of the earth's atmosphere.

Data availability statement

All data used in this study are publicly available. The ERA-5 data set is accessible online at Copernicus Climate Change Service (C3S) Climate Date Store. COSMIC data set is available at http://cdaacwww.cosmic.ucar.edu. MLS data set is available at http/:eaetrthdata.nasa.gov/earth-observation-data/near-real time/download-nrtdata/mls-nrt. Radiosonde data are available at IMD site, and at wyoming University, USA in public domain.

References

[1] M. Salby, Fundamentals of Atmospheric Physics, Academic Press, an imprint of Elsevier, United State of America, 1995.

[2] D. Andrews, An Introduction to Atmospheric Physics, Cambridge University Press, Cambridge, 2010, doi:10.1017/CBO9780511800788.

[3] M.J. Schwartz, et al., Validation of the Aura microwave limb sounder temperature and geopotential height measurements, J. Geophys. Res. 113 (D15) (2003) 11, doi:10.1029/2007JD008783.

[4] S. Abraham, S.K. Dhaka, N. Nath, K.L. Baluja, Ionosphere absorption on October 24, 1995 solar eclipse, Geophys. Res. Lett. 25 (1998) 2945–2947.

[5] S. Abraham, S.K. Dhaka, K.D. Preveen, N. Nath, O.P. Nagpal, K.L. Baluja, Planetary wave effect on ionosphere absorption, J. Atmos. Solar Terr. Phys. 60 (1998) 441–453.

[6] S. Abraham, S.K. Dhaka, K.D. Preveen, N. Nath, O.P. Nagpal, K.L. Baluja, Lower stratosphere coupling with mesosphere through QBO, J. Atmos. Solar Terr. Phys. 59 (1997) 1885–1889.

[7] S.K. Dhaka, B.V.K. Murthy, O.P. Nagpal, R.R. Rao, M.N. Sasi, S. Sundaresan, A study of equatorial waves in the Indian zone, J. Atmos. Terr. Phys. 57 (11) (1995) 1189–1202, doi:10.1016/0021-9169(95)00013-R.

[8] S.L. Vadas, D.C. Fritts, M.J. Alexander, Mechanism for the generation of secondary waves in wave breaking regions, J. Atmos. Sci. 60 (2003) 194–214.

[9] R.A. Anthes, P. Bernhardt, Y. Chen, L. Cucurull, K. Dymond, D. Ector, S. Healy, S. Ho, D. Hunt, Y. Kuo, H. Liu, K. Manning, C. McCormick, T.K. Meehan, W.J. Randel, C. Rocken, W.S. Schreiner, S.V. Sokolovskiy, S. Syndergaard, D.C. Thompson, K.E. Trenberth, N.L. Yen, Z. Zeng, The COSMIC/FORMOSAT-3 mission: Early results, Bull. Am. Meteorol. Soc. 89 (2008) 313–333, doi:10.1175/BAMS-89-3-313.

[10] T.G. Shepherd, The middle atmosphere, J. Atmos. Sol. Terr. Phys. 62 (2000) 1587–1601, doi:10.1016/S1364-6826(00)00114-0.

[11] H. Hersbach, et al., The ERA5 global reanalysis, Q. J. R. Meteorol. Soc. 146 (2020) 1999–2049, doi:10.1002/qj.3803.

[12] W.J. Randel, Middle atmosphere: Zonal mean climatology, in: J.R. Holton (Ed.), Encyclopedia of Atmospheric Sciences, Academic Press, USA, 2003, pp. 1358–1365, doi:10.1016/B0-12-227090-8/00227-X.
[13] M.P. Baldwin, et al., The Quasi-Biennial Oscillation, Rev. Geophys. 39 (2) (2001) 179, doi:10.1029/1999RG000073.
[14] J.R. Holton, H.-C. Tan, The influence of the equatorial Quasi-Biennial Oscillation on the global circulation at 50 mb, J. Atmos. Sci. 37 (10) (1980) 2200–2208, doi:10.1175/1520-0469(1980)037<2200:TIOTEQ>2.0.CO;2.
[15] J.R. Holton, H.-C. Tan, The Quasi-Biennial Oscillation in the Northern Hemisphere lower stratosphere, J. Meteorol. Soc. Japan 60 (1982) 140–147, doi:10.2151/jmsj1965.60.1_140.
[16] M.H. Hitchman, S. Yoden, P.H. Haynes, V. Kumar, S. Tegtmeier, An observational history of the direct influence of the stratospheric Quasi-Biennial Oscillation on the tropical and subtropical upper troposphere and lower stratosphere, J. Meteor. Soc. Japan 99 (2) (2021) 239–267, doi:10.2151/jmsj.2021-012.
[17] H. van Loon, C.S. Zerefos, C.C. Repapis, The Southern Oscillation in the stratosphere, Mon. Weather Rev. 110 (1982) 225–229, doi:10.1175/15200493(1982)110<0225:TSOITS>2.0.CO;2.
[18] G. Fjeldbo, V.R. Eshleman, Atmosphere of Venus as studied with the Mariner 5 Dual Radio-Frequency Occultation experiment, Radio Sci 4 (10) (1969) 879–897.
[19] S.K Dhaka, A. Kumar, O.P. Nagpal, Some studies of tropical/equatorial waves over Indian tropical middle atmosphere: Results of tropical wave campaign, Met. Atmos. Phys. 51 (1) (1993) 25–39, doi:10.1007/BF01080878.
[20] A. Raval, A.H. Oort, V. Ramaswamy, Observed dependence of outgoing longwave radiation on sea-surface temperature and moisture, J. Clim. 7 (1994) 807–821, doi:10.1175/1520-0442(1994)007<0807:ODOOLR>2.0.CO;2.
[21] E. Yulaeva, J.R. Holton, J.M. Wallace, On the cause of the annual cycle in tropical lower-stratospheric temperatures, J. Atmos. Sci. 51 (1994) 169 –169.
[22] V. Kumar, S.K. Dhaka, R.K. Choudhary, S.P. Ho, K.K. Reddy, On the occurrence of coldest temperature in stratospheric region and tropical tropopause stability: A study using COSMIC/FORMOSAT-3 satellite measurements, J. Atmos. Solar-Terr. Phys. 121 (B) (2014) 271–286, doi:10.1016/j.jastp.2014.10.007.

CHAPTER 2

Retrieval of aerosol optical depth from satellite observations: Accuracy assessment, limitations, and usage recommendations over South Asia

Muhammad Bilal[a,b], Alaa Mhawish[b], Md. Arfan Ali[b], Zhongfeng Qiu[b], Gerrit de Leeuw[c,d,e], Manish Kumar[f]

[a]School of Surveying and Land Information Engineering, Henan Polytechnic University, Jiaozuo, China
[b]School of Marine Sciences, Nanjing University of Information Science and Technology (NUIST), Nanjing, China
[c]Royal Netherlands Meteorological Institute (KNMI), R & D Satellite Observations, De Bilt, The Netherlands
[d]Aerospace Information Research Institute, Chinese Academy of Sciences (AirCAS), Beijing, China
[e]School of Environment Science and Spatial Informatics, University of Mining and Technology, Xuzhou, Jiangsu, China
[f]Department of Environmental Science and Analytical Chemistry (ACES), Stockholm University, Stockholm, Sweden

2.1 Introduction

Satellite remote sensing of the atmospheric composition provides information on a global scale that cannot be obtained by any other means. Since the late 1970s, sensors have been developed to measure atmospheric trace gases and aerosols, starting with the Total Ozone Mapping Spectrometer (TOMS) in 1979 [1]. Spectrometers dedicated to the measurement of a wide range of atmospheric trace gases such as the Global Ozone Monitoring Experiment (GOME; launched on ERS-2 in 1995) and the SCanning Imaging Absorption spectroMeter for Atmospheric CHartographY (SCIAMACHY, launched on ENVISAT in 2002) were followed by the Ozone Monitoring Instrument (OMI, launched on Aura in 2004) and TROPOspheric Monitoring Instrument (TROPOMI, launched on the Copernicus Sentinel-5 Precursor satellite in 2017) are examples of the rapid development and improvement of space-based sensors. Although these spectrometers do provide information on aerosols, radiometers with a small number of wavebands over a wide spectral range are much better suited for the retrieval of aerosol properties. The main reason is the smaller footprint, of the order of hundreds of meters, which allows for cloud-free pixels, an absolute condition to obtain reliable aerosol information. Other assets are the wide spectral range, the availability of polarization information, and several viewing angles [2]. The TOMS and the AVHRR series of instruments, starting in the early 1980s, have been used for aerosol retrieval, with varying success. Continuous-time series started with the Along Track Scanning Radiometer (ATSR-2, on ERS-2) and the Advanced ATSR (AATSR, on ENVISAT), together with providing aerosol optical depth (AOD) time series from 1995 until 2012, but the use

Atmospheric Remote Sensing
DOI: https://doi.org/10.1016/B978-0-323-99262-6.00012-2

of aerosol retrieval products really started with the launch of the Moderate Resolution Imaging Spectroradiometer (MODIS) on the Terra and Aqua satellites, providing data from 2000 and 2002, respectively. The evaluation of the AOD time series from these and other sensors by Sogacheva et al. [3] shows the differences between sensors and even between different retrieval algorithms applied to the same sensor.

Sensors like TOMS, AVHRR, and ATSR-2/AATSR are suitable for quantifying aerosols in the atmospheric column [4] whereas other sensors like MODIS, the Polarization and Directionality of Earth Reflectance instrument (POLDER), the Multi-angle Imaging SpectroRadiometer (MISR) and the Visible Infrared Imaging Radiometer Suite (VIIRS) have been designed for that purpose [5–9]. Aerosol particles larger than about 100 nm are optically active, and the total column of AOD can be quantified using visible bands, part of ultraviolet, near-infrared, and short infrared [10–12]. In addition to the diversity of sensors and their improvement over time, new and more accurate aerosol retrieval algorithms have been developed in the last two decades [11–15]. The main aerosol parameter that most of the algorithms retrieve is the AOD which is mainly reported at mid-visible bands such as 550 nm. Some of the instruments can separate aerosol types (e.g., aerosol size, shape, and absorption) and total aerosol loading, such as MIRS [14,16]. In addition, active sensors such as the Cloud-Aerosol Lidar with Orthogonal Polarization (CALIOP) can retrieve the vertical profile of aerosols and differentiate between several aerosol subtypes [17,18].

Aerosol retrieval algorithms applied to passive sensor measurements used the level 1 (L1) calibrated radiances as input to retrieve geophysical data (here AOD), which is known as level 2 data (L2). The L2 data can also be spatially and temporally aggregated to generate level 3 data (L3) with coarser pixel size and/or temporally averaged (monthly, seasonal, and annual). The aerosol retrieval algorithms use different approaches to quantify aerosol properties from the TOA radiances measured by a sensor at different wavelengths. After cloud screening and removing the contributions from the surface reflectance and molecular scattering to the TOA reflectance, the aerosol contribution remains and is used by the retrieval algorithm to retrieve the AOD [19–21]. Therefore, the sensitivity of each algorithm depends on the number of wavebands, on the geometries and the accuracy of the estimated surface reflectance, and on the assumed aerosol optical and microphysical properties in the aerosol models [9,12,22,23]. For example, the L1 measurements from the same sensor can be used by different algorithms and retrieve different AOD values due to different assumptions used by the algorithms. In addition, due to the heterogeneity of land surfaces, the accuracy of aerosol retrieval algorithms can vary spatially and temporally with changing land cover. Therefore, the uncertainty estimation of AOD retrieval under different conditions is very useful for the end-user and the algorithm's developer to improve the algorithm accuracy and sensitivity.

South Asia, which comprises Bangladesh, Bhutan, India, Nepal, Pakistan, and Sri Lanka, is one of the most densely populated regions with around 25% of the world population [24–26]. The rapid economic growth, urbanization, and industrialization in this region have increased air pollution problems [25,27–29]. Air quality monitoring in developing south Asian countries is limited and challenging due to the few air quality monitoring stations which mainly are located in megacities [25,28–30]. Satellite AOD retrievals provide one of the primary data sources that can be used to quantify the air quality over regions with few or no monitoring stations [28,29,31]. Therefore, knowledge of the uncertainty in the satellite aerosol retrieval products is essential in air quality-related applications and the development of environmental policy. This chapter discusses operational aerosol retrieval algorithms and approaches used to evaluate the AOD. Case studies to assess AOD products over South Asia from different retrieval algorithms are presented and the chapter concludes with recommendations for end-users.

2.2 Aerosol optical properties

Atmospheric aerosols are mixtures of particles of different sizes, compositions, shapes, and chemical and physical properties suspended in the atmosphere. Examples of aerosol particles are dust, haze, smoke, sea salt, black carbon, etc. The aerosol particles influence the global climate by changing the radiation energy balance by directly interacting with the incoming solar radiation. Or indirectly by changing the physical and optical properties of clouds that affect the hydrological cycle. Particles with sizes ranging from 0.1 μm to 2.5 μm in the radius are of primary interest for climate, air quality, visibility, and human health research. The aerosol optical properties are determined by the particle size distribution, that is, the number of particles with a certain size, and the chemical composition which determines the complex refractive index. The complex refractive index and particle size together determine the scattering and absorption of incident radiation and all these parameters vary in the atmosphere in response to the ambient relative humidity (RH) due to condensation (at high RH) or evaporation (at low RH) of water vapor. The total extinction of solar radiation at wavelength λ is obtained by the integration of the particle extinction coefficient over the whole particle size distribution and the light path in the atmosphere. The most common aerosol optical property obtained from remote sensing observations is the AOD, that is, the extinction (the sum of absorption and scattering of radiation) of solar radiation by aerosol particles, at a given wavelength (λ), integrated over the atmospheric vertical column.

$$\sigma_{extinction}(\lambda) = \sigma_{scattering}(\lambda) + \sigma_{absorption}(\lambda)$$

Because the aerosol optical properties may change with height above the surface, also the extinction varies with height in response to changes in size and chemical composition of the particles. These changes result in different scattering and absorption properties. Because the AOD is a measure of the total extinction in the vertical column and thus does not discriminate between the extinction at different heights, the AOD is the integral of the aerosol extinction from the surface to the top of the atmosphere (TOA).

$$AOD(\lambda) = \int_0^{TOA} \sigma_{extinction}(\lambda, z)\,dz$$

The AOD is a dimensionless value representing the amount of aerosol in the atmospheric column. So, the clean atmosphere has a low AOD such as 0.01, while in a highly aerosol polluted atmosphere the AOD is high (>0.4) and may reach values larger than 1. Satellite remote sensing measures the TOA reflectance at different wavelengths containing contributions from the land surface and atmospheric components such as gases, aerosols, clouds, and water vapor. The retrieval of the AOD from such measurements is discussed in Section 4.

2.3 Satellite-based remote sensing instruments for aerosol monitoring

Satellite measurements are made using instruments on different satellite platforms, including satellites in a sun-synchronous polar orbit (low earth orbit or LEO satellites) and geostationary satellites. The satellites in sun-synchronous orbits circle the earth from the north pole to the south pole (ascending) or from south to the north (descending) and synchronous with the sun (fixed position relative to the sun), allowing to visit the same spot at roughly the same local time. The elevation of the sun-synchronous (low orbit) satellites varies from 160 km to 1000 km. The geostationary orbit (high orbit or GEO) satellites follow the same earth rotation from west to east and have a rotation period equal to the earth's rotation period (23 hours 56 minutes 4 seconds). This makes satellites in geostationary orbits appear to remain above earth at a constant longitude or stationary over a fixed position. Sun-synchronous satellites cover the whole earth but, depending on the sensor's swath width it may take one or several days to provide global coverage. For instance, the MODIS swath of 2330 km is just too small to cover the whole earth daily, leaving some area near the equator uncovered. In contrast, geostationary satellites view only part of the earth (the earth disk), depending on their position, but with high temporal resolution. Thus, geostationary satellites can provide a time sequence of aerosol observations allowing us to observe the evolution of an air pollution episode.

In the last two decades, several dozens of instruments have been developed to retrieve geophysical parameters for the earth's surface and atmosphere from space. This chapter focuses on the sensors used to retrieve aerosol optical properties, mainly AOD. Sensors measure the radiance at the top of the atmosphere (TOA) at different wavelengths with a relatively small bandwidth (wavebands). These measurements are used as input to retrieval algorithms to derive the geophysical parameters like AOD. Most of the sensors used in aerosol retrieval measure the TOA reflectance at near-UV, visible, near-IR, shortwave IR and TIR bands. Aerosol optical properties have been particularly retrieved by passive satellite sensors. Still, active lidar sensors such as CALIOP onboard of CALIPSO satellite provide valuable information on the vertical distribution of aerosol layer in the atmospheric column during day and night time. Satellite instruments vary in their spectral resolution, spatial resolution, swath width, and the number of scanners. Table 2.1 summarizes the major sensors used in AOD retrieval over the past two decades, including each instrument's main features.

Table 2.1 Summary of the major sensors used for AOD retrieval.

Sensor(s)	Characteristics	Operational period	Swath (km)	Algorithm	Spatial resolution	References
Advanced Very-High Resolution Radiometer (AVHRR)	Bi-spectral, single-view, broad swath radiometer	1989–1991 (NOAA7), 1995–1999 (NOAA14) and 2006–2011 (NOAA18)	2900	Deep Blue/ SOAR, V. 4	8 km	[60,61]
Along Track Scanning Radiometer (ATSR-2) and Advanced ATSR (AATSR)	Dual-view radiometer in the visible and near-infrared; thermal infrared for cloud	1995–2003 (ATSR-2) and 2002–2012 (AATSR)	512	ADV/ASV v2.31	1 km	[4]
Total Ozone Mapping Spectrometer (TOMS)	UV spectrometer	1979–1993 and 1996– 2001	3100	Nimbus-7/TOMS: N7AERUV v. 0.4.3; EP/TOMS: EPAERUV v. 0.1.3	50 km	[62]
Ozone Monitoring Instrument (OMI)	UV spectrometer	2004–2016	2600	OMAERUV v. 1.8.9.1	13 × 24 km	[47,48]
Sea-viewing Wide Field-of-view Sensor (SeaWiFS)	Multispectral, single-view, broad-swath radiometer	1997–2010	1502	Deep Blue/ SOAR V.1	13.5 km	[54]
Polarization and Directionality of the Earth's Reflectances (POLDER) 3	Multispectral, multi-angle polarimeter	Dec 2004–Dec 2013	2100 × 1600	GRASP v.1	6 km	[3,63]
Multi-angle Imaging SpectroRadiometer (MISR)	Multispectral, with four bands, visual–near-infra-red, multi-angle, that is, nine angles, radiometer	2000–present	380	Standard algorithm (SA) V23	4.4 km	[14,16]
Moderate Resolution Imaging Spectroradiometer (MODIS)	Multispectral, single-view, broad-swath radiometer	Terra: 2000–present; Aqua: 2002–present	2300	DT C6.1; DB C6.1; MAIAC V6	3 and 10 km; 10 km; 1 km	[11,12,43]
Visible Infrared Imaging Radiometer Suite (VIIRS)	Multispectral, single-view, broad-swath radiometer	2012–present	3040	DT DB/SOAR EDR		[22,64,65]

2.4 Satellite-based aerosol retrieval algorithms

Aerosol retrieval algorithms were developed to quantify aerosol properties using TOA reflectance measured by satellite instruments in cloud-free conditions, by comparison with TAO reflectance calculated using a radiative transfer model for the same geometry and atmospheric condition, including different aerosol models. Then the best fit aerosol model is used to retrieve AOD, whereas the models are regionally oriented, and in some algorithms, the aerosol models vary seasonally. To reduce processing time and improve accuracy, model-based aerosol climatology can be used as a priori [6]. The TOA reflectance contains reflectance from atmospheric components such as aerosols, atmospheric gases scattering solar light as well as absorbing gases such as water vapor and other trace gases, and reflectance from the earth's surface. The occurrence of clouds is one of the major factors affecting aerosol retrieval because the intense cloud scattering, even from very thin clouds, overwhelms the aerosol signal. Therefore, the first step in all aerosol retrieval algorithms is to mask the cloudy pixels by applying different cloud detection methods [32], sometimes followed by a postprocessing step to remove undetected clouds [33]. The other contributor to the TOA reflectance is the reflectance from the land surface, which varies spatially and temporally and has a very high impact on AOD accuracy, especially over bright surfaces and during low aerosol loading conditions [8,9,27,29,34–40]. Therefore, accounting for the land surface reflectance to the TOA is one of the most important steps in AOD retrieval. Algorithms use different methods to estimate the land surface reflectance. For example, the Dark Target algorithm utilizes the empirical relationship between visible bands (0.47 and 0.66) to shortwave infrared (SWIR, 2.12 μm) surface reflectance (VIS-to-SWIR relationship) for retrieving AOD. DB uses multiple methods depending on the season, location, and land cover types. For example, over bright surfaces, it uses a static seasonal database of spectral surface reflectance, and over vegetated surfaces, it uses a statistical spectral ratio as a function of NDVI similar to the DT algorithm. After correcting for the surface reflectance contribution, the remaining part that needs to be removed from the TOA reflectance is the path radiance contribution from gases. The path radiance contribution to the TOA is estimated using atmospheric pressure and temperature to estimate the Rayleigh scattering. After separating the molecular and surface contributions from the measured TOA reflectance, the aerosol part is retained. Then the aerosol reflectance is compared with that calculated using a radiative transfer model (RTM), for the geometric conditions (sun, satellite, surface) during each observation and different aerosol models. The aerosol model providing the best fit to the measured (retained) aerosol reflectance is selected as the solution to the retrieval problem and provides the AOD. Such procedure is applied because there is no exact solution, that is, the retrieval is underdetermined because there are more variables than independent measurements.

2.4.1 Overview of operational retrieval algorithms

Several algorithms have been developed to retrieve aerosol optical properties using multiple wavelengths sensor measurements from LEO or GEO satellites. Table 2.1 shows an overview of operational aerosol retrieval algorithms for each sensor and the output of aerosol products. The following discussion presents the functional aerosol retrieval algorithms which have been developed and applied to one or more sensors, possibly with modifications required for different sensor characteristics.

2.4.1.1 Dark Target (DT) algorithm

The DT aerosol retrieval algorithms consist of two different algorithms for application over land [12,21] or ocean [41]. The DT algorithms were developed to retrieve aerosol optical properties over dark surfaces like vegetated land or ocean, where at visible wavelengths the aerosols tend to be brighter over the dark background. For each retrieval box (MODIS: 10 × 10 km and 3 × 3 km; VIIRS: 6 × 6 km), the measured TOA reflectance is screened to remove pixels containing cloud and inland water, and also the brightest and darkest pixels at the visible band are discarded. The reflectance of the remaining pixels, which are expected to represent the cloud-free dark surface pixels, is averaged. Then, the algorithm uses the relationship between the TOA reflectance in the short wave infrared (SWIR_ and visible (VIS) bands to determine the surface reflectance in the VIS from the SWIR observations, which in turn is used to effectively separate the surface and aerosol contributions to the TOA reflectance. The AOD is determined using aerosol models saved in look-up tables (LUTs) created using RTMs for different geometries and aerosol models. The model which best fits the measured TOA reflectance is used to derive the AOD. The models are regional and seasonal oriented and contain a combination of fine-mode and coarse mode dominated aerosols. The retrieval algorithm allows for interpolation between the discrete models used in the LUTs. These LUTs are used to speed up the algorithms by saving computer time running the RTMs.

The DT algorithm is applied to MODIS on Terra and Aqua and to VIIRS on Suomi NPP to retrieve AOD over dark land surfaces. MODIS DT AOD is reported at two spatial resolutions, including 3 km and 10 km, while the VIIRS DT AOD is reported at 6 km spatial resolution. One of the limitations of the DT algorithm is the moderate spatial resolution (3 km, 6 km, and 10 km) and that AOD is only retrieved over dark surfaces. Therefore, lots of gaps occur in AOD retrievals over arid regions and some urban areas [9,12]. Moreover, the retrieval accuracy over brighter surfaces such as urban regions is low due to high surface reflectance that interferes with the aerosol signals detected by the sensor.

2.4.1.2 Deep Blue (DB) algorithm

The DB algorithm was initially developed to retrieve AOD over bright surfaces such as desert and arid regions to fill the gaps in the DT AOD, which only retrieves AOD over dark surfaces [42]. The second generation of the DB algorithm [11] is enhanced to retrieve AOD over all land surfaces except snow/ice under cloud-free conditions [9]. DB algorithms have been applied to several sensors measurements, including MODIS, SeaWiFS, AVHRR, and VIIRS, and report AOD at different spatial resolutions (10 km, 13.4 km, 8.8 km, and 6 km) depending on the sensor and application. The DB algorithm retrieves AOD at the sensor's nominal resolution and afterward aggregates the retrieved AOD over the retrieval box or final pixel size. The surface reflectance for cloud-screened pixels is determined by three methods depending on the geographical location, season, and land cover types. The three methods are: (1) precalculated surface reflectance dataset for the pixels with NDVI less than 0.1, (2) precalculated surface reflectance dataset for the pixels with NDVI greater than 0.1, and (3) MODIS Land Cover (MCD12C1) products are used to divide pixels into three groups (e.g., arid and semi-arid regions, general vegetation, and urban/built-up and transitional regions). The main advantage of the DB algorithm is that retrieving AOD over all land surfaces except snow/ice provides better spatial coverage by including both bright and dark land surfaces. The main limitation of DB AOD algorithms is that AOD is provided at moderate resolution of 6–13.4 km, which limits its applications in small urban scales.

2.4.1.3 Multi-Angle Implementation of Atmospheric Correction (MAIAC)

MAIAC is a new unconventional algorithm for retrieving aerosol optical properties over land and ocean at a spatial resolution of 1 km [43] using MODIS Terra and Aqua data. The MAIAC algorithm uses dynamic time series analysis and spatial processing to distinct atmospheric and surface contributions and enhances the quality of cloud detection to retrieve aerosol properties and surface characterization. In particular, the MAIAC algorithm accumulates multi-angle observations from 5 days up to 16 days, which retrieves the surface bidirectional reflectance distribution function (BRDF) and separates the spatial and angular inconsistency in the surface spectral reflectance. Afterward, these are stored with surface signature (spectral, spatial, thermal, etc.) for each 1 km grid cell, which enables MAIAC to separate pixels with clear-sky, clouds, or snow cover which improves the accuracy of atmospheric correction and AOD estimation during cloudy days [43]. MAIAC uses a cloud mask test combined with aerosol-type selection (smoke/dust tests). MAIAC uses three aerosol models to retrieve AOD, including two absorbing models, smoke and dust used when the smoke/dust is detected, and background models. The background models contain eight aerosol models prescribed regionally based on long-term aerosol optical and physical properties obtained from AERONET sites. MAIAC retrieves AOD at blue (470 nm) and green (550 nm) bands, along with columnar water vapor (CWV) at 1 km spatial resolution. In addition, MAIAC provides aerosol types (smoke, dust, and background) as a result of the smoke/dust tests and smoke injection height when smoke is detected along with fine mode fraction over the water.

2.4.1.4 MISR aerosol retrieval algorithm

MISR is a passive sensor onboard the Terra satellite operated from early 2000. MISR is a unique sensor because it images the earth from nine different push-broom cameras at different angles (±70.5°, ±60.0°, ±45.6°, ±26.1°, and 0° (nadir)). Each camera has four spectral bands, three visible (0.446 μm, 0.558 μm, and 0.672 μm) and one near-infrared (0.866 μm) with spatial resolution ranging from 275 m to 1.1 km. MISR aerosol retrieval algorithms utilize the nine different viewing angles, which help reduce the assumption of land surface reflectance [44]. Multiple view-angles enable the MISR aerosol retrieval algorithms for land and water to separate aerosol types (e.g., aerosol size, shape, and absorption) and total aerosol loading in the atmospheric column. In particular, the MISR aerosol retrieval algorithm uses a set of 74 aerosol models in the look-up table (LUT) to calculate AOD, size-fractioned AOD, and other aerosol optical properties [14,16]. Several improvements have been applied to the MISR aerosol retrieval algorithm to improve retrieval accuracy and spatial resolution. The latest version of the algorithm is Version 23 (V23), with a higher spatial resolution of 4.4 km than the previous V22 (17.6 km) [16]. MISR provides several optical properties of aerosols as well as the aerosol type [45] over a retrieval region of ca. 17 × 17 km. The narrower swath of 380 km results in full earth coverage every 2–9 days depending on the latitude.

2.4.1.5 OMI aerosol retrieval algorithm

The OMI is a passive sensor onboard the Aura satellite, which was launched on July 15, 2004, passing the equator at 01:45 pm local time (ascending node) [46]. OMI is a spectrometer that measures the backscattered energy from the Earth using ultraviolet (UV) and visible (VIS) bands ranging from 270 nm to 500 nm with a 2600 km wide swath that provides daily global coverage. The OMI UV and visible bands give advantages to several algorithms to retrieve aerosol optical properties utilizing its sensitivity to aerosol absorption in the near-ultraviolet (UV) spectral region and low near-UV surface reflectance [47,48]. In addition to AOD, absorbing properties of aerosols such as the aerosol absorption optical depth (AAOD), single scattering albedo (SSA), and UV absorbing aerosol index (UV-AAI) can also be retrieved utilizing the near-UV spectral bands [47,48]. Two aerosol retrieval algorithms have been developed to retrieve aerosol properties, including the OMI AERosol UV algorithm (OMAERUV) and OMI Multi-wavelength AEROsol (OMAERO). The OMAERUV uses the TOA measurements in the UV at 354nm and 388nm to retrieve AOD, AAOD, and UV-AAI at 388 nm. The AOD and AAOD are retrieved using a standard inversion algorithm that uses the precomputed TOA reflectance for a set of aerosol models (21 models). The aerosol models are composed of three major types: biomass burning carbonaceous aerosols, desert dust, and weakly absorbing sulfate-based aerosols, and each contains seven models for varying aerosol optical and microphysical properties [48]. The OMAERO retrieval algorithm uses a multiple wavelength optimal estimation approach to retrieve AOD in cloud-free conditions using up to 20 wavelengths from 331 nm to 500 nm [47]. The OMAERO algorithm uses 24 aerosol models in the LUT, representing major aerosol types, including desert dust, biomass burning, volcanic, and weakly absorbing aerosol types based on their sources. The model which best fits the measured TOA reflectance is used to derive AOD and SSA.

2.4.2 Uncertainty evaluation of aerosol retrieval products

Quantifying the uncertainty in satellite-based AOD retrievals (AOD validation) is an essential step for several applications such as air quality modeling and climate applications. The satellite-based AOD retrieval Level 2 (L2) products are usually validated against independent reference datasets obtained from Sunphotometer measurements, mainly from the global Aerosol Robotic Network (AERONET) [19] and over the ocean from the Maritime Aerosol Network (MAN) [49]. The most common approach used to validate satellite-based AOD is to collocate the satellite-retrieved AOD with ground-based AOD measurements in space and time. The ground-based measurements represent AOD measured in clear sky conditions at a specific location several times a day, for example, AERONET-based AOD data measured every 15 minutes on average. On the other hand, satellite-based AOD products are snapshot raster data measured at the specific time of the overpass. Each raster's pixel represents a spatial AOD averaged over the pixel area, which varies from one sensor to other. However, the suspended aerosol particles in the atmosphere are constantly moving with airmass, which can be captured by satellite sensors over a spatial window corresponding to the ground-based station, while the ground-based instrument measure AOD during a particular time window. Therefore, the collocations between spatial statistics of satellite-based AOD with temporal statistics from ground-based AOD provide a meaningful comparison between point value (AERONET) with image pixel (satellite). The spatial window represents AOD for a single-pixel (e.g., 1×1) or an averaged AOD for multiple-pixels (e.g., 3×3 or 5×5) centered at the AERONET site, and the temporal window represents an averaged AOD for a specific local time. The spatio-temporal windows can be varied from one region to another due to spatio-temporal variations of aerosols, the spatial resolution of satellite AOD algorithms, and the transportation of air masses. The standard approach suggested by Ichoku [50] is averaging satellite AOD over spatial windows of 5×5 pixels centered at the AERONET site while the AERONET AOD is averaged over a temporal window of one hour centered at the satellite overpass time. Then the collocations are linearly correlated, and several statistical metrics are determined to quantify the uncertainty in AOD retrievals. The standard regression method used to define the line of best fit between ground-based and satellite-based AOD is ordinary least squares (OLS) which assumed that the Y-axis (here satellite data) is measured with error, while the X-axis (ground data) is measured without error. This assumption is arguable since the ground-based AOD also includes some uncertainty [51] which however is small compared to satellite-based AOD. To address this, several studies [10,28,52,53] suggested using the reduced major axis (RMA) regression method, which assumed both the x-axis and y-axis measured with error. Therefore, the line of best fit will be more reliable using RMA than OLS. The intercept and the slope of the best fit line are used as indicators of the uncertainty related to the estimation at the surface reflectance (intercept) and the error associated with the assumptions in aerosol models (slope). The positive intercept indicates that the satellite aerosol retrieval algorithm estimated positive AOD when it is zero, indicating that the algorithm underestimates the contribution of the surface reflectance and vice versa. While a slope greater than unity indicates that the algorithm overestimates the AOD mainly due to nonaccurate assumptions in the aerosol models used by the algorithm. On the other hand, other statistical metrics used to evaluate the satellite AOD products are listed as follows: Person's correlation coefficient (r), which examines the agreement between ground-based and satellite-based AODs. Root mean squared error (RMSE, Eq. 2.1) represents the variability in the data. Mean absolute error (MAE, Eq. 2.2) represents the mean absolute bias between satellite and ground-based AODs. The relative mean bias (RMB, Eq. 2.3) represents the relative bias which indicates the relative deviation of the retrieved AOD from the ground-based AOD values.

$$RMSE = \sqrt{\frac{1}{n}\sum_{i=1}^{n}\left(AOD_{(s)i} - AOD_{(g)i}\right)^2} \tag{2.1}$$

$$MAE = \frac{1}{n}\sum_{i=1}^{n}\left|AOD_{(s)i} - AOD_{(g)i}\right| \tag{2.2}$$

$$RMB = \left(\frac{\overline{AOD_s} - \overline{AOD_g}}{\overline{AOD_g}}\right) \times 100 \tag{2.3}$$

where AOD_s and AOD_g represent AOD from satellite and ground-based AERONET Sunphotometer, respectively.

The satellite product validation teams estimate the expected magnitude of AOD uncertainty, known as expected error. The EE was calculated based on global validation using all collocations between satellite AOD and AERONET AOD worldwide. The EE is the envelope encompassing absolute and relative errors and contains 68.4% ($\pm 1\sigma$) following Gaussian statistics (Eq. 2.4) [12,23,54]. The MODIS DT algorithm reported the EE for the AOD at 10 km resolution as EE= 0.05 × 0.15AOD, while for the DB algorithm the EE is given by EE = 0.05 ± 0.20AOD. In contrast, the MAIAC algorithm reported the EE at 1km spatial resolution as EE = 0.05 ± 0.10AOD based on global assessment of MAIAC AOD retrieval [43] To estimate the percentage of satellite AOD retrievals falling within the EE (wEE%), above the EE (aEE%), and below the EE (bEE%) envelopes are calculated using the following formulas:

$$EE = \pm\left(0.05 + 0.20 \times AOD_g\right) \tag{2.4}$$

$$wEE\% = \left(AOD_g - |EE| \le AOD_s \le AOD_g + |EE|\right) \times 100 \tag{2.5}$$

$$aEE\% = \left(AOD_s \geq AOD_g + |EE|\right) \times 100 \tag{2.6}$$

$$bEE\% = \left(AOD_s \leq AOD_g - |EE|\right) \times 100 \tag{2.7}$$

For the more accurate AOD products, a higher AOD percentage falls within the EE envelope, the RMSE and RMB are lower and the slope is close to unity. A higher slope and RMB indicate that the algorithm overestimates the AOD, which can be confirmed by higher aEE than bEE. When the algorithm underestimates the AOD, the RMB and slope have values lower than one, and the bEE is greater than aEE.

2.5 Case study of AOD retrieval and accuracy assessment over South Asia (SA)

This section provides a case study of AOD evaluation retrieved by different MODIS algorithms, including DT, DB, and MAIAC over SA. As mentioned above, the satellite-based AOD retrieval was evaluated by comparing it with ground-based [20] Sunphotometer AOD mainly obtained from AERONET stations. Data from 15 AERONET stations distributed across SA were used (Fig. 2.1 and Table 2.2) for

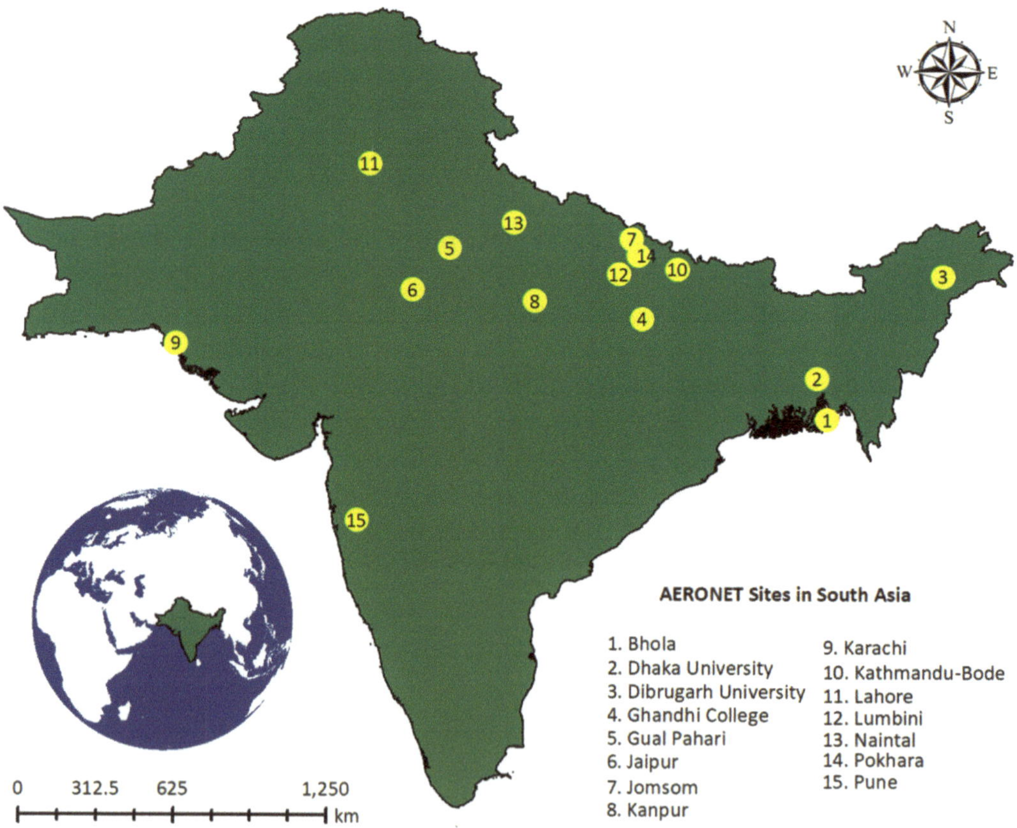

Fig. 2.1 ***Locations of AERONET sites used in the present study.***

Table 2.2 AERONET site information used for evaluating satellite AOD data.

Sr. No.	AERONET site	Country	Latitude/Longitude	Period
1	Bhola	Bangladesh	22.17N/90.75E	2013–2019
2	Dhaka University	Bangladesh	23.73N/90.40E	2012–2020
3	Dibrugarh University	India	27.45N/94.90E	2018–2019
4	Gandhi college	India	25.87N/84.13E	2006–2019
5	Gual Pahari	India	28.43N/77.15E	2008–2017
6	Jaipur	India	26.91N/75.81E	2009–2017
7	Jomsom	Nepal	28.78N/83.71E	2011–2015
8	Kanpur	India	26.51N/80.23E	2001–2019
9	Karachi	Pakistan	24.87N/67.03E	2006–2019
10	Kathmandu Bode	Nepal	27.68N/85.39E	2012–2016
11	Lahore	Pakistan	31.54N/74.32E	2006–2019
12	Lumbini	Nepal	27.49N/83.28E	2013–2019
13	Naintal	India	29.36N/79.46	2008–2012
14	Pokhara	Nepal	28.15N/83.97E	2010–2015
15	Pune	India	18.54N/73.81E	2008–2014

comparison with satellite-retrieved AOD from 2003 to 2017. AERONET Direct Sun algorithm report AOD at four spectral bands 440 nm, 500 nm, 675 nm, and 870 nm. Hence, for evaluating the satellite-based AOD which was reported at 550, AERONET was AOD interpolated to 550 nm using Angstrom Exponent with wavelength pair 440–670 nm applied to Angstrom power law [19,55]. In this case study, we used the latest Version 3 (V3) Level 2 (L2) AERONET AOD data which is cloud screened and quality assured to evaluate the satellite AOD products [56]. In general, AOD retrieval from satellite data is challenging, especially over South Asia, where the land surface is very diverse (urban, semi-urban, cropland, desert, and mountains), and aerosol loading is high and diverse due to emission from diverse sources [29]. The SA is affected by smoke aerosols during the postmonsoon months (October and November), haze in the winter, and dust in the spring (March–May). This diversity of aerosol type makes the AOD retrieval over the region challenging. Therefore, the uncertainty can vary between different aerosol retrieval algorithms applied over this region. Fig. 2.2 shows the spatial distribution of the annual mean AOD for 2017, retrieved by application of the DT, DB, and MAIAC to MODIS Terra and Aqua data collected over South Asia. MAIAC and DB provide higher spatial coverage than DT due to their retrieval ability over all land surfaces except snow/ice, while DT retrieves AOD only over dark surfaces. In addition, the higher spatial resolution of the MAIAC algorithm enables AOD retrieval over complex landscapes such as coastal and hilly areas such as Baluchistan and the Himalayan foot-hills. On the other hand, the inherent coarser spatial resolution of DT and DB cannot retrieve AOD between clouds as well as fine aerosol features such

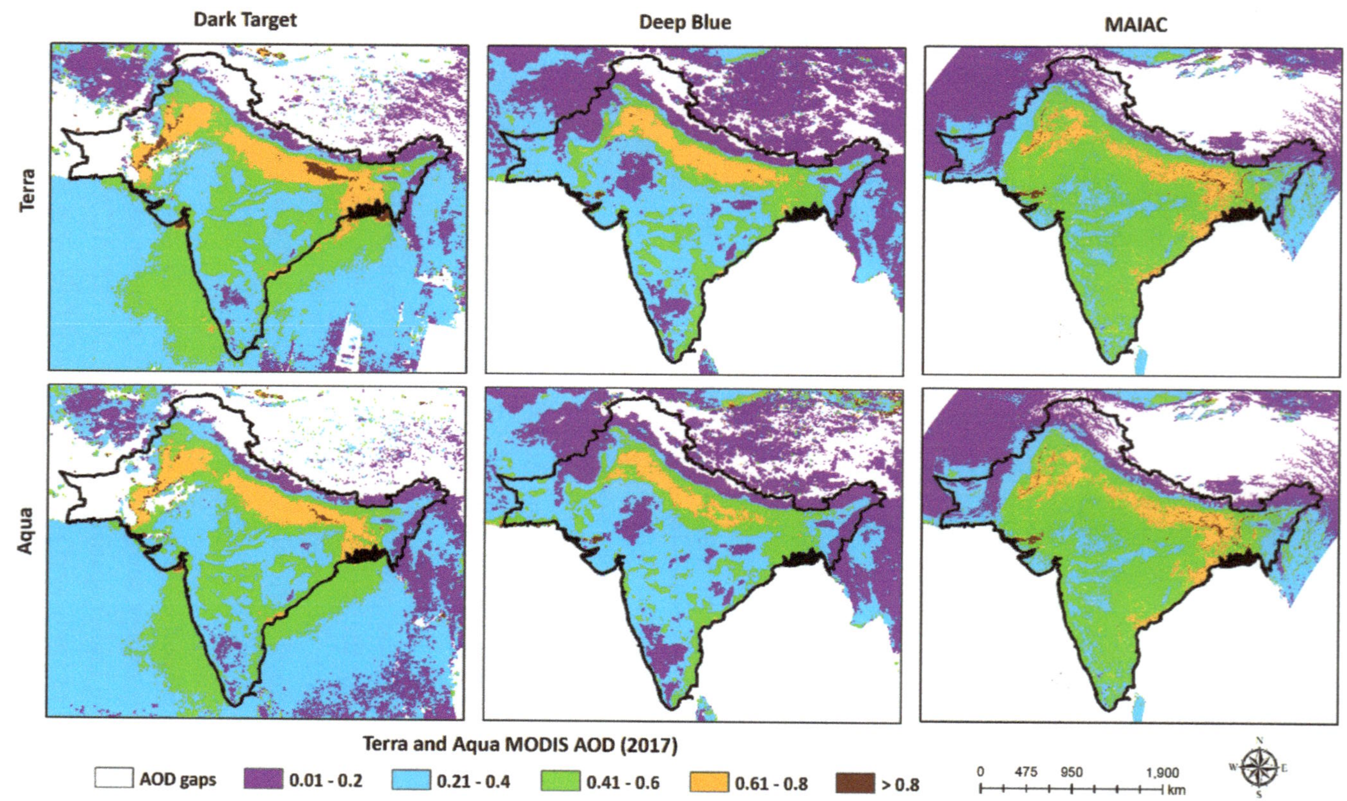

Fig. 2.2 *Annual mean spatial distributions of the AOD retrieved using the Aqua and Terra MODIS DT, DB, and MAIAC algorithms over South Asia for the year 2017.*

as small-scale forest fire [8]. The higher spatial coverage of MAIAC than that from the DT and DB algorithms is manifest in the number of MAIAC-AERONET matching points (Fig. 2.3). The number of match-up points is one of the evaluation metrics in AOD retrieval, and the algorithms providing a larger number of match-ups have a better ability to retrieve AOD over different surfaces with a variety of aerosol loadings and different aerosol types [29,37,57].

The comparison of the AOD retrieved using the MODIS Terra and Aqua algorithms with AERONET data in Fig. 2.3 shows similar performances of the three algorithms. The performance of the MAIAC algorithm is somewhat better than that of the DT and DB algorithms, with a lower retrieval error (RMSE: 0.16–0.17) and slope closer to unity than DT and DB algorithms. The percentage of AOD retrievals within the EE envelope was highest for MAIAC (around 77%). The retrieval accuracy of the DT algorithm is better than that of DB, which significantly underestimated AOD over SA. Around 30% of the DB retrieved AOD is below the EE, indicating that the algorithm overestimates the surface reflectance and/or uncertainty in estimating the aerosol optical and microphysical properties such as single scattering albedo used in aerosol models. Overall, MAIAC outperforms other DT and DB algorithms in terms of a higher number of retrieval as well as retrieval accuracy. Also, DT achieved better retrieval accuracy with 71–73% of retrieval within the EE envelope. However, the slope

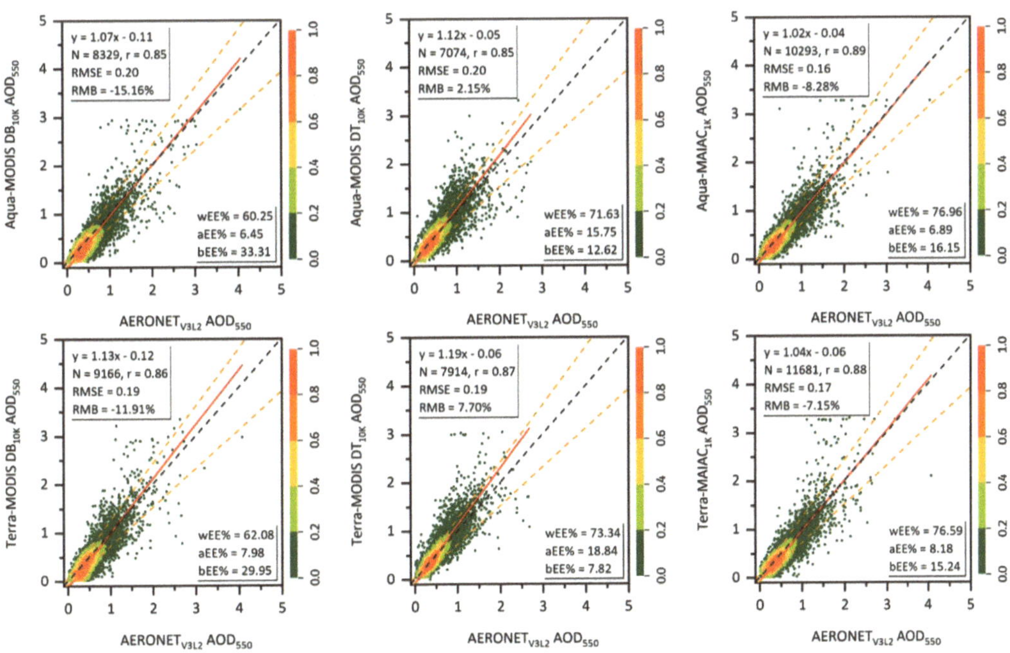

***Fig. 2.3** Validation of Aqua and Terra MODIS DT, DB, and MAIAC AOD retrievals against AERONET measurements from 2003 to 2017.*

of the Terra MODIS retrieved AOD is somewhat larger than 1, indicating that Terra overestimates the AOD by 2–5%.

This comparison provides an overview of the performance of the different algorithms applied to the same satellite measurements over the same area. The algorithm developer groups are continuously improving the quality of the products, and such analysis is very helpful for them to improve the retrieval accuracy of their algorithms. It is also helpful for users to recommend which algorithm to use. However, the user also be aware that the performance of the algorithms varies regionally, and regional analysis is required before applying the product in any application.

2.6 Conclusion

During the last two decades, the number of satellite-based sensors suitable for the retrieval of aerosol properties has increased, with improved technology leading to more accurate results and the availability of a larger suite of aerosol parameters. Some of the new sensors are specifically designed for aerosol retrievals, such as MISR, POLDER, or the Directional Polarization Camera (DPC) flying on the Chinese Geofen-5 satellite, and future sensors such as the EUMETSAT Multi-Viewing Multi-Channel Multi-Polarisation Imaging (3MI) mission or the Multi-Angle Polarimeter instrument SPEXone on the US PACE mission. Not only sensor technology, but also the development and improvement of advanced aerosol retrieval algorithms contribute to the availability of new and improved aerosol products from satellite sensors [6,58,59]. Each retrieval algorithm uses different methods and assumptions for AOD retrieval. The common feature in all algorithms is to separate the aerosol reflectance from the total reflectance at the TOA. Then using this reflectance to estimate the AOD by comparing it with the simulated reflectance in radiative transfer models. However, the two major sources of the uncertainties in AOD retrieval are estimating the surface reflectance and the assumption related to aerosol properties used in aerosol models. Uncertainties in estimating the surface reflectance affect the AOD retrieval, especially for low aerosol loading conditions when the TOA reflectance is dominated by reflectance from the surface. On the other hand, the small uncertainties in aerosol properties, mainly SSA and aerosol size distribution, induce significant error in AOD retrieval, mainly at high aerosol loading conditions when the TOA reflectance is dominated by aerosol reflectance. Therefore, evaluating the uncertainties in AOD retrieval is essential at both global and regional scales. Here, we presented a framework that is commonly used to evaluate the AOD retrieval from passive sensors measurements. In addition, we have presented a case study over South Asia for AOD retrieval from three major aerosol retrieval algorithms applied to MODIS Terra and Aqua measurements and assessed the accuracy of AOD products. We highlighted the major strengths and limitations of each algorithm. Our results show that the high spatial resolution of the MAIAC AOD outperformed DT and

DB AOD products in terms of retrieval accuracy and spatial coverage. DB AOD shows a significant underestimation over SA, while the DT performance is comparable with that of MAIAC, but with the limitation of lower spatial resolution and smaller coverage.

References

[1] J.P. Burrows, P. Borrell, U. Platt, The Remote Sensing of Tropospheric Composition from Space, Springer Nature, Springer-Verlag Berlin Heidelberg, 2011, https://doi.org/10.1007/978-3-642-14791-3.

[2] A.A. Kokhanovsky, G. de Leeuw, Satellite Aerosol Remote Sensing over Land, Springer, Berlin, Heidelberg, 2009.

[3] L. Sogacheva, T. Popp, A.M. Sayer, O. Dubovik, M.J. Garay, A. Heckel, N.C. Hsu, H. Jethva, R.A. Kahn, P. Kolmonen, M. Kosmale, G. de Leeuw, R.C. Levy, P. Litvinov, A. Lyapustin, P. North, O. Torres, A. Arola, Merging regional and global aerosol optical depth records from major available satellite products, Atmos. Chem. Phys. 20 (2020) 2031–2056.

[4] P. Kolmonen, L. Sogacheva, T.H. Virtanen, G. de Leeuw, M. Kulmala, The ADV/ASV AATSR aerosol retrieval algorithm: Current status and presentation of a full-mission AOD dataset, Int. J. Digital Earth 9 (2015) 545–561.

[5] M. Bilal, M. Nazeer, J.E. Nichol, Validation of MODIS and VIIRS derived aerosol optical depth over complex coastal waters, Atmos. Res. 186 (2017) 43–50.

[6] G. de Leeuw, T. Holzer-Popp, S. Bevan, W.H. Davies, J. Descloitres, R.G. Grainger, J. Griesfeller, A. Heckel, S. Kinne, L. Klüser, P. Kolmonen, P. Litvinov, D. Martynenko, P. North, B. Ovigneur, N. Pascal, C. Poulsen, D. Ramon, M. Schulz, R. Siddans, L. Sogacheva, D. Tanré, G.E. Thomas, T.H. Virtanen, W. von Hoyningen Huene, M. Vountas, S. Pinnock, Evaluation of seven European aerosol optical depth retrieval algorithms for climate analysis, Remote Sens. Environ. 162 (2015) 295–315.

[7] R.C. Levy, L.a. Remer, R.G. Kleidman, S. Mattoo, C. Ichoku, R. Kahn, T.F. Eck, Global evaluation of the collection 5 MODIS dark-target aerosol products over land, Atmos. Chem. Phys. 10 (2010) 10399–10420.

[8] A. Mhawish, T. Banerjee, M. Sorek-Hamer, A. Lyapustin, D.M. Broday, R. Chatfield, Comparison and evaluation of MODIS Multi-angle Implementation of Atmospheric Correction (MAIAC) aerosol product over South Asia, Remote Sens. Environ. 224 (2019) 12–28.

[9] A.M. Sayer, L.A. Munchak, N.C. Hsu, R.C. Levy, C. Bettenhausen, M.J. Jeong, MODIS Collection 6 aerosol products: Comparison between Aqua's e-Deep Blue, Dark Target, and "merged" data sets, and usage recommendations, J. Geophys. Res. Atmos. 119 (2014) 13965–13989.

[10] M. Bilal, A. Mhawish, M.A. Ali, J.E. Nichol, G.d. Leeuw, K.M. Khedher, U. Mazhar, Z. Qiu, M.P. Bleiweiss, M. Nazeer, Integration of surface reflectance and aerosol retrieval algorithms for multi-resolution aerosol optical depth retrievals over urban areas, Remote Sen. 14 (2022), 373. https:// doi.org/10.3390/rs14020373.

[11] N.C. Hsu, M.-J. Jeong, C. Bettenhausen, A.M. Sayer, R. Hansell, C.S. Seftor, J. Huang, S.-C. Tsay, Enhanced Deep Blue aerosol retrieval algorithm: The second generation, J. Geophys. Res. Atmos. 118 (2013) 9296–9315.

[12] R.C. Levy, S. Mattoo, L.A. Munchak, L.A. Remer, A.M. Sayer, F. Patadia, N.C. Hsu, The Collection 6 MODIS aerosol products over land and ocean, Atmos. Meas. Tech. 6 (2013) 2989–3034.

[13] M. Bilal, J.E. Nichol, M.P. Bleiweiss, D. Dubois, A simplified high resolution MODIS Aerosol Retrieval Algorithm (SARA) for use over mixed surfaces, Remote Sens. Environ. 136 (2013) 135–145.

[14] R.A. Kahn, B.J. Gaitley, M.J. Garay, D.J. Diner, T.F. Eck, A. Smirnov, B.N. Holben, Multiangle Imaging SpectroRadiometer global aerosol product assessment by comparison with the Aerosol Robotic Network, J. Geophys. Res. 115 (2010) D23209.

[15] X. Su, L. Wang, M. Zhang, W. Qin, M. Bilal, A High-Precision Aerosol Retrieval Algorithm (HiPARA) for Advanced Himawari Imager (AHI) data: Development and verification, Remote Sens. Environ. 253 (2021) 112221.

[16] M.J. Garay, M.L. Witek, R.A. Kahn, F.C. Seidel, J.A. Limbacher, M.A. Bull, D.J. Diner, E.G. Hansen, O.V. Kalashnikova, H. Lee, A.M. Nastan, Y. Yu, Introducing the 4.4 km spatial resolution Multi-Angle Imaging SpectroRadiometer (MISR) aerosol product, Atmos. Meas. Tech. 13 (2020) 593–628.

[17] C.A. Hostetler, K.-P. Lee, R.A. Ferrare, C.R. Trepte, Y. Hu, M.A. Vaughan, D.M. Winker, A.H. Omar, C. Kittaka, R.R. Rogers, R.E. Kuehn, Z. Liu, The CALIPSO automated aerosol classification and lidar ratio selection algorithm, J. Atmos. Ocean. Technol. 26 (2009) 1994–2014.

[18] M.H. Kim, A.H. Omar, J.L. Tackett, M.A. Vaughan, D.M. Winker, C.R. Trepte, Y. Hu, Z. Liu, L.R. Poole, M.C. Pitts, J. Kar, B.E. Magill, The CALIPSO Version 4 automated aerosol classification and Lidar ratio selection algorithm, Atmos. Meas. Tech. 11 (2018) 6107–6135.

[19] B.N. Holben, T.F. Eck, I. Slutsker, D. Tanré, J.P. Buis, A. Setzer, E. Vermote, J.A. Reagan, Y.J. Kaufman, T. Nakajima, F. Lavenu, I. Jankowiak, A. Smirnov, AERONET—A federated instrument network and data archive for aerosol characterization, Remote Sens. Environ. 66 (1998) 1–16.

[20] N. Holben, D. Tanr, A. Smirnov, T.F. Eck, I. Slutsker, W.W. Newcomb, J.S. Schafer, B. Chatenet, F. Lavenu, J. Kaufman, J.V. Castle, A. Setzer, B. Markham, D. Clark, R. Halthore, A. Karneli, N.T.O. Neill, C. Pietras, T. Pinker, K. Voss, G. Zibordi, An emerging ground-based aerosol climatology: Aerosol optical depth from AERONET, J. Geophys. Res. Atmos. 106 (2001) 12067–12097.

[21] Y.J. Kaufman, D. Tanré, L.A. Remer, E.F. Vermote, A. Chu, B.N. Holben, Operational remote sensing of tropospheric aerosol over land from EOS moderate resolution imaging spectroradiometer, J. Geophys. Res. Atmos. 102 (1997) 17051–17067.

[22] R.C. Levy, L.A. Munchak, S. Mattoo, F. Patadia, L.A. Remer, R.E. Holz, Towards a long-term global aerosol optical depth record: applying a consistent aerosol retrieval algorithm to MODIS and VIIRS-observed reflectance, Atmos. Meas. Tech. 8 (2015) 4083–4110.

[23] L.a. Remer, Y.J. Kaufman, D. Tanré, S. Mattoo, D.a. Chu, J.V. Martins, R.-R. Li, C. Ichoku, R.C. Levy, R.G. Kleidman, T.F. Eck, E. Vermote, B.N. Holben, The MODIS aerosol algorithm, products, and validation, J. Atmos. Sci. 62 (2005) 947–973.

[24] K.R. Kumar, N. Kang, Y. Yin, Classification of key aerosol types and their frequency distributions based on satellite remote sensing data at an industrially polluted city in the Yangtze River Delta, China, Int. J. Climatol. 38 (2018) 320–336.

[25] M. Kumar, N. Ojha, N. Singh, Atmospheric aerosols from open burning in South and Southeast Asia. In: R.P. Singh, Asian Atmospheric Pollution, Elsevier, Amsterdam, Netherlands, 2022, pp. 75–96.

[26] N. Ojha, A. Sharma, M. Kumar, I. Girach, T.U. Ansari, S.K. Sharma, N. Singh, A. Pozzer, S.S. Gunthe, On the widespread enhancement in fine particulate matter across the Indo-Gangetic Plain towards winter, Sci. Rep. 10 (2020) 5862.

[27] M.A. Ali, M.M. Islam, M.N. Islam, M. Almazroui, Investigations of MODIS AOD and cloud properties with CERES sensor based net cloud radiative effect and a NOAA HYSPLIT Model over Bangladesh for the period 2001–2016, Atmos. Res. 215 (2019) 268–283.

[28] M. Bilal, A. Mhawish, J.E. Nichol, Z. Qiu, M. Nazeer, M.A. Ali, G. de Leeuw, R.C. Levy, Y. Wang, Y. Chen, L. Wang, Y. Shi, M.P. Bleiweiss, U. Mazhar, L. Atique, S. Ke, Air pollution scenario over Pakistan: Characterization and ranking of extremely polluted cities using long-term concentrations of aerosols and trace gases, Remote Sens. Environ. 264 (2021) 112617.

[29] A. Mhawish, M. Sorek-Hamer, R. Chatfield, T. Banerjee, M. Bilal, M. Kumar, C. Sarangi, M. Franklin, K. Chau, M. Garay, O. Kalashnikova, Aerosol characteristics from earth observation systems: A comprehensive investigation over South Asia (2000–2019), Remote Sens. Environ. (2021) 259.

[30] M.N. Islam, M.A. Ali, M.M. Islam, Spatiotemporal investigations of aerosol optical properties over Bangladesh for the period 2002–2016, Earth Syst. Environ. 3 (2019) 563–573.

[31] A. Mhawish, T. Banerjee, M. Sorek-Hamer, M. Bilal, A.I. Lyapustin, R. Chatfield, D.M. Broday, Estimation of high-resolution PM2.5 over the Indo-Gangetic plain by fusion of satellite data, meteorology, and land use variables, Environ. Sci. Technol. 54 (2020) 7891–7900.

[32] S.A. Ackerman, K.I. Strabala, W.P. Menzel, R.A. Frey, C.C. Moeller, L.E. Gumley, Discriminating clear sky from clouds with MODIS, J. Geophys. Res. Atmos. 103 (1998) 32141–32157.

[33] L. Sogacheva, P. Kolmonen, T.H. Virtanen, E. Rodriguez, G. Saponaro, G. de Leeuw, Post-processing to remove residual clouds from aerosol optical depth retrieved using the Advanced Along Track Scanning Radiometer, Atmos. Meas. Tech. 10 (2017) 491–505.

[34] M. Bilal, M. Nazeer, J.E. Nichol, Z. Qiu, L. Wang, M.P. Bleiweiss, X. Shen, J.R. Campbell, S. Lolli, Evaluation of Terra-MODIS C6 and C6.1 Aerosol Products against Beijing, XiangHe, and Xinglong AERONET Sites in China during 2004-2014, Remote Sens. 11 (2019) 486.

[35] M. Bilal, J.E. Nichol, M. Nazeer, Validation of Aqua-MODIS C051 and C006 operational aerosol products using AERONET measurements over Pakistan, IEEE J. Selected Top. Appl. Earth Observ. Remote Sens. 9 (2016) 2074–2080.

[36] M. Bilal, Z. Qiu, Evaluation of MODIS C6 combined aerosol product at global scale, IGARSS 2018: 2018 IEEE International Geoscience and Remote Sensing Symposium, 2018, pp. 9126–9129.

[37] M. Bilal, Z. Qiu, J.E. Nichol, A. Mhawish, M.A. Ali, K.M. Khedher, G. de Leeuw, W. Yu, P. Tiwari, M. Nazeer, M.P. Bleiweiss, Uncertainty in Aqua-MODIS aerosol retrieval algorithms during COVID-19 lockdown, IEEE Geosci. Remote Sens. Lett. 19 (2022) 1–5.

[38] A. Mhawish, T. Banerjee, D.M. Broday, A. Misra, S.N. Tripathi, Evaluation of MODIS Collection 6 aerosol retrieval algorithms over Indo-Gangetic Plain: Implications of aerosols types and mass loading, Remote Sens. Environ. 201 (2017) 297–313.

[39] A.M. Sayer, N.C. Hsu, C. Bettenhausen, M.-J. Jeong, Validation and uncertainty estimates for MODIS Collection 6 "Deep Blue" aerosol data, J. Geophys. Res. Atmos. 118 (2013) 7864–7872.

[40] A.M. Sayer, N.C. Hsu, C. Bettenhausen, M.J. Jeong, G. Meister, Effect of MODIS Terra radiometric calibration improvements on Collection 6 Deep Blue aerosol products: Validation and Terra/Aqua consistency, J. Geophys. Res. Atmos. 120 (2015) 12157–12174.

[41] D. Tanré, Y.J. Kaufman, M. Herman, S. Mattoo, Remote sensing of aerosol properties over oceans using the MODIS/EOS spectral radiances, J. Geophys. Res. Atmos. 102 (1997) 16971–16988.

[42] N.C. Hsu, S.-C. Tsay, M.D. King, J.R. Herman, Deep Blue retrievals of Asian aerosol properties during ACE-Asia, IEEE Trans. Geosci. Remote Sens. 44 (2006) 3180–3195.

[43] A. Lyapustin, Y. Wang, S. Korkin, D. Huang, MODIS Collection 6 MAIAC algorithm, Atmos. Meas. Tech. 11 (2018) 5741–5765.

[44] D.J. Diner, J.C. Beckert, T.H. Reilly, C.J. Bruegge, J.E. Conel, R.A. Kahn, J.V. Martonchik, T.P. Ackerman, R. Davies, S.A.W. Gerstl, H.R. Gordon, J. Muller, R.B. Myneni, P.J. Sellers, B. Pinty, M.M. Verstraete, Multi-angle Imaging SpectroRadiometer (MISR) instrument description and experiment overview, IEEE Trans. Geosci. Remote Sens. 36 (1998) 1072–1087.

[45] R.A. Kahn, B.J. Gaitley, An analysis of global aerosol type as retrieved by MISR, J. Geophys. Res. Atmos. 120 (2015) 4248–4281.

[46] P.F. Levelt, G.H.J. van den Oord, M.R. Dobber, A. Malkki, V. Huib, V. Johan de, P. Stammes, J.O.V. Lundell, H. Saari, The ozone monitoring instrument, IEEE Trans. Geosci. Remote Sens. 44 (2006) 1093–1101.

[47] O. Torres, C. Ahn, Z. Chen, Improvements to the OMI near-UV aerosol algorithm using A-train CALIOP and AIRS observations, Atmos. Meas. Tech. 6 (2013) 3257–3270.

[48] O. Torres, A. Tanskanen, B. Veihelmann, C. Ahn, R. Braak, P.K. Bhartia, P. Veefkind, P. Levelt, Aerosols and surface UV products from ozone monitoring instrument observations: An overview, J. Geophys. Res. 112 (2007) D24S47.

[49] A. Smirnov, B.N. Holben, D.M. Giles, I. Slutsker, N.T. O'Neill, T.F. Eck, A. Macke, P. Croot, Y. Courcoux, S.M. Sakerin, T.J. Smyth, T. Zielinski, G. Zibordi, J.I. Goes, M.J. Harvey, P.K. Quinn, N.B. Nelson, V.F. Radionov, C.M. Duarte, R. Losno, J. Sciare, K.J. Voss, S. Kinne, N.R. Nalli, E. Joseph, K. Krishna Moorthy, D.S. Covert, S.K. Gulev, G. Milinevsky, P. Larouche, S. Belanger, E. Horne, M. Chin, L.A. Remer, R.A. Kahn, J.S. Reid, M. Schulz, C.L. Heald, J. Zhang, K. Lapina, R.G. Kleidman, J. Griesfeller, B.J. Gaitley, Q. Tan, T.L. Diehl, Maritime aerosol network as a component of AERONET: First results and comparison with global aerosol models and satellite retrievals, Atmos. Meas. Tech. 4 (2011) 583–597.

[50] C. Ichoku, A spatio-temporal approach for global validation and analysis of MODIS aerosol products, Geophys. Res. Lett. 29 (12) (2002) 1616, doi: 10.1029/2001GL013206.

[51] T.F. Eck, B.N. Holben, J.S. Reid, O. Dubovik, A. Smirnov, N.T. O'Neill, I. Slutsker, S. Kinne, Wavelength dependence of the optical depth of biomass burning, urban, and desert dust aerosols, J. Geophys. Res. Atmos. 104 (1999) 31333–31349.

[52] M. Bilal, M. Nazeer, J.E. Nichol, M.P. Bleiweiss, Z. Qiu, E. Jäkel, J.R. Campbell, L. Atique, X. Huang, S. Lolli, A Simplified and Robust Surface Reflectance Estimation Method (SREM) for use over diverse land surfaces using multi-sensor data, Remote Sens. 11 (2019).

[53] W.V. Harper, Reduced major axis regression. In: Wiley StatsRef: Statistics Reference Online, Wiley Online Library, Hoboken, New Jersey, 2016.

[54] A.M. Sayer, N.C. Hsu, C. Bettenhausen, M.-J. Jeong, B.N. Holben, J. Zhang, Global and regional evaluation of over-land spectral aerosol optical depth retrievals from SeaWiFS, Atmos. Meas. Tech. 5 (2012) 1761–1778.

[55] A. Smirnov, B.N. Holben, Y.J. Kaufman, O. Dubovik, T.F. Eck, I. Slutsker, C. Pietras, R.N. Halthore, Optical properties of atmospheric aerosol in maritime environments, J. Atmos. Sci. 59 (2002) 501–523.

[56] A. Sinyuk, B.N. Holben, T.F. Eck, D.M. Giles, I. Slutsker, S. Korkin, J.S. Schafer, A. Smirnov, M. Sorokin, A. Lyapustin, The AERONET Version 3 aerosol retrieval algorithm, associated uncertainties and comparisons to Version 2, Atmos. Meas. Tech. 13 (2020) 3375–3411.

[57] Y. Wang, M.A. Ali, M. Bilal, Z. Qiu, A. Mhawish, M. Almazroui, S. Shahid, M.N. Islam, Y. Zhang, M.N. Haque, Identification of NO2 and SO2 pollution hotspots and sources in Jiangsu province of China, Remote Sens. 13 (18) (2021) 3742, https://doi.org/10.3390/rs13183742.

[58] O. Dubovik, D. Fuertes, P. Litvinov, A. Lopatin, T. Lapyonok, I. Doubovik, F. Xu, F. Ducos, C. Chen, B. Torres, Y. Derimian, L. Li, M. Herreras-Giralda, M. Herrera, Y. Karol, C. Matar, G.L. Schuster, R. Espinosa, A. Puthukkudy, Z. Li, J. Fischer, R. Preusker, J. Cuesta, A. Kreuter, A. Cede, M. Aspetsberger, D. Marth, L. Bindreiter, A. Hangler, V. Lanzinger, C. Holter, C. Federspiel, A comprehensive description of multi-term LSM for applying multiple a priori constraints in problems of atmospheric remote sensing: GRASP algorithm, concept, and applications, Front. Remote Sens. 2 (2021) 706851. doi: 10.3389/frsen.2021.706851.

[59] T. Popp, G. de Leeuw, C. Bingen, C. Brühl, V. Capelle, A. Chedin, L. Clarisse, O. Dubovik, R. Grainger, J. Griesfeller, A. Heckel, S. Kinne, L. Klüser, M. Kosmale, P. Kolmonen, L. Lelli, P. Litvinov, L. Mei, P. North, S. Pinnock, A. Povey, C. Robert, M. Schulz, L. Sogacheva, K. Stebel, D. Stein Zweers, G. Thomas, L. Tilstra, S. Vandenbussche, P. Veefkind, M. Vountas, Y. Xue, Development, production and evaluation of aerosol climate data records from European Satellite Observations (Aerosol_cci), Remote Sens. 8 (5) (2016) 421, https://doi.org/10.3390/rs8050421.

[60] N.C. Hsu, J. Lee, A.M. Sayer, N. Carletta, S.H. Chen, C.J. Tucker, B.N. Holben, S.C. Tsay, Retrieving near-global aerosol loading over land and ocean from AVHRR, J. Geophys. Res. Atmos. 122 (2017) 9968–9989.

[61] A.M. Sayer, N.C. Hsu, J. Lee, N. Carletta, S.H. Chen, A. Smirnov, Evaluation of NASA Deep Blue/SOAR aerosol retrieval algorithms applied to AVHRR measurements, J. Geophys. Res. Atmos. 122 (2017) 9945–9967.

[62] O. Torres, Total Ozone Mapping Spectrometer measurements of aerosol absorption from space: Comparison to SAFARI 2000 ground-based observations, J. Geophys. Res. 110 (2005) D10S18.

[63] O. Dubovik, T. Lapyonok, P. Litvinov, M. Herman, D. Fuertes, F. Ducos, B. Torres, Y. Derimian, X. Huang, A. Lopatin, A. Chaikovsky, M. Aspetsberger, C. Federspiel, GRASP: A versatile algorithm for characterizing the atmosphere, SPIE Newsroom (2014) 1-4, doi: 10.1117/2.1201408.005558.

[64] A.M. Sayer, N.C. Hsu, J. Lee, C. Bettenhausen, W.V. Kim, A. Smirnov, Satellite Ocean Aerosol Retrieval (SOAR) algorithm extension to S-NPP VIIRS as part of the 'Deep Blue' aerosol project, J. Geophys. Res. Atmos. 123 (2018) 380–400.

[65] C. Cao, F.J. De Luccia, X. Xiong, R. Wolfe, F. Weng, Early on-orbit performance of the visible infrared imaging radiometer suite onboard the Suomi National Polar-Orbiting Partnership (S-NPP) satellite, IEEE Trans. Geosci. Remote Sens. 52 (2014) 1142–1156.

CHAPTER 3

Global Navigation Satellite Systems and their applications in remote sensing of atmosphere

Sushil Kumar
School of Information Technology, Engineering, Mathematics and Physics, The University of the South Pacific, Suva, Fiji

3.1 Introduction

Global Navigation Satellite Systems (GNSSs) were originally designed to provide positioning, navigation, and timing (PNT) services including for worldwide civilian applications. GNSS is one of the most important innovations of the last century in the fields of surveying and navigation. Nowadays, GNSSs provide a range of services, positioning, timing, navigation, mapping, public safety surveillance and geographic surveys, time standards, mapping, and weather. In today's technological society, GNSSs have become another pillar along with the internet and mobile communications. The current smart cellphones have integrated the PNT functionality into a single CMOS chip (https://www.gpsworld.com/dual-band-gnss-market-moving-from-insignificant-to-billions-in-less-than-5-years/) at a very reasonable cost. The GNSSs are also used in weather forecasting, natural hazards, climate, environment (e.g., [1–4]) and remote sensing of the atmosphere and ionosphere (e.g., [5–9]).

On October 4, 1957, the first artificial satellite known as *Sputnik*, with a smaller payload, was launched by the Soviet Union to provide information about the space environment and the ionosphere. It carried with it a radio transmitter which transmitted 20–40 MHz signals and provided the first measurements of the ionosphere from above, especially the information about the upper ionosphere [10]. Following the launch of *Sputnik*, the United States in 1958 launched its first satellite called the *Explorer*. US military developed first satellite systems such as Transit, Timation, and then NAVSTAR with basic technology that is used today. The first NAVSTAR was launched in 1989 and the 24th satellite in 1994 and then its full operational capability was declared in April 1995. Today, over 8000 satellites have been sent into space forming the GNSSs and for other purposes of communication and remote sensing.

The PNT technologies play an essential role in our modern society which depends upon the availability of the Global Positioning System (GPS) and the growing family of GNSSs. There are currently four fully operational GNSSs and two regional navigation satellite systems (RNSSs). Two important and well-advanced GNSSs include the Global

Atmospheric Remote Sensing
DOI: https://doi.org/10.1016/B978-0-323-99262-6.00006-7

Positioning System (GPS) of the United States of America and the Global Navigation Satellite System (GLONASS) of the Russian Federation with both systems have 24 satellites. GPS (NAVSTAR) satellites orbit at an altitude of approximately 20,200 km above the earth's surface in three orbital planes, compared with 19,100 km for GLONASS. GNSSs are being modernized for better GNSS-based services and applications which include an increase in the number of transmission frequencies and changes to the signal components. Two other GNSSs are the European Satellite Navigation System (Galileo) of the European Union and the BeiDou Navigation Satellite System of China, currently being developed and deployed, which will enhance the quality of services and increase the number of potential users and applications. Galileo system consists of 26 satellites positioned in three circular medium earth orbit (MEO) planes at 23,222 km altitude. BeiDou is also known as COMPASS which is China's independent regional and global navigation system, having a constellation of 30 MEO satellites at an altitude of 21,150 km and 5 geostationary orbits (GEO) satellites at 36,000 km. At the regional scale, the Indian Regional Navigation Satellite System (IRNSS also called NavIC) and Japan's Quasi-Zenith Satellite System (QZSS), together with several regional augmentation systems are also available. IRNSS has been mainly designed to provide accurate position information service to assist in the navigation of ships in the Indian Ocean waters. IRNSS consists of 3 satellites in geostationary orbit at 34, 83, and 132° east longitude and 4 satellites in geosynchronous orbit placed at an inclination of 29 degrees with longitude crossing at 55 and 111 degrees east. IRNSS uses L5(1191.795 MHz) and S(2491.005 MHz) signals for positioning services. QZSS is designed such that there is at least one satellite near zenith over Japan at any given time. The system uses three satellites in elliptical and inclined GEO at 120° apart and passing over the same ground track. It aims to work in combination with GPS and Galileo to improve its accuracy and services. For the position determination by GNSS systems, a user requires to see (i.e., have a direct line of sight to) at least four satellites simultaneously at any one time. For more detailed information about different GNSSs and RNSSs reader is referred to the "Interoperable Global Navigation Satellite Systems Space Service Volume" [11] (https://www.unoosa.org/res/oosadoc/data/documents/2018/stspace/sfro QZSStspace75_0_html/st_space_75E.pdf) and individual chapters of the book "Position, Navigation, and Timing Technologies in the 21st Century" edited by Morton et al. [12].

3.2 Performance of the GNSS

Table 3.1 gives the summary of frequency bands, carrier to noise C/No (dBHz) ratio, civilian position accuracy, coding (modulation) used by GNSSs position accuracy and services provided. The C/No is the ratio of carrier power to the noise power mixed with the signal, in a 1 Hz bandwidth which defines a limit for the GPS receiver sensitivity and varies with satellite elevation. C/No in any GNSS receiver is used as a measure

Table 3.1 Summary of properties and services of GNSS systems.

Constellation	User frequency bands (MHz)	C/No (dBHz)	Civilian position accuracy (meter)	User coding and modulation	Services
GPS/ NAVSTAR	L1 = 1575.420 L2 = 1227.6 L5 = 1176.45	37–45	7.1	CDMA BPSK	SPS on L1 PPS on L1 and L2
GLONASS	G1 = 1602.0 G2 = 1246.0 G3 = 1204.704	20–45	7.5	FDMABPSK+ CDMA on GLONASS-K1	Open on E1 with 24.552 MHz BW Open on E5a+E5b
Galileo	E1 = 1575.42 E6 = 1278.75 E5 = 1191.795 E5a = 1176.45 E5b = 1207.14	30–55	4	CDMA BOC and BPSK	Open on E1 and E5a+E5b
BeiDou	B1 = 1561.098 B2 = 1207.14 B3 = 1268.52	31–48 on B1 signal	10	QPSK and BOC	Open on B2
QZSS	L1 = 1575.420 L2 = 1227.6 L5 = 1176.45 LEX = 1278.75 (for differential GPS	35–53	1.6	CDMA, BPSK, BOC	Public regulated services
IRNSS	L5 = 1191.79 S = 2491.005 C = 3400	44.79 on L5		BPSK, BOC, CDMA	SPS and PPS

CDMA, code division multiple access; *FDMA*, frequency division multiple access; *BPSK*, binary phase shift keying; *QPSK*, quadrature phase shift keying; *BOC*, binary offset carrier; *SPS*, standard positioning service; *PPS*, precision positioning service.

of signal quality and also to determine the lock of the carrier and code tracking loops. In a typical GNSS receiver, the signals from multiple satellites are weighted according to their C/No ratios such that the computed position is most reliant on the signal with the best C/No which tells the user about the sufficient strength that can be measured with a given instrument to form a usable measurement. The position measurements need to be corrected for the errors (or biases) related to each satellite measurement due to ionospheric and tropospheric delay errors, Multipath (reflections, diffraction and scattering of the incident radio waves from structure of the surrounding environments), RF interference, receiver noise and clock offset, orbital errors, etc. Ionospheric delay is one of the largest sources of error in the GNSSs and RNSSs for the signal for a static receiver.

The C/No given in Table 3.1 are from specific cases reported in the literature and have a large variability and comparatively less variability in position accuracy. The performance of the GNSS systems can be deteriorated under severe space weather conditions and more frequently under the high solar activity period due to ionospheric scintillations. Goswami et al. [13] have studied the relative performance of different navigational satellite constellations (GPS, GLONASS, Galileo) under adverse ionospheric conditions associated with equatorial scintillations for the equinoctial months of 2014 and March 2016 at stations Calcutta (22.58°N 88.38°E geographic; 32°N magnetic dip) and for September 2016 from Siliguri (26.72°N, 88.39°E; 39.49°N magnetic dip), located near the northern crest of equatorial ionization anomaly in the Indian longitude sector. They reported C/No ratio fluctuations of 36.6 to 60% more over limited ionospheric conditions at both stations.

The percentage of signal availability and maximum outage duration (MOD) are the other two important parameters that tell the user about the usability of a particular GNSS system. MOD is the longest duration that a user can expect to be without a signal and is measured in minutes. The percentage of signal availability for at least one signal (L1/E1/B1) for four GNSSs (GPS, GLONASS, Galileo, BeiDou) is about 100% and 99.6% for QZSS, whereas for four or more signals it varies from 79.4% to 99.8%. The MOD for one signal is zero minute for GPS, GLONASS and BeiDou, and 11 minutes for Galileo and 197 minutes for QZSS. However, for four or more than four signals MOD can vary from 0 (BeiDou) to 60 minutes (Galileo). For further details about the global performance estimates of the percentage of time for five GNSS signals (GPS, GLONASS, Galileo BeiDou, QZSS) and MOD, a reader is referred to [11] (page 20, Table 5.2 and full volume on https://www.unoosa.org/res/oosadoc/data/documents/2018/stspace/sfro QZSStspace75_0_html/st_space_75E.pdf).

GNSS-based technology continues to develop as a tool for international cooperation among the satellite operators of the current and planned systems. The growing need to locate objects accurately and reliably has wide-ranging implications for the management of the environment and disaster warning and emergency response, among other areas. Within the United Nations International Committee on GNSS (ICG), there is an initiative underway to ensure that GNSS signals within the Space Service Volume (SSV) are available and interoperable across all international global constellations and regional augmentations https://www.unoosa.org/res/oosadoc/data/documents/2018/stspace/stspace75_0_html/st_space_75E.pdf). This initiative is being carried out within the ICG Working Group B (WG-B) on "Enhancement of GNSS Performance, New Services and Capabilities." For further full details of GNSSs and RNSSs, the reader is referred to the book "Position, Navigation, and Timing Technologies in the 21st Century," edited by Morton et al. [12].

3.3 Technical concepts of GPS

The GPS is a satellite-based radio navigation system that provides accurate position information, the velocity of the moving receiver, and time continuously under all weather conditions around the globe. The whole system is owned and operated by the US Department of Defense (DoD) and was developed mainly for US military applications, however, it is now used widely in civilian communities in most countries with a wide range of applications. The GPS can be divided into three segments:

3.3.1 The space segment

Various constellation architectures such as NTS-1 and NTS2 were considered and finally, GPS constellation placed satellites in six circular orbits with MEO at an altitude of 20,200 km having at least 24 satellites which recently has been to 30 or 31 satellites. Each satellite weighs approximately 2000 pounds and is 17 feet across with the solar panels extended, lasting far beyond their 7.5-year design life. The width including wingspan is 38.025 feet. The six orbits are inclined at 55° up from the geographic equator and are spaced 60° apart. The orbital period of satellites is 11 hours and 58 minutes which is about half a sidereal day, meaning that each day satellites complete about two full earth revolutions [14].

The satellites transmit two radio signals. These signals consist of a C/A (coarse/acquisition) code at 1.023 MHz and a P (precision) code at 10.23 MHz bandwidths, respectively. The P-code is pseudo-random and 37 weeks long protecting interfering signals transmitted by a possible enemy and the P-code can be transmitted encrypted. The GPS transmitted two carrier frequency signals L1 = 1575.42 MHz (wavelength = 19.04 cm) and L2 = 1227.60 MHz (wavelength = 24.44 cm). A second civilian signal channel L2 Civil (L2C) was started in 2006 under further development of NAVSTAR GPS. The third new civilian signal called L5 has been allocated with a carrier frequency of 1176.45 MHz and bandwidth of 20.46 MHz for increased demand of safety of life transportation and other high performance of applications. Every GPS user can access the C/A code, whereas the P-code is only accessible to authorized users. Both L1 and L2 use code division multiple access (CDMA) as the multiplexing process and binary phase shift keying (BPSK) as the modulation method. The signals transmitted by GPS satellites are controlled by an atomic clock with an accuracy of 20–30 ns. The carrier signals L1 and L2 are biphasic modulated by codes to provide satellite clock readings and other information like orbital parameters to the receiver. The codes consist of a sequence or string of binary values of 0 or 1, while the biphasic modulation is performed by a 180° shift in the carrier phase whenever a change in the code state occurs [15]. The navigation message is modulated using the two carriers (L1 and L2) at a chipping rate of 50 bits per second (bps) which contains information on the satellite's pseudo-random code/number (PRN), ephemeris data and almanac data. PRN is a unique identification code that identifies which satellite is transmitting the information.

Ephemeris data gives information about the functioning of the satellite (whether it is working properly or not), current date and time. This is essential in determining the position. The almanac data notifies the receiver where each GPS satellite should be at any time throughout the day and each satellite transmits almanac data showing the orbital information for that satellite and every other satellite in the system [15].

3.3.2 The control segment

Control segment of GPS is responsible for its proper operation which has a master control station, monitor stations, and ground antennas. The master control station is located at Schreiber Air Force Base in the State of Colorado, USA. The motion and positions of satellites are controlled and corrected by the base stations using measurements collected by the monitor stations to predict the behavior of each satellite's orbit and clock. Five base stations are located around the globe: Ascension Island (Atlantic Ocean), Cape Canaveral, Florida, Diego Garcia (Indian Ocean), Hawaii and Kwajalein (both in the Pacific Ocean). Currently, the network is composed of 16 stations, 6 from the Air Force plus 10 from the National Geospatial-Intelligence Agency (NGA). In the event of some catastrophic failure, there are two backup master control stations, one located in Sunnyvale, California, and in Rockville, Maryland. The function of the control segment is to ensure that the space segment is operating within specifications and to provide adjustments where necessary. Control segment and the GPS satellites communicate via an S-band uplink from one of the four ground station antennas.

3.3.3 The user segment

The user segment is responsible for the determination of pseudorange and consists of five modules namely antennas, receiver, signal processing, data processing capabilities input/output device such as a control display unit, and a power supply. The pseudo ranges from three different satellites simultaneously are used to obtain user coordinates by triangulation. It solves the navigation equations to obtain coordinates of a location. Ground receivers passively receive the radio signals from each visible satellite and measure the time it takes the signal to travel to the receiver. Distance is then computed by formula $D = C \times t$, where C is the velocity of electomagnetic wave in free space (3×10^8 ms^{-1}). To get the actual distance the ionospheric and tropospheric range corrections need to be applied. Fig. 3.1 below shows all three segments of the GPS.

GPS offers two levels of navigation and positioning; standard positioning service (SPS) and precise positioning service (PPS). The SPS gives 100 m horizontal accuracy, 156 m vertical accuracy, and 167 ns time accuracy. With PPS the accuracy increases to 17.8 m horizontal accuracy, 27.7 m vertical accuracy, and 100 ns time accuracy [14]. Recently, civilian uses of GPS have far outpaced the US DoD's purpose of implementing it. GPS is being used in land, sea, air, and space navigation, search and rescue missions, surveying recreation as well as in intelligent vehicle highway systems (IVHS). A

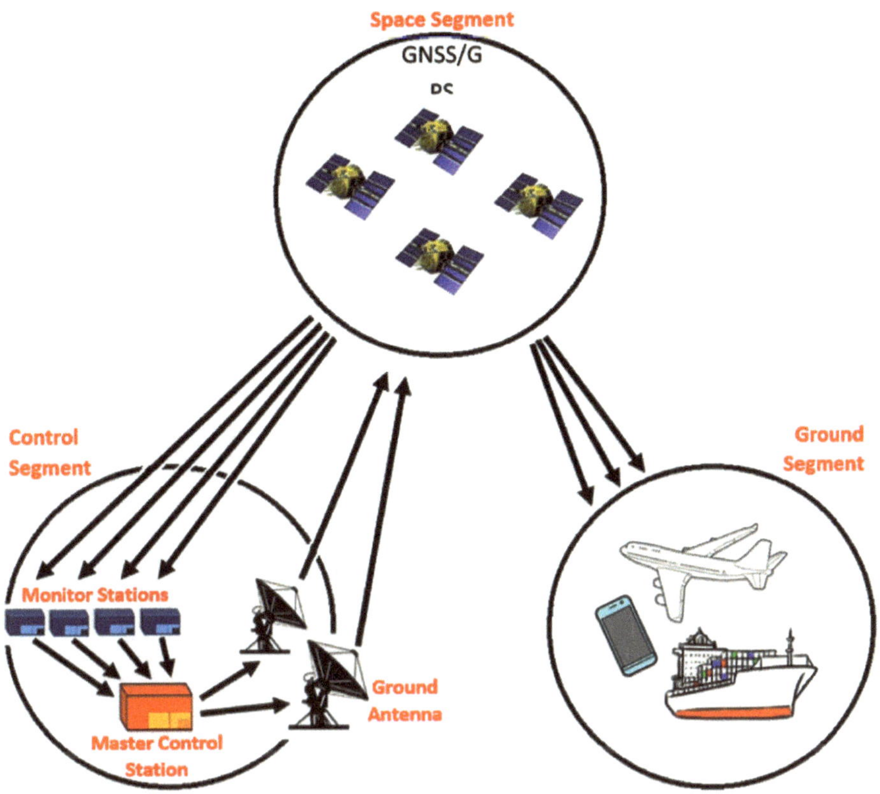

Fig. 3.1 *A block diagram of three segments of the GPS/GNSS system.*

modification to the receiver with Kalman Filter, the dual-frequency GPS receivers has been used to record the ionospheric parameters like total electron content (TEC) and ionospheric L band scintillations [7,16] under normal and space weather conditions.

3.4 Receivers and GPS antenna

The fundamental design of a GNSS receiving system is shown in Fig. 3.2. GPS receiving system consist of three basic components: an antenna, a receiver-processor unit which converts the radio signal to a useable navigation solution, a control/display unit to display the positioning information. A variety of GNSS receivers are used for different purposes such as navigation, surveying and ionospheric studies. GNSS receivers have evolved over the time for highly specialized and customized GNSS practical and scientific applications. GNSS receivers estimate the code phase, carrier frequency, and carrier phase with respect to locally generated reference signals. The carrier phase measurements are used for precise positioning services where minimization of errors is vital.

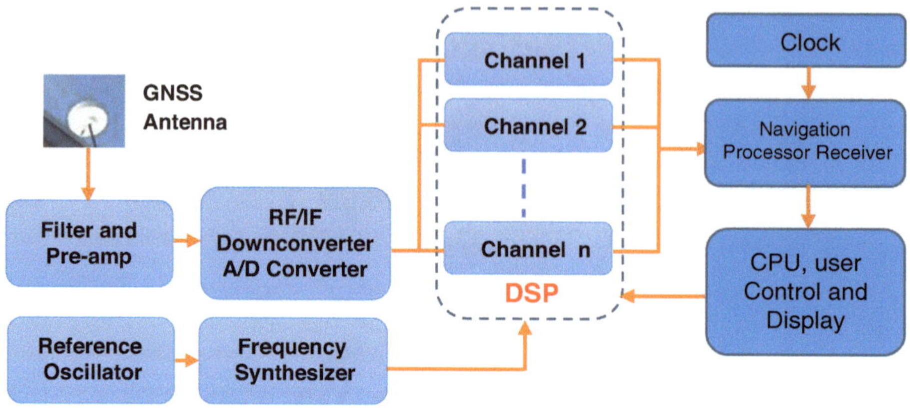

Fig. 3.2 ***Block diagram of a typical GNSS receiver.*** [Concept adapted from Chen et al. [21], Antennas for Global Navigation Satellite Systems, Chapter 2, Figure 1.9, https://onlinelibrary.wiley.com/doi/book/10.1002/9781119969518].

The PNT receivers in design using radio wave signals vary from other digital receivers such as used in ionospheric studies. The carrier phase can suffer the cycle ambiguities which need to be corrected for better positioning services. The real-time kinematic (RTK) receivers are used for high accuracy of millimeter-level carrier-phase-based relative positioning.

The GSV4004B receiver has been the most widely used for ionospheric studies (e.g., [16–19]). The GSV 4004B has 24 channels; 22 channels are used to receive L1 and L2 band signals from the satellites simultaneously and have been specifically designed for recording of TEC and ionospheric scintillations on L1 frequency (1575.42 MHz). The 23rd channel is used for the measurement of local background noise (C/No) and scintillation index S_4 correction factors. The 24th channel is configured to track a satellite-based augmentation system (SBAS) which measures phase and amplitude at a sampling rate of 50 Hz and code/carrier divergence at 1 kHz for each satellite being tracked on L1 [20]. The receiver can track up to 11 GPS satellites transmitting L1 and L2 band frequency signals. Septentrio PolaRx5S receiver is another highly capable of computing some of the most important ionospheric parameters such as TEC, change in TEC (dTEC), L1, L2 and L5 band amplitude and phase scintillations indices (S_4, σ_Φ) and scintillation index (*SI*). Septentrio PolaRx5S receiver is a multi-frequency and multi-constellation receiver designed for space weather monitoring and applications. The receiver works with a user graphical interface (UGI) logging tool (RX Tools) for continuous monitoring and recording of the data. The data is recorded in an ASCII file having a range of parameters at a sampling rate of 50 Hz.

GPS signals are right-hand circularly polarised (RHCP) electromagnetic waves. Therefore, a receiver antenna should be designed to sense RHCP waves and to reject reflected multipath signals. The GNSS signals while propagating from satellites to the ground receivers suffer attenuation, atmospheric (ionosphere and troposphere) delays, multipath and interference. Fundamental issues in antenna design for GNSS applications are almost the same as they are related to position accuracy, ionospheric frequency-dependent phase shift (group delay) due to dispersion and tropospheric delays by the refraction of the radio waves. The narrowband helical, spiral, and microstrip patch antennas could be used for practical GNSS applications. Mobile phones are most widely used for position and navigation nowadays. Due to the limitation of size GPS antennas on mobile phones are of planar inverted-F antennas (PIFAs) or inverted-F antennas (IFAs). These antennas do have lower effective antenna gain which is compensated by using low-noise amplifier (LNA) together with the antenna. The integrated GNSS receivers and broadband (wideband) CP receiving antennas are required for different GNSSs and RNSSs. For further details about GNSS antennas reader is referred to the book "Antennas For Global Navigation Satellite Systems" by Chen et al. [21] and in particular the chapter "Fundamental Considerations for GNSS Antennas." GPS 702 antenna shown in Fig. 3.3A is an active antenna designed by NovAtel (GSV4004B) to receive the GPS L1 and L2 signals. Septentrio PolaNt Choke Ring B3/E6 antenna, shown in Fig. 3.3B, is an aluminum-based high precision geodetic multi-frequency, multi-constellation choke ring antenna which supports the full range of GNSS signals, that is, GPS, GLONASS, Galileo, BeiDou, IRNSS, QZSS, and SBAS constellations. The major advantage of using this antenna is that its choke ring characteristic provides slow phase-center variation and better multi-path rejection [22] and the inclusion of low-noise amplifiers provides accurate and reliable data recording. The major advantage

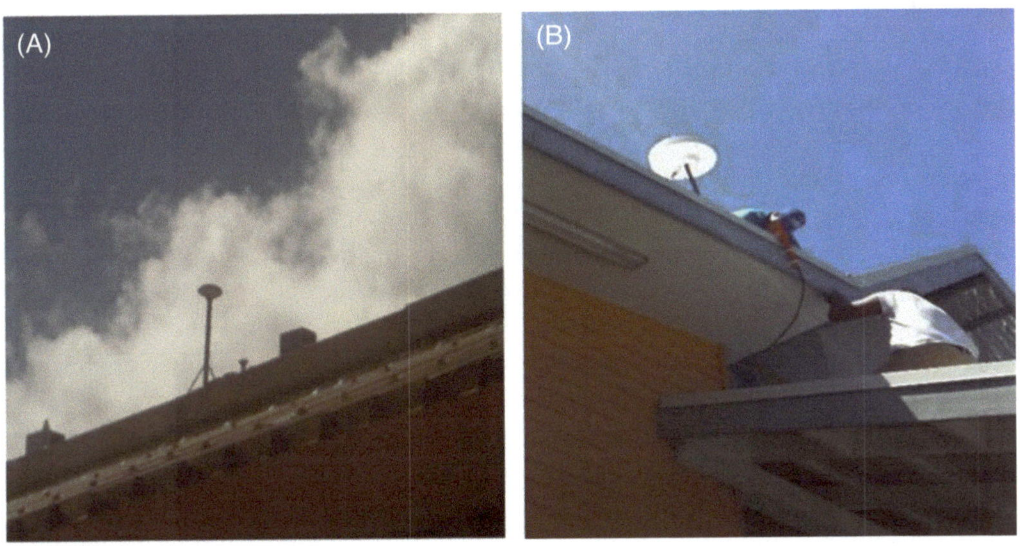

Fig. 3.3 (A) GPS 702 antenna installed at a USP site, Fiji, (B) Choke Ring B3/E6 antenna at USP site in Tarawa, Kiribati (Geog. lat. 1.33°N, long. 173.01°E).

of using this antenna is that its choke ring characteristic provides slow phase-center variation and better multi-path rejection [22] and the inclusion of low-noise amplifiers provides accurate and reliable data recording.

A simple GPS receiver only makes use of one global navigation satellite system, while multi-constellation GNSS receivers receive a multitude of GNSS signals at the same time. The multi-frequency GNSS receivers have several advantages including removal of ionospheric errors, radiofrequency interference robustness, better multipath rejection, better accuracy, additional spoofing detection etc. Ground-based receivers utilized by International GNSS Service (IGS) network do provide the continuous monitoring of the tropospheric vapour content and the ionospheric electron content. The GNSS receiver for atmospheric sounding (GRAS) instrument flown on the first European operational weather satellite MetOp [23] provided GPS occultation measurements and navigation data based on which bending angles and meteorological data are derived. The GRAS has provided a wealth of rising and setting occultation measurements [24]. The integrated GPS and occultation receiver (IGOR) (https://www.climate-policy-watcher.org/radio-occultation/igor-receiver-and-gps-antennas.html, https://www.researchgate.net/figure/IGOR-GPS-receiver-TerraSAR-X-flight-unit_fig1_265568568) manufactured by Broad Reach Engineering has flown on numerous research satellites.

3.5 GNSS limitations

The atmospheric delay consisted of tropospheric and ionospheric propagation delays can seriously affect the high accuracy of the GNSS which needs to be mitigated. The maximum signal delay producing error comes from the ionosphere. The tropospheric error of about 2.5 m in the zenith direction for a site located at sea level is typically small compared to ionospheric error. The troposphere (free of charged particles) is the lowermost region of the earth's atmosphere whereas the ionosphere extends over an altitude range of 50–1000 km and consists of sufficient ionization (ions and electrons) to refract the radio waves. The tropospheric delay does not depend on signal frequency whereas ionospheric delay varies roughly with the inverse square of the signal frequency. The major limiting factor in the accuracy of GPS is ionospheric refraction. The GPS uses an ionospheric correction model to account for this refraction. However, due to the dynamic nature of the ionosphere, this model does not compensate for all the parameters and as a result the ionospheric refraction influxes an error of up to 30 m in space coordinates [14] due to a delay in the code information. The ionospheric delay also depends upon the location of the user in the equatorial, mid-latitude, and high latitude belts. The US DoD started measuring the ionospheric group path delay in the early 1980s on the two carrier frequencies L1 (1575.45 MHz) and L2 (1227.60 MHz) to correct the ionospheric time delay for better accuracy. The two frequencies transmitted by the GPS are the 154th and 120th harmonics of 10.23 MHz. This 10.23 MHz is bi-phase modulated on both carriers with a pseudorandom code resulting in a [(sin x)/x]2 shaped spectrum with 20 MHz width to the first nulls [25]. It has been found that at night the ionospheric delay is approximately five to ten times less than that in the daytime [26]. Multipath is another significant source of measurement error which occurs due to the satellite trajectories and the environment surrounding the antenna. High-accuracy systems do provide corrections to users to improve their PNT performance beyond that provided by a GNSS service. Apart from the propgational errors, the othere sources of errors having implications in the PNT services are satellite clock error, ephemeris error, receiver noise, etc. [27]. The satellite-based augmentation system (SBAS) and the ground-based augmentation system (GBAS), are two of such systems designed to improve GNSS service such that the augmented service meets the requirements of air navigation such as flight takeoff, ascent, enroute, terminal area, and approach. Navigation systems designed for use in aviation are judged from accuracy, integrity, continuity, and availability, and the four parameters are considered in designing a navigation system. GBAS systems have been installed at airports of several countries such as the United States, Europe, Australia, India, Brazil, and Russia. For further details about GBAS and SBAS the reader is referred to chapters 12 and 13 of the book "Position, Navigation, and Timing Technologies in the 21st Century".

3.6 GPS/GNSS meteorology

The atmospheric remote sensing using both ground-based and satellite-based methods have opened unique opportunities for scientists to address the issues such as weather, climate, global warming, environment, and sustainability. The availability of GNSS signals globally and under all weather conditions has provided a popular passive sensing tool for the earth surface, the troposphere (neutral atmosphere), ionosphere under normal and severe terrestrial and space weather conditions. Though this section aims to describe how tropospheric error could be used for GPS/GNSS meteorology, the ionospheric error has also been described briefly. In the early 1990's, scientists developed techniques to utilize tropospheric delays to determine the tropospheric parameters such as temperature, pressure, and amount of integrated water vapor (IWV), which is also known by other names in the literature, such as precipitable water vapor (PWV), precipitable water (PW), integrated precipitable water vapor (IPWV), and so forth. The use of the effect of the atmosphere on the propagation of the GPS radio signals to derive information on the state of the atmosphere (neutral) created a new science called *GPS/GNSS Meteorology*. GNSS meteorology comprises three branches: GNSS reflectometry (GNSS-R), GNSS refractometry (known as GNSS radio occultation or in short GNSS-RO) and GNSS ground-based measurements. The GNSS-RO, an active microwave limb sounding technique [28,29] was first developed at the Jet Propulsion Laboratory, USA and used by Stanford University to study the planetary atmospheres and ionospheres. It utilizes the GNSS receivers installed on low earth orbit (LEO) satellites to measure the refracted radio signals coupled with ground-based atmospheric sounding techniques that use continuously operating reference station (CORS) networks for the determination of water vapor (WV) variability forms the basis for GNSS meteorology [30]. Yu et al. [30] have presented an overview of atmospheric remote sensing using the GNSS technique and addressed the relevant challenging issues associated with GNSS remote sensing. A radio occultation event (ROE) occurs when a GPS satellite is setting or rising behind the earth's limb as viewed by a receiver onboard a low earth orbit LEO satellite as shown in Fig. 3.4.

During an ROE, the GPS signals propagating to the LEO satellite are delayed due to the intervening atmosphere and ionosphere in the path, and their path is bent by gradients in the refractivity field. This delay can be converted into a profile of bending angles and consequently processed to produce a refractivity index profile. GPS RO provides global coverage and all-weather sounding capability with high vertical resolution, high precision, and accuracy. RO measurements provide temperature data for the stratosphere and upper troposphere, and WV data for the troposphere. The RO data have a high vertical resolution of 0.5–1 km, but a low horizontal resolution of ~200 km. The GNSS-RO brings significant benefits to the southern hemisphere because of limited access to ground-based observations.

The LEO satellites such as CHAllenging Minisatellite Payload (CHAMP), Gravity Recovery and Climate Experiment (GRACE), and Constellation Observing System for Meteorology, Ionosphere, and Climate (COSMIC) have contributed significantly to the GNSS atmosphere remote sensing from LEO satellite orbit heights down to the earth's surface [31,32]. However, due to the relatively limited number of available LEO satellites, the RO measurements in a region of specific interest with good temporal coverage are still rather sparse.

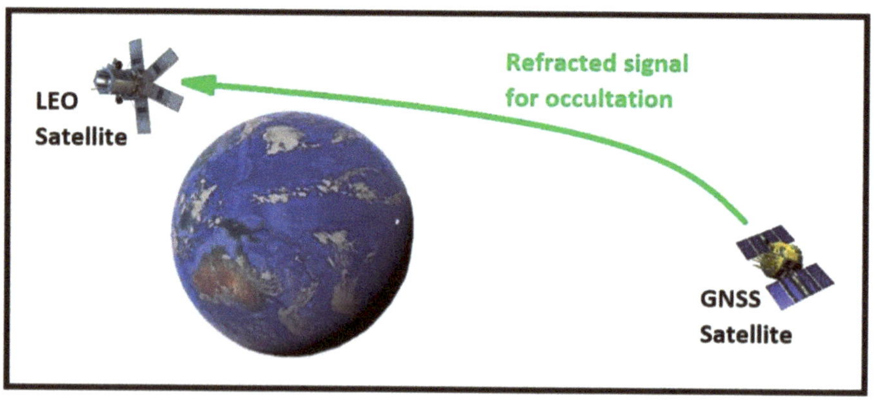

Fig. 3.4 *A typical display of a radio occultation event.*

Bevis et al. [5] were the first to put forward a suggestion of GPS meteorology (remotely sensing WV) before even GPS was opened for the public. In the ground-based GNSS meteorology, the L-band radio signals transmitted by the GNSS satellites are received by ground-based GNSS receivers. While propagating to the ground-based receiver the signals are affected by both the ionosphere and neutral atmosphere (tropospheric) producing propagation delays due to both ionosphere and neutral atmosphere. From the total propagation delay, the delay due to troposphere can be estimated as a by-product of the geodetic data processing which is used to obtain IWV in the zenith direction above each GPS station between receiver and GPS satellites. The basis for GNSS meteorology is the CORS networks for the determination of WV variability. The details of GNSS RO and GPS/GNSS meteorology for sounding atmospheric WV have been described in a review paper by Yu et al. [30]. GNSS ground-based CORS networks provide IWV with high temporal and spatial resolution, whilst GNSS RO profiles provide vertical profiles of WV. The CORS are expanding becoming more denser providing an opportunity to study the spatial and temporal features of WV. The different GNSS methodologies to retrieve atmospheric WV content such as tomography, conversion of GNSS tropospheric delay to water vapor estimates, analyses of errors, and combinations of GNSS with other sources to enhance water vapor information have been described in the recent review by Vaquero-Martinez and Anton [33] on the role of GNSS meteorology in monitoring WV for atmospheric physics. They have also briefly described the published results on the spatio-temporal observations from different regions, China, Japan, Sumatra, South Korea, India, Thailand, European Countries, Antarctica, and some other regions globally.

The neutral atmosphere is a mixture of dry gases and water vapor. It is nondispersive for GNSS band frequencies. The GPS signals while propagating from the satellite to the receiver take excess time due to the neutral atmosphere through which the signal propagates compared to the time taken through the empty space due to a decrease in the speed of propagation of the GNSS signal carrier phase in the neutral atmosphere. The time delay can be converted into the path delay called slant total delay (STD). The STD can be further converted into zenith total delay (ZTD) which excluding hydrometeors is further divided into hydrostatic and wet components called zenith hydrostatic delay (ZHD) and zenith wet delay (ZWD), respectively, so $ZTD = ZHD + ZWD$. The ZHD delay is due to the total mass of the dry atmosphere above the GPS antenna, and ZWD is due to the total amount of water vapor along the GPS signal path.

Let us consider L1 (f_1 = 1.57542 GHz) and L2 (f_2 = 1.22760 GHz) band signals transmitted by GPS. Neglecting the noise and multipath effects, the simplified mathematical model of phase measurements by a receiver R for L1 and L2, on satellite i (expressed in cycles) can be written as follows:

$$\varnothing^i_{R,L1} = \frac{f_1}{c}\left(D^i_R + T^i_R - I^i_{R,L1} + c(\Delta t^i - \Delta t_R\right) + N^i_{R,L1} \tag{3.1}$$

$$\varnothing^i_{R,L2} = \frac{f_2}{c}\left(D^i_R + T^i_R - I^i_{R,L2} + c(\Delta t^i - \Delta t_R\right) + N^i_{R,L2} \tag{3.2}$$

where D_R^i is the geometric distance between receiver R and satellite i, T_R^i the tropospheric error, $I_{R,L1}^i$ and $I_{R,L2}^i$ the dispersive ionospheric errors on L1 and L2 signals, Δt^i the satellite clock synchronization error, Δt_R the receiver clock synchronization error, $N_{R,L1}^i$ and $N_{R,L2}^i$ the phase ambiguity of L1 and L2 signals carrier frequency and c is the speed of electromagnetic waves.

The ionospheric range error ($I_{R,L1}^i$, $I_{R,L2}^i$) is dominant and is due refractive effect from electrons in the ionosphere. The ionosphere is weakly ionized inhomogeneous, collisional, and anisotropic plasma. The complex refractive index n_{ion} of waves in the ionosphere is given by the famous Appleton-Hartree formula [34]:

$$n_{ion} = 1 - \frac{X}{1 - iZ\left[\frac{Y_T^2}{2(1-X-iZ)}\right] \pm \left[\sqrt{\frac{Y_T^4}{4(1-X-iZ)} + Y_L^2}\right]} \tag{3.3}$$

where X, Y_T, Y_L, and Z are dimensionless magneto-ionic parameters, defined as:

$$X = \left(\frac{\omega_P}{\omega}\right)^2, Y_T = \frac{\omega_H}{\omega}\sin\theta = Y\sin\theta, Y_L = \frac{\omega_H}{\omega}\cos\theta = Y\cos\theta, \text{and } Z = \frac{\nu_e}{\omega},$$

where ω_p, ω_H, and ω are the angular plasma frequency, electron gyrofrequency, and wave frequency, respectively. ν_e is the electron-neutral collision frequency and θ is the angle between the propagation direction and the earth's magnetic field vector. ω_p is defined as: $\omega_p = \sqrt{n_e e^2 / \varepsilon_o m_e}$, where n_e is the electron density, e (= −1.601 × 10^{-19} C) is the charge on an electron, ε_o (= 8.854 × 10^{-12} F/m) is the permittivity of free space, and m_e (= 9.109 × 10^{-31} kg) is the mass of an electron.

An approximate form of the ionospheric refractive index is given by

$$n_{ion} \approx 1 - \frac{1}{8\pi^2}\frac{n_e}{f^2}\frac{e^2}{\varepsilon_0 m_e} \tag{3.4}$$

where n_e is the electron density per meter cubed, e is the electron charge, ε_0 is the permittivity of free space in farads per meter, and m_e is the electron mass in kilograms and f is the frequency of the GNSS signal carrier in hertz. The ionospheric refractivity (N_{ion}) ignoring absorption using N_{ion} = (n_{ion} − 1) × 10^6 can be expressed as

$$N_{ion} = 40.3\frac{n_e}{f^2} \tag{3.5}$$

The tropospheric error $\left(T_R^i\right)$ is the total slant path delay of the neutral atmosphere to which main contribution (HD) comes from the thickness and the density of the neutral atmosphere (about 90% from tropospheric layer) which depends on the altitude of the station and the fields of pressure and temperature and second contribution (WD) comes from integrated water vapor. A third small contribution to tropospheric error can be of hydrometeors (ZHmD) which can cause high variability of 3% in T_R^i. For a satellite i located (at azimuth α_i and elevation ϵ_i) visible from a station R at time t, T_R^i can be converted to zenith tropospheric error ($T_R^{izenith}$) above a GNSS station R using a mapping function MF by relationship, $T_R^{zenith}(t) = T_R^i(t) / MF(t, \alpha_i, \varepsilon_i)$ which is also referred to as zenith total delay (ZTD) given as

$$ZTD = ZHD + ZWD + ZHmD \tag{3.6}$$

Due to a variable refractive index of the atmospheric layer, the GPS signals from satellite i to receiver R along the ray path travel slower than they would in a vacuum and in a curved path instead of in a straight line. This produces a delay in signal arrival time which in terms of an equivalent increase in travel path length [5] is given by

$$\Delta L = \int_L [n(s) - 1] ds + [S - G] \tag{3.7}$$

where n(s) is the refractive index as a function of position s along the curved ray path L, S is the path length along L and G is the straight-line path length through the neutral atmosphere (the path that would occur if the atmosphere was replaced by a vacuum).

The first term of Eq. (3.7) is due to the slowing propagation effect of electromagnetic waves in the neutral atmosphere and the second term $[S - G]$ is due to bending of waves due to refraction by the neutral atmosphere. The bending term $[S - G]$ is much smaller, about 1 cm or less, for paths with elevations greater than about 15°. Using the relation between atmospheric refractivity N and index of refraction n, $N = 10^6(n - 1)$, (Eq. 3.7) is written as

$$\Delta L = 10^{-6} \int_L N(s) ds + [S - G] \tag{3.8}$$

The refractivity of the atmosphere ($N(s) = N_{atmos}$) is a function of its temperature, pressure, and water vapor pressure. Thayer [35] suggested the following relationship

$$N_{atmos} = k_1 (P_d / T) Z_d^{-1} + k_2 (P_v / T) Z_v^{-1} + k_3 (P_v / T^2) Z_v^{-1} \tag{3.9}$$

where refractivity coefficients $k_1 = (77.606 \pm 0.04)$ K mbar^{-1}, $k_2 = (64.79 \pm 0.08)$ mbar^{-1} and $k_3 = (3.776 \pm 0.004) \times 10^5$ K mbar^{-1}, T is the temperature, P_d the partial pressure of dry air (in millibars), P_v the partial pressure of water vapor, and Z_d^{-1} and Z_v^{-1} are the inverse compressibility factors for dry air and water vapor, respectively. The contribution of water vapor to refractivity above the troposphere is insignificant which can be neglected and refractivity can be determined from just the pressure and temperature (Eq. 3.9). With an equation of state (e.g., the ideal gas law) and the additional assumption of hydrostatic equilibrium, the pressure and temperature profiles can be derived by vertical integration of the refractivity [36].

The ZHD is the largest component of ZTD and includes the contribution of the total density of all gas in the neutral atmosphere (due to nondipole component of water vapor refractivity). Elgered et al. [37] suggested a model to obtain ZHD in millimeters is given as

$$ZHD = (2.2779 \pm 0.0024)P_s / f(\lambda, H) \tag{3.10}$$

where P_S is the total pressure (in millibars) at the earth's surface, and term

$$f(\lambda, H) = (1 - 0.00266\cos 2\lambda - 0.00028H) \tag{3.11}$$

accounts for the variation in gravitational acceleration with latitude λ and the height H of the surface above the ellipsoid (in kilometers). ZHD represents the contribution of the total density of all gas in the neutral atmosphere to ZTD (including water vapor density). The troposphere accounts for about 75% of the total ZHD. The subtraction of ZHD from the ZTD will give ZWD which can be used to estimate the presence of water vapor because it is the only atmospheric component with a permanent dipolar moment.

The water vapor content over a site makes another specific additional contribution (due to the dipole component of water vapor refractivity) to *ZTD* known as *ZWD* given as [37]:

$$ZWD = 10^{-6}\left[k_2' \int (P_v / T)\, dz + \int k_3 \left(P_v / T^2\right) dz\right] \tag{3.12}$$

where $k_2' = (17 \pm 10)$ K mbar^{-1} and the integral is along the zenith path and P_v is the partial pressure of water vapor. These expressions can be evaluated from profiles of P_v and T provided by instruments such as radiosondes.

Almost all of the *ZWD* occurs in the troposphere, and most of it occurs in the lower troposphere. The delays along paths at arbitrary elevation angles can be obtained from zenith path delays using the mapping functions [5] which depend upon elevation angle (ε). Most expressions for the delay along a path of arbitrary elevation consist of the zenith delay multiplied or divided by a mapping function. The mapping function describes the dependence of the delay on elevation angle (e.g., [38]). A simple mapping function is $1/\sin(\varepsilon)$ which generally does not work realistically at low elevation angles. For low elevation and the error caused by straight-line approximation, more complicated mapping functions have been developed by several researchers (e.g., [39–42]).

The hydrometeors, that are liquids (e.g., cloud and water rain) and solid water (e.g., pristine ice, snow, hailstones' and graupel) give ZHmD which is expressed as

$$ZHmD = 10^{-6} \int_{z0}^{\infty} (N_{lw} - N_{ice})\, dz \tag{3.13}$$

where N_{lw} and N_{ice} are the atmospheric refractivity of liquid and solid water, respectively. The hydrometeor delay can be significant during extreme terrestrial weather events such as heavy rainfall associated with tropical cyclones at temperate latitudes and snowfall at mid-and high latitudes, otherwise, it is small and can be neglected.

3.7 Estimating ZWD and water vapor using GNSS

There are various ground-based equipment (e.g., radiosonde, a balloon-borne instrument, sun photometer, LiDAR, spectrometer), in situ methods (remote sensing and airborne instruments), and remote sensing (IR, visible and microwave sensors) techniques to obtain the water vapor retrieval but they are limited in temporal and spatial coverages and operationality. The GNSS radio occultation (RO) technique is relatively a low-cost technique with global coverage at a high temporal frequency (sub-hourly) operating in all weather conditions. The science behind RO was developed in the 1960s to study the atmosphere of planets and their moons [43,44]. More recently, scientists have begun to apply RO to obtain atmospheric refractivity using GPS signal phase measurements as they are occluded by earth's atmosphere. GNSS-RO is a remote sensing technique for earth's atmosphere and ionosphere in which GNSS (e.g., GPS) signals are recorded by a receiver onboard by LEO satellites providing global coverage. It provides high accuracy and vertical resolution of GNSS-RO data and this technique is ideally suited to study, weather forecasting and atmospheric processes, climate monitoring, model verification, etc. The GNSS-RO for atmospheric measurement is based on the Doppler shift measurements at a given frequency and removing the Doppler effects caused by satellite motion. From the total GNSS signal bending both due to ionosphere and atmosphere the bending of the signal due to the atmosphere is recovered. The precision of bending angle, the altitude range, and the vertical resolution are customized to obtain the desired bending angle profile which is then used to derive atmospheric parameters such as refractivity, temperature, pressure, and humidity. GNSS RO applications are in the continuous expansion phase first by the FORMOSAT-3/COSMIC mission which was a microsatellite mission for weather forecast, climate monitoring, and atmospheric, ionospheric and geodesy research. The initial results from the FORMOSAT-3/COSMIC along with CHAMP, GRACE have been presented by Wickert et al. [45] (2009). The FORMOSAT-3/COSMIC, a collaborative Taiwan-USA science experiment to deploy a constellation of six microsatellites in low earth orbits, had been extended to FORMOSAT-7/COSMIC-2 mission. FORMOSAT-7/COSMIC-2, is a joint program between NSPO (Taiwan) and NOAA (USA), with a constellation of six satellites launched in 2019. FORMOSAT-7/COSMIC-2 satellites provide over 4000 radio occultation soundings daily, which reveal data including temperature, pressure, and water vapor between 50 degrees north and south latitudes. The amount of sounding data is about 3 to 4 times larger than that of FORMOSAT-3 (the mission from 2006 to 2019), and substantially increased meteorological and ionospheric observation data in low-latitude regions. The detailed coverage of GNSS RO is beyond the scope of this chapter and the reader is referred to chapter 33 "GNSS Radio Occultation" of book "Position, Navigation, and Timing Technologies in the 21st Century."

The water vapor content is a key parameter for the understanding and forecasts of the meteorological process (i.e., initiation of large-scale convection, mesoscale deep convection' and humidity of the lower troposphere) because it is the only constituent that possesses a dipole moment contribution to its refractivity.

An analytical expression for ZTD in L-band radio signal propagation (Position, Navigation, and Timing Technologies in the 21st Century, chapter 30, pp. 860–861) is given as

$$ZTD = 10^{-6} \int_{z0}^{\infty} N(z)\, dz = ZHD + ZWD + ZHmD \tag{3.14}$$

where the z is the vertical altitude and z_0 is the altitude of the GNSS site. The refractivity N_{atmos} of the atmosphere can be expressed as

$$N_{atmos} = N_{hydrostatic} + N_{wet} + N_{hydrometers} \tag{3.15}$$

The three components N_{atmos} ($N_{hydrostatic}$, N_{wet}, $N_{hydrometers}$) are defined by ZTD (ZHD, ZWD, $ZHmD$). The contribution from the hydrometeors is typically small and neglected. The tropospheric refractivity due to gas (dry gas and water vapor) can be written as $N_{gas} = N_{hydrostatic} + N_{wet}$.

The combined refractivity (N) of the atmosphere and ionosphere (N_{ion}) is given as

$$N = N_{gas} + N_{hydrometers} + N_{ion} \tag{3.16}$$

N_{ion} as given in Eq. (3.5) varies as $1/f^2$ and represents the refractive effect from electrons in the ionosphere. Since the ionosphere is dispersive, the ionospheric error or ionospheric contribution to total refractivity from GNSS satellite to receiver can be isolated through the use of two GNSS frequencies giving neutral atmosphere (troposphere) error or refractivity.

The expression for ZHD expression (Position, Navigation, and Timing Technologies in the 21st Century, chapter 30, pp. 860–861) is given as

$$ZHD = 10^{-6} \int_{z0}^{\infty} k_1\, R_d\, \rho\, dz = 10^{-6} \int_{z0}^{\infty} k_1 \frac{P}{T_v}\, dz \tag{3.17}$$

where ρ is the density of air, P the pressure; R_d the specific molar gas constant for dry air (287.0586 ± 0.0055 J/(kmol. K), T_V the virtual temperature; and k_1 an atmospheric refractivity coefficient.

A specific additional contribution by the water vapor content (due to the dipole component of water vapor refractivity) over a site to ZTD known as ZWD is given as

$$ZWD = 10^{-6} \int_{z0}^{\infty} \left(k_2\, R_w - k_1\, R_d + \frac{k_3}{T} R_w \right) \rho_{wv}\, dz = 10^{-6} \int_{z0}^{\infty} \left(k_2' \frac{P_v}{T} + k_3 \frac{P_v}{T^2} \right) dz \tag{3.18}$$

where $k_2' = k_2 - k_1 \dfrac{R_d}{R_w}$, where ρ_{wv} is the density of water vapor and R_w the specific molar gas constant for water vapor (461.525 ± 0.013 J/(kmol. K)), and P_v the partial pressure of water vapor.

Different estimations of constants k_1, k_2, and k_3 have been reported in the literature [35,46–48]. Davis et al. [38] reported that the uncertainties in the constants of

(Eq. 3.18) limit the accuracy in atmospheric refractivity to 0.02%. Bevis et al. [46] revised data used by Hasegawa and Stokesberry [47] and provided mean values for k_1 = (0.7760 ± 0.0005) Pa/K, k_2 = (0.704 ± 0.022) Pa/K, k_3 = (373900 ± 1200) Pa/K, and $k_2^{'}$ = (0 2213 ± 0 0220) Pa/K. In the troposphere, the dipole component of the refractivity is about 20 times larger than the nondipole component [46] and wet delay is always much smaller than the hydrostatic delay but it is usually far more variable. For temperate latitudes (low altitudes) the daily variability of the wet delay can exceed that of the hydrostatic delay [37].

To estimate integrated water vapor (IWV) from water vapor density (ρ_{wv}), Hogg et al. [49] reported an expression as given

$$IWV = \int_{z0}^{\infty} \rho_{wv}\, dz = \kappa\,.\,ZWD \tag{3.19}$$

and Askne and Nordius [50] proposed an expression for κ:

$$\kappa = \frac{10^8}{R_w\left(\frac{k_3}{T_m} + k_2^{'}\right)} \tag{3.20}$$

where T_m is the mean temperature of the column above the GNSS site. The analytical expression for T_m suggested by Davis et al. [39] and Askne and Nordius [50] is given

$$T_m = \frac{\int_{z_o}^{\infty} \frac{P_v}{T}\, dz}{\int_{z_o}^{\infty} \frac{P_v}{T^2}\, dz} \tag{3.21}$$

A simple relation between T_m and surface temperature T_s (in K) was proposed by Bevis et al. [46] given as

$$T_m = 70.2 + 0.72T_s \tag{3.22}$$

For meteorological applications, the term precipitable water (PW) is used rather than the IWV. The IWV (expressed in kg/m^2) is used when we refer to the mass of vapor per unit area and PW (expressed in millimeters) to the height of an equivalent column of water. The PW is defined as the liquid equivalent of the total water vapor contained in an air column above a site and can be obtained from IWV using $PW = IWV/\rho$, where ρ is the density of water. To retrieve IWV or PW from an observed ZWD using (Eq. 3.19) the value of constant κ should be estimated using a value for T_m which depends upon the specific area and season. The T_m could be obtained based on the observations in the area and season of concern or from meteorological models or other estimation techniques using GPS data processing software packages and techniques stated by Davis et al. [38]. GPS receivers offer the possibility of observing the horizontal distribution of IWV and PW with unprecedented coverage.

Eq. (3.19) for GNSS estimate can be written as

$$IWV_{GNSS} = \kappa\,.\,ZWD_{GNSS} \tag{3.23}$$

Nowadays GNSS is a key and most widely used technique for meteorological observations. The accuracies of IWV_{GNSS} extractions from ZWD_{GNSS} have been reported to be very good within accuracies of less than 1.5 kg/m^2 and 0.01 m, respectively. Van Malderen et al. [51] using more than 15 years of ZWD_{GNSS} retrievals at 28 worldwide sites compared IWV_{GNSS} with coincident satellite IWV observations and with sun photometer and/or radiosonde observations. They found mean biases in IWV due to the different techniques compared to the GNSS estimates varying between −0.3 and 0.5 kg/m^2.

3.8 Summary

Both GNSS ground-based and RO techniques have potential in detecting and predicting severe terrestrial and space weather events. Bonafoni et al. [52] have presented a comprehensive review of GNSS contributions to studies of extreme weather and climate events. Water vapor is the most variable and the major constituent of the atmosphere. Despite being relatively low in concentration, it contributes more than any other component of the atmosphere to the greenhouse effect and also plays a critical role in the vertical stability of the atmosphere and the evolution of the tropospheric storm systems. The dynamics of water vapor have a strong influence on weather and climate due to the large energy and heat transfers in the hydrological processes on local, regional, and global scales. Global warming will cause systematic changes in the total water vapor content of the atmosphere. Long-term, global measurement of IWV could provide vital information on regional and global climate change. Bernet et al. [53] analyzed IWV trends from a ground-based microwave radiometer in Bern, from a Fourier transform infrared (FTIR) spectrometer at Jungfraujoch, from reanalysis data (ERA5 and MERRA-2) and Swiss ground-based GNSS stations and found IWV generally increasing by 2% per decade to 5% per decade, with deviating trends at some GNSS stations. This is consistent with global warming on a regional scale and stresses the importance of the water vapor feedback in climate change.

References

[1] B.W. Crowell, Y. Bock, D. Melgar, Real-time inversion of GPS data for finite fault modeling and rapid hazard assessment, Geophys. Res. Letts. 39 (2012), L09305. doi:10.1029/2012GL051318.

[2] G. Guergana, J. Jones, J. Douša, G. Dick, S. de Haan, E. Pottiaux, O. Bock, R. Pacione, G. Elgered, H. Vedel, M. Bende, Review of the state of the art and future prospects of the ground-based GNSS meteorology in Europe, Atmos. Meas. Tech. 9 (2016) 5385–5406.

[3] P. Segall, J.L. Davis, GPS applications for geodynamics and earthquake studies, Ann. Rev. Earth Planet. Sci. 25 (1997) 301–336.

[4] P. Tregoning, T. van Dam, Atmospheric pressure loading corrections applied to GPS data at the observation level, Geophys. Res. Letts. 32 (2005) 153.

[5] M. Bevis, S. Businger, A.H. Thomas, C. Rocken, A.A. Richard, R.H. Ware, GPS meteorology' remote sensing of atmospheric water vapor using the global positioning system, J. Geophys. Res. 97 (D14) (1992) 15787–15801.

[6] N. Jakowski, V. Wilken , S. Schlueter, S.M. Stankov, S. Heise, Ionospheric space weather effects monitored by simultaneous ground and space based GNSS signals, J. Atmos. Sol. Terr. Phys. 67 (12) (2005) 1074–1084.
[7] S. Kumar, Sarvesh Kumar, Equatorial ionospheric TEC and scintillations under the space weather events of 4–9 September 2017: M-class solar flares and a G4 geomagnetic storm, J. Atmos. Sol. Terr. Phys. 209 (2020) 105421. doi:10.1016/j.jastp.2020.105421.
[8] I. Zakharenkova, E. Astafyeva, I. Cherniak, GPS and GLONASS observations of large-scale traveling ionospheric disturbances during the 2015 St. Patrick's Day storm, J. Geophys. Res. Space Phys. 121 (12) (2016) 12–138.
[9] D.M. Tralli, T.H. Dixon, S.A. Stephens, Effect of wet troposheric path delays on estimations of geodetic baselines in the Gulf of California using Global Positioning System, J. Geophys. Res. 93 (1988) 6545–6557.
[10] V. Kuznetsov, V. Sinelnikov, S. Alpert, Yakov Alpert: Sputnik-1 and the first satellite ionospheric experiment, Adv. Space Res. 55 (12) (2015) 2833–2839. doi:10.1016/j.asr.2015.02.033.
[11] The Interoperable Global Navigation Satellite Systems Space Service Volume, (2018), *United Nations Publication,* Sales No. E.19.IV.1, ISBN 978-92-1-130355-1, eISBN 978-92-1-047440-5.
[12] Y.T.J. Morton, F. van Diggelen, J.J. Spilker, Jr., B.W. Parkinson, Position, Navigation, and Timing Technologies in the 21st Century, 1, IEEE Press, Piscataway, NJ, 2021.
[13] S. Goswami, A. Paul , S. Haldar, Study of relative performance of different navigational satellite constellations under adverse ionospheric conditions, Space Weather 16 (2018) 667–675. doi:10.1029/2017SW001762.
[14] G.T. French, Understanding the GPS, GeoResearch Inc, Bethesda, MD, 1996.
[15] V.E. Kuzminykh, Global Navigation Satellite Systems Monitoring of Ionosphere, Russian State Hydrometeorological University, Saint Petersburg, 2010.
[16] R. Prasad, S. Kumar, P.T. Jayachandran, Receiver DCB estimation and GPS vTEC study at a low latitude station in the South Pacific, J. Atmos. Sol. Terr. Phys. 149 (2016) 120–130.
[17] A.O. Adewale, E.O. Oyeyemi, A.B. Adeloye, C.N. Mitchell, J.A. Rose, P.J. Cilliers, A study of L-band scintillations and TEC at an equatorial station Lagos, Nigeria, Radio Sci 47 (2012) RS2011. doi: 10.1029/2011RS004846.
[18] M.S. Bagiya, H.P. Joshi, K.N. Iyer, M. Aggarwal, S. Ravindran, B.M. Pathan, TEC variations during low solar activity period (2005–2007) near the Equatorial Ionization Anomaly Crest region in India, Ann. Geophys. 27 (2009) 1047–1057, doi:10.5194/angeo-27-1047-2009.
[19] P.V. RamaRao, S. Gopi Krishna, K. Niranjan, D.S. Prasad, Temporal and spatial variations in TEC using simultaneous measurements from the Indian GPS network of receivers during the low solar activity period of 2004-2005, Ann. Geophys. 24 (2006) 3279–3292. doi:10.5194/angeo-24-3279-2006.
[20] GPS Silicone Valley, GSV4004B GPS-Ionospheric Scintillation and TEC Monitor (GISTM), User's Manual, GPS Silicone Valley, Los Altos, 2007.
[21] X. Chen, C.G. Parini, B. Collins, Y. Yao, M. Ur Rehman, Antennas for Global Navigation Satellite Systems, John Wiley & Sons Ltd, UK, 2012.
[22] V. Giralo, D. Eddy, S. D'Amico, Development and verification of the Stanford GNSS Navigation Testbed for spacecraft formation flying, Proceedings of the 2018 International Technical Meeting of The Institute of Navigation, Reston, VA, Stanford University, 2017, pp. 837–856. doi:10.33012/2018.15544.
[23] O. Montenbruck, Y. Andres, H. Bock, T. van Helleputte, J. van den Ijssel, M. Loiselet, C Marquardt, P. Silvestrin, P. Visser, Y. Yoon, Tracking and orbit determination performance of the GRAS instrument on MetOp-A, GPS Solution 12 (2008) 289–299.
[24] M. Loiselet, N. Stricker, J. Christensen, C. Carrascosa, In orbit verification results from GRAS receiver on MetOp-A satellite, Proceedings of Joint 2007 EUMETSAT Meteorological Satellite Conference and the 15th American Meteorological Society (AMS) Satellite Meteorology and Oceanography Conference, Amsterdam, 24–28 September 2007.(2007)
[25] J.A. Klobuchar, Modern total electron content measurement techniques, in: C.H. Liu (Ed.), World Ionosphere/Thermosphere Study, ICSU Scientific Committee on Solar Terrestrial Physics, Chicago, Illinois, 1989, pp. 137–147.
[26] A. Komjathy, Global ionospheric total electron content mapping using the Global Positioning Systems. In: Technical Report Submitted in Partial Fulfillment of the Requirements for the Degree

of Doctor of Philosophy in the Department of Geodesy and Geomatics Engineering, University of New Brunswick, Fredericton, 1997.
[27] S. Kumar, A.K. Singh, R.P. Singh, Probing the tropospheric water vapor using GPS. In: GPS and GNSS Technology in Geosciences, Elsevier Ltd, London, UK. 2021, pp. 119–134.
[28] W. Melbourne, et al., The application of space borne GNSS to atmospheric limb sounding and global change monitoring, JPL Publ. 147 (1994) 1–26.
[29] R. Ware, M. Exner, D. Feng, M. Gorbunov, K. Hardy, B. Herman, Y. Kuo, T. Meehan, W. Melbourne, C. Rocken, W. Schreiner, S. Sokolovskiy, F. Solheim, X. Zou, R. Anthes, S. Businger, K. Trenberth, GPS sounding of the atmosphere from low earth orbit: Preliminary results, Bull. Am. Meteor. Soc. 77 (1996) 19–40. doi:10.1175/1520-0477.
[30] K. Yu, C. Rizos, D. Burrage, A.G. Dempster, K. Zhang, M. Markgraf, An overview of GNSS remote sensing, J. Adv. Signal Process. 2014 (2014) 134.
[31] Y.-A. Liou, A.G. Pavelyev, S.-F. Liu, A.A. Pavelyev, N. Yen, C.-Y. Huang, G.-J. Fong, FORMOSAT-3/COSMIC GPS radio occultation mission: preliminary results, IEEE Trans. Geosci. Remote Sens. 45 (11) (2007) 3813–3826.
[32] A.G. Pavelyev, K. Zhang, S.S. Matyugov, Y.Y-A Liou, C-S Wang, O.I Yakovlev, I.A Kucherjavenkov, Y. Kuleshov, Analytical model of bistatic reflections and radio occultation signals, Radio Sci. 46 (RS1009) (2011). doi:10.1029/2010RS004434.
[33] J. Vaquero-Martinez, M. Anton, Review on the role of GNSS meteorology in monitoring water vapor for atmospheric physics, Remote Sens. 13 (2021) 2287. doi:10.3390/rs13122287.
[34] K.G. Budden, Radio Waves in the Ionosphere, Cambridge University Press, England, 1961.
[35] D. Thayer, An improved equation for the radio refractive index of air, Radio Sci. 9 (1974) 803–807.
[36] G.A. Hajj, E.R. Kursinski, L.J. Romans, W.I. Bertiger, A technical description of atmospheric sounding by GPS occultation, J. Atmos. Sol. Terr. Phys. 64 (4) (2002) 451–469. doi:10.1016/s1364-6826(01)00114-6.
[37] G. Elgered, J.L. Davis, T.A. Herring, I.I. Shapiro, Geodesy by radio interferometry: Water vapor radiometry for estimation of the wet delay, J. Geophys. Res. 96 (1991) 6541–6555.
[38] J.L. Davis, T.A. Herring, I.I. Shapiro, A.E. Rogers, G. Elgered, Geodesy by radio interferometry: Effects of atmospheric modeling errors on estimates of baseline length, Radio Sci. 20 (1985) 1593–1607.
[39] H.D. Black, An easy implemented algorithm for the tropospheric range correction, J. Geophys. Res. 83 (1978) 1825–1828.
[40] J. Boehm, A.E. Niell, P. Tregoning, H. Schuh, Global mapping function (GMF): A new empirical mapping function based on numerical weather model data, Geophys. Res. Letts. 33 (L07) (2006) 304. doi:10.1029/2005GL025546.
[41] J.W. Marini, Correction of satellite tracking data for arbitrary atmospheric profile, Radio Sci. 7 (1972) 223–231.
[42] S.M. Yionoulis, Algorithm to compute tropospheric refraction effects on range measurements, J. Geophys. Res. 75 (1970) 7636–7637.
[43] G. Fjeldbo, V.R. Eshelman, The atmosphere of Mars analysed by integral inversion of the Mariner IV occultation data, Planet. Space Sci. 16 (1968) 123–140.
[44] G.F. Lindal, G.E. Wood, G.S. Levy, J.D. Anderson, D.N. Sweetnam, H.B. Hotz, B.J. Buckles, D.P. Holmes, P.E. Doms, V.R. Eshleman, G.L. Tyler, T.A. Croft, The atmosphere of Jupiter: An analysis of the Voyager radio occultation measurements, J. Geophys. Res. 86 (1981) 8721–8727.
[45] J. Wickert, G. Michalak, T. Schmidt, G. Beyerle, C.-Z. Cheng, S.B. Healy, S. Heise, C.-Y. Huang, N. Jakowski, W. Kohler, C. Mayer, D. Offiler, E. Ozawa, A.G. Pavelyev, M. Rothacher, B. Tapley, C. Arras, GPS radio occultation: Results from CHAMP, GRACE and FORMOSAT-3/COSMIC, Terrestrial Atmos. Ocean. Sci. 20 (1) (2009) 35–50.
[46] M. Bevis, S. Businger, S. Chiswell, GPS meteorology: Mapping zenith wet delays onto precipitable water, J. Appl. Meteorol. 33 (1994) 379–386.
[47] S. Hasegawa, D. Stokesberry, Automatic digital microwave hygrometer, Rev. Sci. Instrum. 46 (1975) 867–873.
[48] E. Smith, S Weintraub, The constants in the equation for atmospheric refractive index at radio frequencies, Proceedings of IRE, Boulder Collarado, US, 1953, pp. 1035–1037.
[49] D.C. Hogg, F.O. Guiraud, M.T. Decker, Measurement of excess transmission length on Earth Space paths, Astron. Astrophys. 95 (1981) 304–307.

[50] J. Askne, H. Nordius, Estimation of tropospheric delay for microwaves from surface weather data, Radio Sci 22 (3) (1987) 379–386.
[51] R. Van Malderen, H. Brenot, E. Pottiaux, et al., A multi-site intercomparison of integrated water vapour observations for climate change analysis, Atmos. Meas. Tech. 7 (2014) 2487–2512. doi:10.5194/amt-7-2487-2014.
[52] S. Bonafonia, R. Biondib, H. Brenotc, R. Anthes, Radio occultation and ground-based GNSS products for observing, understanding and predicting extreme events: A review, Atmos. Res. 230 (2019) 104624. doi:10.1016/j.atmosres.2019.104624.
[53] L. Bernet, E. Brockmann, T. von Clarmann, N. Kämpfer, E. Mahieu, C. Mätzler, G. Stober, K. Hocke, Trends of atmospheric water vapour in Switzerland from ground-based radiometry, FTIR and GNSS data, Atmos. Chem. Phys. 20 (2020) 11223–11244.

CHAPTER 4

Estimation of ionospheric total electron content (TEC) from GNSS observations

Gopi Krishna Seemala
Indian Institute of Geomagnetism, Navi Mumbai, India

4.1 Introduction

In the early days of radio, mathematical physicists theoretically proved by that even for very long waves it will be impossible to receive radio signals at long distances beyond curvature of earth because of the diffractive attenuation. In 1882, Balfour Stewart in his studies of earth's magnetism [1], discovered that diurnal variation of magnetic field can be explained by currents in the upper atmosphere. However, in 1901 Guglielmo Marconi demonstrated the reception of long-distance radio signals across the Atlantic Ocean, that is, around an appreciable portion of the curved earth. That stimulated widespread interest in the studies of this phenomenon. The following year, Kennelly [2] and Heaviside [3] independently postulated that the earth is surrounded by an ionized atmospheric layer which acts as a reflector and so prevents the wave from escaping into space; to explain the long-distance radio transmission at a time when others were investigating diffraction effects as an explanation. The first direct experimental evidence for the existence of an ionized region was obtained by Appleton and Barnett [4], with the "frequency change" experiments using the BBC transmitter at Bournemouth. They measured the arrival angle of waves from a distant transmitter and showed that they arrived at an elevated angle, that is, that there is a "sky-wave." Experiments of this kind indicated that the heights of reflection were of the order of 100–200 km. Subsequent work has demonstrated the existence of several well-defined ionized regions, and of rather complicated layer structure within each. The name "ionosphere" for these ionized regions of atmosphere was suggested by Watson-Watt, and was universally adopted.

Earlier, the ionospheric research was mostly carried out by radio engineers as they are motivated by the need to improve long-range radio communication. Then, subsequently, the focus shifted to understanding the effects of ionosphere on satellites as they orbit in this ionized medium. Today, we have vast scientific knowledge of the ionosphere owing to the amount of data generated from satellite remote sensing, and was enhanced by progress in theoretical and laboratory experimental techniques.

Thus, the ionosphere is a part of earth's upper atmospheric shell where the electrons and ionized atoms and molecules, that spans from an altitude of about 50 km to more than 1000 km. In the ionosphere, the charged particles or electrons and ions are

Atmospheric Remote Sensing
DOI: https://doi.org/10.1016/B978-0-323-99262-6.00022-5

created by the photo-ionization due to solar radiation incidence on neutral atoms and molecules of air. This ionization occurs primarily due to ultraviolet radiation (UV) and X-rays from the sun. The ionosphere can conduct electrical currents as well as reflect, deflect, and scatter radio waves. In general, ionosphere has a lower altitude limit of about 50–70 km and there is no distinct upper limit, but in most studies and for application purposes the upper limit is kept at 2000 km. At higher altitudes of ionosphere, the region is called magnetosphere which extends from about several hundred kilometers from earth's surface to tens of thousands of kilometers into space. In this region, the charged particle's behavior is strongly influenced by the magnetic fields of both earth and sun. The maximum electron density of the ionosphere is usually observed at an altitude of about 300 km, where only a tiny fraction (<1%) of neutral atmospheric gas is ionized.

4.1.2 Ionospheric layers

It is well known that atmospheric density reduces with height due to the gravity of the earth, and this vertical gradient is much larger than the horizontal gradient. The simplest type of ionized layer that can be considered theoretically is the so-called "Chapman layer," [5]. Suppose ultra-violet or X-ray radiation of a wave-length with enough photon energy for photo-ionization of an atmospheric gas is incident on atmosphere, whose density and absorption coefficient, vary exponentially with altitude. As the radiation goes deeper into the atmosphere, it gets absorbed more and the radiation intensity is decreased. At a sufficiently low height the greater part of the radiation is absorbed, and a lower bound is formed to the region of ionization produced by the radiation. This results in a region with maximum density of ionization called a Chapman layer. The actual ionization density present at any given time will represent a balance between the rates of production, accumulation, and loss.

Practically, the ionosphere has more complicated structure than a single layer of ionization, as there are several gases present in the atmosphere and their concentrations vary differently with height. The most abundant gas molecules that are ready for ionization due to incident solar radiation are molecular oxygen (O_2) and nitrogen (N_2) below 200 km; atomic oxygen (O) above 200 km, and hydrogen (H) and Helium (He) above 600 km altitude (for more details refer to Fig. 1.2 in [6]). The different gases need different wavelengths of radiation to get ionized. The resultant plasma density is a balance between the rate at which electrons are produced and the rate at which they are lost. The electrons and ions are lost mostly due to recombination and diffusion. These rates of recombination and diffusion change with altitude. This altitudinal variation of electron production and recombination rates creates a density profile with altitude, and results in several distinct layers (D, E, and F). Now it is known that these layers are not so distinct as earlier thought. Edward V. Appleton, who studied ionosphere extensively by radio probing it, is the one who brought this nomenclature. Initially, when he studied

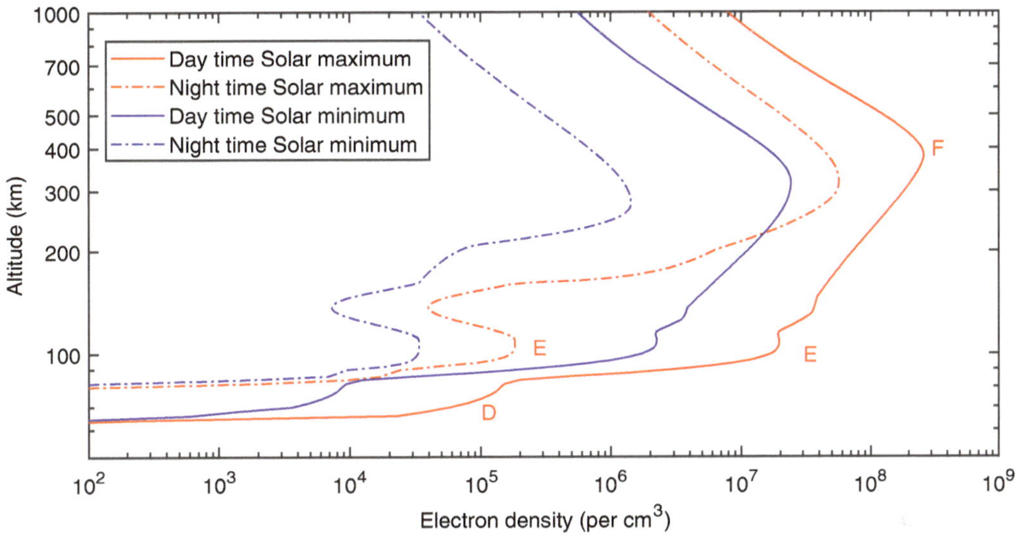

Fig. 4.1 ***Typical variation of ionospheric density values, calculated using IRI 2016 model, during high and low solar active times.***

the first layer where electric field of wave got reflected, he used the symbol "E" to describe wave's electric field. After that, at higher altitudes, a second layer was identified by him, then "F" was used as symbol for that reflected wave. Then he used the symbol "D" for a layer that he suspected at lower altitude. With time, these letters were associated with the ionospheric layers rather than electric field of those reflected waves.

The Fig. 4.1 shows the typical schematic of ionospheric layers and electron density profile. The electron density profiles for typical day and night for schematic representation are derived using the IRI model [7] available at http://irimodel.org/, for the high and low solar active times. As shown for each layer there is an altitude of maximum electron density, above and below which the ionization density tends to drop off. The ionospheric layers vary with respect to chemical composition, and particle concentration. The lowest ionospheric region is D layer whose ionization is present only during daylight hours and ranges with altitudes of about 70–90 km. The E region density extends from 90 km to about 160 km. After sunset, the electron density of E-region is diminished by a factor of 10 or more in a short time before reaching a night-time density. The F region extends upward from an altitude of about 160 km, during the day time the F-region is further distinguished into two regions F1 and F2. The F1 region has an altitude peak near 200 km and during night time this layer is absent or not distinct. The F2 region has a maximum density near 300 km during the day time and this shifts to higher altitudes at night.

The electron density of ionosphere varies with altitude and also spatially with latitude. The electron density also varies temporally with time of the day (diurnal variation), time of the year (seasonal variation) and solar cycle (11-year variation). It is obvious that the amount of ionization in the ionosphere changes with the flux of radiation received from the sun. The ionosphere varies dramatically with geomagnetic latitudes; in particular there are three different ionospheric regions due to earth's geographic and magnetic field; they are high latitude, mid latitude and low latitude regions. The mid latitude ionosphere has obvious diurnal and seasonal variations, and this region is most modeled by many ionospheric models. At high latitudes, the ionosphere is directly coupled to the magnetospheric tail by the magnetic field lines and this connection has important consequences for high latitude ionosphere. The low latitude ionosphere region is dynamic and sensitive to plasma instabilities, and responds to the magnetospheric ring currents. The ionospheric disturbances in general are associated directly or indirectly with the events on the sun.

4.1.3 Ionosphere and its effects on Global Navigation Satellite System (GNSS)

The radio wave/signals traversing through the ionosphere experience group delay on the signal/code or information due to its encounter with the free electrons in the ionosphere. The other effects of ionosphere on the radio wave are phase advance, Doppler shift, angular refraction or bending of path, Faraday rotation of linear polarized waves, distortion of pulses transmitted. Another significant effect is due to ionospheric density irregularities that cause amplitude and phase scintillations [8]. Except ionospheric scintillation effects, all the ionospheric effects experienced by the radio signal are proportional, at least to a first order to the total number of electrons encountered by it while traversing through ionosphere [8,9]. The total electron content (TEC) is defined as the number of free electrons in a column of unit cross-sectional area along the ray path from transmitter to the receiver end [39]. Slant TEC is defined as the electron column density along an (oblique) line of sight, and slant TEC equals vertical TEC only if the line of sight goes through zenith. The measurement unit of TEC is TECU (1 TECU = 10^{16} el/m^2).

In the recent times, owing to the ever-increasing applications in space, and dependence on satellite-based navigation and such other trans-ionospheric communication systems; the study of ionosphere characteristics particularly during the space weather events such as geomagnetic storms and ionospheric scintillations has gained importance. The quality of the received signals at the GNSS receiver may be severely degraded by the density gradients in ionosphere and scintillations, affecting the accuracy and reliability of the position, velocity, and time (PVT) solution. On the other hand, the TEC and its derivatives, as well as the scintillation indices extracted from the GNSS observations, serve as the signals of opportunity for understanding and mitigating the effects in the real-time applications.

A satellite-based navigation system uses satellites to provide autonomous geo-spatial positioning. This system allows the receivers to determine their location coordinates to a good accuracy (within a few centimeters to meters) by using time stamped radio signals transmitted by satellites. The receiver using these time stamped signals can calculate the current time to a very high precision that enables its use also for time synchronization. A satellite-based navigation system having global coverage is called as GNSS. Global coverage of such systems can be achieved by a satellite constellation of 18–30 medium earth orbit (MEO) satellites which are spread across several orbital planes. These satellites usually have orbital inclinations of >50° and about 12 hours orbital periods, so that at least 4 satellites are in the field of view at any point on globe to enable calculation of the position of the user/receiver. GNSS term is used for any satellite constellation that provides positioning, velocity, and time (PVT) services on a global or regional basis. While Global Positioning System (GPS) is one the most popular GNSS, other GNSS systems emerged to provide complementary and independent PVT capability. The other GNSS systems are BeiDou Navigation Satellite System (BDS), Galileo, GLONASS (Globalnaya Navigazionnaya Sputnikovaya Sistema), Indian Regional Navigation Satellite System (IRNSS), Quasi-Zenith Satellite System (QZSS). More information on current GNSS can be sought on webpage at link https://en.wikipedia.org/wiki/Satellite_navigation.

The signals of GNSS satellites must travel through the earth's ionosphere on their way to receivers on or near the earth's surface. Therefore, for the GNSS signals to have the precise positional accuracies, these signals must be corrected for the added ionospheric group delay and carrier phase advance. Clearly, the ionospheric TEC parameter is of great importance for maintaining accuracy in satellite ranging systems. There are many methods to determine TEC like differential Doppler, Faraday rotation, differential carrier phase, and group delay techniques. If the dual frequency GNSS receivers are used, then the ionospheric effect on the signals can be almost eliminated by taking advantage of the ionosphere's dispersive nature. On the other hand, we can also determine the error induced by the ionosphere, and reduce ionosphere parameters like TEC and irregularity strength. Using observations from GNSS receivers, the TEC estimation is more accurate than others methods, cost effective and have greater spatial coverage due to a greater number of satellites available. Thus, GNSS has become one of the powerful tools available to the ionospheric science community for studying the temporal and spatial variations of the ionosphere. In the following sections, we use GPS L1 (1575.42 MHz) and L2 (1227.60 MHz) frequencies to explain the derivation for TEC from GPS observables, as the method is very similar for any GNSS system observations using two frequencies.

4.2 Estimation of ionospheric total electron content

The effect of the ionosphere on radio waves cannot be described by simple dispersion. The ionospheric refractive index is a function of the frequency of radio wave, electron density, and also a minor influence from intensity of earth's magnetic field. The integrated phase and group refractive indices for trans-ionosphere signal will result in a range that is different from the true geometric range between satellite and receiver. This difference in range is called ionospheric range error or ionospheric delay (in time). For carrier phase observations, this ionospheric error is negative (phase advance, shorter than geometric range), and it is positive (group delay, longer than geometric range) for the pseudorange or code observations. The ionosphere delay $\delta\rho$ experienced by the trans-ionospheric signals traversing a distance ρ, is expressed by the refractive index η integrated along the line of sight from the satellite to the receiver:

$$\delta_\rho = \int_{Rx}^{Sat} c\frac{dl}{\upsilon} - \rho = \int_{Rx}^{Sat} (\eta - 1)\, dl \tag{4.1}$$

where c is the speed of light in vacuum or space, and υ is the actual trans-ionospheric signal propagation velocity, Rx = receiver position and Sat = satellite position.

The equation of complex refractive index of the ionosphere is often referred to as the Appleton-Hartree formula, whose detailed derivation can be found in Davies [10]. Assuming that there is cold, collision less, magnetized plasma, the phase refractive index, η_p, can be expressed for both ordinary waves (+ sign) and extraordinary (– sign) waves as:

$$\eta_p^2 = 1 - \frac{X}{1 - \frac{Y_T^2 \sin^2\theta}{2(1-X)} \pm \left[\frac{Y_T^4 \sin^4\theta}{4(1-X)^2} + Y_L^2 \cos^2\theta\right]^{\frac{1}{2}}} \tag{4.2}$$

$$X = \frac{\omega_p^2}{\omega^2} = \frac{e^2 N_e}{4\pi^2 f^2 m\varepsilon}, Y = \frac{\omega_g}{\omega} = \frac{eB}{2\pi fm} \tag{4.3}$$

where $\omega = 2\pi f$ is the circular frequency of the carrier frequency f. The ω_p and ω_g are plasma and gyro frequencies associated with the free electrons of the ionosphere. $Y_T = Y\sin\theta$, and $Y_L = Y\cos\theta$, θ is the angle between the propagation direction of electromagnetic wave and the magnetic field B. And $m = 9.109 \times 10^{-31}$ kg is mass of an electron; $e = 1.602 \times 10^{-19}$ C, is charge of electron; $\varepsilon = 8.854 \times 10^{-12}$ F/m is the electric permittivity in free space (vacuum); N_e = number density of free electrons.

For the signals at GNSS frequencies, the frequency $\omega >> \omega_p$ (and hence $\omega >> \omega_g$), the Eq. (4.2) can be expanded into a second-order Taylor approximation and keeping the terms only up to 4th inverse power of frequency [11].

$$\eta_p = 1 - \frac{1}{2}X \pm \frac{1}{2}XY_L - \frac{1}{8}X^2 - \frac{1}{4}XY^2\left[1 + \cos^2\theta\right] \tag{4.4}$$

In the above Eq. (4.4), the positive and negative signs represent expressions for ordinary and extraordinary waves. The expression for phase refractive index η_p can be obtained for the extraordinary signals by substituting the values of X, Y, Y_L as expressed in Eq. (4.3) in the above Eq. (4.4), we get:

$$\eta_p = 1 - \frac{e^2}{8\pi^2 m\varepsilon} \cdot \frac{N_e}{f^2} - \frac{e^3}{16\pi^3 m^2\varepsilon} \cdot \frac{N_e B\cos\theta}{f^3} - \frac{e^4}{128\pi^4 m^2 \varepsilon^2} \cdot \frac{N_e B^2\left(1+\cos^2\theta\right)}{f^4} \quad (4.5)$$

The 2nd term in the above Eq. (4.5) accounts for more than 99.9% of the ionospheric delay calculated at GNSS frequencies [11,12,40]. At GNSS frequency bands, the 3rd, 4th, and 5th terms have the magnitude which is about 3, 5, and 6 orders less compared to the 2nd term at GNSS frequencies. Therefore, we usually use only the 1st and 2nd terms of Eq. (4.5) to get the refractive index of the ionosphere by neglecting the higher order terms; so, we get:

$$\eta_p = 1 - \frac{e^2}{8m\varepsilon\pi^2} \cdot \frac{N_e}{f^2} \quad (4.6)$$

After substituting the physical constants in the Eq. (4.6), we get:

$$\eta_p = 1 - \frac{40.3 \times N_e}{f^2} \quad (4.7)$$

Using the relation between the phase and group velocities [10], the group refractive index is expressed as:

$$\eta_g = 1 + \frac{40.3 \times N_e}{f^2} \quad (4.8)$$

after inserting Eqs. (4.7 or 4.8) in Eq. (4.1), that is, integrating along the ray path we get the delay due to ionosphere in meters as:

$$\delta_\rho = \frac{40.3}{f^2} \int_{Rx}^{Sat} N_e dl = \frac{40.3}{f^2} TEC \quad (4.9)$$

In Eq. (4.9), the term TEC is the total electron content along the ray path from receiver to satellite in a square meter cross-sectional area. For example, consider a signal from a system operating at frequency of 1 GHz traverses through TEC of 100 TEC units or 10^{18} electrons/m^2, a value (of slant TEC) that is more commonly exceeded with lower elevation angle. The 100 TEC units will produce an additional time delay of about 134 ns or 40.2 m of range error in case of satellite navigation. At a lower frequency like 100 MHz, same TEC will result in range error of above 4 km. Thus, the radio signals while traversing ionosphere experience different delays for different frequencies given by Eq. (4.9) as it is a dispersive medium. At GPS L1 frequency (1575.42 MHz), a range delay of 1 m corresponds to 6.15 TECU, or about 16.2 m range delay for 100 TEC units as per above example. Using a single frequency GPS receiver, it is difficult to estimate the range delay without knowledge of TEC. With single frequency GPS receivers, the ionospheric models are used to correct for the ionospheric range errors to an extent. The conventional ionosphere models cannot accurately estimate TEC variations with local time, season, location, and the level of solar and magnetic disturbance. The day-to-day TEC varies from a minimum of few TEC units during night time to a day time maximum that ranges from few tens to about 100 TECU or more, depending on the location of the station, season, and solar activity ([13] and references therein). Apart from the diurnal and seasonal variations, the ionosphere density varies by substantial amounts from its average behavior during the geomagnetic storms and space weather events. The slant TEC values can go up higher depending on the elevation of the satellite.

4.2.1 TEC calculation from GPS observations

All the GPS satellites transmit the same carrier frequencies. The code modulation that is different for each satellite is used to reduce the signal interference [41,42]. These codes are known as pseudo random noise (PRN) codes, these random codes are generated using a mathematical algorithm. Since each satellite broadcasts using its own unique code combination, the satellite can be identified by its unique PRN sequence number. The raw data recorded from GNSS/GPS receiver is in various data formats. The popular format is **R**eceiver **In**dependent **Ex**change (RINEX or rinex) is an ASCII (i.e., **A**merican **S**tandard **C**ode for **I**nformation **I**nterchange) format, which is receiver independent and is mostly used for sharing data between various groups or publishing the data. Please see the link (https://igs.org/formats-and-standards/) for rinex data format definition and standards. To measure the TEC, both carrier phase and pseudorange/code measurements from two frequencies transmitted by GNSS/GPS, that is, four observables are required. TEC is measured from GNSS/GPS using two methods, (1) from group delay or code measurements, and (2) carrier phase measurements. The TEC computation using code measurements is simple and robust [14], but the code measurements are noisier compared to carrier phase measurements. Using the difference in range delay between the two frequencies f_1 (1575.42 MHz) and f_2 (1227.6 MHz) transmitted by GPS from the same satellite to the same receiver, we can exclude all the other atmospheric delays and errors that are common for both frequencies, leaving the residue of ionospheric differential delay (Δρ). Thus, TEC from code (pseudorange) measurements (C_1 and C_2 corresponding to frequencies f_1 and f_2) is given by

$$C_2 - C_1 = \Delta_\rho = \delta_{\rho 2} - \delta_{\rho 1} = \frac{40.3\left(f_1^2 - f_2^2\right)}{f_1^2 f_2^2} TEC \tag{4.10}$$

$$TEC_G = \frac{f_1^2 f_2^2}{40.3\left(f_1^2 - f_2^2\right)} \left(C_2 - C_1\right) \tag{4.11}$$

Similarly, we get TEC from phase measurements (L_1 and L_2) given by

$$TEC_P = \frac{f_1^2 f_2^2}{40.3\left(f_1^2 - f_2^2\right)}\left(L_1 - L_2\right) \tag{4.12}$$

The calculation of TEC from code measurements (TEC_G) are absolute estimates but noisy, of course they are biased by differential code biases (which will be addressed in the next section). And the measurements using carrier-phase are about two orders more precise compared to code measurements but these are only relative TEC (TEC_P) due to an unknown offset as the actual number of cycles of phase is not known. Therefore, usually combination of both code and phase measurements are used to overcome phase ambiguity and noise problems for an improved TEC estimate. The phase TEC measurements (TEC_P) are levelled using code TEC measurements (TEC_G) to eliminate phase ambiguity.

4.2.2 Cycle slip correction

The phase observables (L_1 and L_2) of the single PRN arc must be corrected for any cycle slips before they are used in Eq. (4.12) for derivation of phase TEC. Also seldom, researchers do the correction of cycle slips in the differential phase or phase TEC (TEC_P), wherein a cycle slip [43] is defined as a jump in differential phase TEC of more than or equal to fixed threshold TEC units (usually 1 to few TECU) in the interval defined by mostly data sampling [15,16]. Ionospheric density irregularities impact the phase of the signal; the spatially irregular phases emerging from the medium can cause constructive or destructive interference leading to increased or decreased signal amplitude [17]. These varying electron densities can cause Doppler shift that can exceed the bandwidth of the receiver's phase lock loop (PLL), which is used to continuously track the phase. In the event of phase shifts causing PLL to exceed its bandwidth, the receiver loses its lock on to signal, leading to cycle slips [18,19]. Also, during ionospheric scintillation, the destructive interference causing lower signal power can cause the receiver to lose of lock on the signals and several losses of lock events prevent will cause discontinuity in the TEC variation. Cycle slip correction requires removing the discontinuities in different phase connected arcs and producing a continuous curve. In general, the most cycle slip correction methods use either a kind of curve fitting or simple arithmetic to detect and correct the slip. The curve fitting methods are accurate but fail when there are multiple cycle slips, or during the ionospheric disturbances and/or rapid ionospheric variations. The arithmetic algorithms assume that ionosphere does not change rapidly in the data sampling interval time to correct for the offset and does a decent job at handling multiple slips. The readers can refer to a review of the popular cycle-slip correction methods and their theoretical performance, which are reported in [20].

To detect correct the cycle slips in phase TEC, the author of this chapter has adopted a simple arithmetic algorithm which is described here. The simple arithmetic progression difference with a tolerance of standard deviation of the last 10 data samples is used to detect the slip in phase TEC. The slip is corrected with an offset of the running mean of the previous 5 samples difference, given by the following equation:

$$Slip = \begin{cases} no, & (x_i - x_{i-1}) \leq \sigma \\ yes, & (x_i - x_{i-1}) > \sigma \end{cases} \tag{4.13}$$

where *xi* is the phase TEC data sequence, and σ is standard deviation of the previous 10 data samples. This method has an advantage that there is no fixed threshold and it changes dynamically with data and so is suitable for any sampling interval. The false trigger of slip detection is rare as the difference with previous sample is compared against standard deviation of last 10 data values. Thus, with the occurrence of cycle slip, the difference will become significantly larger due to contribution from artificial jump as result of cycle slip. The difference in phase TEC caused by cycle slips is linearly proportional to the size of the cycle slip. This method has computational advantage compared to curve fit methods, but at times this method may produce error which is again most likely less than fraction of TEC unit, as the observed TEC with consecutive data of the satellite arc does not change more than a fraction of a TEC unit often at 30 second data interval that is popularly used in most GNSS/GPS receivers. The Fig. 4.2 shows

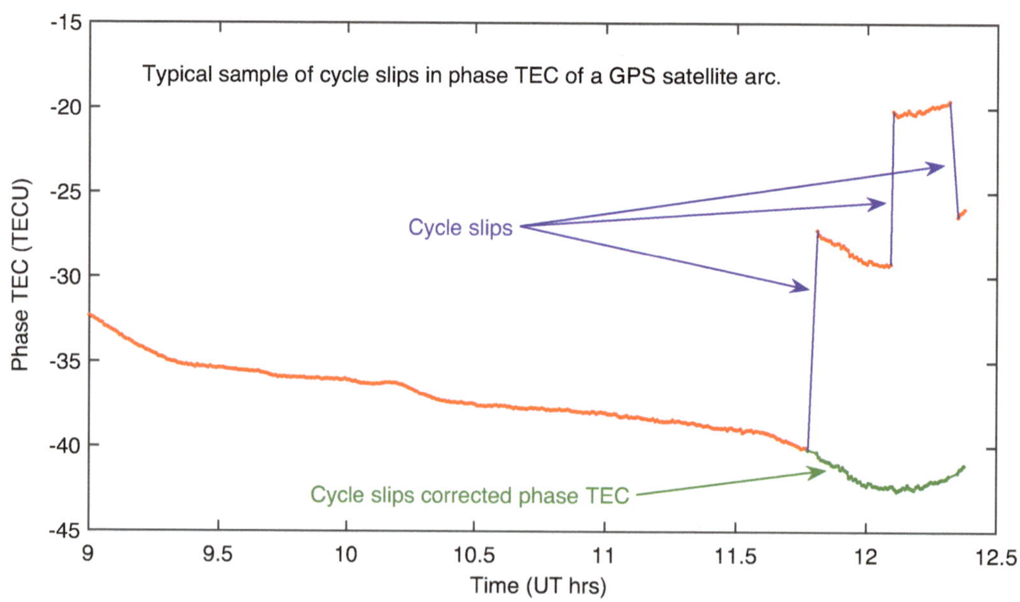

Fig. 4.2 ***A typical example showing phase TEC (in red color) with cycle slips indicated (blue color), and the phase TEC corrected for cycle slips (green color).***

a typical example for phase TEC derived from GPS satellite arc that has experienced cycle slips and the sampling rate is 30 seconds. The un-corrected phase TEC is drawn in red color with cycle slips (i.e., the jumps in phase TEC) are indicated in blue color. The phase TEC (the section with cycle slips) that is corrected for cycle slips is shown in green color line, which results in single continuous arc of phase TEC without any jumps or slips. The author has adopted and used this slip correction method in program GPS_TEC tool developed by him, that calculates TEC from rinex data [21,22] and it can be downloaded from webpage https://seemala.blogspot.com.

4.2.3 TEC leveling

The phase TEC levelling process is based on computing the baseline for each connected arc of phase measurements from a single PRN as shown in Fig. 4.3. To the level the phase TEC to the absolute and noisy group TEC, the arithmetic difference (offset) between all the TEC_P and TEC_G at each of the epochs of single satellite (or PRN) is calculated, for elevation angles >20° that are continuous in time and that form single arc. If there is a time gap, that is, the values that are from same satellite, considered as a separated arc and the same procedure is repeated for that set separately. At lower

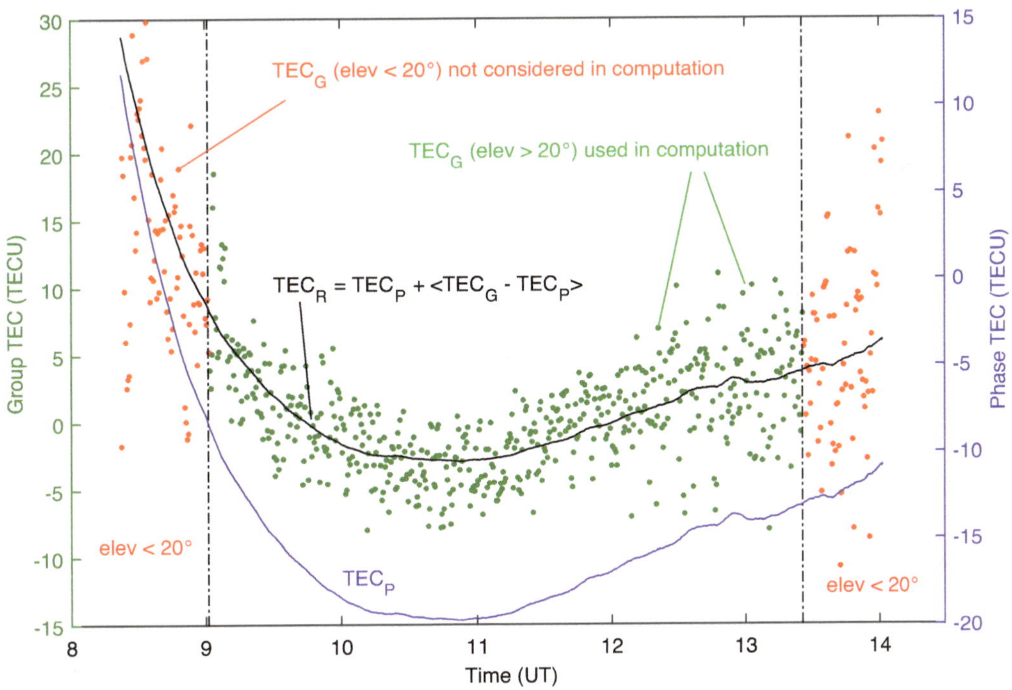

Fig. 4.3 ***Estimation of absolute TEC (REC_R) by levelling using the combination of code TEC (TEC_G) and the phase TEC (TEC_P).***

elevation angles, the group TEC values could be noisier due to multipath signals coming from horizon, ionospheric irregularities, and tropospheric scatter etc. The average offset is calculated after removing any outliers from the calculated offsets/differences. The outliers are defined as elements that are more than two local scaled median absolute deviation (MAD) [23] from the local median of selected window length (here, 1 hour is considered), as given by the following equation:

$$|y_i| = \begin{cases} x_i, & |x_i - \mu| < 2\sigma \\ outlier, & |x_i - \mu| \geq 2\sigma \end{cases} \tag{4.14}$$

where x_i is the data sequence of offsets ($TEC_G - TEC_P$) that belongs to selected window length of one hour, the μ is mean of the selected window length, and σ is its standard deviation. When the difference between the value x_i and the local mean is greater than two times the local standard deviation, then it is considered as an outlier and the data point is omitted from further calculation. The average of all the offset values from single arc of PRN, is the computed baseline value as given by the second term in the following equation is added to the phase TEC (TEC_P) values to get the slant TEC (TEC_R).

$$TEC_R = TEC_P + \sum_{i=1}^{N} (TEC_{Gi} - TEC_{Pi}) / N \tag{4.15}$$

where N is the number of data points in the single arc of the PRN. In the above equation, the ambiguity or calculated offset (second term) is considered constant for all phase TEC values for the duration of satellite pass or arc, so that it can be used to fit or level to code TEC [24,25].

The Fig. 4.3 depicts the TEC leveling process of GPS PRN 2, recorded from station "IISC," which is a part of International GNSS Service (IGS) network (https://igs.org/) for the date March 31, 2021, it was downloaded from ftp site ftp://data-out.unavco.org/pub. In the figure, code TEC (TEC_G) values are plotted as green dots for elevation angles >20° and red dots for elevation angles ≤20° which are not considered for offset baseline calculation. The phase TEC (TEC_P) values are plotted as a blue line, after adding the offset as per Eq. (4.15), the slant TEC (TEC_R) is plotted as a black line. It can be clearly seen that the final TEC_R values (shown in black line) are more precise as phase TEC and absolute without any ambiguity as code TEC; yet these slant TEC values must be processed for receiver and satellite differential code biases.

4.2.4 Vertical TEC conversion

The computed TEC_R values Eq. (4.15) shown in Fig. 4.3 as black line, are referred to as slant TEC, these values dependent on elevation angle of satellite but does not represent TEC at a particular location. The kind of unique TEC that characterizes the location will be obtained when the satellite is exactly overhead or above the location, this zenith value is called vertical TEC (VTEC). Since the GPS (in general GNSS) satellites move in different parts of the sky and the electron content in the ionosphere varies both spatially and temporally, the slant TEC cannot used to characterize the ionospheric variation of that location; therefore, it is often desirable to calculate VTEC. During the geomagnetic storms and ionospheric irregularities, when the spatial density gradients are high, the accurate VTEC values are useful in studying the ionospheric variations which are independent of satellite elevation angles. In the conversion of slant TEC to vertical TEC, it is often assumed that the ionosphere and the plasmasphere are horizontally stratified and are spatially uniform. Further, the ionosphere is simplified to a thin layer at fixed altitude of 350–400 km above the earth's surface. This height is the effective height or of the centroid of electron density of the ionosphere, which is usually called as ionospheric pierce point height (IPP) or altitude of ionospheric intersection of the user line-of-sight to the tracked satellite as shown in the Fig. 4.4. The Fig. 4.4 illustrates the schematic of the conversion of slant TEC to vertical TEC, the assumed altitude of IPP, the geometry of the ray path from satellite to ground through the ionospheric pierce point. This method is called the thin shell (approximation) model and is widely used [26–28]. The vertical equivalent (VTEC) of measured STEC at the ionospheric piercing point (IPP) height can be determined using the ionospheric thin shell mapping function given by:

$$VTEC = \left[STEC - \left(DCB_R + DCB_{Si}\right)\right] \times S(E) \tag{4.16}$$

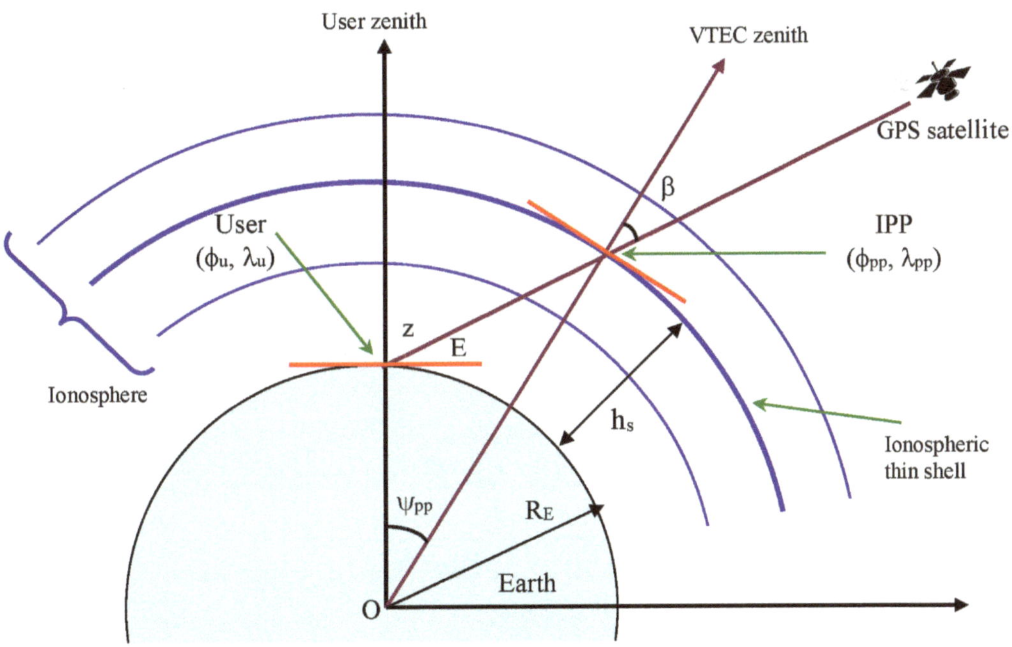

Fig. 4.4 ***Schematic of Slant TEC to Vertical TEC conversion.***

where DCB_R and DCB_{Si} are the differential code biases of the receiver and "i"th satellite observed, these will be discussed in the following section. S(E) is the thin shell mapping function, whose mathematical expression is presented below:

$$S(E)=\left\{1-\left(\frac{R_E\times\cos(E)}{R_E+H_{ipp}}\right)^2\right\}^{\frac{1}{2}} \tag{4.17}$$

Here R_e is the radius of the earth (R_e = 6378 km) and E is the elevation angle of satellite with respect to receiver. H_{ipp} is the altitude of ionospheric piercing point (IPP), as mentioned it is usually assumed in the range 350–400 km, apparently based on the maximum electron density altitudes in the ionosphere.

This effective height (or IPP altitude) influences the conversion of the measured slant TEC to vertical TEC and vice-versa, it is defined by the obliquity factor that depends on the elevation angle of satellite. In the equatorial and low latitude regions, where the phenomenon like equatorial ionization anomaly (EIA) crests appear; the electron densities and the gradients are high, and the determination/assumption of a single suitable IPP altitude is more difficult. However, from the studies by Rama Rao et al. [29] the VTEC computed from STEC for different IPP altitudes ranging from 250 km to 750 km in the Indian low latitude region has revealed that resultant VTEC does not change significantly with the IPP altitude as long as the elevation angle of the satellite is greater >50°. Whereas, in the case of satellites with lower elevation angles (<50°), there is a significant departure in the computed VTEC using different IPP altitudes from both the methods. Therefore, the commonly assumed IPP altitude of 350 km can safely be used for higher elevation angles, otherwise for lower elevation angles, the local/regional average F-region peak density altitude may be used.

Using the GPS receiver's geographic latitude and longitude (ϕ_u, λ_u), the IPP location coordinates can be computed using the observed azimuth (A) and elevation angle (E) to the tracked satellite, using the single-layer ionospheric model [30]. The latitude (ϕ_{pp}) of IPP location is given by:

$$\phi_{pp}=\sin^{-1}\left(\sin\phi_u\cdot\cos\psi_{pp}+\cos\phi_u\cdot\sin\psi_{pp}\cdot\cos A\right) \tag{4.18}$$

where Ψ_{pp} is the angle subtended at the center of the earth between the user position vector and the earth projection of the pierce point, it is computed by expression:

$$\psi_{pp}=\frac{\pi}{2}-E-\sin^{-1}\left(\frac{R_E}{R_E+H_{IPP}}\cos E\right) \tag{4.19}$$

and the longitude (λ_{pp}) of the IPP is given by:

$$\lambda_{pp}=\lambda_u+\sin^{-1}\left(\frac{\sin\psi_{pp}\sin A}{\cos\phi_{pp}}\right) \tag{4.20}$$

Thus, the VTEC calculated from the measured slant TEC represents the TEC value at the IPP location (ϕ_{pp}, λ_{pp}) determined, and thus can be used in the navigation applications to study the TEC spatial and temporal behavior with a coverage limited by the elevation angle. The elevation "E" and azimuth "A" of the satellite at that epoch, with respect to the receiver location can be calculated using the position of the satellite at that time which is estimated from the ephemeris downloaded from any IGS processing center (for example: ftp://data-out.unavco.org/pub/rinex/nav/, ftp://garner.ucsd.edu/pub/nav/, http://ftp.aiub.unibe.ch/CODE/, etc.), or using the navigation file recorded by the receiver. The navigation message or the ephemeris is global to all the receivers and so the navigation file of any station can be used as long as it contains ephemeris for all satellites. For example, the program written by the author uses the station navigation file if it is present, otherwise it downloads a navigation file of predetermined station "chpi" or "cro1" and uses it for calculation of satellite's elevation and azimuth angles. The calculation of satellite's position, elevation and azimuth is beyond the scope of this chapter and the readers may refer to Strang and Borre [31].

4.2.5 Receiver and satellite differential code biases (DCBs)

As discussed in Section 4.2.3, code/pseudorange measurements are assumed to be absolute, but noisy. For this reason, usually we use combination of pseudorange and phase measurements to eliminate noise and phase ambiguity in the resultant TEC. Yet this absolute slant TEC (from Eq. 4.15) is not truly absolute. All effects such as distance of receiver to satellite, tropospheric delay, clock offsets; that are common to both frequencies are removed, but the frequency-dependent effects, like differential instrumental biases in the satellite and the receiver, and multipath are still present. The effects of multipath can be avoided with an elevation mask of 30° [32,33]. Thus, the STEC (from Eq. 4.15) still needs to be corrected for differential instrumental/code biases at satellite and receiver.

The differential code biases (DCBs), is the delay related with the different, frequency dependent times for signals in RF paths/channels at satellites and receivers. This delay bias in each signal of the two GPS/GNSS frequencies is also known as differential instrumental bias. Several meters of error can occur if the effect of the DCBs is ignored, and the estimated TEC values can even become nonphysical negative TEC [34]. For example, a 1 ns delay due to DCB causes an error of 2.85 TECU (at GPS frequencies) in TEC estimation. The DCB biases are known to vary systematically. Seasonal variations in estimated TEC errors (up to ~20 TECU) associated with receiver DCBs are observed. These variations can be related to variations in the receiver environment, like temperature [35]. Thus, the DCBs must be accounted for precise estimate of slant TEC in the Eq. (4.16) before calculation ofVTEC.

The DCBs are derived by various methods; some methods are developed to calculate biases by considering more than one station and model the TEC on the double differences of GPS observations. But, for the calculation of TEC and DCBs using single station observation, in general there are two basic approaches. The first method, the TEC is modeled by polynomial coefficients including DCBs; these unknown coefficients and DCBs in the form of linear system of equations are solved by least squares method [25,27,36,37]. In the second method, the DCB value is chosen based on searching for the true value with a constraint condition through minimizing the standard deviation [36,38]. The calculated DCBs of satellite and IGS station receiver can be directly obtained from internet from IGS analysis centers, namely, Jet Propulsion Laboratory (JPL) Pasadena, CA, USA; Center for Orbit Determination in Europe (CODE) University of Berne, Switzerland; European Space Operations Center (ESOC) of European Space Agency (ESA), Germany; and gAGE/UPC of Polytechnical University of Catalonia, Spain.

In this section, a similar approach to standard deviation minimization [36] for calculation receiver DCB using single station is described, assuming that satellite DCBs are known (or downloaded from one of IGS centers). In this technique, for a selected period, the TEC values calculated from all satellites are assumed to be close. Preferably the local post mid-night hours are chosen when the TEC is spatially uniform; so that TEC values calculated from all satellites (in that duration) will be close, as the ionosphere is mostly spatially uniform then. For this dataset, the optimum receiver DCB is selected by searching or using trail DCB values in TEC calculation and the value that minimizes the standard deviation is chosen as the DCB for that receiver/station. The minimization of standard deviation method assumes that the calculated VTEC from all satellites in view are equal satellite zenith are same. For example, if the ionosphere is horizontally homogeneous and DCBs are properly removed, the calculated VTECs from all the visible satellites must be identical. In reality, it is known that ionosphere has a horizontal gradient and vertical structure. However, if the correct DCB values are used for calculatingVTEC, the scattering ofVTEC values from all satellites can be safely assumed to be minimum. To reduce multipath effects an elevation mask angle of 30° is used on the dataset and this will also ensure that ionospheric spatial homogeneity to an extent as assumed in this method.

The receiver DCB is estimated by using a series of trail bias candidates and then selecting the one that has a minimum deviation of TECs with respective mean values. Mathematically, for a given range of DCB*(k)* for receiver bias to test, the standard deviation $\sigma_k(t)$ of VTECs with their mean (as discussed below) is calculated for all the 'M' satellites at each observation epoch "t," the expression is given by:

$$\sigma_k(t) = \sqrt{\frac{1}{M_n} \sum_{j=1}^{M_t} (VTEC_k^j(t) - \overline{VTEC_k}(t))^2} \tag{4.21}$$

where M_t represents the total number of satellites at epoch "t." In the Eq. (4.21), $\overline{VTEC_k}(t)$ denotes mean of all VTEC values from M_t satellites for that epoch or selected interval. The total of standard deviations $\sigma_{Total}(k)$ is obtained for the entire day with VTEC values calculated using DCB(k) is given as

$$\sigma_{Total}(k) = \sum_{i=1}^{N} \sigma_k(t) \tag{4.22}$$

where $\sigma_k(t)$ is the sum of standard deviation values for "k"th DCB at an epoch "t," N is the number of epochs in the measurement period or selected data.

The range of DCB test values that are converted into TEC units (so to add to STEC) with a step size of 0.1 TEC units, as input for the minimization of standard deviation. Then for each of these DCB(k) values, the $VTEC_k$ values are computed using Eqs. (4.16 and 4.17), and total standard deviation $\sigma_{Total}(k)$ are calculated from above Eqs. (4.21 and 4.22). When the minimum $\sigma_{Total}(k)$ value is found, then corresponding DCB(k) is the correct estimate of receiver DCB.

This method needs a priori search range and the computation time increases with increase in search range. Also, if the given DCB range is narrower, the value of correct DCB may not be within this small range leading to an incorrect estimate of DCB. In general, most of the receiver DCB values fall within the range of −30 to 30 ns or about −90 to 90 TEC units. But, to include any peculiar receiver with high DCB value, a range of −500 to 500 TEC units will be used as trail DCB range, that could easily increase the computation time. To reduce the computation time, the variable step size was used as shown in the Fig. 4.5. Initially, a step size of 50 TECU was used for −500 to 500 TECU range (20 iterations) see Fig. 4.5A. After selecting the first minimum $\sigma_{Total}(k1)$, the range DCB(k1) ± 50 TECU with step size of 10 TECU is used again for second time (10 iterations). As shown in Fig. 4.5B, with $\sigma_{Total}(k2)$ being the second time minima selected, the third range of DCB(k2) ± 10 TECU with step size of 1.0 TECU is used for third time (20 iterations). Finally, with $\sigma_{Total}(k3)$ being the third minima (in Fig. 4.5C), another range of DCB(k3) ± 1 TECU with final step size of 0.1 TECU is used (20 iterations). The resultant selected minima $\sigma_{Total}(k4)$ will give the optimum receiver bias value DCB(k4) as shown in Fig. 4.5D. The usage of variable step size in 4 stages results in a reduced total number of iterations from 10,000 iterations to just 70 iterations instead of using uniform step size of 0.1 TECU for entire range of -500 to 500 TEC units for DCB values in Eq. (4.22).

Thus, with this usage of variable step size, the computational time is reduced by more than two orders magnitude. In addition, the data that is usually sampled at 1 to 30 s is decimated to 3 minutes, leveraging at least another one order magnitude reduction for 30 s data rate in computation time as the decimation does not the effect the estimation of optimal bias with the assumption that receiver DCB is constant throughout the day or 24 hours' time. After the estimation of optimal receiver DCB, it is used with the dataset of original sampling rate and final bias corrected STEC and VTEC values are calculated.

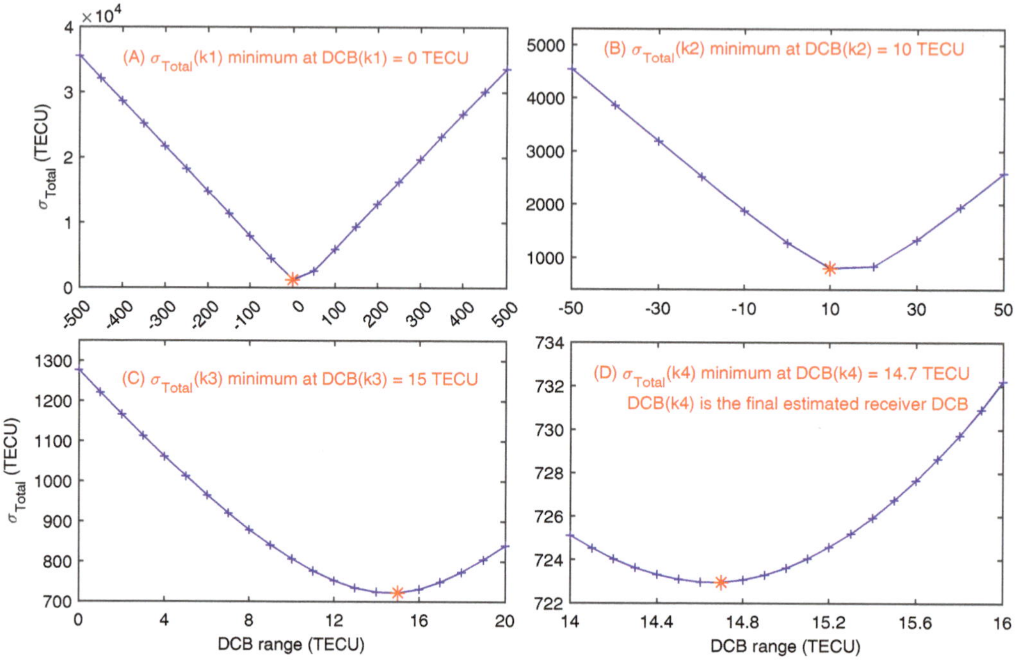

Fig. 4.5 ***Typical values σ_{Total} calculated for all epochs of day at different ranges/steps of receiver DCB ranges shown on x-axis, calculated for "IISC" station on March 31, 2021.*** This figure showcases how the minimum number of variable steps were used to reach the optimal minimum of σ_{Total}.

The distribution of vertical TEC arcs at different receiver DCB bias conditions is depicted in Fig. 4.6 by using actual data set. The data from "IISC" an IGS station is chosen, which is located at (13.02 N, 77.57 E) for the date of March 31, 2021. The absolute values of TEC are obtained by including the differential satellite biases published by the University of Bern as already mentioned in this section. The Fig. 4.6 shows VTEC values in three panels, the applied satellite biases were identical, but receiver DCBs in each of the case are different as, panel (A) the receiver DCB is over corrected or with DCB value of −20.0 TECU, we can see that VTEC (blue color) values spread throughout the day including night time and satellite arcs of hyperbolic shape; panel (B) the receiver DCB is under corrected with DCB value of 40.0 TECU, which also results in spread of VTEC (red color) values and all the satellite arcs are of parabolic shape; panel (C) the receiver DCB is optimally corrected with estimated DCB value of 14.7 TECU, we can observe that VTEC (green color) values of different satellites/PRNs are closer during night time hours and overall the most arcs are closer except at noon time where the horizontal gradients are higher.

This method of minimizing the standard deviation with variable step size and down-sampled VTEC values (to achieve 4 orders of computation efficiency) for receiver DCB estimate is currently used in GPS_TEC program tool developed by the author (Gopi

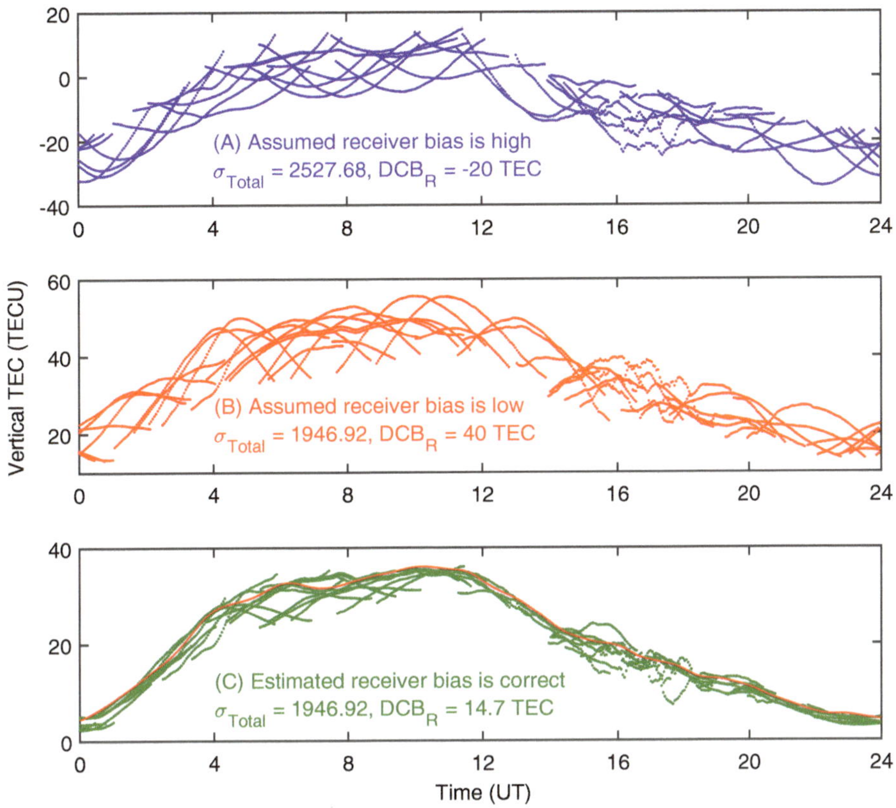

Fig. 4.6 The calculated VTEC values from observed STEC at station "IISC" for the date March 31, 2021 for different receiver bias corrections (A) when receiver bias is over corrected (B) when receiver bias is under corrected (C) the receiver bias applied is correct. The sum of the standard deviation of VTECs to their mean at all epochs of the day, is shown in each panel. The data of IISC, IGS station of date March 31, 2021 is used for VTEC derivation.

Seemala) available at webpage https://seemala.blogspot.com. Here, the VTECs are calculated by using satellite DCBs published by the University of Bern available at the link http://ftp.aiub.unibe.ch/CODE/ and receiver DCB is estimated as discussed in this section. Earlier, the first method for single station mentioned in this section, was used by the author in the program GPS_TEC, using satellite DCBs published by the University of Bern and the receiver bias is calculated by using least squares method for selected data of local night hours 02:00 to 06:00 LT [13,22]. Both these methods are practically feasible and can be used offline with an exception of assuming known satellite DCBs or that were downloaded from one of the IGS centers.

The program took a time of about 200–300 milliseconds to complete the above task of estimating the receiver DCB for 30 second daily rinex file. The total time taken by the program to complete the task of estimation ofVTEC values, which includes reading one day rinex data (30 second sampling), estimating STEC and receiver DCB, reading navigation file, calculating IPP coordinates and conversion of STEC to VTEC, and also saving the processed ASCII data and VTEC figure files, is usually less than or about 2 seconds on a simple intel processor with 2 GHz clock speed. The program is highly optimized to handle large number of rinex files at one go with built-in batch processing options, automatic download of required navigation and satellite DCB files.

The two-sigma-mean method is used to represent the mean diurnal variation of TEC at the station. This TEC diurnal variation curve is shown as red line in the Fig. 4.6C on top of individual satellite arcs of the day that are drawn in green color. The two-sigma-mean is described as follows. The mean and standard deviation of VTEC values from all satellites is calculated for all the epochs/data points in 1 minute window. Then, the values that are outside of mean ± standard deviation are eliminated. And the same process is repeated for second time, to further eliminate data points that are outside standard deviation from the mean. Now, with the remaining TEC data points in 1 minute window, the mean is calculated, this mean data value obtained over this selected window is called two-sigma-mean. The entire process is repeated for all the epochs with selected time windows, and the resultant is continuous diurnal mean TEC values with 1 minute sampling representing TEC variation for that station. These 1 minute two-sigma-mean samples are further smoothened using a used a low pass filter (with cutoff of 4 hours period) to smooth the final diurnal variation TEC values. From the Fig. 4.6C, the two-sigma-mean TEC variation in comparison with VTEC values for different satellite arcs, it can be clearly seen that the mean TEC is well represent preserving the diurnal pattern without yielding to pattern of multiple satellite arcs.

4.3 Summary

The ionospheric effect on radio signals, and the estimation of ionospheric TEC using dual frequency signals from GNSS are explained with theory. The instrumental biases or DCBs of satellite and receiver must be considered for GPS precise positioning and in calculation of TEC. The GPS data from dual frequency receiver was used to demonstrate the estimation of receiver DCB and calculation of VTEC from raw GPS observables. The chapter also explained the implementation of efficient method for the receiver DCB estimation, which is faster and sufficiently accurate for the data from GPS/GNSS receivers installed across the globe, as there is no selection of data is needed with respect to local time to ensure night time homogeneity in ionosphere. This process of TEC estimation can be extended to any GNSS raw dataset. The author has also discussed the program developed by that is available at webpage http://seemala.blogspot.com/ that has used the methods described in this chapter and that is being used by several researchers to estimate TEC data from rinex data recorded by the dual frequency GNSS receivers.

Acknowledgements

Author acknowledges IRI team for making the IRI model available online, and the IGS data archival services provided by UNAVCO, Inc. The author acknowledges Patricia Doherty, Institute for Scientific Research, Boston College for sharing the Fortran program (rinex_11.for). The author also acknowledges Institute for Scientific Research, Boston College, USA; where the author got the opportunity and initiated the work on GPS_TEC program for TEC data analysis.

References

[1] D. Martyn, Location of the currents causing the solar and lunar diurnal magnetic variations, Nature 160 (1947) 535–537. doi:10.1038/160535b0.
[2] A.E. Kennelly, On the elevation of the electrically conducting strata of the earth's atmosphere, Electr. World Eng. 39 (1902) 473.
[3] O. Heaviside, Telegraphy, Britannica 33, 10th ed., Encyclopedia Britannica, Chicago, 1902, pp. 213–218.
[4] E.V. Appleton, M.A.F. Barnett, On some direct evidence for downward atmospheric reflection of electric rays, Nature 115 (1925) 333.
[5] S. Chapman, The absorption and dissociative or ionizing effect of monochromatic radiation in an atmosphere on a rotating earth, Proc. Phys. Soc. 43 (1931) 26, doi.org/10.1088/0959-5309/43/1/305.
[6] M.C. Kelley, The Earth's Ionosphere: Plasma Physics and Electrodynamics. In: The Earth's Ionosphere: Plasma Physics and Electrodynamics, second ed., Elsevier, New York, 2009, doi.org/10.1088/0959-5309/43/1/305.
[7] D. Bilitza, B.W. Reinisch, International Reference Ionosphere 2007: Improvements and new parameters, Adv. Space Res. 42 (4) (2008) 599–609. doi:10.1016/j.asr.2007.07.048.
[8] L. Wanninger, Effects of the Equatorial Ionosphere on GPS, GPS World, Cleveland, 1993, pp. 48.
[9] J.A. Klobuchar, Ionospheric effects on GPS, in: B.W. Parkinson, J.J. Spilker (Eds.), Global Positioning System: Theory and Applications, Progress in Astronautics and Aeronautics, Volume 164, vol. 2, American Institute of Aeronautics and Astronautics INC., Washington, 1996, pp. 485–515.
[10] K. Davies, Ionospheric Radio. In: Ionospheric Radio, Peter Pergrinus Ltd., London, 1990.
[11] S. Bassiri, G. Hajj, High-order ionospheric effects on the global positioning system observables and means of modeling them, Manuscr. Geodaet. 18 (1993) 280–289.
[12] N. Jakowski, F. Porsch, G. Mayer, Ionosphere-induced ray-path bending effects in precision satellite positioning systems, Zeitschrift fur Satellitengestutzte Positionierung, Navigation und Kommunikation 1/94 (1994) 6–13.
[13] P.V.S. Rama Rao, S. Gopi Krishna, K. Niranjan, D.S.V.V.D. Prasad, Temporal and spatial variation in TEC using simultaneous measurements from the Indian GPS network of receivers during the low solar activity period of 2004–2005, Ann. Geophys. 24 (2006) 3279–3292. doi:10.5194/angeo-24-3279-2006.
[14] F. Arikan, C.B. Erol, O. Arikan, Regularized estimation of vertical total electron content from Global Positioning System data, J. Geophys. Res. 1 (18) (2003) 1469–1480 2003.
[15] Z. Liu, A new automated cycle slip detection and repair method for a single dual-frequency GPS receiver, J. Geod. 85 (2011) 171–183. doi:10.1007/s00190-010-0426-y.
[16] S. Sharma, N. Dashora, P. Galav, R. Pandey, Cycle slip detection, correction and phase leveling of RINEX formatted GPS observables, Curr. Sci. 100 (2) (2011) 205–212.
[17] P.M. Kintner, B.M. Ledvina, E.R. de Paula, GPS and ionospheric scintillations, Space Weather 5 (2007) S09003. doi:10.1029/2006SW000260.
[18] G. Blewitt, An automatic editing algorithm for GPS data, Geophys. Res. Lett. 17 (3) (1990) 199–202 .
[19] P.V.S. Rama Rao, S. Gopi Krishna, K. Niranjan, D.S.V.V.D. Prasad, Study of spatial and temporal characteristics of L-band scintillations over the Indian low latitude region and their possible effects on GPS navigation, Ann. Geophys. 24 (6) (2006) 1567–1580. doi:10.5194/angeo-24-1567-2006.

[20] S. Banville, R.B. Langley, S. Saito, T. Yoshihara, Handling cycle slips in GPS data during ionospheric plasma bubble events, Radio Sci 45 (2010) RS6007. doi:10.1029/2010RS004415.
[21] D. Okoh, G. Seemala, B. Rabiu, J.B. Habarulema, S. Jin, K. Shiokawa, et al., A neural network-based ionospheric model over Africa from Constellation Observing System for Meteorology, Ionosphere, and Climate and Ground Global Positioning System observations, J. Geophys. Res. Space Phys. 124 (2019) 10512–10532. doi:10.1029/2019JA027065.
[22] G.K. Seemala, C.E. Valladares, Statistics of total electron content depletions observed over the South American continent for the year 2008, Radio Sci 46 (2011) RS5019. doi:10.1029/2011RS004722.
[23] C. Leys, et al., Detecting outliers: Do not use standard deviation around the mean, use absolute deviation around the median, J. Exp. Social Psychol. 2013 (2013). doi:10.1016/j.jesp.2013.03.013.
[24] A. Komjathy, Global Ionospheric Total Electron Content Mapping Using the Global Positioning System (PhD thesis), Technical Report No. 188, Department of Geodesy and Geomatics Engineering, University of New Brunswick, Fredericton, New Brunswick, Canada, 1997.
[25] G.E. Lanyi, T. Roth, A comparison of mapped and measured total ionospheric electron content using Global Positioning System and beacon satellite observations, Radio Sci. 23 (4) (1988) 483.
[26] L. Ciraolo, P. Spalla, Comparison of ionospheric total electron content from the Navy Navigation Satellite System and the GPS, Radio Sci. 32 (1997) 1071–1080.
[27] D. Coco, C. Coker, S.R. Dahlke, J.R. Clynch, Variability of GPS satellite differential group delay biases, IEEE Trans. Aerospace Electr. Syst. 27 (1991) 931–938.
[28] B.D. Wilson, A.J. Mannucci, Instrumental-biases in ionospheric measurements derived from GPS data, Proceedings of ION GPS-93, Institute of Navigation, Washington, D.C., 1993.
[29] P.V.S. Rama Rao, K. Niranjan, D.S.V.V.D. Prasad, S. Gopi Krishna, G. Uma, On the validity of the ionospheric pierce point (IPP) altitude of 350km in the Indian equatorial and low-latitude sector, Ann. Geophys. 24 (2006) 2159–2168. doi:10.5194/angeo-24-2159-2006.
[30] J.A. Klobuchar, Ionospheric time-delay algorithm for single frequency GPS users, IEEE Trans. Aerospace Electr. Syst. AES-23 (3) (1987) 321–331.
[31] G. Strang, K. Borre, Linear Algebra, Geodesy, and GPS, Wellesley-Cambridge Press, Cambridge, 1997. https://klein.mit.edu/~gs/books/gps.html.
[32] M. Hernandez-Pajares, J.M. Juan, J. Sanz, Medium-scale traveling ionospheric disturbances affecting GPS measurements: Spatial and temporal analysis, J. Geophys. Res. 111 (2006) A07S11. doi:10.1029/2005JA011474.
[33] J.A.R. Rose, J.R. Tong, D.J. Allain, C.N. Mitchell, The use of ionospheric tomography and elevation masks to reduce the overall error in single-frequency GPS timing applications, Adv. Space Res. 47 (2) (2011) 276–288. doi:10.1016/j.asr.2010.05.030.
[34] E. Sardón, A. Rius, N. Zarraoa, Estimation of the transmitter and receiver differential biases and the ionospheric total electron content from Global Positioning System observations, Radio Sci 29 (3) (1994) 577–586. doi:10.1029/94RS00449.
[35] A.A. Mylnikova, Yu.V. Yasyukevich, V.E. Kunitsyn, A.M. Padokhin, Variability of GPS/GLONASS differential code biases, Results Phys. 5 (2015) 9–10. doi:10.1016/j.rinp.2014.11.002.
[36] G. Ma, T. Maruyama, Derivation of TEC and estimation of instrumental biases from GEONET in Japan, Ann. Geophys. 21 (2003) 2083–2093. doi:10.5194/angeo-21-2083-2003.
[37] R. Warnant, Reliability of the TEC computed using GPS measurements: The problem of hardware biases, Acta Geodaet. Geophys. Hungarica 32 (1997) 451–459. doi:10.1007/BF03325514.
[38] F. Arikan, H. Nayir, U. Sezen, O Arikan, Estimation of single station interfrequency receiver bias using GPS-TEC, Radio Sci 43 (2008) RS4004. doi:10.1029/2007RS003785.
[39] J.A. Klobuchar, Ionospheric effects on GPS, GPS World 2 (4) (1991) 48–51.
[40] M.M. Hoque, N. Jakowski, Estimate of higher order ionospheric errors in GNSS positioning, Radio Sci 43 (RS5008) (2008), doi:10.1029/2007RS003817.
[41] R.B. Langley, Why is the GPS signal so complex? GPS World1 (3) (1990) 56–59.
[42] B. Hofmann-Wellenhof, H. Lichtenegger, J. Collins, Satellite signal, Global Positioning System, 4th ed., Springer, New York (1997).
[43] I. Horvath, S. Crozier, Software developed for obtaining GPS-derived total electron content values, Radio Sci. 42 (20) (2007), doi.org/10.1029/2006RS003452.

CHAPTER 5

Remote sensing data extraction and inversion techniques: A review

Arjun Pratap Shahi[a], Praveen Kumar Rai[b], Rabi-ul-Islam[c], Varun Narayan Mishra[d]
[a]Independent Researcher, Varanasi, India
[b]Department of Geography, K.M.C. Language University, Lucknow, India
[c]Haryana Space Applications Centre, Hisar, India
[d]Amity Institute of Geoinformatics & Remote Sensing (AIGIRS), Amity University, Noida, Gautam Buddha Nagar, India

5.1 Introduction

The amalgamation of space exploration and remote sensing have helped in continuous and robust observation of earth surface. The earth observation satellites have provided us with satellite images which collect information in various bands. Remote sensing is a science or art, an activity or function, tool or technique of accruing information about an object, area, or phenomenon in different wavelength (ultraviolet, visible, infrared, and microwave) regions of the electromagnetic spectrum by means of instruments such as scanners and cameras mounted on mobile platforms, such as aircraft or spacecraft, and the analysis of acquired information by means of photo interpretive techniques, image interpretation, and state-of-the-art image processing systems [1–3]. Data extraction techniques have laid foundation for successful exploitation of embedded information in satellite image. To provide desired thematic information, data extraction techniques are not only required to be accurate and reliable, but also the ability to integrate with various physical and prediction models. There are various information extraction techniques such as classification, change detection, extraction of indices, feature identification, etc. The systematic and structured way of information extraction from remote sensing data is accomplished through classification. Recent studies have showed that data extraction through artificial intelligence [4], deep learning [5], convolution neural network (CNN) [6], and machine learning algorithms like support vector machine (SVM) and random forest (RF) is widely used for wetland classification [7], to assess the changes in agricultural irrigated areas [8], for MODIS global burned area classification [9], coral reef change detection in remote pacific islands [10], land-use change in coal mining [11], for crop classification [12], for object-based urban land use/land cover classification [13], assessment of Antarctic moss health [14], and flood mapping [15].

This chapter deals in detail about classification techniques, band ratioing, different indices, principal component analysis, visual image interpretation method, data handling techniques, and fundamental file formats of remotely sensed data.

Atmospheric Remote Sensing
DOI: https://doi.org/10.1016/B978-0-323-99262-6.00021-3

5.2 Remote sensing data extraction

5.2.1 Classification

Image classification is a process of automatically assigning pixel in an image to meaningful classes or themes [16]. It is the basic operation for information extraction and feature detection. Multispectral data is necessary to perform classification process. Pixel is considered as the basic unit of an image which contains spectral signature of features/objects in several bands. Different features have different combination of digital numbers (DNs) based on their spectral reflectance and emittance properties. These spectral patten present within each pixel in different band is used as numerical basis for classification [17]. In a classified satellite image, pixels will differ spectrally from one category to another category of the same image [18]. Therefore, each theme or category have different area and the final digitally classified map is a mosaic of these different themes represented with different color or symbol also known as thematic maps [3]. Lu and Weng [65] proposed six major steps of image classification that are: (1) choice of a suitable classification system, (2) selection of training samples, (3) image preprocessing and feature extraction, (4) selection of suitable classification methods, (5) postclassification processing, and (6) accuracy assessment. Major steps involved during different classification is shown Fig. 5.1.

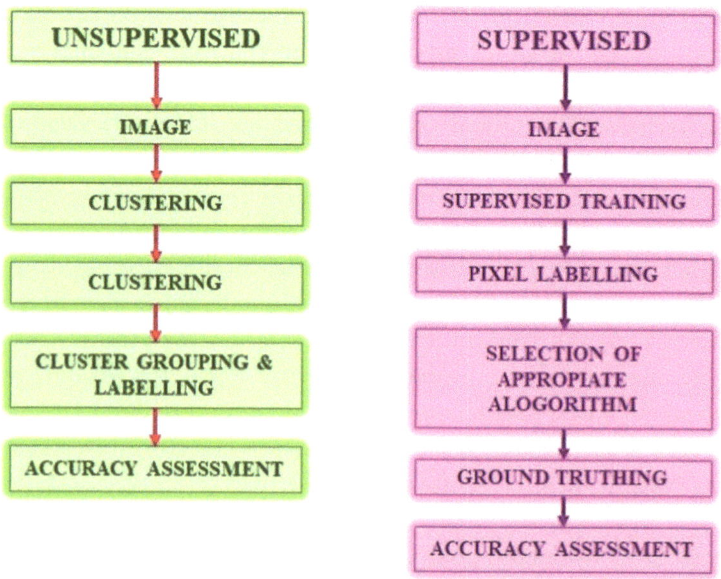

Fig. 5.1 ***Steps followed during unsupervised and supervised classification of optical multispectral satellite images.***

The image classifiers which form the backbone of multispectral image classification are:

1. Supervised classification
2. Unsupervised classification

5.3 Supervised classification

In supervise classification, an image analyst based on his image interpretation proficiency, "supervises" the operation of selecting various suitable sample sites also known as "training area." These training area or sample site are used to assemble numerical descriptors and/or numerical interpretation key that represent spectral characteristics for each feature type of interest. Each pixel of the image is compared numerically to each class within interpretation key and tagged with the name of class it matches most. There are three major steps involved in the classic supervised classification, (1) training: the user identifies representative training areas or samples and develops a numerical description of the spectral attributes (spectral signature) of each land cover class of interest in the image, (2) classification: each pixel in the image is classified into the land cover class based on its resemblance to the input training pixel and if the pixel is not matching to any predefined class signature then it is classified as unknown or unclassified, and (3) accuracy assessment: the classified thematic image is compared with reference image or ground reference data to check the accuracy of the classification [19]. Numerous mathematical classifier strategies have been developed for comparison of unknown pixel to training area pixels.

5.3.1 Minimum-distance-to-mean classifier

It is one of the simplest classification techniques. This classifier evaluates spectral mean/average in each band for each class and these values contain mean vector indicated with "+" sign (Fig. 5.2) [20,21]. Unknown pixels are classified by determining distance between the unknown pixel and class mean (in the two-coordinate system to reduce dimensionality of data and increase the computational efficiency). After evaluating distance, the unknown pixel is assigned to the nearest class. The pixel is classified as unknown if the computed distance from all the class mean is greater than analyst defined threshold. This classifier is mathematically simple and computationally efficient but also, insensitive to variance in the spectral domain. The thematic map generated using this classification in Fig. 5.6B.

5.3.2 Parallelepiped classifier

This classifier uses range of values in each category to classify pixels. The range is defined by the highest and lowest values of DNs in each band for each class. This range appears as rectangular area and also known as "parallelepipeds" (Fig. 5.3) [3].

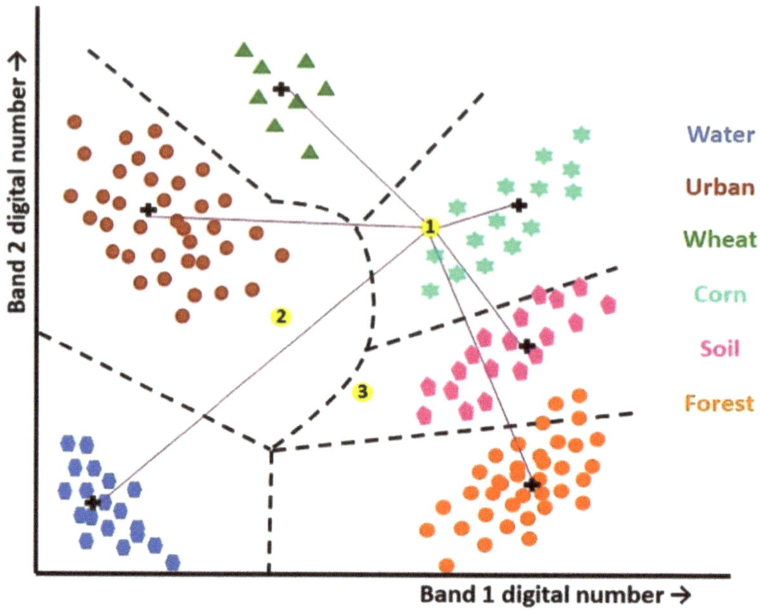

Fig. 5.2 ***Minimum-distance-to-mean classifier strategy*** *(Modified from: [45]).*

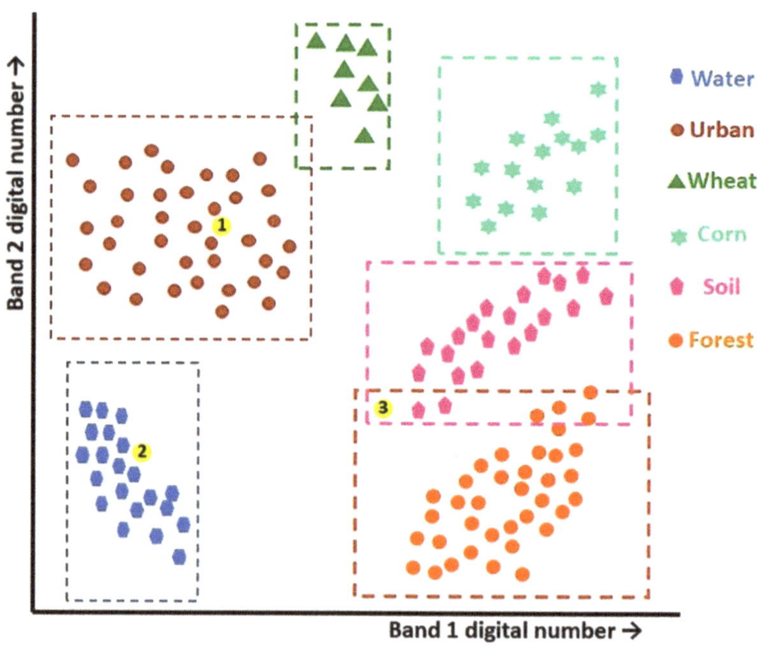

Fig. 5.3 ***Parallelepiped classifier strategy*** *(Source: [46]).*

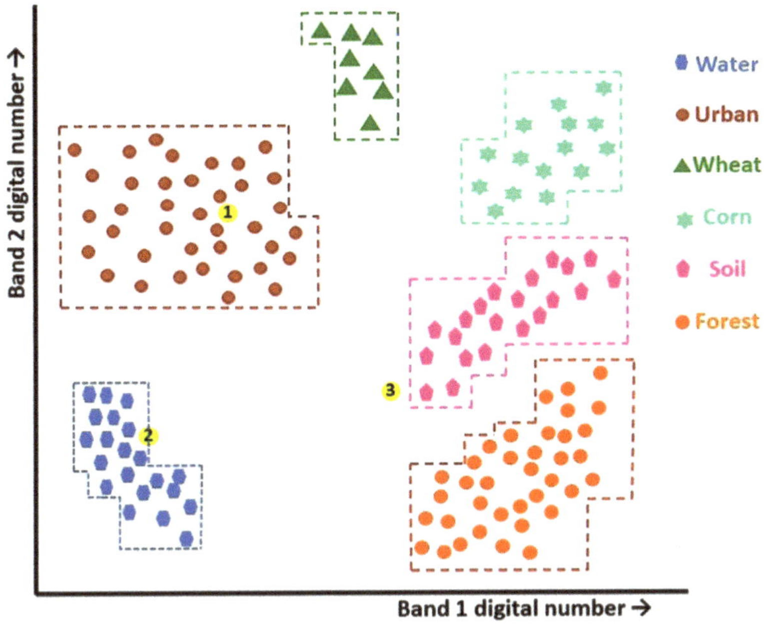

Fig. 5.4 ***Parallelepiped classifier strategy with stepped decision region boundaries*** *(Source: [3]).*

This classifier introduces sensitivity to class variance through range of values in each category training set. An unknown pixel is assigned to the class according to decision range or class range and labelled as the class within which it falls or else it is labelled as "unknown" if it lies outside all the decision ranges. Sometimes overlapping of decision region occurs and the pixel placed in those area is either labelled as "not sure" or assigned to any one of the overlapping classes [3]. Overlap is mainly caused due to class distribution exhibiting high covariance or correlation (Fig. 5.3). Due to high covariance spectral values have tendency to vary similarly in two bands. The rectangular decision area fits class training data very poorly, thus creating confusion for parallelepiped classifier where spectral response is highly correlated and have high covariance [21]. Classification of such pixels and allotting these pixels to their correct class is of great importance, as overlapping parallelepipeds are common in remotely sensed data analysis. Lillesand, Kiefer and Chipman [3] suggested another method of using a series of rectangles with stepped borders in place of the single rectangle (Fig. 5.4). The output of the classification is shown in Fig. 5.6C.

5.3.3 Maximum likelihood classifier

This classifier quantitatively evaluates both the variance and covariance of the class spectral signature when categorizing unknown pixel [21]. This classifier assume that point data cloud derived from training data set is normally distributed [22]. Due to this

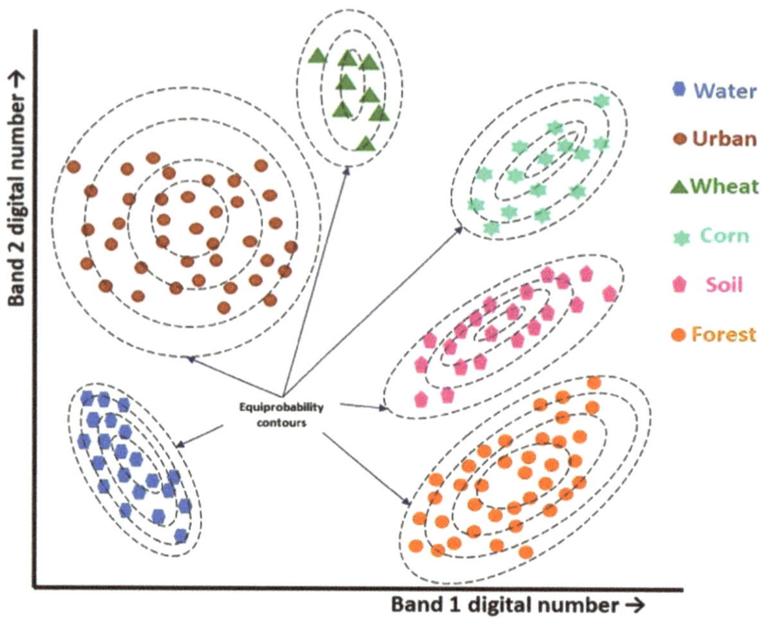

Fig. 5.5 ***Equiprobability contours defined by maximum likelihood classifier (MLC)*** *(Source: [47]).*

assumption spectral response of each class can be viewed in terms of mean vector and covariance matrix. The statistical probability for a given pixel value is computed. Each class has a bell-shaped surface in a gaussian plane also known as "probability density function" (PDF). The PDFs are used to classify an unidentified pixel by calculating the probability of the pixel value belonging to each class [3]. After evaluating the probability in each class, the pixel would be assigned to the most likely class (highest probability value) or be labelled "unknown" if the probability values are below user defined threshold. In short, maximum likelihood classifier technique form an equiprobability contours and these decision regions for each class is shown in Fig. 5.5. The output obtained implementing this classification is shown in Fig. 5.6D.

5.4 Unsupervised classification

Unsupervised classification does not require point cloud data or spectral signature of classes within each band of training data set to classify unknown. In this classification techniques different algorithms are employed to investigation of unknown pixels of an image. The basic logic is that the reflectance pixel values within a given cover type should be close together in the measurement space while values should be well separated in different classes. These algorithms (K mean and ISODATA) examine natural grouping or clustering in the pixel values and these pixel values are labelled into number of classes after identification of substantially separated clusters [18]. These classes are

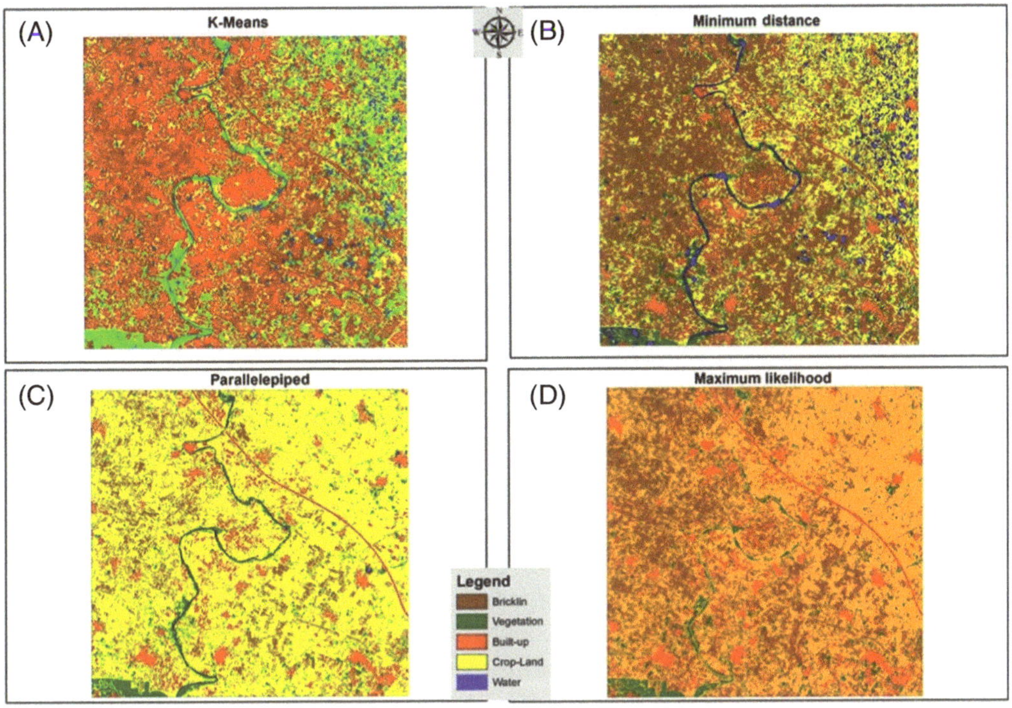

Fig. 5.6 The LULC map obtained using different classifier technique (A) K-mean, (B) minimum-distance-to-mean, (C) parallelepiped, and (D) maximum likelihood, over of Ghaziabad, U.P., India. The multispectral high resolution (10 m) Sentinel-2A (Band: Blue, Green, Red, and NIR) satellite image is used for classification.

known as spectral classes because they are result of natural spectral grouping present in an image. At first the identity of these classes is not known to the image analyst. After spectral grouping, the analyst identifies the obtained classes to some form of reference data. Thus, in unsupervised classification procedure: (1) training datasets are not required and (2) only a specification of number of classes is required by the user. Several clustering algorithms exist that can be used to verify the natural spectral clusters present in the image. The most popular classifiers which use this algorithm are K-means and iterative self-organizing data analysis technique (ISODATA) [3].

5.4.1 K-mean

K-mean algorithm is common form of clustering algorithm that tries to split-up the dataset into "K" predefined distinct nonoverlapping clusters [21]. K-mean is an unsupervised, nondeterministic, numerical, iterative method of clustering [23]. It assigns data points to a cluster whose mean vector is closest (arithmetic mean of all the data points that belong to that cluster is at the minimum). After classifying all the pixels in this manner new mean vectors are computed for each cluster. The new mean vector is used as the basis to reclassify the image pixels. The process continues until there is no significant change in the location of class mean vectors between successive iterations of the algorithm. The less variation we have within clusters, the more homogeneous (similar) the data points are within the same cluster [3]. The output of the classification is shown in Fig. 5.6A.

5.4.2 ISODATA

It is one of the widely used clustering algorithm of unsupervised classification. ISODATA stands for iterative self-organizing data analysis technique [24]. This algorithm frequently changes the number of clusters from one iteration to the next, by merging, splitting, and deleting clusters. The general process is similar to the K-mean algorithm. However, the image analyst has to define threshold values for parameters and the number of iterations the process has to continue. After assigning pixels to the cluster, the statistics of each cluster is revaluated. The standard deviation within each cluster, and the distance between cluster centers is calculated. A single cluster is split in two, if standard deviation (in any one dimension) is greater than the user-defined threshold. On the other hand, two clusters are merged, if the distance between them is less than the predefined minimum distance [3]. Finally, the process repeats, until either there is no substantial change in the cluster statistics or some maximum number of iterations is reached.

5.5 Hybrid classification and artificial neural networks (ANNs)

Hybrid classification technique employs amalgamation of elements from both supervised as well as unsupervised algorithms to improve classification accuracy procedures. Hybrid classification approach exhibits better results for data sets having inconsistent spectral signature within information classes. The iterative guided spectral class rejection (IGSCR) is one such hybrid classifier [18]. This classifier requires user defined set of rejection criterion and large number of training pixels to perform spectral clustering of comparable pixels into two or three classes. Thereafter, with the help of successive iterations the spectral class is accepted if it meets predefined inclusion threshold otherwise rejected it if it does not. The sequence of iteration goes on until convergence threshold is met. In the end supervised classification is used to classify these pure classes.

ANNs are one the most significant technological development which has been implemented vigorously for information extraction from satellite images. ANNs are structured to mimic the philosophy of progressive learning of human brain by establishing relationship between input and output data with the help of computer programs or set of rules [25]. The processing node is the basic element of ANNs. The input values are received and summed up at each processing node. These summed up values are passed

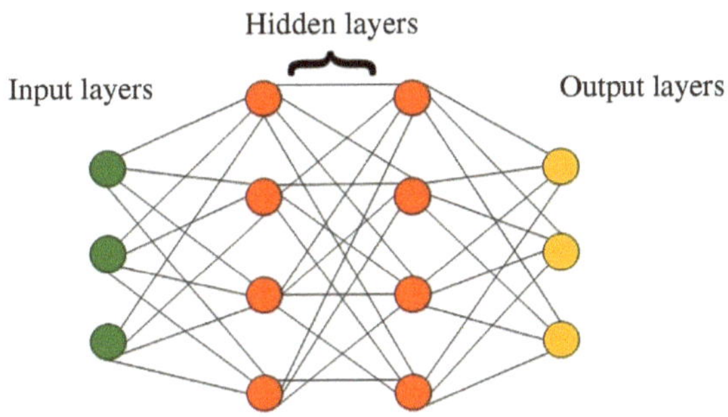

Fig. 5.7 ***A feed-forward neural network, showing the input layer, one hidden layer and the output layer*** *(Source:[25]).*

through an activation function. The output value of each node forms one of the input values to other processing node in the next layer of ANNs (Fig. 5.7).

ANNs are composed of three elements (Fig. 5.7):

(1) An input layer: multispectral satellite imagery with related ancillary data forms the source data. (2) The output layer: contains user defined classes and training data set in which relation between output classes and input data is well recognized. The processing runs smoothly if the number of output classes are small in comparison to the number input channels. (3) Training phase: during training phase, an ANN forms a relationship between input and output data by assigning weights within one or more hidden layers.

In the field of remote sensing, the most popular ANN model has been the multilayer perceptron (MLP), also known as feedforward neural network. MLP-type networks work in a feed-forward direction where information progresses from an input layer to an output layer in the learning phase. Such networks contain an extra layer termed the hidden layers to overcome the problems of the perceptron. Various classification techniques have been implemented to extract information from multispectral spectral high resolution remotely sensed imageries (Table 5.1).

5.6 Elements of visual image interpretation

Shape, size, texture, tone patter, and association are very important [26]. Shape refers to general form, configuration, or outline of individual object. Also, in stereoscopic pairs height of object is often defined by its shape. Relative size among the objects on image of the same scale must be taken into consideration. Pattern refers to its spatial arrangement of objects. Reptation forms relationship with characteristics of objects (artificial or natural) origin. Tone is relative brightness or color of an object in an image. Say if we encounter soil of light tone which refers to lack of moister content however dark tone soil means presence of abundance of moisture. Without tonal difference the shape,

Table 5.1 Different classifiers used for data extraction from remotely sensed imageries.

Study area	Classifier	Application/classes	Satellite image	Result	References
Jaipur, Rajasthan	Mahalanobis, maximum likelihood (MLC), minimum distance, and parallelepiped classifier	Land use	Landsat MSS, TM, and ETM+	Minimum distance classifier has higher accuracy than other three classifiers for Landsat MSS image however MLC classifier produces higher accuracy than other three classifier for Landsat TM and ETM+.	[17]
Jaipur, Rajasthan	Maximum likelihood classifier (MLC)	LU/LC; 5 (urban, wasteland, vegetation, hills, and water bodies)	Landsat ETM+	MLC is a robust classifier and has very slim chances of misclassification. This classifier achieved overall accuracy of 93.75%, producer accuracy 94%, user accuracy 96.09% and overall Kappa accuracy 90.52%.	[19]
Ghent in Belgium	Ensemble classifiers: decision tree, artificial neural networks and random forest	Urban land cover mapping; 9 (water, grass, trees, buildings with dark red and bright roof, roads, other artificial objects, shadow)	IKONOS	Ensemble classifiers generate significantly higher accuracies than a single classifier.	[48]
NCR, Delhi	K-mean clustering algorithm	LULC change detection; 6 (agricultural land, water body, built-up, vegetation, open/fallow land, and ridge)	Landsat TM and OLI/TIRS	Result show that during 1990–2018, built-up and open/fallow land area increased by 326% and 44% respectively, while vegetation cover and agricultural land area decreased by 34% and 12% respectively, of the total area of study.	[49]
Pico da Vara Natural Reserve, Archipelago of the Azores, Portugal	SVM, ANN, MhD, and MXL	Vegetation mapping; 8 (forestry production species, aggressive alien invasive species (CL and NN), native species, bare soil areas, clouds, natural pasture areas, and shadows of clouds)	IKONOS	Regardless of poor separability of some vegetation type, MXL, SVM and ANN classifications have achieved good overall accuracies and per class accuracies (overall accuracy >75% and Kappa Index Agreement >0.6).	[50]

Khorezmprovince, Uzbekistan	Maximum likelihood classifier (MLC)	Land use change detection; 7 (agricultural land, built up, bare land, lowland, saline land, sand, and waterbody)	Landsat TM and OLI	Results show significant increase in area of agricultural land, built up and saline land at an expanse of decrease in area of bare land and lowland.	[51]
Sahibganj, Jharkhand; Malda and Murshidabad, West Bengal	Random forest (RF), SVM, ANN, fuzzy ARTmap, SAM and Mahalanobis distance (MD)	LULC change mapping; 6 (agricultural land, built-up area, sand bar, fallow land, vegetation, and river and wetlands)	Landsat OLI	Result depicts that Random Forest algorithm is the best machine-learning LULC classifier, among the six examined algorithms.	[52]
Bhutan–Bengal foothill	Maximum likelihood classifier (MLC)	land use and land cover (LULC) change monitoring; 6 (vegetation, agriculture, plantation, built-up area, barren land, and waterbody)	Landsat MSS, TM, and ETM+	Over three decades built-up area (6.44%), barren land (2.71%) and waterbody (1.29%) showed increasing trend however slight decline in vegetation (−2.93%), agriculture (−6.955%), and plantation (−0.5%) area were observed.	[53]
Brandenburg, Germany	Knowledge-based methods	Forest types; 5 (pine, larch, beech, robinia, and oak)	QuickBird	Results show a good separability with approximately 80% to 90% overall accuracy for the tree species beech, oak, robinia, larch, and pine.	[54]
Place Glacier Basin, Western Canada	Supervised Maximum likelihood classifier (MLC)	wet-snow line mapping; 3 (wet snow, glacier ice, and bedrock)	ERS-1 SAR	Kappa coefficients for two images are 0.49 ± 0.02 and 0.48 ± 0.02, respectively, at the 95% confidence level. This represents a classification accuracy of about 50% for each of the SAR images.	[55]

pattern, and texture of object cannot be differentiated. Texture refers to frequency of tonal change on an image. It is a product of individual shape, size, pattern, and tone very small features. On basis of texture objects with similar spectral reflectance can be identified. Association means occurrence of certain features in relation to others.

5.7 Band/spectral rationing

Band rationing is technique in which DN values in one spectral band is divided from the corresponding values in another band. The resultant image is commonly known as ratioed image. Band ratioing specially enhances spectral characteristics of features in image, irrespective of variation in scene illumination. Ratio images are often used in discerning very minute spectral variation which are masked due to brightness variation. Band rationing magnifies the anomaly of the feature under study, subsequently, stretching the result or rescaling the values, and then displaying the new values as an image [27,28]. The idea behind band ratioing is that ratio values should be constant for pixels representing features of similar spectral properties and should vary for some (not necessarily all) pixels depicting features of differing spectral properties [29]. In other words, this technique magnifies the anomaly of the feature under study, subsequently, stretching the result or rescaling the values, and then displaying the new values as an image [27,28]. Ratioing can be used for change detection when applied to multitemporal data. The possible number of ratios combination that can be obtained from n bands of data is n(n–1) for example: for Landsat MSS 12 combinations are possible [3]. A quantitative criterion has been proposed to assist in selecting three ratio combinations to include in colour composite is known as "optimum index factor (OIF)" [30,31]. With the help of OIFs best three band combination is picked up for generating FCC which relays very minute spectral difference with ease and have optimum information [32]. The underlying logic in selection procedure is dependent on total variance in each ratio and degree of correlation between bands. The three-band combination with the highest value of OIF must always have high variance and low correlation [16].

5.8 Principal component analysis (PCA)

Data extraction often becomes difficult with the availability of too many combinations of ratios. It also reduces computational efficiency due exhaustive inter-band correlation. Principal component transformation is a mathematical technique where it reduces the dimensionality of data. That is, it is a robust technique where it confines all the necessary in fewer number of bands than original number of bands and distinguishable information can be accumulated in few bands [16,33–35]. Eklundh and Singh [36], suggested that principal component transformation is completed in three steps which are as follow:

Step1: with the help of input data set (original image) covariance or correlation matrix is calculated.

Step2: we derive eigenvalues and eigenvector which are also known as transformation coefficients

Step3: calculation of principal components.

Often, PCA increases the computational efficiency of classification process. Generally, this procedure is followed when little no information is available about the feature in an image prior to classification or image interpretation.

5.9 Normalized difference index

Abundance of ratio combinations sometimes presents challenges in information extraction. These limitations are addressed by mathematical variants or normalized difference index. One of the most recognised mathematical variants is normalized difference vegetation index (NDVI), which is based on the difference of reflectance in the near-infrared and red bands. Satellite image pixels with higher NDVI value (brighter pixels) depict densely vegetated land whereas pixels with lower NDVI value (darker pixels) nonvegetated area such as water sources, river channels and bare sand etc. often, NDVI values are corroborated with vegetation monitoring and assessment, with crop biomass accumulation, with leaf chlorophyll levels, with leaf area index values, and with crop canopy activity level [3]. The normalized difference indices range from −1 to +1. Thus, they shun scaling difficulties present by simple ratios. Many indices have been suggested for different feature identification and information extraction (Table 5.2).

Table 5.2 Description of various indices.

Sr. No.	Indices acronym	Description	Formulation	References
1	NDVI	Normalized difference vegetation index	$\frac{NIR - Red}{NIR + Red}$	[56]
2	GNDVI	Green normalized difference vegetation index	$\frac{NIR - Green}{NIR + Green}$	[57]
3	NDWI	Normalized difference water index	$\frac{Green - NIR}{Green + NIR}$	[58]
4	NDBI	Normalized difference built-up index	$\frac{SWIR1 - NIR}{SWIR1 + NIR}$	[59]
5	NMDI	Normalized multiband difference index	$\left[\frac{NIR - (SWIR1 + SWIR2)}{NIR + (SWIR1 + SWIR2)}\right]$	[60]
6	NDSI	Normalized difference snow index	$\frac{VIS - SWIR}{VIS + SWIR}$	[61]

GNDVI is used to determine water and nitrogen uptake into the plant canopy. GNDVI is a reliable indicator of crop health condition. NDWI is used to differentiate water from the dry land or rather most suitable for water body mapping, NDBI is applied for built-up area extraction, NMDI is used to extract the dry soil or bare land and NDSI is used to delineate the presence of snow/ice.

5.10 File format

Satellite images are digital representation of different features on the earth's surface. This imagery is a systematic arrangement of raster cells. Each cell of raster data also known as pixel [37]. These pixels are laid out in a grided format and contains a digital number between a certain range which is dependent upon radiometric resolution of detector system. The pixel values can vary from 0 to 127 (7-bit image) or 0 to 255 (8-bit image) and represents information like temperature, height values or reflected electromagnetic radiation. The sequential arrangement of pixels, representing a digital image in a computer compatible storage medium is known as image file format. There are three basic file formats which are as follow:

A. Band interleaved by pixel (BIP): The BIP format stores the first pixel for all bands in sequential order, followed by the second pixel for all bands, followed by the third pixel for all bands, and so forth, interleaved up to the number of pixels [38]. This format provides optimum performance for spectral (z) access of the image data. Standard TIFF images are in BIP format.

B. Band interleaved by line (BIL): The BIL format stores the data in line-major order, that is, the first line of the first band, followed by the first line of the second band, and so on, interleaved up to the number of bands. Subsequent lines for each band are interleaved in similar fashion [38]. It is the recommended file format for most ENVI processing tasks. SPOT data is in BIL format.

C. Band sequential (BSQ): In BSQ format, each band is stored as a separate file and each individual band uses the same raster order, that is, where each line of the data is followed immediately by the next line in the same spectral band [38]. This format is optimal for spatial (x, y) access of any part of a single spectral band. TM (thematic mapper) scenes are in BSQ format.

Other file formats which are essential for any remote sensing image analyst are as follow:

Hierarchical data format (HDF): HDF file format supports n-dimensional datasets. It is portable with no vendor lock-in, and self-describing file format, that is, everything (all data as well as metadata) can be passed along in one file. Thus, show big data handling capacity as well as fast input/output performance. It can easily run on a range of computational platforms hence supports cross platform criterion. HDF is the prescribed format for data products that are derived from EOS missions. Usually, HDF data format is used to store data collected from EOS satellites: Terra, Aqua and Aura.

Network common data form (netCDF): netCDF support the creation, access, and sharing of array-oriented scientific data. It includes information about the data it contains. It can be simultaneously accessed at the same instance by multiple readers. It is highly portable, scalable, appendable as well as archivable. netCDF support cross platform access. Numerous portions of the netCDF format have been endorsed by standards bodies such as integrated ocean observing system (IOOS) and climate and forecast (CF) conventions. It is the main data format for atmospheric models and marine Geophysics data which contains information attributing to gravity, magnetics, and bathymetry.

Geographical tagged image file format (GeoTIFF): It is an extension of the regular TIFF format. GeoTIFF data format uses a small set of reserved TIFF tags to store a broad range of georeferencing information. GeoTIFF commonly use generic image format that can handle geolocation information. It is available in all standard remote sensing and GIS packages. GeoTIFF file format covers Landsat data, national elevation dataset (NED), digital orthophotoquadrangle (DOQ) which is a computer-generated image of an aerial photograph, SRTM data etc.

Geometric and radiometric correction: In remote sensing applications precise geometric and radiometric is a required not only in improvements in classification accuracy but are a prerequisite in any multitemporal change detection [39]. Geometric and radiometric corrections can be performed independently one after another [40].

5.11 Geometric correction

Geometric distortions are anomalies in the relative position of pixels of a scene with respect to their absolute position within some defined map projection [41]. Geometric distortions may arise due varied reasons such as panoramic distortion, earth curvature, atmospheric refraction, relief displacement, and nonlinearities in the sweep of a sensor's IFOV [42]. These distortions are broadly classified into two types, viz. systematic and random distortion. Those error which are known and can be easily predicted as well as corrected using mathematical models are known as systematic distortion. However, errors which arises due to unknown reason and can only be corrected using ground control points (GCP), taken from stable area such as highway intersection or shore lines are known as random distortion [3]. Irregularities in pixel position due to changes in attitude is difficult to model mathematically so a procedure called image rectification is used instead. Image rectification is process in which geometrically distorted satellite images are registered to map projection such that satellite images are made planimetric for accurate area, distance, and direction measurements. The rectification of satellite images is achieved with the help of geometric transformation and resampling techniques.

5.12 Radiometric correction

This data handling step is very important for image analyst to perform accurate and quantitative interpretation of remotely sensed data [40,43,44]. Radiometric distortion in optical satellite imagery, which is related to sensor are calibration and de-striping while which is related to scene are atmospheric effects, topography effects, illumination effects, view angle effects, and target reflectance characteristics [44]. Variations in illumination and viewing geometry between images (for optical sensors) can be corrected by applying a sun elevation correction and an earth–sun distance correction. The sun elevation correction process is performed to compensate the seasonal variation of the sun with respect to the earth. During this process the optical image data is normalized by dividing each pixel value in a scene by the sine of the solar elevation angle (or cosine of the solar zenith angle) for the particular time and location of imaging. The earth–sun distance correction is applied to normalize for the seasonal changes in the distance between the earth and the sun. Suspended particles in atmosphere cause scattering which attenuate energy illuminating the surface or signal propagating from the target to the sensor. This introduces "haze" in the imagery and reduces image contrast. The process which is performed to compensate "haze" in an optical image is referred to as dark object subtraction. Noise in an image may be due to irregularities the sensor. Common forms of noise are striping and dropped lines. The corrective process made a relative correction among the six sensors to bring their apparent values in line with each other. Dropped lines are normally "corrected" by replacing the line with the pixel values in the line above or below, or with the average of the two. Another quantitative radiometric correction process is of transformation of DN in absolute radiance value.

5.13 Conclusion

In this chapter we have discussed in detail about different types of classification methods (supervised, unsupervised, and hybrid classification). We have explained the working principle of each classification algorithms (minimum-distance-to-mean, parallelepiped, maximum likelihood, K-mean, ISODATA and artificial neural networks) in different classification technique. We have laid emphasis on elements (shape, size, texture, tone patter, and association) of visual image interpretation. Arithmetic operations like band ratioing which can enhance minute details of features masked due to variation in scene illumination has been explained. Data extraction becomes challenging when number of bands increases in multispectral satellite images. We have discussed the basic concept and steps involved in PCA. Principal component transformation accumulates distinguishable in fewer number of bands. Normalised difference indices avoid scaling difficulties of simple ratios. We have explained different types of normalised difference indices and their application (Table 5.2). Furthermore, we have highlighted fundamental file formats of remotely sensed data and various other formats which are machine independent, self-explanatory, easily handled and widely recognized. The two most important process of data handling technique, viz. geometric and radiometric correction are discussed in detail.

References

[1] P.R. Luney, H.W. Dill, Uses, potentialities, and needs in agriculture and forestry. Remote Sensing with Special Reference to Agriculture and Forestry, (1970) 1–34

[2] J. Fussell, D. Rundquist, J.A. Harrington, On defining remote sensing, Photogramm. Eng. Remote Sens. 52 (9) (1986) 1507–1511.
[3] T.M. Lillesand, R.W. Kiefer, J.W. Chipman, Remote Sensing and Image Interpretation, John Wiley & Sons, Hoboken, NJ, 2008.
[4] W. Shi, M. Zhang, R. Zhang, S. Chen, Z. Zhan, Change detection based on artificial intelligence: State-of-the-art and challenges, Remote Sens. 12 (10) (2020) 1688.
[5] A. Abdollahi, B. Pradhan, N. Shukla, S. Chakraborty, A. Alamri, Deep learning approaches applied to remote sensing datasets for road extraction: A state-of-the-art review, Remote Sens. 12 (9) (2020) 1444.
[6] S. Bera, V.K. Shrivastava, Analysis of various optimizers on deep convolutional neural network model in the application of hyperspectral remote sensing image classification, Int. J. Remote Sens. 41 (7) (2020) 2664–2683.
[7] S. Banks, L. White, A. Behnamian, Z. Chen, B. Montpetit, B. Brisco, J. Pasher, J. Duffe, Wetland classification with multi-angle/temporal SAR using random forests, Remote Sens. 11 (6) (2019) 670.
[8] F. Traoré, J. Bonkoungou, J. Compaoré, L. Kouadio, J. Wellens, E. Hallot, B. Tychon, Using multitemporal Landsat images and support vector machine to assess the changes in agricultural irrigated areas in the Mogtedo region, Burkina Faso, Remote Sens. 11 (12) (2019) 1442.
[9] R. Ramo, E. Chuvieco, Developing a random forest algorithm for MODIS global burned area classification, Remote Sens. 9 (11) (2017) 1193.
[10] J.J. Gapper, H. El-Askary, E. Linstead, T. Piechota, Coral reef change detection in remote Pacific islands using support vector machine classifiers, Remote Sens. 11 (13) (2019) 1525.
[11] S.K. Karan, S.R. Samadder, Improving accuracy of long-term land-use change in coal mining areas using wavelets and support vector machines, Int. J. Remote Sens. 39 (1) (2018) 84–100.
[12] P. Kumar, D.K. Gupta, V.N. Mishra, R. Prasad, Comparison of support vector machine, artificial neural network, and spectral angle mapper algorithms for crop classification using LISS IV data, Int. J. Remote Sens. 36 (6) (2015) 1604–1617.
[13] S.E. Jozdani, B.A. Johnson, D. Chen, Comparing deep neural networks, ensemble classifiers, and support vector machine algorithms for object-based urban land use/land cover classification, Remote Sens. 11 (14) (2019) 1713.
[14] D. Turner, A. Lucieer, Z. Malenovský, D. King, S.A. Robinson, Assessment of Antarctic moss health from multi-sensor UAS imagery with random forest modelling, Int. J. Appl. Earth Observ. Geoinform. 68 (2018) 168–179.
[15] S. Mahdavi, B. Salehi, W. Huang, M. Amani, B. Brisco, A PolSAR change detection index based on neighborhood information for flood mapping, Remote Sens. 11 (16) (2019) 1854.
[16] N. Patel, B. Kaushal, Classification of features selected through Optimum Index Factor (OIF) for improving classification accuracy, J. Forestry Res. 22 (1) (2011) 99–105.
[17] P.S. Sisodia, V. Tiwari, A. Kumar, Analysis of supervised maximum likelihood classification for remote sensing image, International Conference on Recent Advances and Innovations in Engineering (ICRAIE-2014), IEEE, 2014, pp. 1–4.
[18] S.D. Jawak, P. Devliyal, A.J. Luis, A comprehensive review on pixel oriented and object oriented methods for information extraction from remotely sensed satellite images with a special emphasis on cryospheric applications, Adv. Remote Sens. 4 (2015) 177–195.
[19] P.S. Sisodia, V. Tiwari, A. Kumar, A comparative analysis of remote sensing image classification techniques, 2014 International Conference on Advances in Computing, Communications and Informatics (ICACCI), IEEE, 2014, pp. 1418–1421.
[20] A. Akgün, A.H. Eronat, N. TürkIstanbul, Turkey, Comparing different satellite image classification methods: An application in Ayvalik District, Western Turkey, The 4th International Congress for Photogrammetry and Remote Sensing, 2004.
[21] R Eastman, IDRISI Kilimanjaro: Guide to GIS and Image Processing, Clark Labs, Clark University, Worcester, 2003.
[22] J.T. Tou, R.C. Gonzalez, Pattern recognition principles. Addison-Wesley Publishing Company, (1974) 377 p.
[23] J. Yadav, M. Sharma, A review of K-mean algorithm, Int. J. Eng. Trends Technol. 4 (7) (2013) 2972–2976.

[24] G.H. Ball, D.J. Hall, ISODATA, A Novel Method of Data Analysis and Pattern Classification, Stanford Research Institute, Menlo Park, CA, 1965.
[25] T. Kavzoglu, An Investigation of the Design and Use of Feed Forward Artificial Neural Networks in the Classification of Remotely Sensed Images (Doctoral dissertation), University of Nottingham, 2001.
[26] C.E. Olson, Elements of photographic interpretation common to several sensors, Photogramm. Eng. Remote Sens. 26 (4) (1960) 651–656.
[27] J.M. Abrams, Remote sensing for porphyry copper deposits in Southern Arizona, Econ. Geol. 78 (1983) 591–604.
[28] D.A. Mouat, G.G. Mahin, J. Lancaster, Remote sensing techniques in the analysis of change detection, Geocarto Int. 8 (2) (1993) 39–50.
[29] R.E. Crippen, The dangers of underestimating the importance of data adjustments in band ratioing, Remote Sens. 9 (4) (1988) 767–776.
[30] P.S. Chavez, G.L. Berlin, B. Sowers, Statistical method for selecting Landsat MSS ratios, J. Appl. Photogr. Eng. 8 (1982) 23−30.
[31] P.S. Chavez, S.C. Guptill, J.A. Bowell, Image processing techniques for thematic mapper data, Proc. ASPRS-ACSM Tech. Paper 2 (1984) 728−742.
[32] S.K. Saha, M. Kudrat, Selection of spectral band combination for land cover/land use classification using a brightness value overlapping index (BVOI), J. Indian Soc. Remote Sens. 19 (3) (1991) 141–147.
[33] J. Estornell, J.M. Martí-Gavilá, M.T. Sebastiá, J. Mengual, Principal component analysis applied to remote sensing, Modell. Sci. Educ. Learn. 6 (2013) 83–89.
[34] B.B. Jackson, Multivariate Data Analysis: An Introduction, Irwin, Homewood, IL, 1983.
[35] C. Munyati, Use of principal component analysis (PCA) of remote sensing images in wetland change detection on the Kafue Flats, Zambia, Geocarto Int. 19 (3) (2004) 11–22.
[36] L. Eklundh, A. Singh, A comparative analysis of standardized and unstandardized principal components analysis in remote sensing, Int. J. Remote Sens. 14 (7) (1993) 1359–1370.
[37] J.R. Parker, Extracting vectors from raster images, Comput. Graph. 12 (1) (1988) 75–79.
[38] Q. Ding, M. Khan, A. Roy, W. Perrizo, The P-tree algebra, Proceedings of the 2002 ACM Symposium on Applied computing, 2002 426–431.
[39] K.I. Itten, P. Meyer, Geometric and radiometric correction of TM data of mountainous forested areas, IEEE Trans. Geosci. Remote Sens. 31 (4) (1993) 764–770.
[40] R.J. Woodham, T.K. Lee, Photometric method for radiometric correction of multispectral scanner data, Can. J. Remote Sens. 11 (2) (1985) 132–161.
[41] M. Kumar, R.K Singh, Digital image processing of remotely sensed satellite images for information extraction, Conference on Advances in Communication and Control Systems (CAC2S 2013), 2013.
[42] T. Toutin, Geometric processing of remote sensing images: models, algorithms and methods, Int. J. Remote Sens. 25 (10) (2004) 1893–1924.
[43] R.N. COLWELL, Manual of Remote Sensing, American Society of Photogrammetry, Falls Church, VA, 1983.
[44] P.M. Teillet, Image correction for radiometric effects in remote sensing, Int. J. Remote Sens. 7 (12) (1986) 1637–1651.
[45] L.M. Goncalves, C.C. Fonte, E.N.B.S. Júlio, M. Caetano, Evaluation of remote sensing images classifiers with uncertainty measures. In: Spatial Data Quality from Process to Decisions (Editors R. Devillers, H. Goodchild), CRC Press, Boca Raton, 2009, pp. 163–177.
[46] S.A. El-Rahman, Performance of spectral angle mapper and parallelepiped classifiers in agriculture hyperspectral image, Int. J. Adv. Comput. Sci. Appl. 7 (5) (2016) 55–63.
[47] L. Bruzzone, D.F. Prieto, Unsupervised retraining of a maximum likelihood classifier for the analysis of multitemporal remote sensing images, IEEE Trans. Geosci. Remote Sens. 39 (2) (2001) 456–460.
[48] J.C.W. Chan, L. Demarchi, T. Van de Voorde, F. Canters, Binary classification strategies for mapping urban land cover with ensemble classifiers, IGARSS 2008-2008 IEEE International Geoscience and Remote Sensing Symposium, 3, IEEE, 2008, pp. III–1004.
[49] M.W. Naikoo, M. Rihan, M. Ishtiaque, Analyses of land use land cover (LULC) change and built-up expansion in the suburb of a metropolitan city: Spatio-temporal analysis of Delhi NCR using Landsat datasets, J. Urban Manag. 9 (3) (2020) 347–359.

[50] A. Gil, Q. Yu, A. Lobo, P. Lourenço, L. Silva, H. Calado, Assessing the effectiveness of high resolution satellite imagery for vegetation mapping in small islands protected areas, J. Coastal Res. (2011) 1663–1667.
[51] S. Rakhmonov, U. Umurzakov, K. Rakhmonov, I. Bozarov, O. Karamatov, Land use and land cover change in Khorezm, Uzbekistan, E3S Web of Conferences, 227, EDP Sciences, 2021, pp. 01002.
[52] S. Talukdar, P. Singha, S. Mahato, S. Pal, Y.A. Liou, A. Rahman, Land-use land-cover classification by machine learning classifiers for satellite observations—a review, Remote Sens. 12 (7) (2020) 1135.
[53] M. Chamling, B. Bera, Spatio-temporal patterns of land use/land cover change in the Bhutan–Bengal foothill region between 1987 and 2019: Study towards geospatial applications and policy making, Earth Syst. Environ. (2020) 1–14.
[54] A. Clasen, A. Frick, B. Kleinschmit, Knowledge based methods for the classification of forest parameters with QuickBird data, Center Remote Sens. Land Surf. 28 (2006) 30.
[55] S. Adam, A. Pietroniro, M.M. Brugman, Glacier snow line mapping using ERS-1 SAR imagery, Remote Sens. Environ. 61 (1) (1997) 46–54.
[56] T.S. Purevdorj, R. Tateishi, T. Ishiyama, Y. Honda, Relationships between percent vegetation cover and vegetation indices, Int. J. Remote Sens. 19 (18) (1998) 3519–3535.
[57] D.A.G. Cárdenas, J.A.R. Valencia, D.F.A. Velásquez, J.R.P. Gonzalez, Dynamics of the indices NDVI and GNDVI in a rice growing in its reproduction phase from multi-spectral aerial images taken by drones. In: International Conference of ICT for Adapting Agriculture to Climate Change, Springer, Cham, 2018, pp. 106–119.
[58] B.C. Gao, NDWI: A normalized difference water index for remote sensing of vegetation liquid water from space, Remote Sens. Environ. 58 (3) (1996) 257–266.
[59] Y. Zha, J. Gao, S. Ni, Use of normalized difference built-up index in automatically mapping urban areas from TM imagery, Int. J. Remote Sens. 24 (3) (2003) 583–594.
[60] L. Wang, J.J. Qu, NMDI: A normalized multi-band drought index for monitoring soil and vegetation moisture with satellite remote sensing, Geophys. Res. Lett. 34 (20) (2007) 1–5.
[61] A.G. Riggs, K.D. Hall, O.M. Roman, VIIRS snow cover algorithm theoretical basis document (2015) https://modis-snow-ice.gsfc.nasa.gov/uploads/VIIRS_snow_cover_ATBD_2015.pdf.
[62] https://desktop.arcgis.com/en/arcmap/latest/manage-data/raster-and-images/supported-raster-dataset-file-formats.htm.
[63] https://www.unidata.ucar.edu/publications/factsheets/current/factsheet_netcdf.pdf.
[64] https://www.unidata.ucar.edu/software/netcdf/docs/.
[65] D. Lu, Q. Weng, A survey of image classification methods and techniques for improving classification performance. Int. J. Remote Sens. 28 (5) (2007) 823–870.

CHAPTER 6

Appraisal of radiative transfer model 6SV for atmospheric correction of multispectral satellite image towards land surface temperature retrieval

Prashant K. Srivastava[a], Nishita Jaiswal[a], Swati Suman[a], Smrutisikha Mohanty[a,b], Mona Sharma[a,b]
[a]Remote Sensing Laboratory, Institute of Environment and Sustainable Development, Banaras Hindu University, Varanasi, Uttar Pradesh, India
[b]Department of Environmental Studies, School of Interdisciplinary and Applied Sciences, Central University of Haryana, Haryana, India

6.1 Introduction

Satellite-derived land surface temperature (LST) has a wide range of applications, especially in hydro-meteorological such as in urban heat island, crop management, evapotranspiration, fire monitoring, etc. [1]. LST is important to monitor burned land surfaces on both local and overall scales in case of forest fire (deforestation). The intense heat of active fires can be measured by using thermal instruments along with satellite-derived LST to monitor the fire regime and burned area mapping [2]. Integrating both satellite-derived and model-estimated LST gives better results with higher accuracy than the current LST products [3]. Although the forcing data is significant for the analysis, parameterizations play a key role in driving the simulations. For example, variations in modelled cloud properties affect the LST values. Such uncertainties in the simulations may contribute to the energy partitioning in terms of sensible and latent heat flux, which may create large differences in the model-simulated and satellite-derived LST. LST obtained from earth observing satellites has been used to study the rock types and geothermal anomalies [4], monitor the water in crops by taking the relationship between canopy temperature and transpiration into account, and soil moisture. Using thermal inertia (TI), defined as the resistance of an object to temperature variations depending upon its material density, thermal conductivity, and specific heat, soil moisture availability can be evaluated. As compared to dry soil higher TI is observed in water bodies that show a lower diurnal temperature fluctuation [5]. The costal/aerosol band of Landsat 8 operational land imager (OLI) image is used to reduce the existence of aerosol particles in the atmosphere. They concluded that the radiometric corrected image shows better performance. There are various LST retrieval methods are used such

Atmospheric Remote Sensing
DOI: https://doi.org/10.1016/B978-0-323-99262-6.00014-6

as radiative transfer equation method (RTE), single channel algorithm (SCA), and split window algorithm (SWA), which is mainly dependent on land surface emissivity (LSE) models [6,7].

In past, many satellites were used for retrieving LST such as ASTER, AVHRR, MODIS, and LANDSAT [8–10]. Importance to be given to the atmospheric correction when measuring indices related to vegetation by using reflectance in the red and near-infrared (NIR) spectral areas of the electromagnetic spectrum. When there is a reduction in visible and NIR reflections due to atmospheric phenomena such as clouds, water vapor, and particles, the normalized difference vegetation index (NDVI) values become lower than their true values. [11,12] presented the work on LST using 6SV algorithm on Landsat 8 OLI sensors and thermal infrared sensor (TIRS) dataset over Walnut Creek region, they estimated the noises through WRF and these noises are removed by applying the 6SV algorithm on Landsat satellite image. LST retrieval from corrected satellite image was compared against LST field validation datasets obtained during Soil Moisture Experiments in 2002 (SMEX02), where they found that a higher accuracy in LST estimates can be obtained with 100 hPa corrected satellite image.

In purview of the above, this study has been focused on (1) to evaluates the atmospheric correction of Landsat-8 OLI image by using 6SV radiative transfer model, (2) to estimate LST retrieval using atmospherically corrected Landsat-8 OLI image, and (3) to evaluate corrected LST estimated from Landsat-8 OLI image using 6SV by using the ground-based measurements of LST.

6.2 Study area

The study area is Mirzapur district (23.52 - 25.32° N and 82.7 - 83.33° E), situated in Uttar Pradesh in the Northern India. The selected site is surrounded by the Varanasi district in north, Chandauli district in east, Sonbhadra district in south, and Allahabad district in the northwest direction. It covers an area of around 4521 km^2. The average elevation of the study site is 80 meters (265 feet). Indian Standard Time is calculated from Mirzapur based on of 82.5° E longitude. In summer season, from March to June and the climax month of May, the temperature can reach to maximum. Monsoon season, from July to September, has been characterized by heavy irregular rainfalls. The weather during this period is wet and humid. In winter season from December to February, climate of Mirzapur mild and enjoyable with temperatures ranging from 12°C to 20°C. In Fig. 6.1, the geographical location of the study site is shown along with the false color composite. In this map of Mirzapur red color indicate vegetation, black color shows the water body, white color indicates the sand, dark grey color indicates the settlement, and light sky-blue color shows the fellow land.

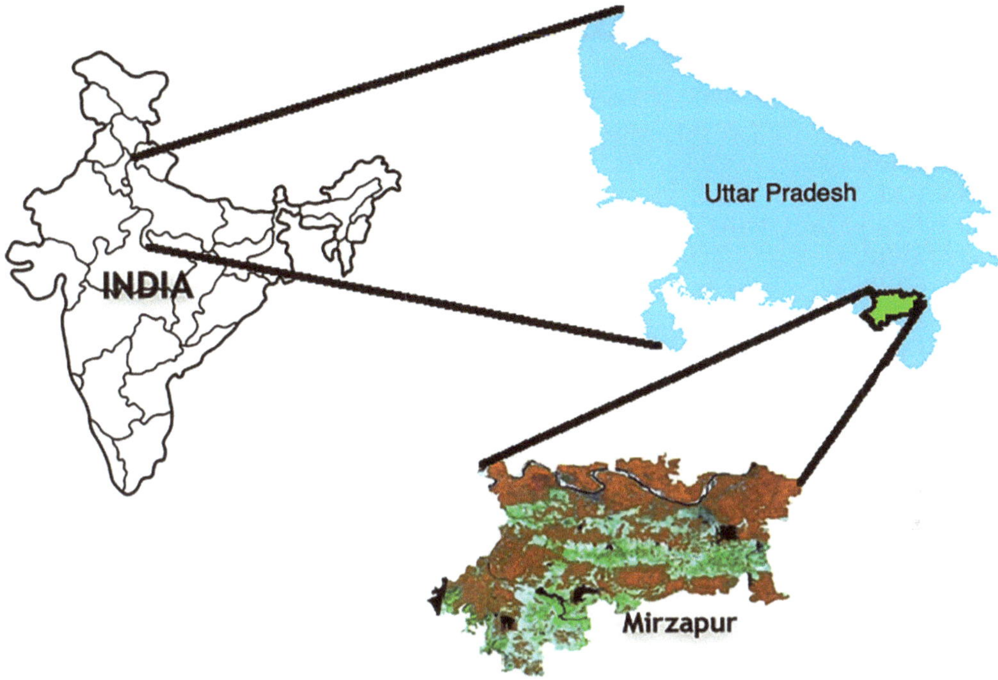

Fig. 6.1 ***Study area: map of Mirzapur, U.P., India.***

6.3 Datasets and methodology

6.3.1 Landsat 8

In 1972, the first satellite of Landsat series was launched along with two Earth-viewing imagers namely a return beam vidicon (RBV) and an 80-meter 4-band multispectral scanner (MSS). In 1975 and 1978, the Landsat 2 and Landsat 3 was launched respectively. Landsat 4 was launched in 1984 with the MSS along with the thematic mapper (TM). The Landsat 8 was launched in February 11, 2013 from Vandenberg Air Force Base, California. It has two-sensor payload named as the OLI and the thermal infrared sensor (TIRS), which is built by the Ball Aerospace & Technologies Corporation and the NASA Goddard Space Flight Center (GSFC), respectively. The information obtained from this satellite is used in diverse fields such as education, business, science, and national security. The purpose of Landsat program is to collect moderate-resolution multispectral data of the earth's surface on a global level in a systematic and repeatable manner. Datasets are freely available to public for at a higher spatial resolution over earth's surface [13].

Table 6.1 Landsat-8 OLI and TIRS bands (μm).

Bands	Wavelengths	Resolution
Band 1: Coastal aerosol	0.43–0.45	30
Band 2: Blue	0.45–0.51	30
Band 3: Green	0.53–0.59	30
Band 4: Red	0.64–0.67	30
Band 5: Near infrared (NIR)	0.85–0.88	30
Band 6: SWIR 1	1.57–1.65	30
Band 7: SWIR 2	2.11–2.29	30
Band 8: Panchromatic	0.50–0.68	15
Band 9: Cirrus	1.36–1.38	30
Band 10: TIRS 1	10.6–11.19	100
Band 11: TIRS 2	11.50–12.51	100

In this study, Landsat 8 OLI was used because it is easily available, free and affordable. It has a revisit time of 16 days. The Landsat 8 OLI satellite imagery datasets can be obtained from (https://earthexplorer.usgs.gov) for the date February 21, 2017 (Table 6.1). The path and row of satellite image are 143 and 42 respectively with cloud cover is 0.01, sun elevation is 46.40560977, earth–sun distance is 0.989. The atmospheric correction of Landsat 8 was performed by using 6SV. The datasets over this study region like water vapor (g/cm^2) was obtained from (https://rda.ucar.edu/datasets/ds083.2/) and ozone (Dobson) was obtained from (https://neo.sci.gsfc.nasa.gov). The visibility of sampling day was obtained from (http://www.timeanddate.com/weather/india/mirzapur/historic). The detailed methodology for LST retrieval is provided in Fig. 6.2.

6.3.2 Second simulation of satellite signal in the solar spectrum-vector (6SV)

In atmosphere some gases are responsible for atmospheric noise in the satellite images due to which researchers are facing problems to obtain the accurate information from the satellite images. The noises are created by different gases that are present in the solar spectrum such as oxygen (O_2), water vapor (H_2O), carbon dioxide (CO_2), ozone (O_3), methane (CH_4) and nitrous oxide (N_2O).

For atmospheric correction the accurate and advanced 6SV code is used which is applicable for all optical satellites [14,15]. The 6S code was developed to make atmospheric correction in shorter wavelength. The equation of reflectance is used as given below [14]:

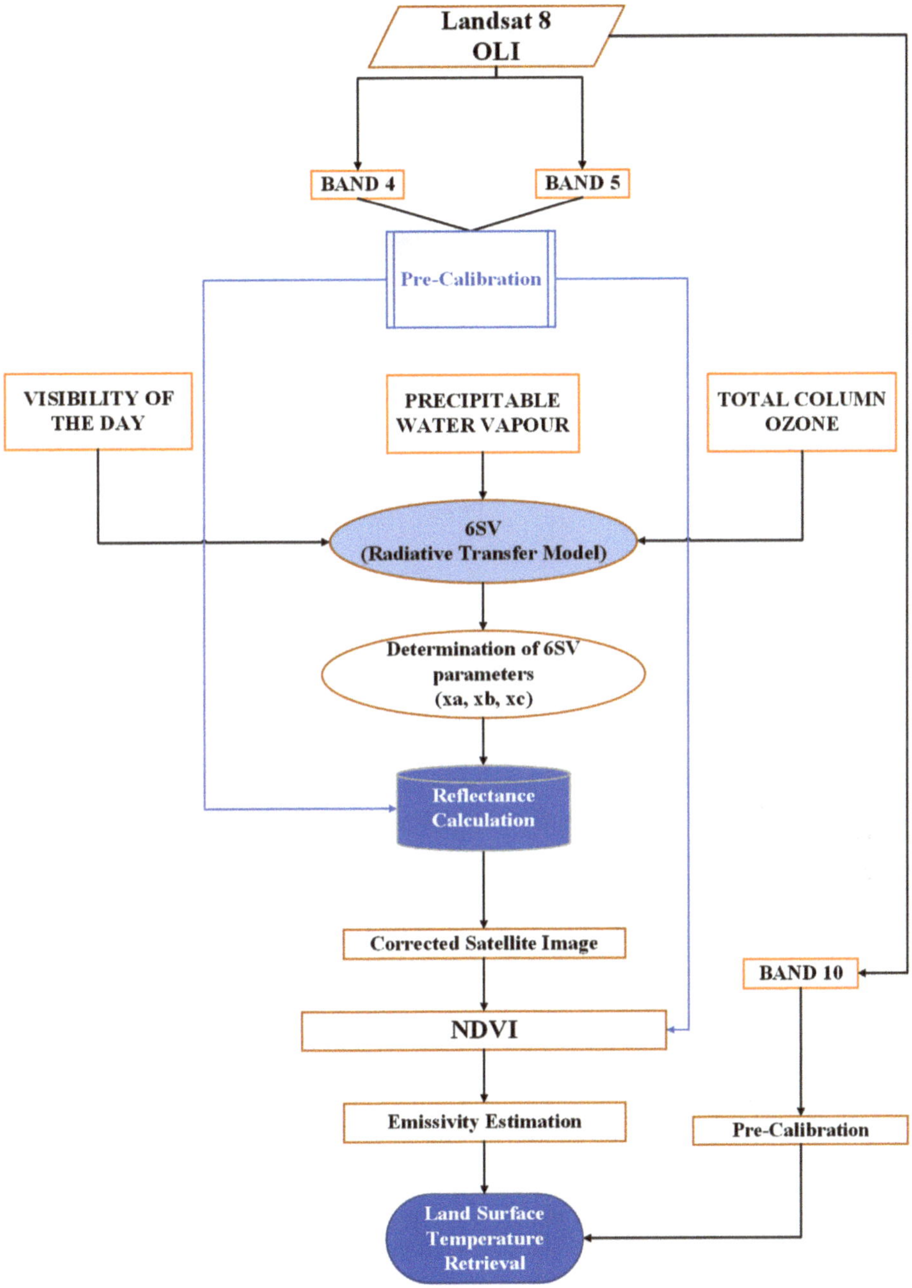

Fig. 6.2 ***Flowchart depicting the land surface temperature retrieval from corrected and uncorrected satellite images.***

$$\rho(\theta_s,\theta_v,\phi_s-\phi_v,z_t)=T_g(\theta_s,\theta_v,z_t)\left[\rho_r(z_t)+\rho_a+\frac{\rho_r}{1-S(z_t)\rho_t}T(\theta_v,z_t)T(\theta_s,z_t)\right] \tag{6.1}$$

where θ_s is the sun zenith angel, θ_v is the zenith angel, ϕ_s is the sun angels, ϕ_v is the azimuthal angel, z_t target the altitude, ρ_r is Rayleigh reflectance, ρ_t is the reflectance of the target, ρ_a is the atmospheric reflectance, T_g is gaseous transmittance, S is spherical albedo.

To consider the uncertainty between vertical distribution of aerosol and water vapor, top-of-atmosphere reflectance value is computed individually [14].

$$\underset{\rho TOA}{i=}1.3(\theta_s,\theta_v,\varphi_s-\varphi_v)=T_g^{OG}(\theta_s,\theta_v)\left[\rho_r+(\rho_{r+a}-\rho_r)T_g^{H_2^O}\left(\theta_s,\theta_v\frac{i-1}{2}U_{H_2^O}\right)+T^{\downarrow}(\theta_s)T^{\uparrow}(\theta_v)\frac{\rho_s}{1-S_{\rho_s}}T_g^{H_2^O}\left(\theta_s,\theta_v U_{H_2O}\right)\right] \tag{6.2}$$

where U_{H_2O} is the absorber amount, T_g^{OG} refers to the gaseous transmission for gases other than water vapor, $T_g^{H_2O}$ refers to H_2O absorption and $(\rho_{r+a}-\rho_r)$ is an estimate of the aerosol intrinsic reflectance, $T(\theta_s)$ is the total transmittance.

Physically based 6S model was chosen because it is not just for one object type. To use the 6S code, the below mentioned information can be needed: (1) geometrical condition, (2) atmospheric models, (3) aerosol model type, (4) spectral condition, and (5) ground reflectance. Using the above-mentioned input data, the 6S code was used to obtain the conversion equation from the observed radiance to actual reflectance (acr). In this code, acr is calculated as Eq. (6.4):

$$\mathrm{Y}=\mathrm{xa}^*(\mathrm{R})-\mathrm{xb} \tag{6.3}$$

$$\mathrm{acr}=\mathrm{y}/(1+\mathrm{xc}^*\mathrm{y}) \tag{6.4}$$

where R is the observed radiance [W/(m^2 Sr Am)]; xa is the inverse of the transmittance; xb is the scattering term of the atmosphere; and xc is the reflectance of the atmosphere for isotropic light entering the base of the atmosphere [12].

6.3.3 Normalized difference vegetation index

From spectral bands, band 5 (red) and band 4 (near infrared) were used to calculate vegetation indices. Clouds, water vapor, and aerosols make it hard to see the distinction of visible verses near infrared (NIR) reflections, which lower the NDVI values as compared to its real values [16].

Formula of NDVI is given below

$$NDVI = (NIR - IR)/(NIR + IR) \quad (6.5)$$

The band math is a mathematical expression that is used for NDVI calculation in eq. (6.5). To calculate indices ENVI band math was used. The range of NDVI lies between (−1) and (+1), (−1) is the minimum NDVI value which indicates the unhealthy condition of vegetation and (+1) is maximum value of NDVI indicates the healthy condition of plant [17]. Out the Landsat 8 OLI sensor sensitivity may have degraded by as much as 20% from the prefight calibration. In this study, the conversion of digital values of band 4 and 5 to reflectance at top-of-atmosphere level is with the help of radiometric calibration and band math. 6S gives xa, xb, and xc values by running 6S code in Linux software. In ENVI software 6S code was used for the atmospheric corrections ofTOA measurements to calculate the reflectance of band 4 and 5 which is further used for LST retrieval.The atmospheric model identity of this code is the user defined water vapor, which was 1.88 g/cm^2 with visibility of 4 km and ozone value of 0.251 DU.

6.3.4 Land surface temperature retrieval

LST for a satellite image is the average of the temperature of the surface type in each pixel weighed by their fractional cover [12].This is quantified as the radiation emitted by surface and is measured as the amount of signal acquired by the sensors in electromagnetic spectrum called radiance. Radiance is believed to vary with the surface albedo of the material. To study the surface and atmospheric properties from space surface albedo play an important role. To infer the surface properties, the spectral and angular dependence of reflected insolation is used. Albedo varies in different surfaces. Low albedo observed in oceans, lakes, and forests as small fractions of the incident sunlight is reflected by them, while in case of snow, sea ice, and deserts albedo is higher. The LST of a densely vegetated ground is the vegetation canopy surface temperature whereas for sparsely vegetated areas it is the average temperature of the vegetation canopy, the vegetation body, and the surface temperature.

If the area surrounding a target is assumed to be the same as (or similar to) the target and the target are assumed to be Lambertian and uniform, the reflectance at the target can be expressed conveniently as [14].

$$\rho = \frac{\pi (L_t - L_p)}{(E_{dir} + E_{diff})\tau + \pi S (L_t - L_p)} \quad (6.6)$$

where ρ is reflectance; L_t is the satellite-based radiance; S is the reflectance of the atmosphere; L_p is the atmospheric path radiance; E_{dir} is the direct irradiance at the surface; E_{diff} is the diffuse irradiance at the surface; and τ is the total diffuse transmittance from the ground to the top of the atmosphere in the view direction of the satellite. The relationship between LST and band integrated radiance (T_s), $L_{\lambda(T_s)}$ may be approximated as:

$$L_{\lambda(T_s)} = gain*DN + offset \tag{6.7}$$

where T_s is LST in Kelvin; (T_s) $L_{\lambda(T_s)}$ is integrated band radiance obtained in (W m^{-2} Sr^{-1} μm^{-1}); and *k1* and *k2* are "calibration" constants chosen to optimize the approximation for the band pass of the sensor. For LANDSAT 8, *k1* = 774.89 W m^{-2} Sr^{-1} μm^{-1}, *k2* = 1321.08 W m^{-2} Sr^{-1} μm^{-1} (http://ltpwww.gsfc.nasa.gov/IAS/handbook/handbook_toc.html).

Emissivity is controlled by things like the content, chemical composition, structure, and roughness of a surface, as well as how it looks. It is possible for emissivity to vary greatly between plant species, areal density, and growth stages of vegetated surfaces. However, in the meanwhile, emissivity is determined by the wavelength of light, and this property is often known as "spectral emissivity." [18]. However, in this study the NDVI based emissivity method was applied by using the following equations [19–21]

$$\varepsilon = \varepsilon_v P_v + E_v\left(1 - P_v\right) + d_\varepsilon \tag{6.8}$$

$$d_\varepsilon = \left(1 - \varepsilon_s\right)\left(1 - P_v\right) F \varepsilon_v \tag{6.9}$$

$$\varepsilon = \left(\varepsilon_v - \varepsilon_s - (1 - \varepsilon_s\right) F \varepsilon_v P_v + \left(\varepsilon_s + (1 - \varepsilon_s\right) F \varepsilon_v \tag{6.10}$$

Where, d_ε is effect of geometrical distribution of the natural surfaces, which is negligible for plain surface; F is a shape factor, ε_s and ε_v are the emissivity of soil and vegetation respectively. For fully vegetated area the emissivity value is assumed to be 0.99 and for soil it is around 0.97 [21].

P_v is the vegetation portion calculated from corrected satellite images and uncorrected using the relationship [17] by Eq. (6.11)

$$P_v = \left[\frac{\left(NDVI - NDVI_{\min}\right)}{\left(NDVI_{\max} - NDVI_{\min}\right)}\right]^2 \tag{6.11}$$

$NDVI$, $NDVI_{min}$, $NDVI_{max}$ in Eq. (6.11) are calculated from corrected satellite images and uncorrected one. The emissivity(ε)of thematic mapper band 10(TM10) is determined from corrected and uncorrected satellite images using the relation by Eq. (6.12).

$$\varepsilon_{TM6} = 0.004 P_v + 0.986 \tag{6.12}$$

Finally, the emissivity-corrected land surface temperatures (LST) were computed from corrected and uncorrected satellite images using the relation by the eq. (6.13)

$$LST = \frac{T_s}{1 + \left(\lambda T_s / \rho\right) \ln \varepsilon} \tag{6.13}$$

where λ is the wavelength of emitted radiance (for which the peak response and the average of the limiting wavelengths (λ = $10.9*10^{-6}m^2$) will be used; $\rho = h \times c / \sigma (1.43 \times 10^{-2}$ m K), σ is Boltzmann's constant (1.38×10^{-23} JK^{-1}), h is Planck's constant (6.626×10^{-34} Js), and c is the velocity of light (2.998 × 108 ms^{-1}).

6.3.5 Performance analysis

Atmospherically corrected and uncorrected satellite images were used here for LST retrieval which is compared against the field-based observed datasets. Among the available statistics, root-mean-square error (RMSE), standard deviation (SD) and correlation (r) were used in this study. Taylor diagram [22] represents the correlation between RMSE and SD by using R statistical package tool. It interprets graphically summarizing the close association of two observations based on their correlation, centered root-mean-square difference, and variation amplitude which is represented by their standard deviations).

$$RMSE = \sqrt{\left(\frac{1}{n}\sum_{i-1}^{n}[y_i - x_i]\right)^2} \tag{6.14}$$

$$r = \frac{n\left(\sum xy\right) - \left(\sum x\right)\left(\sum y\right)}{\sqrt{\left[n\left(\sum x^2\right) - \left(\sum x\right)^2\right]\left(n\left(\sum y^2\right) - \left(\sum y\right)^2\right)}} \tag{6.15}$$

where n indicates the number of observations; x is observed; and y is simulated values.

6.4 Result and discussions

6.4.1 6SV output for atmospheric correction

An array of advanced algorithms is now a days available for satellite image corrections however selection of the suitable algorithm is important. This study investigates usage and applicability of one of the advanced image correction algorithms, second simulation of satellite signal in solar spectrum-vector (6SV), capable of accounting for radiation polarization. The algorithm is also sometimes called as a modified version of the previous scalar 6S. this subsection of the report briefly summarizes the results obtained after the using 6S algorithm on Landsat 8 images for atmospheric correction. While analyzing the 6S code output, values of xa considered as the inverse of the transmittance, while xb is scattering of atmosphere and xc represents reflectance of the atmosphere for isotropic light entering the base of the atmosphere are used for calculating the reflectance of uncorrected satellite image of band 4 and band 5. These reflectance values were then further used for arithmetical calculations on satellite images for atmospheric correction to get atmospheric corrected satellite image and Land surface temperature retrieval. The values of xa, xb, xc for bands 4,5 and other parameters obtained after running the 6SV code in Linux software is shown in Table 6.2. Also, raw and corrected image showing

Table 6.2 The output of different bands.

Bands	Wavelength	Visibility	xa	xb	xc	Irradiance at ground level	Radiance at satellite level	Reflectance at satellite level
						Direct atmospheric environment Solar diffused irradiance Irradiance	Atmospheric environment target Intrinsic radiance Radiance	Atmospheric environment target Intrinsic reflectance Reflectance
1	0.636–0.673	4	0.086	0.454	0.062	10.630 47.028 0.317	5.273 0.274 0.336	0.048 0.002 0.003
2	0.851–0.879	4	0.082	0.246	0.051	17.934 35.899 0.770	2.994 0.682 0.000	0.040 0.009 0.000

reflectance for band 4 and 5 of the study area Mirzapur are shown in Figs. 6.3 and 6.4, respectively.

From the above figure, it can be noticed that the expansion of the all maps is clearer in the corrected images than the raw images for both bands. This can be explained by the fact that atmospheric correction is dependent on the surface type, and this affect the variation between low and high reflective surfaces [23]. Atmospheric Rayleigh scattering occurred in the visible spectral reflectance and in between atmospheric corrected and uncorrected spectral bands, which reduces the reflectance values of near-infrared band. A change in reflectance values mainly reduction was observed in the near-infrared

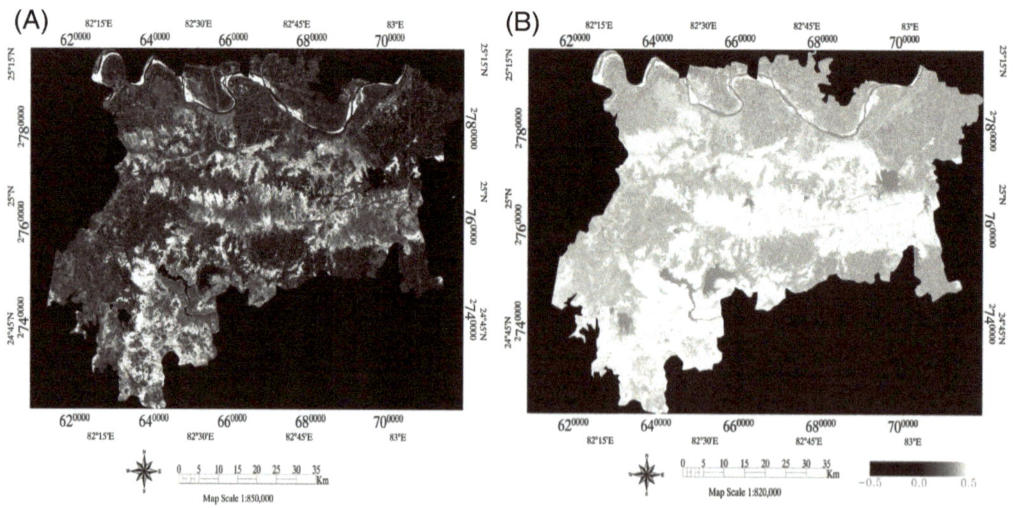

Fig. 6.3 Reflectance of band 4 of Mirzapur from (A) atmospherically uncorrected satellite image (B) atmospherically corrected satellite image.

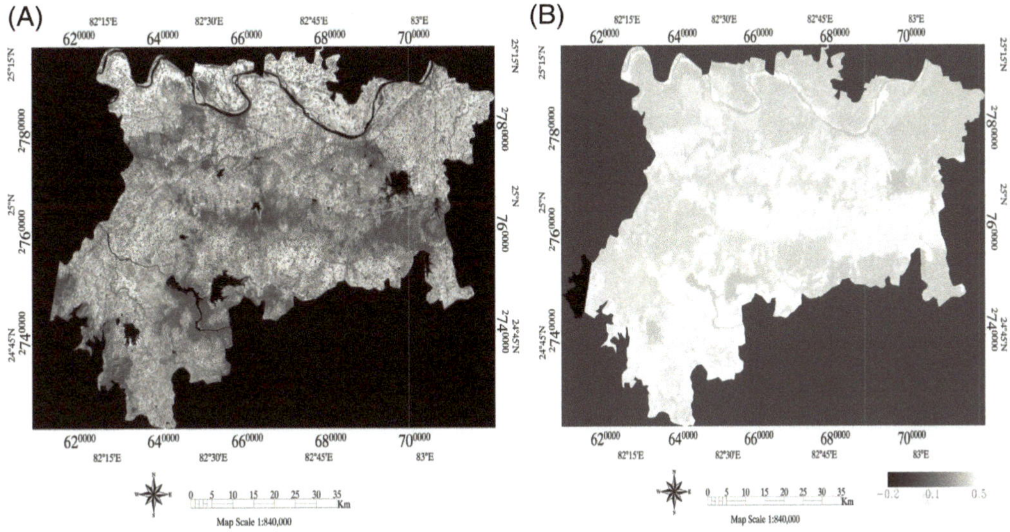

Fig. 6.4 Reflectance of band 5 from (A) atmospherically uncorrected satellite image (B) atmospherically corrected satellite image.

band which is mainly attributed to contribution of atmospheric Rayleigh scattering in the visible spectral reflectance and atmospherically corrected versus uncorrected spectral bands. The lesser impact of air particles, as reported by other researcher, may be one explanation for the higher NIR reflectance levels [23]. In these bands, the greater reflectance clearly demonstrates the impact of aerosol scattering. Gas and water vapor absorption is clearly indicated by reduced reflectance values of TOA in the NIR range. Fig. 6.5a and b represents the correlation of NDVI before 6S correction and after atmospheric correction over crop land respectively. The results show that the NDVI has lesser value before correction, but after correction a broad range is observed. In all datasets there is a continuous increase in NDVI range can be observed.

6.4.2 LST retrieval using Landsat 8 OLI

After the atmospheric correction of the satellite images, next step in the analysis was to quantify the variation of LST with different land cover classes. The analysis for this was performed in a step wise procedure mentioned in the LST retrieval methodology explained in Fig. 6.2 of previous section using Exelis Visual Information Solution (ENVI, 5.1 version), a software widely used for geo-spectral analysis and spectral image processing. Using ENVI, LST values were extracted for each cover classes and their variation is presented in figure below. Fractal analysis of LST values for cover classes used for image classification show highest values of LST for Sand cover followed by urban, agricultural, plantation and least for water body.

Fig. 6.5 shows the NDVI map generated after analysis performed in ENVI on the uncorrected and corrected satellite images of the study area. The map illustrates NDVI

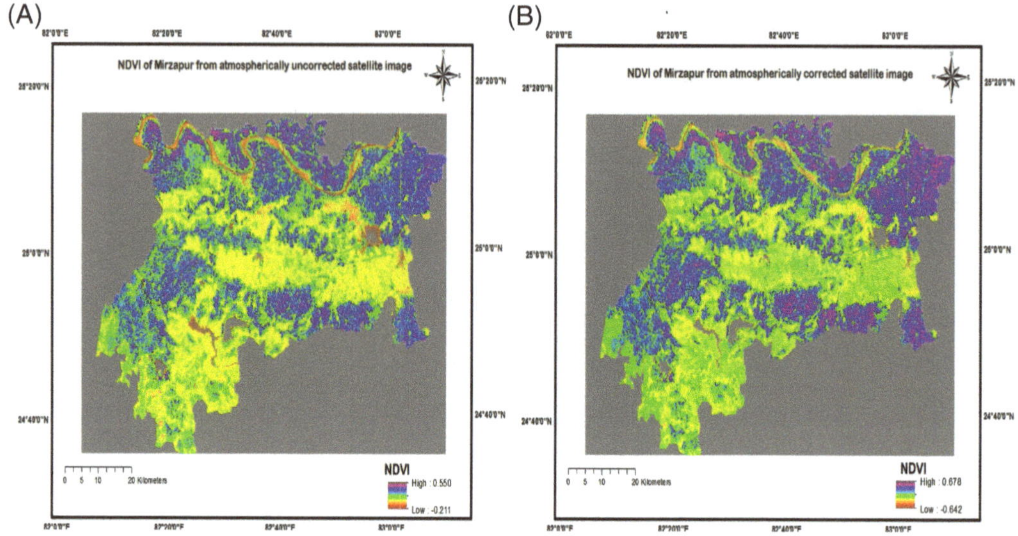

Fig. 6.5 NDVI calculated from (A) atmospherically uncorrected satellite image (B) atmospherically corrected satellite image.

distribution with the help of color distribution with different intensities explained in the map key.

The NDVI map shows higher or maximum NDVI values in the areas having healthy and dense vegetation as compared to unhealthy or sparsely vegetated area with lower NDVI values. In this atmospherically uncorrected satellite image, maximum value NDVI obtained was 0.550 whereas minimum value recorded was –0. 211. Fig. 6.5a represents the upper north-east side of the study area having a healthy vegetation while the central part of the region having unhealthy vegetation cover.

The Fig. 6.5b shows atmospherically corrected image, in which maximum value of NDVI was reported as 0.678 while the minimum value was –0.642. Map clearly shows that in the upper side of north-east, healthy vegetation is maximum.

The Fig. 6.6 above depicts the plant fraction obtained for the region. The plant fraction values were maximum in the areas having maximum vegetation cover or NDVI. On other hand plant fraction was also reported to be low in the area with no vegetation cover or low NDVI value. In this atmospherically uncorrected satellite (Fig. 6.6a) image maximum value of Pv is 1.00 and minimum value is 1.729. Map clearly shows that upper side of north-east, vegetation cover is maximum and lowest at south-east and south direction.

Fig. 6.6b shows the plant fraction for the atmospherically corrected satellite image of the study site. Above map shows same trend in results without atmospheric correction but its statics value is more due to its high reflectance value. In this atmospherically corrected satellite image maximum value of Pv was reported 1.00 and a minimum value

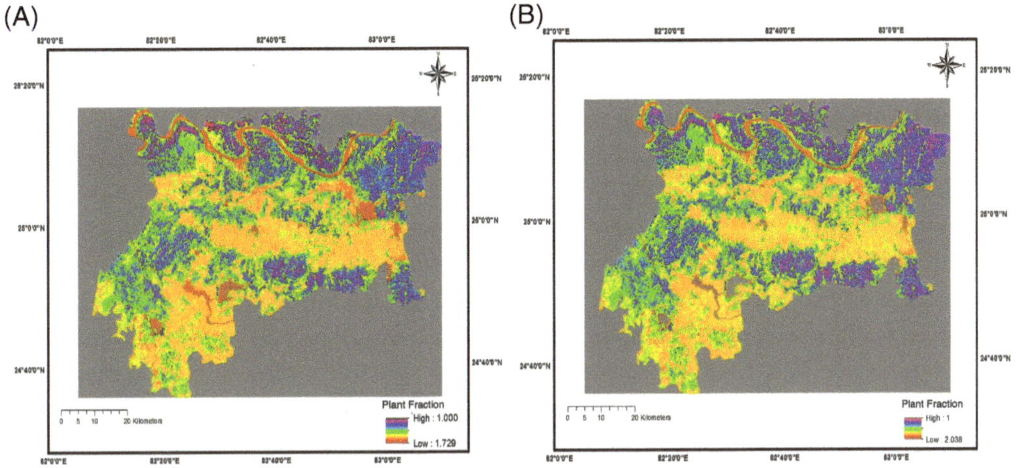

Fig. 6.6 Plant fraction of Mirzapur from (A) atmospherically uncorrected satellite image (B) atmospherically corrected satellite image.

was found as 2.038. Map clearly shows that upper side of north-east, vegetation cover is maximum and lowest at south-east and south direction.

Fig. 6.7 shows the emissivity of the study site from the uncorrected (Fig. 6.7a) and corrected satellite images (Fig. 6.7b). Fig. 6.7a represents the area with maximum emission of radiation, on other hand low emissivity values were reported in the area with low emissivity of radiation. In this atmospherically uncorrected satellite image maximum value of ε was found as 0.990 and a minimum value 0.986 was reported. Map clearly shows that upper side of north-east, emission of radiation is maximum and

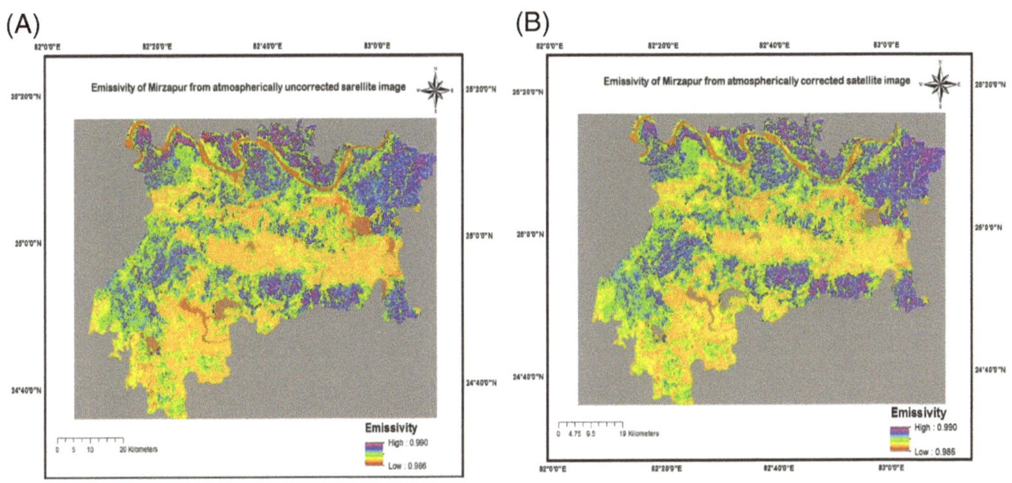

Fig. 6.7 Emissivity estimated from (A) atmospherically uncorrected satellite image (B) atmospherically corrected satellite image.

lowest at south-east and south direction. In Fig. 6.7b, in the atmospherically corrected satellite image a maximum value of ε (0.990) is reported and a minimum value of 0.986 was obtained. Map clearly shows that upper side of north-east, emissivity is maximum and lowest at south-east and south direction.

In above map where LST value is maximum, area shows high land surface temperature on other hand where LST value is lowest, area shows low land surface temperature. In this atmospherically uncorrected satellite image maximum value of LST is 38.429 and minimum value is 20.671. It is noticed from the Map that east and south direction have maximum LST and lowest at north-east, west, and north direction. In Fig. 6.8B atmospherically corrected satellite image maximum value of LST is 38.068 and minimum value is 20.329. Map clearly shows that east and south direction have maximum LST and lowest at north-east, west, and north direction. The RMSE value between LST of uncorrected satellite image and ground measure is 3.39, on other hand the value between corrected satellite image and ground measure is 3.38.

The study showed a better performance in case of atmospherically corrected image as compared to the uncorrected image. For an instance, the statistics of reflectance values were reported better after applying atmospheric correction by 6SV algorithm than the raw images. This study highlights the role of noise present in the atmospheric layer, responsible for scattering and absorption of the radiation before it reaches the satellite. During study, NDVI proved to be a useful tool because of its simple calculation process and inverse relation with vegetation in both the red and infrared region. The model used for LST retrieval provided good estimation of LST up to ~1–2 cm depth. Nonetheless, the results also show the changes of land use and land cover type on LST distribution. The vegetated area has diverse LST due to area-based density of vegetation

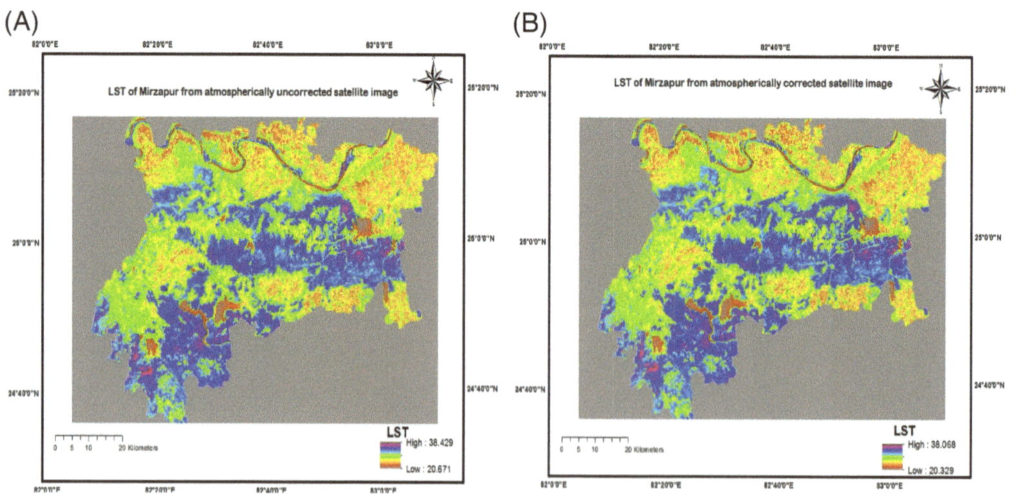

Fig. 6.8 LST retrieval from (A) Atmospherically uncorrected satellite image (B) atmospherically corrected satellite image.

which means low LST values with area having high vegetation density and high values for sparsely vegetated and bare regions. When it compares with ground truth LST, we found atmospheric corrected LST is more accurate as compare to atmospheric uncorrected satellite data. According to the findings of this study, undoubtedly LST is a credible indication of large-scale NDVI fluctuation after 6SV atmospheric correction.

6.5 Conclusion

Satellite images provide a broad prospect of viewing Earth surface from space which is crucial in visualizing and understanding the driving factors on earth system functioning. In an attempt to study the role of satellite image in observing various factors affecting land surface temperature variation and retrieval algorithms, this study investigates the comparative performances of raw and atmospherically corrected images. The study here explores the possibility of long-term monitoring of temperature variation and changes on a regular temporal basis at larger spatial coverages with help of remote sensing technologies. This research employs Landsat 8 OLI images for land surface temperature retrieval and also to evaluate the efficiency of the retrieval procedure after atmospheric correction algorithm, that is, 6SV. The study also analyses the effects of noise on satellite images and also brief on highlighting role of atmospheric correction algorithms in reducing effects of noises. The study concluded that the NDVI value is inversely proportional to the LST and vice versa as bare or rocky areas have more temperature as compared to vegetated area and after correction by using 6SV a better retrieval of LST has been obtained. Further findings of this study are believed to contribute to the future research in the investigations of areas related to soil moisture estimation at deeper soil layers, whose effects on LST can be estimated by using an improved technological approach and by reforming the mathematical expression in band math.

References

[1] C.J. Tomlinson, et al., Remote sensing land surface temperature for meteorology and climatology: A review, Meteorol. Appl. 18 (3) (2011) 296–306.
[2] L. Vlassova, et al., Analysis of the relationship between land surface temperature and wildfire severity in a series of Landsat images, Remote Sens. 6 (7) (2014) 6136–6162.
[3] D.P. Roy, et al., Landsat-8: Science and product vision for terrestrial global change research, Remote Sens. Environ. 145 (2014) 154–172.
[4] K. Watson, Geologic applications of thermal infrared images, Proc. IEEE 63 (1) (1975) 128–137.
[5] Z. Zhan, et al., The application of LST/NDVI index for monitoring land surface moisture in semiarid area, Proceedings of 2004 Geoscience and Remote Sensing Symposium, IGARSS'04, IEEE International, 2004.
[6] Y. Jiang, W. Lin, A comparative analysis of retrieval algorithms of land surface temperature from Landsat-8 data: A case study of Shanghai, China, Int. J. Environ. Res. Publ. Health 18 (11) (2021) 5659.
[7] A. Sekertekin, E. Zadbagher, Simulation of future land surface temperature distribution and evaluating surface urban heat island based on impervious surface area, Ecol. Indicators 122 (2021) 107230.
[8] W. Zhengming, J. Dozier, Land-surface temperature measurement from space: Physical principles and inverse modeling, IEEE Trans. Geosci. Remote Sens. 27 (3) (1989) 268–278.

[9] X. Ye, et al., Land surface temperature estimate from Chinese Gaofen-5 satellite data using split-window algorithm, IEEE Trans. Geosci. Remote Sens. 55 (10) (2017) 5877–5888.
[10] H. Ren, et al., Improving land surface temperature and emissivity retrieval from the Chinese Gaofen-5 satellite using a hybrid algorithm, IEEE Trans. Geosci. Remote Sens. 56 (2) (2017) 1080–1090.
[11] Y.J. Kaufman, B.-C. Gao, Remote sensing of water vapor in the near IR from EOS/MODIS, IEEE Trans. Geosci. Remote Sens. 30 (5) (1992) 871–884.
[12] P.K. Srivastava, et al., Estimation of land surface temperature from atmospherically corrected LANDSAT TM image using 6S and NCEP global reanalysis product, Environ. Earth Sci. 72 (12) (2014) 5183–5196.
[13] J. Storey, et al., Landsat 8 Operational Land Imager on-orbit geometric calibration and performance, Remote Sens. **6** (11) (2014) 11127–11152.
[14] E. Vermote, et al., Atmospheric correction of visible to middle-infrared EOS-MODIS data over land surfaces: Background, operational algorithm and validation, J. Geophys. Res. Atmos. 102 (D14) (1997) 17131–17141.
[15] K. Tachiiri, Calculating NDVI for NOAA/AVHRR data after atmospheric correction for extensive images using 6S code: a case study in the Marsabit District, Kenya, ISPRS J. Photogramm. Remote Sens. 59 (3) (2005) 103–114.
[16] B. Pinty, M. Verstraete, GEMI: A non-linear index to monitor global vegetation from satellites, Plant Ecol. 101 (1) (1992) 15–20.
[17] T.N. Carlson, D.A. Ripley, On the relation between NDVI, fractional vegetation cover, and leaf area index, Remote Sens. Environ. 62 (3) (1997) 241–252.
[18] Q. Weng, et al., Estimation of land surface temperature–vegetation abundance relationship for urban heat island studies, Remote Sens. Environ. 89 (4) (2004) 467–483.
[19] J. Sobrino, et al., Multi-channel and multi-angle algorithms for estimating sea and land surface temperature with ATSR data, Int. J. Remote Sens. 17 (11) (1996) 2089–2114.
[20] J. Sobrino, et al., A comparative study of land surface emissivity retrieval from NOAA data, Remote Sens. Environ. 75 (2) (2001) 256–266.
[21] J.A. Sobrino, et al., Land surface temperature retrieval from LANDSAT TM 5, Remote Sens. Environ. 90 (4) (2004) 434–440.
[22] K.E. Taylor, Summarizing multiple aspects of model performance in a single diagram, J. Geophys. Res. Atmos. 106 (D7) (2001) 7183–7192.
[23] A.R. Sharma, et al., Comparison of ground reflectance measurement with satellite derived atmospherically corrected reflectance: A case study over semi-arid landscape, Adv. Space Res. 43 (1) (2009) 56–64.

CHAPTER 7

Spatio-temporal variation of biomass burning fires over Indian region using satellite data

Darga Saheb Shaik[a,b], Yogesh Kant[a], M. Sateesh[c], Vikram Sharma[d], Deependra Singh Rawat[e], H.C. Chandola[f]
[a]Marine & Atmospheric Sciences Department, Indian Institute of Remote Sensing, ISRO, Dehradun, India
[b]National Atmospheric Research Laboratory, Department of Space, Gadanki, India
[c]Climate Change Centre, King Abdullah University of Science and Technology, Thuwal, Saudi Arabia
[d]Department of Geography, Banaras Hindu University, Varanasi, Uttar Pradesh, India
[e]Department of Allied Sciences (Physics), Graphic Era Hill University, Bhimtal Campus, Nainital, India
[f]Department of Physics (UGC-Centre of Advanced Study), Kumaun University, Nainital, India

7.1 Introduction

Biomass burning (BB), worldwide is either natural or anthropogenic, causing adverse impact on environmental, economic, and ecological systems at regional to global scale [1–3]. The distribution of BB (either forest fires and/or crop residue burning) around the world are mainly caused by anthropogenic activities (90%) compared to natural causes (10%). Fire has been used as a tool to manage land clearance as well as nutrient recovery in the soil for many centuries. Reasons for forest fire are due to either artificial or natural but the advancement depends on environmental factors like weather, fuel, and topographic conditions [4] and persist days to weeks whereas fire due to crop residue burning (CRB) is purely anthropogenic in nature and persist to few hours. Hence, the emissions and causative factors of BB may differ region to region.

Several studies have reported the BB and its impacts in different environmental regimes. In Asia, total of 730 Tg biomass was burnt during 1999–2000, of which 45% comes from forest fires, 34% from crop residue fires, and 20% from grassland and savanna fires and the proportion of BB is highest in China (25%) followed by India (18%) [5]. According to the UN-International Forest Fire News (IFFN) 2006 report, in Indian region the major environment and economic consequences are coming from forest fire occurrences such as loss of fuel wood and valuable timber products (Teak, Deodar, Sal, Chir-pine, Rosewood, etc.), bio-diversity, wildlife habitat, soil erosion, and reduction of forest cover [6]. Venkataraman et al. [7] estimated the BB over Indian region during 1995–2000 and found the annual CRB is 116 (58–289) Tg yr^{-1}, forest and shrub burning is 32 (16–61) Tg yr^{-1} and corresponding total aerosol emissions including particle matter (PM) and its constituents black carbon (BC), organic carbon (OC), etc. are about 360 (16–924) Gg yr^{-1} and 836 (86–2393) Gg yr^{-1} for crop residue and forest fires respectively. Recent

Atmospheric Remote Sensing
DOI: https://doi.org/10.1016/B978-0-323-99262-6.00009-2

study over India reports annual 488 Tg of total agricultural waste was generated and about 24% (116 Tg) was burnt in the field while the resulting annual emissions are 824, 239, 58, and 211 Gg for particulate matter ($PM_{2.5}$), organic carbon, elemental carbon, and greenhouse gases, respectively [8]. Apart from the regional emissions, BB also affects the global climate system [9–11]. Jacobson [12] reported a net global warming of ~0.4 K caused by BB based on 20 years modelling-based simulations. In addition, Bali et al. [13] reported the Himalayan glacier snow cover decreased by 5–20% during Himalayan forest fire period of long term mean of 14 years (2003–2016). Hence, quantification and measurement of BB on regional scale needs more attention to understand the cause-effect scenarios.

In this aspect, space-borne sensors are capable of addressing the problem with well-defined scientific strength. Satellite based remote sensing techniques emerged as a powerful tool to measure daily active fire counts and generate BB frequency, forest fire risk maps, regional fire trend analysis, and air pollution emission scenarios [7,14–18]. The present study analyzes the spatial and temporal variation of BB over Indian region based on multi-satellite data sets (MODIS and VIIRS) during 2003–2018. We have analyzed the seasonal and interannual variations of fire counts over Indian region along with state level statistical trends. Furthermore, we identified the climatic BB hotspot regions based on fire density map and fire type (forest fire or crop residue fire) with respect to MODIS standard land cover type classification.

7.2 Datasets and methodology

Moderate resolution imaging spectroradiometer (MODIS) is a satellite sensor, mounted on aboard NASA's Terra and Aqua satellites and it consists of 36 spectral bands ranging from 250 m to 1 km spatial resolution. MODIS is widely used sensor in various fields of applications like agriculture, forestry, meteorology, urban, etc. across the world [17,19,20]. MODIS data products are freely available and can be downloaded from the NASA's Earthdata web portal. More details about the design, specifications, data products, algorithms, and calibration are described elsewhere [21–24]. The MCD14 product provides the thermal anomaly and fire location information, which can be derived from the MODIS 4 μm and 11 μm spectral radiance [25]. MCD14 fire product can be freely download from the Fire Information for Resource Management System (FIRMS) portal (https://firms.modaps.eosdis.nasa.gov/). Several researchers have used MCD14 product as a proxy for ground forest fire assessment and analysis of crop residue burning occurrences [2,17,26]. The present study, used monthly fire location product (MCD14ML) with high confidence level (80–100%) data for analyzing the fire occurrences over Indian region during the January 2003–December 2018 period.

The visible infrared imaging radiometer suite (VIIRS) sensor was launched on aboard the NASA's Suomi National Polar-orbiting Partnership (Suomi-NPP) satellite in October 2011 [27]. Currently, VIIRS/S-NPP active fire products are available at two resolutions: one at 375 m [28] and another at 750 m [29] in different user-friendly formats and archived in FIRMS portal. More details pertaining to algorithms, data products and validation are reported in earlier studies [28–31]. VIIRS active fire detection product (VNP14IMGTDL) has been providing information on the fire occurrence across the world on a daily basis with a spatial resolution of 375 m and these data can be downloaded from the FIRMS website. The VIIRS product has also been derived from the spectral radiance of 4 μm and 11 μm. The algorithm of fire detection for both MODIS and VIIRS is similar for derivation of the respective fire products [29]. The advantage of VIIRS is high spatial resolution (375 m) and wider swath coverage (3060 km) as compared to MODIS (1 km spatial resolution and swath length 2330 km) and VIIRS is more sensitive to detect smaller fires as well as delineate the fire perimeters [28]. Several approaches have been taken to quantify errors in fire detection, including comparison of satellite data with co-located field data, model simulations and other satellite observations. The overall accuracy of detection rates of active fires for MODIS is 82% [32] while for VIIRS is 80% [33], however, the accuracies are varied by region.

In order to delineate and segregate the fire hotspot regions, we have used the standard MODIS land cover type (MCD12Q1) product over Indian region. MODIS Terra + Aqua Land Cover Type Yearly L3 Global 500 m SIN Grid product consists of six classification schemes, which was derived from 8 day observed data of MODIS Terra and Aqua satellites. MODIS land cover product supplies global maps of land cover at annual time steps at 500 m spatial resolution from 2001 to till date. The International Geosphere Biosphere Programme (IGBP), land cover schema was used in the present study. Details of the MODIS IGBP global land cover product can be found in [34]. MODIS land cover product is widely used for regional land use and land cover classifications/assessments over Indian region [17,35–37]. On all India basis, the classification accuracy of MODIS land cover type is 84% for cropland, 76% forest cover and pasture and barren lands are 64% and 21%, respectively compared with Indian ministry of agriculture report during 2002–2005 [35]. State-wise comparison of land use and land cover (LULC) classification shows the high degree of correlation in cropland (R^2~0.9) followed by forest cover and pasturelands in India [35]. In addition, Roy et al. [38] found the overall accuracy of MODIS global land cover is 67.10% with kappa value of 0.656 over Indian region during 2005. For segregation of fire type (forest fire or crop residue fire), we have classified the LULC types in 3 classes namely forests, croplands, and urban and others.

7.3 Results

7.3.1 Spatial and seasonal variation of fire points over India

Climatological BB density maps were generated over Indian region using the MODIS active fire product during 2003–2018 as well as the VIIRS active fire product during the years 2012–2018 (Fig. 7.1). For generating the spatial BB maps, the Indian region

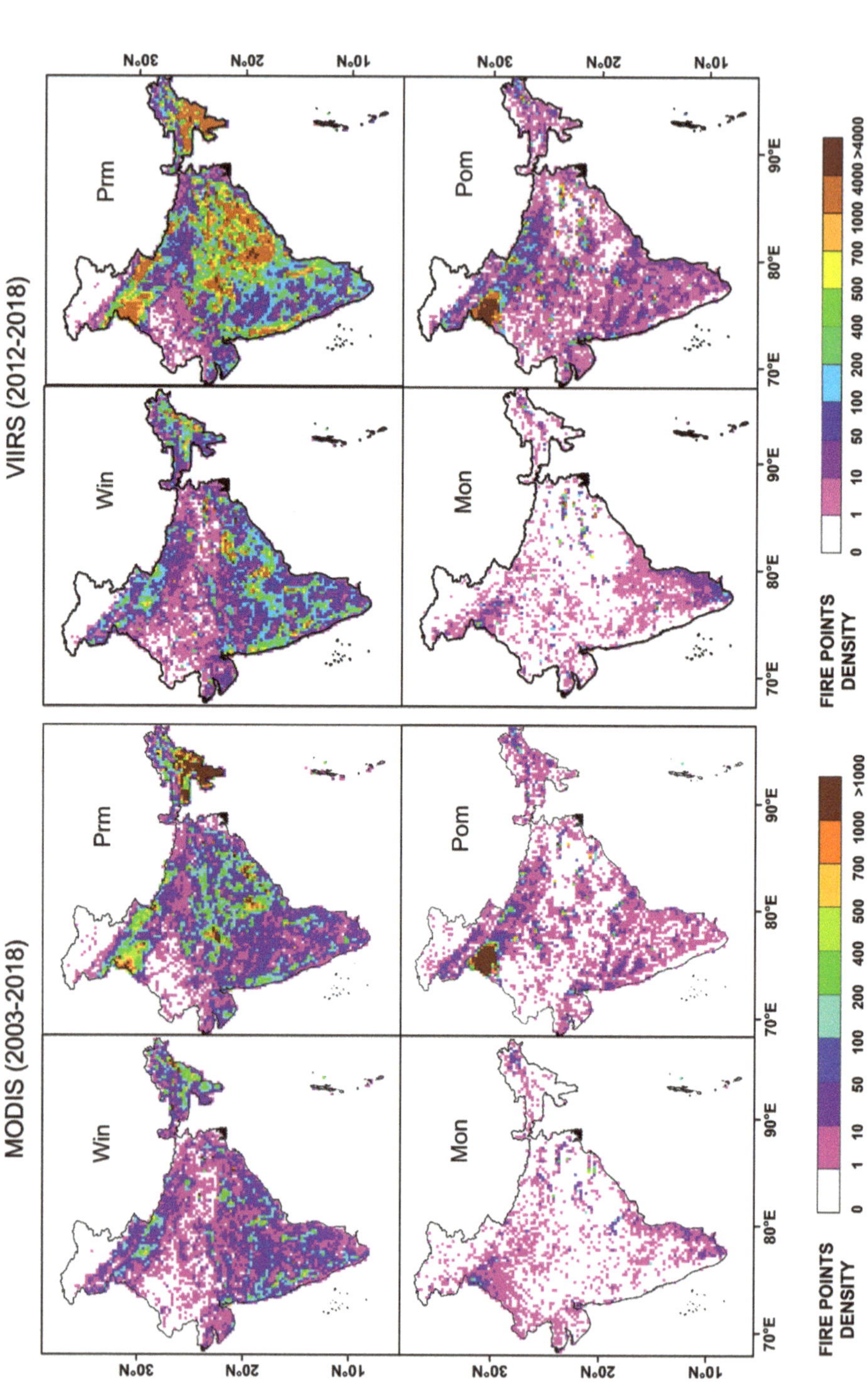

Fig. 7.1 Seasonal fire density (no. fires per grid) maps with spatial resolution 0.25° × 0.25° grid from (A) MODIS during 2003–2018 and (B) VIIRS during 2012–2018.

was divided into 0.25° × 0.25° grids and the number of fire counts falling in each grid were aggregated and subsequently averaged on the basis of seasons, that is, winter (Win), premonsoon (Prm), monsoon (Mon), and postmonsoon (Pom). The results revealed that the BB starts from winter (February) and continue till premonsoon (June). BB over India exhibits average minimum of 917 fires occurrences during Mon to average maximum of 43,053 fires occurrences during Prm as observed from MODIS whereas observation from VIIRS reveals average minimum of 7559 fires in Mon to average maximum of 34,8672 fires occurrence in Prm. Overall, data shows an annual 75,786 and 574,381 fires occurrences detected over India from MODIS and VIIRS sensors respectively. Moreover, both sensors show similar spatial pattern and variability of BB fires over Indian region with different magnitudes. Magnitudes of fire counts differ by the sensor's capabilities/resolutions [18]. During Prm and Pom, BB shows higher density (>1000 per grid) in regions, of northwest IGP states (Punjab and Haryana), foothills of Himalayas (Uttarakhand) as well as central India (Madhya Pradesh and Chhattisgarh) and north-east India (Mizoram and Manipur). In India, crop residue fires mostly occur during Prm and Pom seasons whereas the forest fires occur in Prm season only. The major crop residue burning regions are western IGP states such as Punjab, Haryana, and Uttar Pradesh, whereas the major forest fire regions are north eastern states namely Mizoram, Assam, Manipur, and Nagaland along with Uttarakhand in northern India. Combination of forest fire and CRB are observed in Andhra Pradesh, Madhya Pradesh, and Chhattisgarh due to the dominance shifting cultivation practices [39].

Fig. 7.2 shows the average monthly variation of fire occurrences over Indian region observed (a) from MODIS during 2003–2018 and (b) from VIIRS during 2012–2018. From the figure it is observed that annual BB has two peaks, one is during Prm season (March–April) and other during Pom (October–November) season. On an average, the annual contribution of fire occurrences during Prm is 57% from MODIS and 61% from VIIRS while during Pom it is 24% and 19% for MODIS and VIIRS, respectively. The maximum fire counts observed are in the month of March (29% and 27%) followed by April (18% and 21%), October (13% and 7%), November (11% and 12%), and May (9% and 11%) for MODIS and VIIRS, respectively. These results are comparable with previous study over India during 2010 which reported the fire season extends from October to June during which more than 70% of total annual fires are recorded with peak months March (41.2%), April (19.7%), and May (6.03%) [39]. In general, over Indian region the forest fires are from February to May whereas CRB is in two phases (1) March–April (wheat residue burning) and (2) October–November (Paddy residue burning), these patterns are mostly observed in western Indo-Gangetic plain region [7,40].

7.3.2 Inter annual variations of active fire occurrences

Fig. 7.3 shows annual variation of fire count observed from (a) MODIS during 2003–2018 and (b) from VIIRS during 2012–2018. A significant inter annual variability of

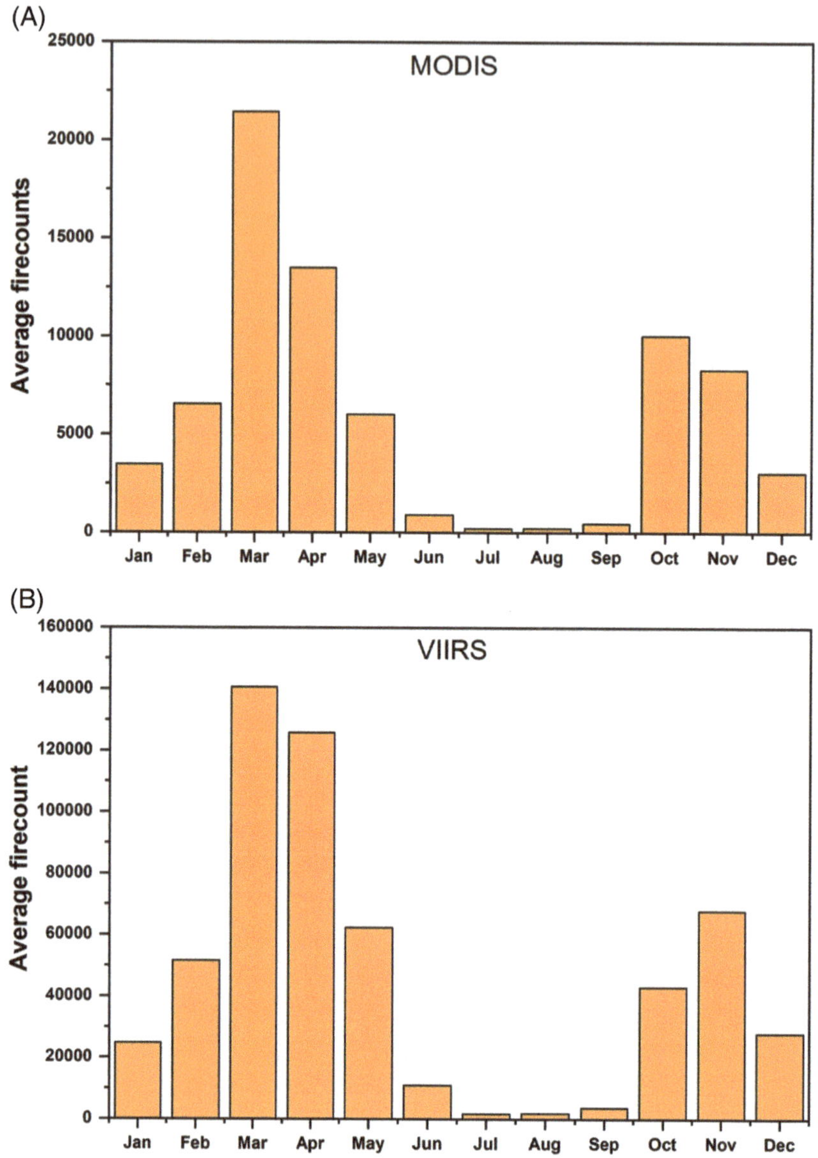

Fig. 7.2 Monthly variation of fire counts over Indian region detected from (A) MODIS and (B) VIIRS sensors.

fire counts observed in both data sets with maximum peaks in 2012 (93,570), 2018 (91,531), 2009 (90,696), 2016 (89,378) and minimum in 2003 (56,177) from MODIS 16 years record while in 7 years records of VIIRS, the maximum peak is observed in 2016 (678,682) followed by 2018 (646,340) and minimum 2015(481,448). Inter annual analysis show that BB is maximum in the year 2012 followed by 2009 and 2016, this is

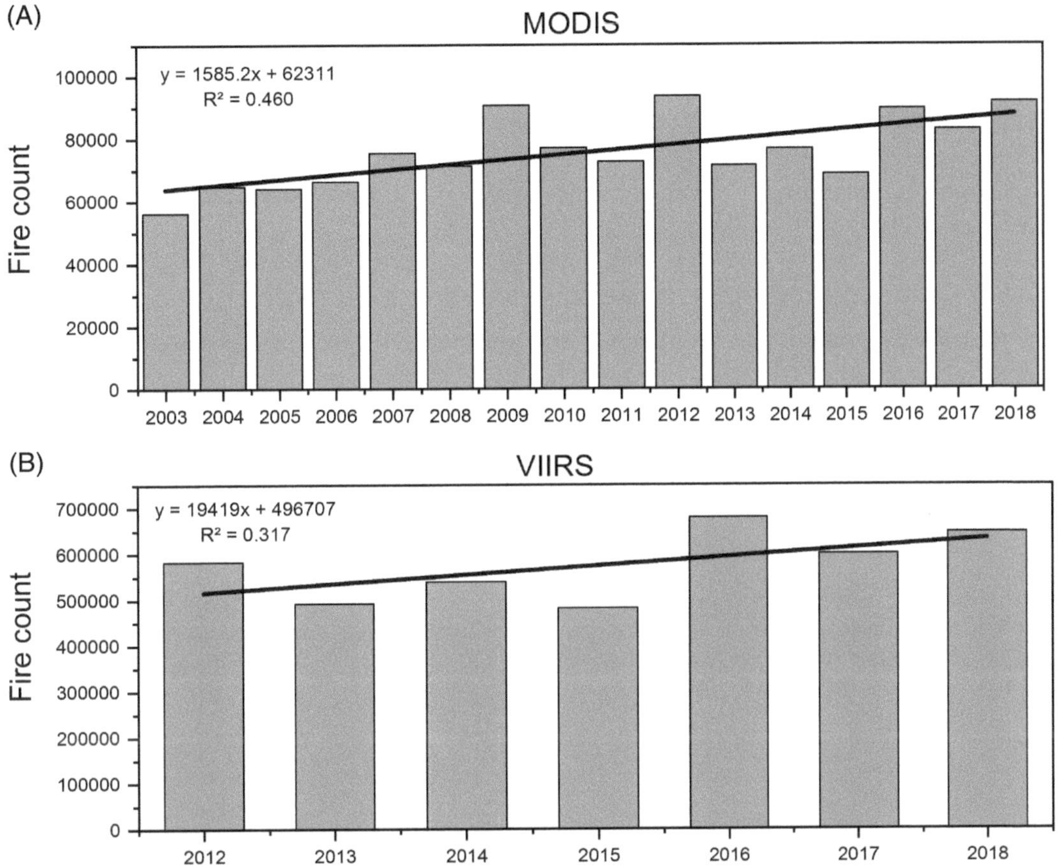

Fig. 7.3 Inter annual variation of active fire counts over Indian region observed from (A) MODIS during 2003–2018 and (B) VIIRS during 2012–2018.

justified by the reports mentioned that 2006, 2009 and 2012 are drought years [41,42]. The observed variability in fire count could be attributed to the extended dry season and drought conditions over India. The long-term average fire counts over India were found to be ~75,786 for MODIS (2003–2018) and ~57,4381 for VIIRS (2012–2018). VIIRS fire counts are 7.5 times high than observed MODIS fire counts. The reason is than VIIRS capability of capturing smaller and cooler fires [18].

The nonparametric Mann-Kendall test has been applied on annual fire count data. The M-K test is suitable for evaluating the long-term trend with non-normal distribution in data samples [43]. The fire counts over Indian region shows significant (95% of confidence level) increasing trend with a rate of 2.7% yr^{-1} and 3.4% yr^{-1} for MODIS and VIIRS, respectively. The observed increase in fire counts may be due to several factors: (1) increase in crop sown area and agriculture residue and farmers burn the agriculture residue in the field due to high labor costs and less time span for the next crop [44], (2) dry seasons has increased in recent years causing increase in forest fires [45], (3) rainfall patterns and intensive droughts leave the vegetation dryer and are more prone to burning [46], and (4) anthropogenic forest fires increases with rising human and cattle population as their dependence and demand for fodder and forest products increases [39].

7.3.3 State level analysis of fire counts over Indian region

The monthly average fire count observed from MODIS over different states of India during the years 2003–2018 is shown in Fig. 7.4 (since considering the long-term observations from MODIS instead of VIIRS). It is evident that almost all the states show a large intra-annual variability such as the BB periods differ across states. Almost in all the states, BB period extends to 6 months in dual phase (1) February to May and (2) November to December except Punjab and Haryana where the burning period is 4 months, that is, April–May and October–November. The seasonality of BB over India is mainly determined by the harvesting practices such as burning of agricultural residue and seasonal land clearance [15]. For the estimation of long-term trends in each state, we have applied the Mann-Kendall statistics with Sen's slope estimation for the period of 16 years (2003–2018) MODIS data. Statistical parameters of each state are reported in Table 7.1 and spatial trends are shown in Fig. 7.5. Maximum trends were observed

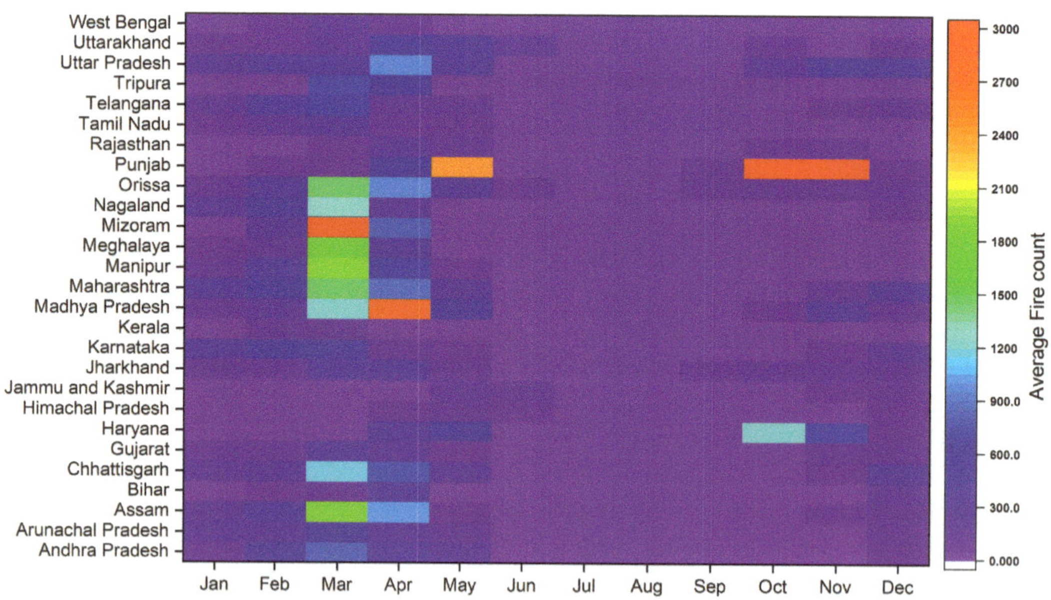

Fig. 7.4 ***Time series monthly average fire count for the different states in India.***

Table 7.1 The statistical parameters of MODIS fire count data over different states of India during the years 2003-2018. Trend analysis was done with non-parametric Mann-Kendall test.

State	Intercept	Average fire count	Z	Slope	Trend (%)	Significance
Andhra Pradesh	1336	2306	1.88	76	5.68	+
Arunachal Pradesh	1592	1843	−0.49	−29.5	−1.85	NS
Assam	2858	3912	0.49	44.83	1.57	NS
Bihar	386	791	3.56	50.76	13.15	***
Chhattisgarh	2282	3244	2.77	217	9.51	**
Goa	23	25	0.69	0.4	1.74	NS
Gujarat	700	1206	1.19	18.57	2.65	NS
Haryana	2159	3102	1.98	80.07	3.71	*
Himachal Pradesh	435	430	0.2	4.6	1.05	NS
Jammu and Kashmir	291	571	0.99	16.9	5.81	NS
Jharkhand	683	1686	2.87	91.45	13.39	**
Karnataka	1073	2071	2.47	52.75	4.91	*
Kerala	172	316	0.64	5.5	3.2	NS
Madhya Pradesh	1348	5350	2.87	429.62	31.87	**
Maharashtra	3366	4134	1.78	114.76	3.41	+
Manipur	3104	3257	−1.78	−73.8	−2.38	+
Meghalaya	1869	2336	0	−3.7	−0.2	NS
Mizoram	5056	4909	−2.27	−269.08	−5.32	*
Nagaland	2271	2359	−1.33	−48.46	−2.13	NS
Orissa	3076	3790	3.16	193.66	6.29	**
Punjab	14655	16715	0.69	155	1.05	NS
Rajasthan	250	678	3.95	63.83	25.53	***
Sikkim	6	16	1.89	1	16.67	+
Tamil Nadu	628	852	0.49	5.7	0.90	NS
Telangana	999	1796	3.36	112.45	11.25	***
Tripura	1562	1299	−2.96	−80.28	−5.13	**
Uttar Pradesh	2304	3640	1.58	126.75	5.5	NS
Uttarakhand	1461	1559	−0.1	−9	−0.61	NS
West Bengal	208	428	2.63	31.09	14.95	**

Note: *** for 99.9% confidence level of significance, ** for 99% confidence level of significance, * for 95% confidence level of significance, + for 90% confidence level of significance and NS for no significance.

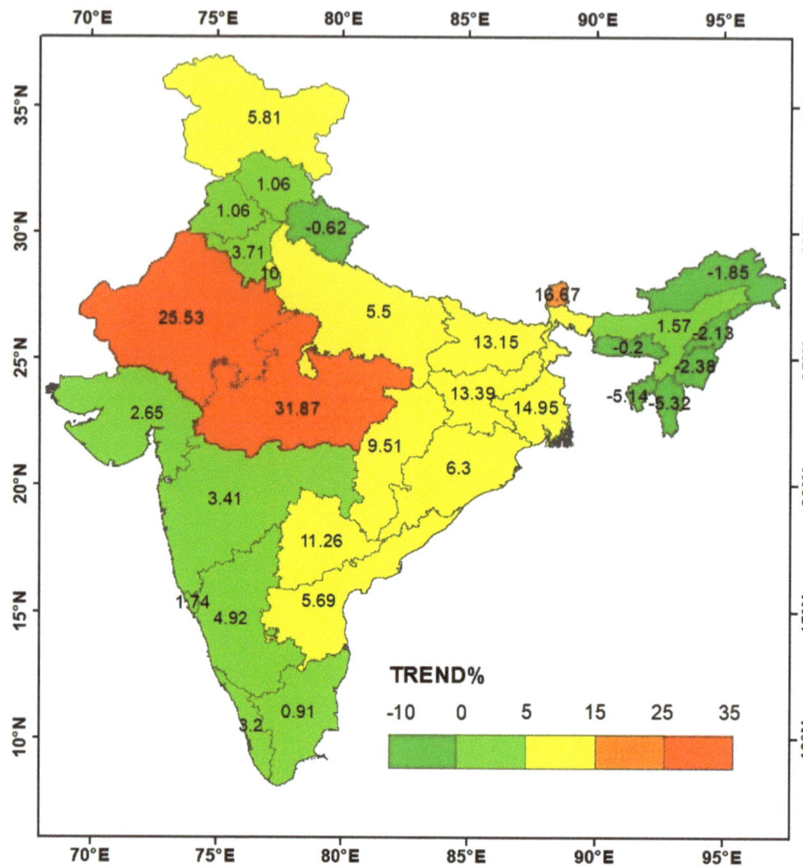

Fig. 7.5 ***State-wise long-term annual trends of fire occurrences in India during 2003–2018.***

in Madhya Pradesh (31.87 yr^{-1}), Rajasthan (25.53 yr^{-1}) and Sikkim (16.67 yr^{-1}) while minimum trend is observed in Mizoram (–5.32 yr^{-1}), Tripura (–5.13 yr^{-1}), and Manipur (–2.38 yr^{-1}). Based on the annual fire count data, the increasing trend of 11–31% yr^{-1} with 95% confidence level of significance were found in Madhya Pradesh, Rajasthan, Sikkim, West Bengal, Jharkhand, Bihar, and Telangana states respectively. On the other hand, the decreasing trends of 2–5% yr^{-1} were found for the states of Mizoram, Tripura, Manipur, Nagaland, and Arunachal Pradesh in decreasing order, respectively. However, no significant trends were observed in Tamil Nadu, Meghalaya, and Uttarakhand. Long-term MODIS fire data over the entire India region show an increasing trend with a rate (10–31 yr^{-1}), which covers about 75% of total area of India except north-east states. Interestingly, the north-east states are showing decreasing trend (2–5% yr^{-1}) where the deciduous forest is majorly covered. These results are consistent with earlier researchers and found declining trend of forest fires in Nilgiri Reserve, north-east states and some other parts of India [47].

7.3.4 Identification of fire hotspot regions

For quantification of fire types (forest fire or crop residue burning) and identification of fire hotspot regions we have used MODIS annual land cover product over Indian region during 2003–2018. Fig. 7.6A shows the example of reclassified LULC map of India for the year of 2017 derived from MODIS IGBP classification and Fig. 7.6B shows the MODIS active fire density map for the 16 years data (2003–2018). Based on the long-term data there were seven frequently BB occurring (hotspot) regions identified over Indian region (Table 7.2). From the Fig. 7.6B, it is observed that most of the agricultural fires are from Punjab, Haryana as well as some parts of the Uttar Pradesh (Regions 1 and 2). Very dense forests in the north-east states along with foot hills of Himalaya states (Uttarakhand and Himachal Pradesh) show higher forest fire occurrences (Region 3). According to ISFR-2017 report, forests cover nearly 24.39% of total country area, which predominantly is occupied by deciduous forests and about 50% forest is prone to fires annually. More number of forest fires occurred in Odisha state (Region 4), Andhra Pradesh, Telangana, and Chhattisgarh states (Region 5) are attributed due to livelihood dependence of tribal people and regional shifting cultivation. BB shows increasing trend in Madhya Pradesh and Maharashtra (Region 6 and 7) due to several factors like meteorological conditions, postharvest burning and land use practices.

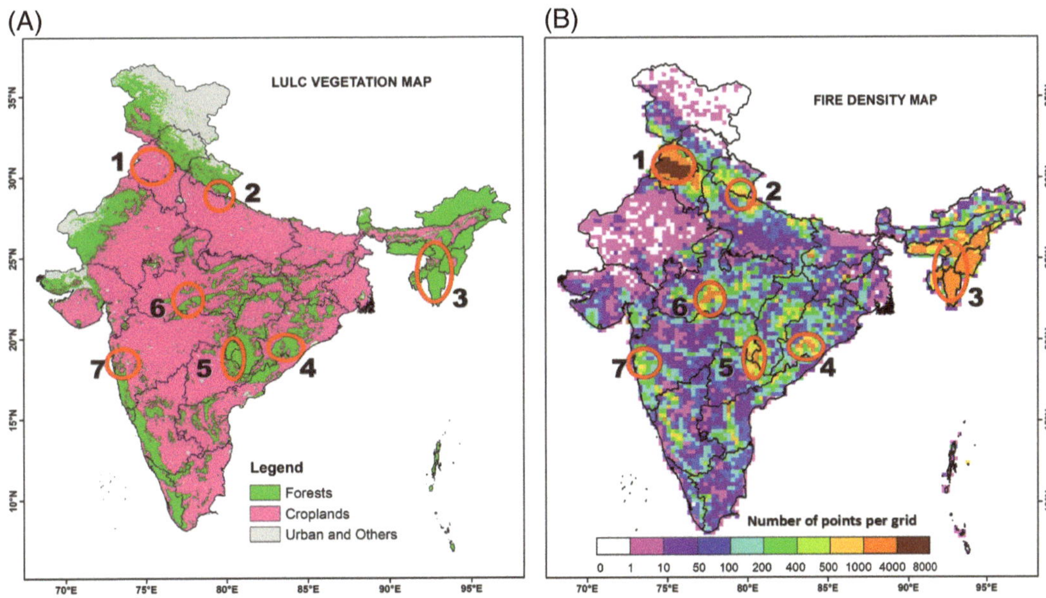

Fig. 7.6 (A) LULC reclassified map over Indian region generated from IGBP land cover type classification during 2017 and (B) MODIS fire density (no. of fires per grid) over Indian region during 2003–2018.

Table 7.2 Climatological hotspot regions of fire occurrences over Indian region.

Region	Land type	BB fire period	States (districts)
Region 1	Cropland	Dual phase (April–May and October–November)	Punjab (all districts), Haryana (Fatehabad, Karnal, Kaithal, and Sirsa)
Region 2	Forest and cropland	Single phase (February–May)	Uttarakhand (Nainital, Pauri Garhwal, Tehri Garhwal, and Udham Singh Nagar), Uttara Pradesh (Lakhimpur Kheri and Pilibhit)
Region 3	Forest	Single phase (February–May)	Assam (Karbi Anglong, Nalbari, and North Cachar Hills), Manipur (Churachandpur, Tamenglong, and Ukhrul), Meghalaya (East Garo Hills, Ri-Bhoi, and West Khasi Hills), Mizoram (Aizawl, Champhai, Mamit Lawngtlai, and Lunglei), and Nagaland (Kohima, Mokokchung, Mon, and Tuensang)
Region 4	Forest	Single phase (February–May)	Orissa (Gajapathi, Kandhamal, and Rayagada)
Region 5	Forest	Single phase (February–May)	Chhattisgarh (Bastar, Dantewada, Kanker, and Raipur) Maharashtra (Chandrapur and Garhchiroli), Telangana (Khammam and Warangal)
Region 6	Forest and cropland	Dual phase (April–May and October–November)	Madhya Pradesh (Harda, Hoshangabad, Raisen, and Sehore)
Region 7	Forest	Single phase (February–April)	Maharashtra (Pune, Raigarh, Ratnagiri, Satara, and Thane)

Agriculture fires are more in the states of Punjab and Haryana (contributing 88.13% of total crop residue fires in IGP region). These agriculture residue fires are major cause of air pollution in entire IGP region [2]. Most of the forest fires occurred in the months of April and May. Chirpine and deciduous forests are more prone to fire during premonsoon season [7]. Based on LULC annual classification, annual total fires detected over Indian region is reported in Table 7.3. Results suggested that the fraction of all type of irrigated cropland fires ranged from 39% to 55% on an all-India basis, in contrast, the all type of forest fires range from 40% to 57% and had a strong regional and seasonal variation. Similar results obtained over Indian region in the year of 2010, they found that all irrigated croplands have high fire counts (59%) followed by closed to open shrub-land category (13.88%) and closed to open broad leaf evergreen or semi-deciduous forest (13.63%) [39].

Table 7.3 Total annual active fires detected by MODIS over India during 2003–2018 and the number of fires detected per year and percentage of fire occurrences of forest and crop lands derived based on the MODIS annual land cover map aggregated into three broad land cover classes during 2003–2018.

Year	Total fire (count)	Forest fires (count)	Fire occurrence (%)	Cropland fires (count)	Fire occurrence (%)	Urban and other fires (count)	Fire occurrence (%)
2003	56,177	30,638	54.54	24,481	43.58	1058	1.88
2004	64,845	37,912	58.47	25,688	39.61	1245	1.92
2005	64,066	30,091	46.97	31,883	49.77	2092	3.27
2006	66,353	37,948	57.19	26,268	39.59	2137	3.22
2007	75,448	39,217	51.98	34,071	45.16	2160	2.86
2008	71,266	30,765	43.17	38,186	53.58	2315	3.25
2009	90,696	52,235	57.59	35,998	39.69	2463	2.72
2010	76,983	40,248	52.28	33,890	44.02	2845	3.70
2011	72,567	30,317	41.78	39,419	54.32	2831	3.90
2012	93,570	46,599	49.80	44,568	47.63	2403	2.57
2013	71,422	30,659	42.93	38,166	53.44	2597	3.64
2014	76,741	33,729	43.95	40,281	52.49	2731	3.56
2015	68,635	28,506	41.53	37,493	54.63	2636	3.84
2016	89,378	37,620	42.09	48,854	54.66	2904	3.25
2017	82,893	33,965	40.97	45,257	54.60	3671	4.43
2018	91,531	36,963	40.38	50,445	55.11	4123	4.50

7.4 Conclusion

Spatial patterns of satellite derived fire occurrences over Indian region have been analyzed using MODIS (combined Terra and Aqua) and VIIRS fire count data during 2003–2018. MODIS land cover type product (MCD12Q1) has been used for identification of fire hotspot regions and segregate the fire type (forest fire or crop residue burning). Present study highlights the spatial variability of BB and hotpot regions which are important in addressing the regional fire mitigation and emission issues in India. The important conclusions of the present research are:

(i) From MODIS and VIIRS, maximum fire occurrences s observed are in two phases: premonsoon (combination forest fires and crop residue burning) and postmonsoon period (crop residue burning).

(ii) On an annual average, the contribution of fire occurrences during premonsoon is observed to be 57% of total fires from MODIS and 61% from VIIRS while during postmonsoon the fire occurrences was observed to be 24% and 19% for MODIS and VIIRS, respectively.

(iii) Monthly fire data show the highest number events of BB during premonsoon as compared to monsoon. Maximum percentage of contribution to total annual fire count observed is in March (29% and 27%), April (18% and 21%), October (13% and 7%), November (11% and 12%), and May (9% and 11%) for MODIS and VIIRS, respectively.

(iv) A significant inter annual variability of fire counts can be observed over Indian region. Long-term average fire counts were found as ~75,786 from MODIS (2003–2018) and ~574,381 from VIIRS (2012–2018).

(v) The fire occurrences over India show a significant (95% confidence level) increasing trend with a rate of 2.7% yr^{-1} and 3.4% yr^{-1} from MODIS and VIIRS, respectively.

(vi) Based on the state-wise analysis, most of the states (8 states) shows the increasing trend (10–30% yr^{-1}) except north-east states (6 states), where the decreasing trend (2–5% yr^{-1}) was observed.

(vii) From long-term (16 years) of MODIS active fire count density, 7 frequently BB occurrence regions were identified over Indian region and annual fraction of forest fires and crop residue burning ranged from 40% to 57% and 39% to 55%, respectively on pan India basis and had a strong regional and seasonal variation.

Acknowledgements

The current research work was carried out under ISRO-GBP project on Aerosol Radiative Forcing over India and authors are grateful to ISRO for financial support. Dr. Shaik Darga Saheb acknowledges ISRO-GBP for providing fellowship to carry out the study at IIRS, Dehradun. The authors gratefully acknowledge the constant encouragement received from Head, MASD and Director, IIRS. The authors would like to thank MODIS and VIIRS science data teams for providing active fire data sets.

References

[1] D.G. Kaskaoutis, S. Kumar, D. Sharma, R.P. Singh, S.K. Kharol, M. Sharma, A.K. Singh, S. Singh, A. Singh, D. Singh, Effects of crop residue burning on aerosol properties, plume characteristics, and long-range transport over northern India, J. Geophys. Res. Atmos. 119 (2014) 5424–5444, doi:10.1002/2013JD021350.

[2] K. Vadrevu, K. Lasko, Fire regimes and potential bioenergy loss from agricultural lands in the Indo-Gangetic plains, J. Environ. Manag. 148 (2015) 10–20, doi:10.1016/j.jenvman.2013.12.026.

[3] K. Vadrevu, K. Lasko, L. Giglio, C. Justice, Vegetation fires, absorbing aerosols and smoke plume characteristics in diverse biomass burning regions of Asia, Environ. Res. Lett. 10 (10) (2015) 105003, doi:10.1088/1748-9326/10/10/105003.

[4] P.S. Roy, Forest fire and degradation assessment using satellite remote sensing and geographic information system, Satellite Remote Sensing and GIS Applications in Agricultural Meteorology. Proceedings of a Training Workshop held 7–11 July 2003 in Dehra Dun, India, 2004 361–400.

[5] D.G. Streets, K.F. Yarber, J.H. Woo, G.R. Carmichael, Biomass burning in Asia: Annual and seasonal estimates and atmospheric emissions, Global Biogeochem. Cycles 17 (4) (2003) 1099, doi:10.1029/2003GB002040.

[6] M.S.R. Murthy, K.V.S. Badarinath, B. Gharai, G. Rajshekha, P.S. Ro, Int. Forest Fire News 34 (2006) 72–77.

[7] C. Venkataraman, G. Habib, D. Kadamba, M. Shrivastava, J.F. Leon, B. Crouzille, O. Boucher, D.G. Streets, Emissions from open biomass burning in India: Integrating the inventory approach with high-resolution Moderate Resolution Imaging Spectroradiometer (MODIS) active-fire and land cover data, Global Biogeochem. Cycles 20 (2006) GB2013, doi:10.1029/2005GB002547.
[8] K. Ravindra, T. Singh, S. Mor, Emissions of air pollutants from primary crop residue burning in India and their mitigation strategies for cleaner emissions, J. Clean. Prod. 208 (2019) 261–273, doi:10.1016/j.jclepro.2018.10.031.
[9] A.L. Daniau, P.J. Bartlein, S.P. Harrison, I.C. Prentice, S. Brewer, P. Friedlingstein, T.I. Harrison-Prentice, J. Inoue, K. Izumi, J.R. Marlon, S. Mooney, M.J. Power, J. Stevenson, W. Tinner, M. Andrič, J. Atanassova, H. Behling, M. Black, O. Blarquez, K.J. Brown, C. Carcaillet, E.A. Colhoun, D. Colombaroli, B.A.S. Davis, D. D'Costa, J. Dodson, L. Dupont, Z. Eshetu, D.G. Gavin, A. Genries, S. Haberle, D.J. Hallett, G. Hope, S.P. Horn, T.G. Kassa, F. Katamura, L.M. Kennedy, P. Kershaw, S. Krivonogov, C. Long, D. Magri, E. Marinova, G.M. McKenzie, P.I. Moreno, P. Moss, F.H. Neumann, E. Norstrm, C. Paitre, D. Rius, N. Roberts, G.S. Robinson, N. Sasaki, L. Scott, H. Takahara, V. Terwilliger, F. Thevenon, R. Turner, V.G. Valsecchi, B. Vannière, M. Walsh, N. Williams, Y. Zhang, Predictability of biomass burning in response to climate changes, Global Biogeochem. Cycles 26 (4) (2012) GB4007, doi:10.1029/2011GB004249.
[10] M. Keywood, M. Kanakidou, A. Stohl, F. Dentener, G. Grassi, C.P. Meyer, K. Torseth, D. Edwards, A.M. Thompson, U. Lohmann, J. Burrows, Fire in the air: Biomass burning impacts in a changing climate, Crit. Rev. Environ. Sci. Technol. 43 (1) (2013) 40–83, doi:10.1080/10643389.2011.604248.
[11] J.R. Marlon, P.J. Bartlein, C. Carcaillet, D.G. Gavin, S.P. Harrison, P.E. Higuera, F. Joos, M.J. Power, I.C. Prentice, Climate and human influences on global biomass burning over the past two millennia, Nat. Geosci. 1 (10) (2008) 697–702, doi:10.1038/ngeo313.
[12] M.Z. Jacobson, Effects of biomass burning on climate, accounting for heat and moisture fluxes, black and brown carbon, and cloud absorption effects, J. Geophys. Res. 119 (2014) 1–8, doi:10.1002/2013JD021860.
[13] K. Bali, A. Kumar Mishra, S. Singh, Impact of anomalous forest fire on aerosol radiative forcing and snow cover over Himalayan region, Atmos. Environ. 150 (2016) 264–275, doi:10.1016/j.atmosenv.2016.11.061.
[14] E. Chuvieco, R.G. Congalton, Application of remote sensing and geographic information systems to forest fire hazard mapping, Remote Sens. Environ. 159 (29) (1989) 147–159, doi:10.1016/0034-4257(89)90023-0.
[15] L.K. Sahu, V. Sheel, K. Pandey, R. Yadav, P. Saxena, S. Gunthe, Regional biomass burning trends in India: Analysis of satellite fire data, J. Earth Syst. Sci. 124 (7) (2015) 1377–1387, doi:10.1007/s12040-015-0616-3.
[16] S. Sarkar, R.P. Singh, A. Chauhan, Crop residue burning in northern India: Increasing threat to greater India, J. Geophys. Res. Atmos. 123 (13) (2018) 6920–6934, doi:10.1029/2018JD028428.
[17] K.V. Suresh Babu, A. Roy, P. Ramachandra Prasad, Forest fire risk modeling in Uttarakhand Himalaya using TERRA satellite datasets, Eur. J. Remote Sens. 49 (August) (2016) 381–395, doi:10.5721/EuJRS20164921.
[18] K. Vadrevu, K. Lasko, Intercomparison of MODIS AQUA and VIIRS I-band fires and emissions in an agricultural landscape-implications for air pollution research, Remote Sens. 10 (7) (2018) 978, doi:10.3390/rs10070978.
[19] D.S. Shaik, Y. Kant, D. Mitra, S.S. Babu, Assessment of aerosol characteristics and radiative forcing over northwest Himalayan region, IEEE J. Selected Top. Appl. Earth Observ. Remote Sens. 10 (12) (2017) 5314–5321, doi:10.1109/JSTARS.2017.2749481.
[20] A.J. Stern, P.C. Doraiswamy, E. Raymond Hunt, Comparison of different MODIS data product collections over an agricultural area, Remote Sens. Lett. 5 (1) (2014) 1–9, doi:10.1080/2150704X.2013.862600.
[21] C. Ichoku, D. A. Chu, S. Mattoo, Y. J. Kaufman, L. A. Remer, D. Tanré, I. Slutsker, A spatio-temporal approach for global validation and analysis of MODIS aerosol products, Geophysical Research Letters 29 (12) (2002) 1616, doi:10.1029/2001GL013206.
[22] R.C. Levy, S. Mattoo, L.A. Munchak, L.A. Remer, A.M. Sayer, F. Patadia, N.C. Hsu, The collection 6 MODIS aerosol products over land and ocean, Atmos. Meas. Tech. 6 (11) (2013) 2989–3034, doi:10.5194/amt-6-2989-2013.

[23] J.T. Morisette, J.L. Privette, C.O. Justice, A framework for the validation of MODIS land products, Remote Sens. Environ. 83 (83) (2002) 77–96.
[24] J.R.G. Townshend, C.O. Justice, Towards operational monitoring of terrestrial systems by moderate-resolution remote sensing, Remote Sens. Environ. 83 (1–2) (2002) 351–359, doi:10.1016/S0034-4257(02)00082-2.
[25] I. Csiszar, L. Denis, L. Giglio, C.O. Justice, J. Hewson, Global fire activity from two years of MODIS data, Int. J. Wildland Fire 14 (2) (2005) 117–130, doi:10.1071/WF03078.
[26] E.H. Chowdhury, Q.K. Hassan, Development of a new daily-scale forest fire danger forecasting system using remote sensing data, Remote Sens. 7 (3) (2015) 2431–2448, doi:10.3390/rs70302431.
[27] R.E. Wolfe, G. Lin, M. Nishihama, K.P. Tewari, J.C. Tilton, A.R. Isaacman, Suomi NPP VIIRS prelaunch and on-orbit geometric calibration and characterization, J. Geophys. Res. Atmos. 118 (20) (2013) 11508–11521, doi:10.1002/jgrd.50873.
[28] W. Schroeder, P. Oliva, L. Giglio, I.A. Csiszar, The new VIIRS 375m active fire detection data product: Algorithm description and initial assessment, Remote Sens. Environ. 143 (2014) 85–96, doi:10.1016/j.rse.2013.12.008.
[29] I. Csiszar, W. Schroeder, L. Giglio, E. Ellicott, K.P. Vadrevu, C.O. Justice, B. Wind, Active fires from the Suomi NPP visible infrared imaging radiometer suite: Product status and first evaluation results, J. Geophys. Res. 119 (2) (2014) 803–816, doi:10.1002/2013JD020453.
[30] E. Vermote, C. Justice, I. Csiszar, Early evaluation of the VIIRS calibration, cloud mask and surface reflectance earth data records, Remote Sens. Environ. 148 (2014) 134–145, doi:10.1016/j.rse.2014.03.028.
[31] D. Wang, S. Liang, T. He, Y. Yu, Direct estimation of land surface albedo from VIIRS data: Algorithm improvement and preliminary validation, J. Geophys. Res. Atmos. 118 (22) (2013) 12577–12586, doi:10.1002/2013JD020417.
[32] T.J. Hawbaker, V.C. Radeloff, A.D. Syphard, Z. Zhu, S.I. Stewart, Detection rates of the MODIS active fire product in the United States, Remote Sens. Environ. 112 (5) (2008) 2656–2664, doi:10.1016/j.rse.2007.12.008.
[33] P. Sofan, D. Bruce, W. Schroeder, E. Jones, J. Marsden, Assessment of VIIRS 375 m active fire using tropical peatland combustion algorithm applied to Landsat-8 over Indonesia's peatlands, Int. J. Digital Earth 13 (12) (2020) 1695–1716, doi:10.1080/17538947.2020.1791268.
[34] M.A. Friedl, D. Sulla-Menashe, B. Tan, A. Schneider, N. Ramankutty, A. Sibley, X. Huang, MODIS collection 5 global land cover: Algorithm refinements and characterization of new datasets, Remote Sens. Environ. 114 (1) (2010) 168–182, doi:10.1016/j.rse.2009.08.016.
[35] P. Acharya, M. Punia, Comparison of MODIS derived land use and land cover with Ministry of Agriculture reported statistics for India, J. Appl. Remote Sens. 7 (1) (2013) 073524, doi:10.1117/1.jrs.7.073524.
[36] V. Dheeravath, P.S. Thenkabail, G. Chandrakantha, P. Noojipady, G.P.O. Reddy, C.M. Biradar, M.K. Gumma, M. Velpuri, Irrigated areas of india derived using MODIS 500 m time series for the years 2001–2003, ISPRS J. Photogramm. Remote Sens. 65 (1) (2010) 42–59, doi:10.1016/j.isprsjprs.2009.08.004.
[37] S. Paul, S. Ghosh, R. Oglesby, A. Pathak, A. Chandrasekharan, R. Ramsankaran, Weakening of Indian summer monsoon rainfall due to changes in land use land cover, Sci. Rep. 6 (December 2015) (2016) 1–10, doi:10.1038/srep32177.
[38] P.S. Roy, A. Roy, P.K. Joshi, M.P. Kale, V.K. Srivastava, S.K. Srivastava, R.S. Dwevidi, C. Joshi, M.D. Behera, P. Meiyappan, Y. Sharma, A.K. Jain, J.S. Singh, Y. Palchowdhuri, R.M. Ramachandran, B. Pinjarla, V. Chakravarthi, N. Babu, M.S. Gowsalya, P. Thiruvengadam, M. Kotteeswaran, V. Priya, K.M.V.N. Yelishetty, S. Maithani, G. Talukdar, I. Mondal, K.S. Rajan, P.S. Narendra, S. Biswal, A. Chakraborty, H. Padalia, M. Chavan, S.N. Pardeshi, S.A. Chaudhari, A. Anand, A. Vyas, M.K. Reddy, M. Ramalingam, R. Manonmani, P. Behera, P. Das, P. Tripathi, S. Matin, M.L. Khan, O.P. Tripathi, J. Deka, P. Kumar, D. Kushwaha, Development of decadal (1985-1995-2005) land use and land cover database for India, Remote Sens. 7 (3) (2015) 2401–2430, doi:10.3390/rs70302401.
[39] K.P. Vadrevu, I. Csiszar, E. Ellicott, L. Giglio, K.V.S. Badarinath, E. Vermote, C. Justice, Hotspot analysis of vegetation fires and intensity in the Indian region, IEEE J. Selected Top. Appl. Earth Observ. Remote Sens. 6 (1) (2013) 224–238, doi:10.1109/JSTARS.2012.2210699.

[40] D.S. Shaik, Y. Kant, D. Mitra, A. Singh, H.C. Chandola, M. Sateesh, S.S. Babu, P. Chauhan, Impact of biomass burning on regional aerosol optical properties: A case study over northern India, J. Environ. Manag. 244 (2019) 328–343, doi:10.1016/j.jenvman.2019.04.025.
[41] V. Mishra, Long-term (1870–2018) drought reconstruction in context of surface water security in India, J. Hydrol. 580 (2020) 124228, doi:10.1016/j.jhydrol.2019.124228.
[42] K. Pangaluru, I. Velicogna, A. Geruo, Y. Mohajerani, E. Ciracì, S. Charakola, G. Basha, S. Vijaya Bhaskara Rao, Soil moisture variability in India: Relationship of land surface-atmosphere fields using maximum covariance analysis, Remote Sens. 11 (3) (2019) 335, doi:10.3390/rs11030335.
[43] D.S. Shaik, Y. Kant, D. Mitra, H.C. Chandola, S. Suresh Babu, Long-term variability of MODIS 3 km aerosol optical depth over Indian region, ISPRS Ann. Photogramm. Remote Sens. Spatial Inform. Sci. 4 (2018) 359–366.
[44] T. Liu, M.E. Marlier, A. Karambelas, M. Jain, S. Singh, M.K. Singh, R. Gautam, R.S. Defries, Missing emissions from post-monsoon agricultural fires in northwestern India: Regional limitations of modis burned area and active fire products, Environ. Res. Commun. 1 (1) (2019) 11007, doi:10.1088/2515-7620/ab056c.
[45] R. Krishan, B.R. Nikam, S.M. Pingale, A. Chandrakar, D. Khare, Analysis of trends in rainfall and dry/wet years over a century in the eastern Ganga canal command, Meteorol. Appl. 25 (4) (2018) 561–574, doi:10.1002/met.1721.
[46] E.R. Ronay, S.F.M. Breitenbach, J.L. Oster, Sensitivity of speleothem records in the Indian summer monsoon region to dry season infiltration, Sci. Rep. 9 (1) (2019) 1–10, doi:10.1038/s41598-019-41630-2.
[47] C. Sudhakar Reddy, K.V. Satish, P.V.V. Prasada Rao, Significant decline of forest fires in Nilgiri biosphere reserve, India, Remote Sens. Appl. Soc. Environ. 11 (October 2017) (2018) 172–185, doi:10.1016/j.rsase.2018.07.002.

CHAPTER 8

Identification of different aerosol types over a semi-arid location in southern peninsular India retrieved from the CALIPSO

Raja Obul Reddy Kalluri, Rama Gopal Kotalo
Aerosol & Atmospheric Research Laboratory, Department of Physics, Sri Krishnadevaraya University, Anantapur, Andhra Pradesh, India

8.1 Introduction

Aerosols play a crucial role in atmospheric dynamics and chemistry, weather and climate and hydrological cycle but still represent one of the most significant uncertainties in the assessment of climate change [1–5]. Aerosol sources, on the other hand, are highly dependent on time and location, and their environmental effects have also been explored in-depth, especially in light of rising industrialization and urbanization [6–8]. Furthermore, the presence of atmospheric aerosols scatters and absorbs solar radiation, causing the air to cool and warm, which is strongly influenced not only by aerosol load but also by their optical properties and chemical composition, and by cloud cover and the albedo of the surface below [9–12]. Therefore, the spatio-temporal distribution of aerosol types is challenging for understanding atmospheric processes, improving aerosol algorithms, and providing useful inputs to regional and global chemical-transport models [13–15]. In this connection, efforts have been made significantly to monitor and predict aerosol sources, sinks, and movements through the use of satellites [16,17]. Furthermore, there has been extensive research on regional aerosol characteristics [18–20], particularly in areas that have a strong influence on climate and pose serious environmental concerns [21–25].

The optical depth of aerosols is a significant quantity, which describes the total attenuation of solar radiation as it passes through the atmosphere. Remote sensing techniques can classify the different types of aerosols based on particle sizes and aerosol loading. Still, if we were able to accurately classify the basic types, such as urban aerosols, biomass burning aerosols, continental aerosols, dust aerosols, and maritime aerosols, that would prove to be a major achievement. Cloud-Aerosol Lidar and Infrared Pathfinder Satellite Observations (CALIPSO) offer extensive vertical information, including backscatter attenuation and depolarization ratios, and vertical feature masks of both cloud and aerosol layers in different conditions such as bright surfaces, lower-lying clouds, as

Atmospheric Remote Sensing
DOI: https://doi.org/10.1016/B978-0-323-99262-6.00016-X

well as clear sky conditions [26,27]. Aerosol layers in the atmosphere are detected by the CALIPSO satellite, and they are classified into six subtypes (dust, marine, clean continental, biomass burning, polluted dust, and polluted continental). Several researchers have validated Cloud-Aerosol Lidar with Orthogonal Polarization (CALIOP) Version-2 chemical characterization of aerosol types using active remote sensing lidars and Aerosol Robotic Network, and they classification showed fairly consistent for coarse mode dust aerosols and reasonably consistent for polluted continental types [28–30]. Researchers also used the CALIPSO aerosol categorization data to identify specific types of aerosol air masses [31,32]. In India, air pollution concentrations are rapidly increasing due to urbanization, industrialization, the use of fossil fuels, and transportation networks (Manoj et al., 2016; [33,55]). Anantapur is a semi-arid environment in southern peninsular India with an estimated 4.3 million people, which is a hotspot for distinct aerosol sources due to the lack of rainfall and dry continental environment. It is necessary to collect more and better aerosol characteristics and loads to increase our scientific understanding of how aerosols relate to the climate in the semi-arid region, as well as the accuracy of the radiation budget model outputs.

The CALIPSO aerosol layer and profile products were utilized to derive seasonal aerosol characteristics over Anantapur semi-arid region during December 2008–November 2017. The remaining sections of this work are divided into sub-sections. The CALIPSO satellite instrumentation and data quality and the site description are described in Section 8.2. Section 8.3 discusses the seasonal variations of various aerosol subtypes, their top layer heights, and the vertical variation of the extinction coefficient during the study period. These findings are discussed in the context of Hybrid Single-Particle Lagrangian Integrated Trajectory (HYSPLIT), The Modern-Era Retrospective analysis for Research and Applications (MERRA-2) winds, and Moderate Resolution Imaging Spectroradiometer (MODIS) fire counts. Finally, Section 8.4 illustrates the main conclusions drawn from the present work.

8.2 Instrument and observational site

8.2.1 CALIPSO

CALIPSO satellite was launched on April 28, 2006, by the National Aeronautics and Space Administration (NASA) and the French Agency, CNES, to study clouds and aerosols impact on the earth's radiation budget. A 16-day repeated cycle was examined when the earth crossed the equator at around 13:30 and 01:30 h local time [27]. Three instruments are on-board the CALIPSO satellite, including CALIOP, the imaging infrared radiometer, and the wide field camera. At 532 and 1064 nm, CALIOP gives daytime and nighttime attenuated backscatter patterns and information on depolarization and color ratios. As part of the CALIOP data analysis package, aerosol and cloud layers have automatically identified and stored as an atmospheric volume description (AVD) flag and a vertical feature mask [34,35]. In addition, the lidar ratio, a critical parameter for extinction retrieval, has recently been revised in CALIOP Version (4.20) for clean marine, dust, clean continental, raised smoke subtypes, and a new "dusty marine" aerosol subtype has been added to the troposphere aerosols. Young et al. [36] provided more detailed information, including exact lidar ratios of aerosol subtypes.

In this study, CALIOP Level 2 (version 4.20), aerosol layer, and aerosol profile products were taken throughout the day. Over the study region, the aerosol layer and profile products are considered grid between 77–78°N and 14–15°E from December 2008 to November 2017. The aerosol layers have been identified at average horizontal resolutions of 5, 20, and 80 km and vertical resolutions of up to 30 m (depending on circumstances and altitude) in the CALIPSO 5-km aerosol layer product [37,38]. The CALIOP 5 km aerosol layer products provide the aerosol optical depth (AOD) parameters in a different layer, aerosol layer top altitude, and Troposphere column AOD at 532 nm. The aerosol subtypes were separated using the CALIOP layer products "Feature Classification Flags." The AOD is assessed using the protocols. (1) Cloud-free conditions. (2) Extinction QC for each aerosol layer for which extinction has been determined. Extinction QC 532 = 0 indicates that each layer's default lidar ratio resulted in a successful extinction solution. Extinction QC = 1 (constrained technique employing two-way transmittance) was utilized in this work. Cloud-aerosol discrimination (CAD) is a routine that distinguishes clouds from aerosols [26,35]. Using a CALIPSO cloud aerosol discrimination algorithm, the CALIPSO provided CAD scores in the range of -100 to 0 for aerosols and 0 to +100 for clouds. The larger absolute value of the CAD score indicates higher confidence in the featured classification. When a feature has a value of 0, it is equally likely to be a cloud or an aerosol. As a result, the no-confidence CALIPSO aerosol retrievals were removed from the analysis. Only CALIPSO aerosol retrievals with a CAD score of ≤ −70 were included in this study. (3) The signal-to-noise ratio was lower during the day, resulting in poor layer detection performance. We have omitted optically thin layers with very low AODs (0.1≤ AOD532 nm ≤ 3.0) for the current study to reduce uncertainty and improve optical depth quality assessment and reliability. Finally, the quality of the CALIPSO data was controlled using "uncertainty of optical depth at 532 nm." We also employed level 2 (4.20) aerosol profile products (APro) files to collect the extinction coefficient as a function of altitude from December 2008 to November 2017. The following data quality control procedures were applied (1) −100 ≤ CAD ≤ −70, (2) Ext_QC = 0, 1. The study excluded data points where extinction uncertainty calculations disagreed (marked by a value of 99.99 km^{-1} for the estimated uncertainty). CALIPSO data quality summaries and the CALIPSO users guide (http://www-calipso.larc.nasa.gov/resources/calypso-users-guide/) are good resources for detailed discussions of CALIOP accuracy, uncertainty, and data quality flags.

8.2.2 HYSPLIT, MERRA-2 winds, and MODIS fire counts

The HYSPLIT model was used to generate backward trajectories to determine where aerosols originate and the transport process [39]. The Air Resources Laboratory (ARL) of the National Oceanic and Atmospheric Administration (NOAA) developed the HYSPLIT model for estimating simple air parcel trajectories. The HYSPLIT model has been widely used to study atmospheric pollutant transport and physical transport processes and identifies emission sources [40,41]. Earlier articles had more information on the cluster analysis [42,43].

Fire Information for Resource Management System (FIRMS) has been used to obtain data on MODIS fire counts (http://firms.modaps.eosdis.nasa.gov/). MODIS (4 mm and 11 mm channels) provides fire count information based on the brightness temperature algorithm. The algorithm categorizes fire pixels into three groups: low confidence (0–30%), nominal confidence (30–80%), and high confidence (>80%). Data with high confidence (>80%) have been considered during the study period. In addition, the UV wind components have been retrieved from NASA's Modern-Era Retrospective Analysis for Research and Applications,Version-2 (MERRA2) re-analysis data during the study period (https://disc.gsfc.nasa.gov/datasets/).

8.2.3 Observational site

Anantapur district lies between 13.4–15.15°N and 76.50–78.30°E, and it is one of the largest in the state of Andhra Pradesh and is located in a semi-arid tropical region over the southern part of the country (Fig. 8.1). The district is a dry, continental region in the southern region of India, and it has the second-lowest rainfall in the country. Several

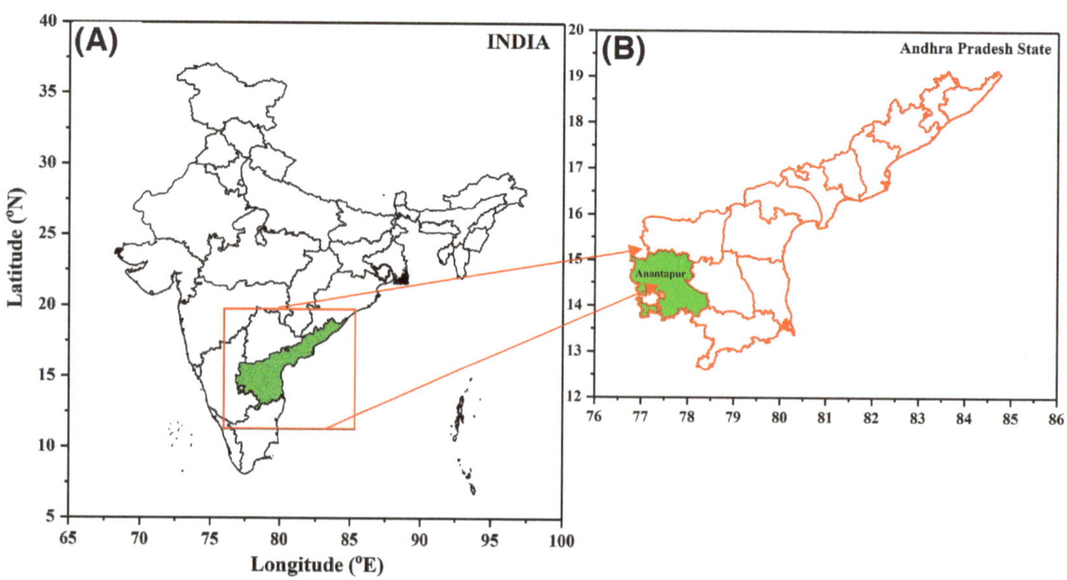

Fig. 8.1 ***Geographical location of Andhra Pradesh (AP) over India represented with a solid circle symbol indicates district map of Anantapur.***

claim polishing industries and dry environmental conditions release dust particles over the study region. Noteworthy that the predominant wind direction is northwesterly/easterly during all seasons (except monsoon), which may carry anthropogenic aerosols to the measurement location from the upper parts of the country. According to the Indian meteorological department, the entire month was separated into four distinct seasons: winter (December–February), hot summer (March–May), south-west monsoon (June–September), and postmonsoon (October–November).

8.3 Results and discussion

8.3.1 Intra-season variability in tropospheric columnar aerosol optical depth

The 9-year time record of intra-seasonally averaged tropospheric columnar aerosol optical depth at 532 nm (AOD) obtained in Anantapur from December 2008 to November 2017 is shown in Fig. 8.2. In order to characterize a truly realistic seasonal cycle, multi-year averages are required. Due to adverse sky circumstances, data gaps in AOD have been noticed. For the whole study period, the intra-annual mean AOD ranged from 0.19 ± 0.09 to 0.38 ± 0.12, with a mean value of 0.31 ± 0.05. The year

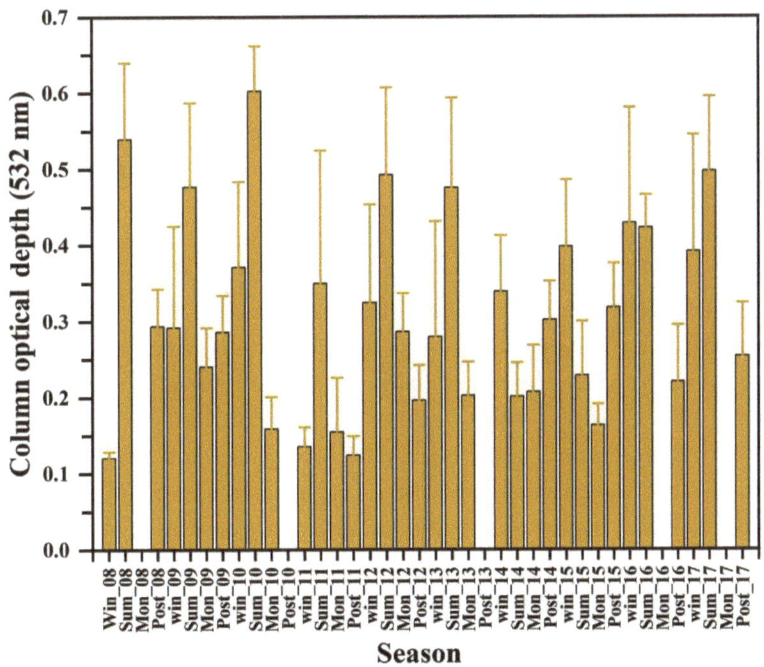

Fig. 8.2 ***Temporal variation of column optical depth observed from the CALIPSO over the study region (error bars correspond to standard deviation).***

2017 had the highest aerosol concentration than the other years, with an AOD value of 0.38 ± 0.12 over the study location. A lower AOD value was observed in 2010, with a mean AOD value of 0.19 ± 0.01. Meanwhile, the intra-seasonal mean ($\pm\sigma$) AOD was 0.31 ± 0.1, 0.43 ± 0.1, 0.20 ± 0.04, and 0.25 ± 0.06 during the winter, summer, monsoon, and postmonsoon, respectively. During summer 2010 recorded the highest AOD value (0.60 ± 0.05), whereas winter 2008 recorded the lowest (0.12 ± 0.01). The observed seasonal pattern of AOD over the measurement site is fairly consistent with ground-based measurements of AOD [9]. Apart from the local sources, the continental air masses generally originated from the Indo Gangetic Plain (IGP) region during the winter and postmonsoon season over the observation site. The IGP region was characterized by vast amounts of anthropogenic aerosols, primarily generated from vehicular emissions and industrial units that exist in the region. An increase in AOD during the summer results from the increased availability of small particles generated by agricultural burning activities [44]. The highest air surface temperatures also facilitate lifting loose soil particles during the summer. In addition to dust particles from local sources, northwest winds carried dust particles from the Thar Desert to the measurement site [45]. On the other hand, there are low AOD values during the monsoon because precipitation is high and anthropogenic activities are low [24].

8.3.2 Seasonal variation of different aerosol subtypes

Fig. 8.3 illustrates the frequency distribution of aerosol layer optical depth (LAOD) from December 2008 to November 2017 over Anantapur. Fig. 8.3 shows that polluted dust optical depths were mostly recorded (LAOD < 0.6), with relative frequencies was about 56%, 38%, and 39% of total seasonal observations throughout summer, monsoon, and postmonsoon, respectively. Similarly, throughout the monsoon, aerosol dust layers were seen (with LAOD < 0.9), and their relative frequencies was about 26% of total seasonal observations, respectively (Fig. 8.3C). The polluted continental layer optical depths were detected (LAOD < 0.5) in the winter and postmonsoon, accounting for 50% and 28% of total seasonal observations, respectively (Fig. 8.3A and D).

As shown in Table 8.1, polluted dust is the predominant aerosol type, representing the mix of smoke and dust. Their percent contributions varied greatly, with the highest in summer (69%), followed by monsoon (53.3%), postmonsoon (44.7%), and finally winter (24.1%). Polluted continental is prevalent in winter and postmonsoon, and their percent contributions varied greatly, with the highest percentages in the winter (63.6%) and postmonsoon (37.8%) seasons. In winter, the winds originate from northeasterly, which brings anthropogenic emissions from human activities and combustion of fossil fuels and biofuels/biomass. Smoke aerosol layers are prevalent in all seasons, with nearly identical percentages of smoke aerosols in the winter (11.4%), summer (8.5%), monsoon (7.1%), and postmonsoon (15.1%) seasons. Similarly, dust layers played a significant role, accounting for 32%, 11.6% of the total in the monsoon and summer. There is a

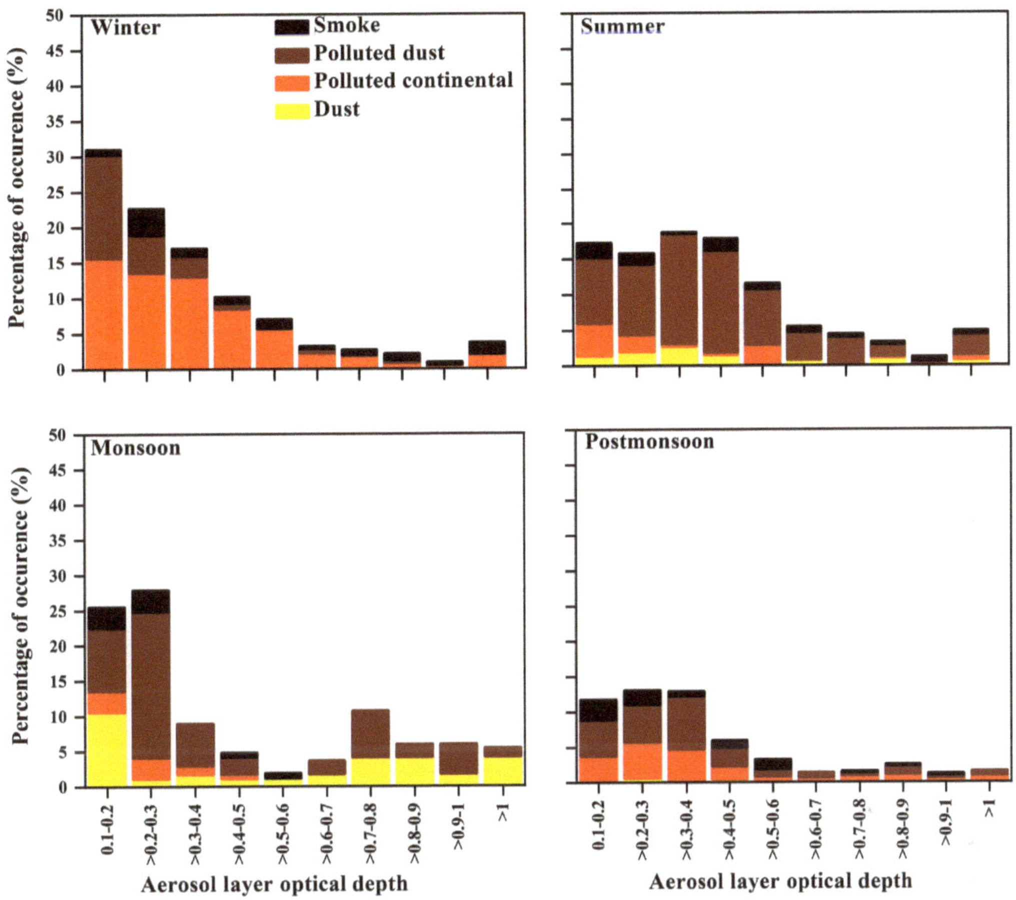

Fig. 8.3 ***Frequency distributions of aerosol layered optical depths obtained at the wavelength of 532 nm for different seasons over Anantapur.***

predominant north-westerly and south-westerly wind direction at the station during summer and monsoon seasons. Dust aerosol originated from the Saharan deserts provides a mechanism for injecting large amounts of material into the free troposphere and transporting it across great distances. As a result, dust and soot from complicated aerosol loading in the troposphere are substantially combined (summer).

8.3.3 Percentage contribution of aerosol top layer heights

Fig. 8.4 depicts frequency distributions of top layer heights for different aerosol types during the study period. These are the only layers examined since the remaining layers have relatively low percentages. As shown in Fig. 8.4A, the surface-attached polluted continental aerosol layer accounted for the highest proportion of aerosol loads at the

Table 8.1 Relative seasonal abundance (%) of representative aerosol subtypes obtained at Anantapur. Dust, polluted continental (PC), polluted dust (PD), and smoke.

	AOD bins	0.1-0.2	0.2–0.3	0.3–0.4	0.4–0.5	0.5–0.6	0.6–0.7	0.7–0.8	0.8–0.9	0.9–1.0	>1.0	Total
Winter	**Dust**	0.2	0.5	0.2	–	–	–	–	–	–	–	**0.9**
	PC	15.6	13.1	12.9	8.6	5.8	2.3	1.9	0.9	0.5	2.1	**63.6**
	PD	14.5	5.2	2.8	0.7	–	0.5	–	0.3	–	–	**24.1**
	Smoke	0.7	3.7	1	0.9	1.2	0.3	0.7	0.9	0.3	1.6	**11.4**
Summer	**Dust**	1.47	2.06	2.79	1.62	0.29	0.88	0.29	1.18	0.15	0.88	**11.6**
	PC	4.56	2.35	0.29	0.29	2.65	–	–	0.15	–	0.59	**10.9**
	PD	9.12	9.85	15.44	14.26	7.79	3.82	3.68	1.62	0.44	2.94	**69**
	Smoke	2.06	1.47	0.15	1.62	0.74	0.74	0.44	0.29	0.59	0.44	**8.5**
Monsoon	**Dust**	10.7	1.2	1.8	1.2	1.2	1.8	4.1	4.1	1.8	4.1	**32**
	PC	3	3	1.2	0.6	–	–	–	–	–	–	**7.7**
	PD	8.9	20.7	5.9	2.4	–	1.8	6.5	1.8	4.1	1.2	**53.3**
	Smoke	3	3	–	0.6	0.6	–	–	–	–	–	**7.1**
postmonsoon	**Dust**	0.3	1.2	0	0.9	–	–	–	–	–	–	**2.4**
	PC	6.6	9.4	8.8	3.3	1.5	1.2	1.8	2.1	1.2	1.8	**37.8**
	PD	9.4	9.7	13.6	4.8	1.8	1.2	0.6	2.1	0.6	0.9	**44.7**
	Smoke	5.1	3.6	1.2	1.8	2.4	–	0.3	0.3	0.3	–	**15.1**

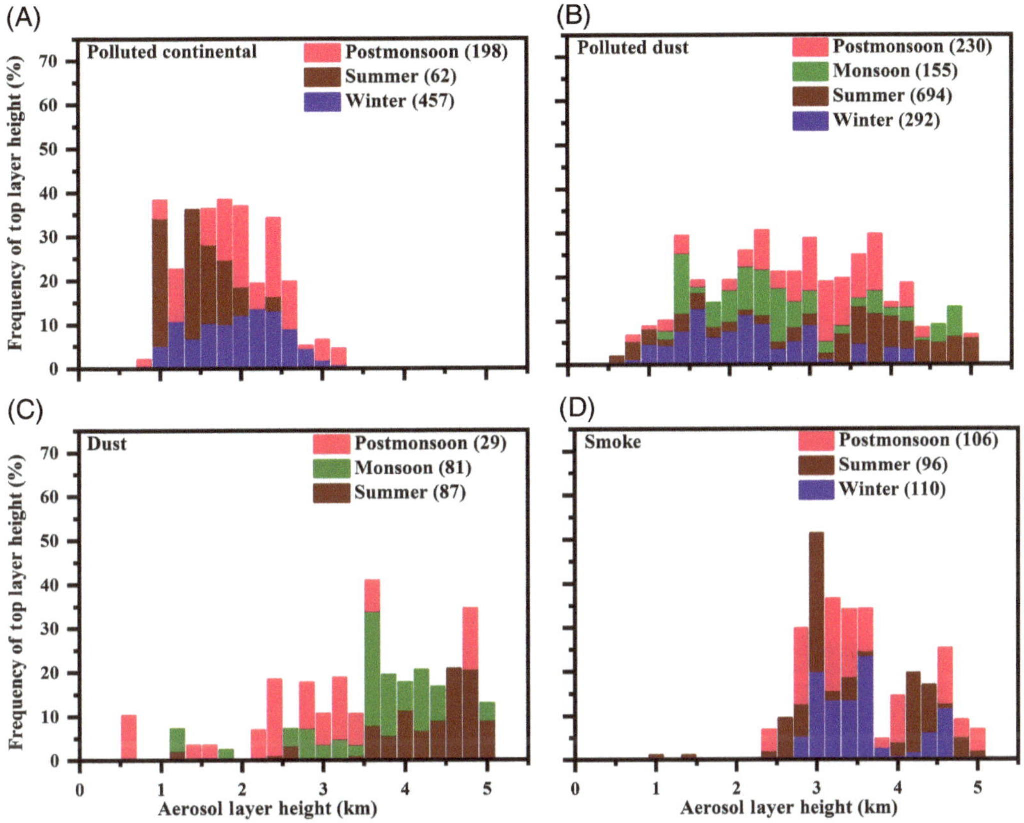

Fig. 8.4 ***Frequency distributions of aerosol top layer altitudes at Anantapur.*** The labels in parenthesis correspond to the number of observations.

lowest altitudes during all seasons. During winter and postmonsoon, around 80% of polluted continental aerosol layers can be found below the boundary layer (2.4 km), owing to a mixture of fine aerosols released from small-scale industries, local vehicles, and pollution transport from northern India. According to Bridhikitti et al. [46], polluted continental top layers below 2 km, indicating afternoon mixing heights over Bangkok. Significant amounts of polluted dust are visible above two kilometers and just a little below two kilometers, regardless of the season. Maximum vertical frequency (>60%) of polluted dust layers above 2 km altitude in all seasons, mainly because of convection induced by localized heating and stronger vertical mixing (Fig. 8.4B). According to Fig. 8.4C, the height of aerosol dust layers was greater than the height of other aerosol subtypes. Monsoon conditions generated convection produced by localized heating and vertical mixing, causing dust to occur more frequently (>70%) above 3-km intervals.

8.3.4 Intra-seasonal variation of the aerosol extinction coefficient

In addition to containing information about aerosols, the extinction coefficient can reflect atmospheric aerosol loads by reflecting the aerosol type and aerosol concentration; consequently, the extinction profile was utilized to examine the vertical aerosol distribution measured at 532 nm in different seasons over Anantapur (Fig. 8.5). During the winter, the obtained extinction coefficient showed the highest (>0.45 km^{-1}) near the surface and which showed a declining trend as altitude increased (Fig. 8.5A). During the winter, high BC mass concentrations resulted in the highest surface aerosol extinction coefficient and shallow the mixed layer height. During the summer, the greatest aerosol extinction coefficient (>0.2 km^{-1}) have been reported in the 0–2.5 km range,

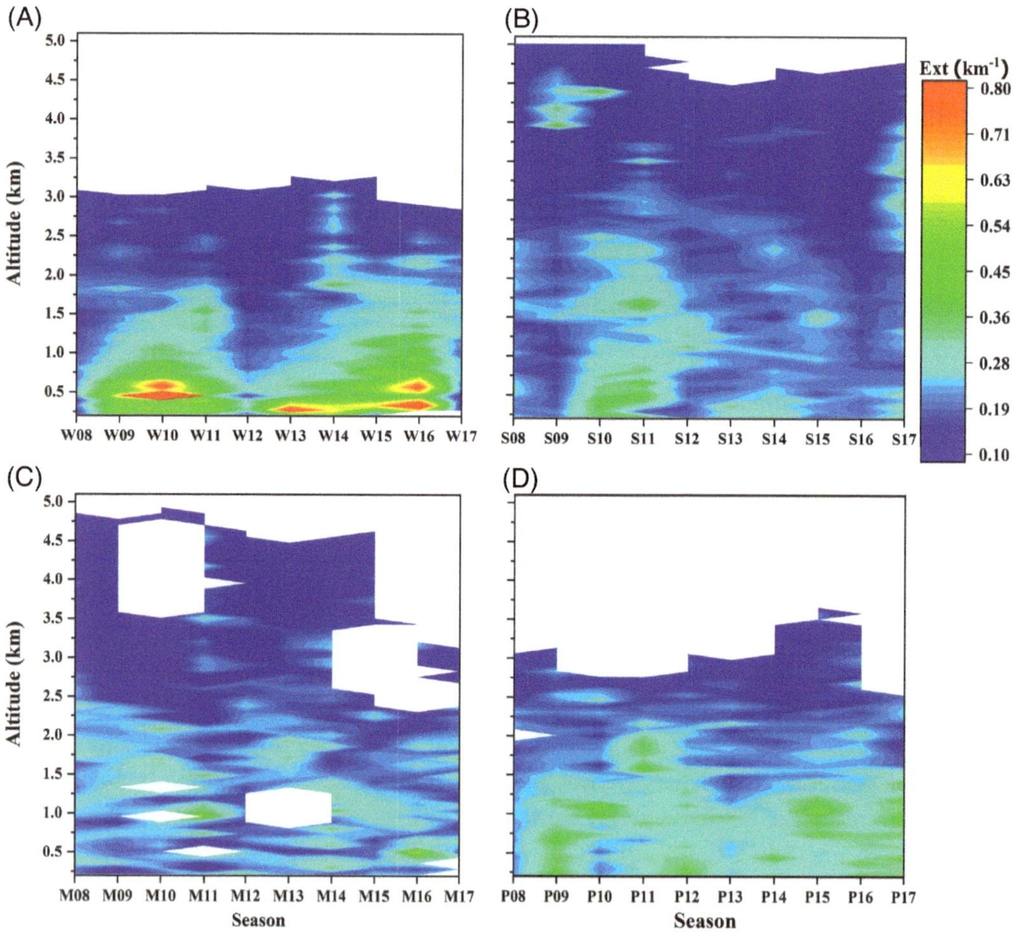

Fig. 8.5 ***Seasonally averaged time series of the vertical extinction coefficient (km^{-1}) at the wavelength of 532 nm during the study period.***

which could be due to strong convection and vertical mixing process boosted particles high altitudes (Fig. 8.5B). We also observed the highest extinction values above the mixed layer height in the summer, mainly due to strong northwest winds carrying dust particles from distant sources. Anantapur dominates mostly northwesterly winds and consistently rising temperatures during the summer. High convection and surface heating conditions are therefore capable of causing dust particles from the soil and those transported onto the surface of the mixture to climb to the top layers and remain suspended for extended periods. In general, during the summer and southwest monsoon seasons, the extinction coefficient in the altitude range 2.5–5 km is >50% greater than during the other two seasons. Postmonsoon aerosol extinction coefficients have similar vertical gradients to those in winter, but near the surface, they are not as high as in winter (Fig. 8.5D). According to Moorthy et al. [47], a campaign conducted throughout the Indian subcontinent and the surrounding marine regions in summer found that most aerosol layers were located between 2 km and 5 km in height throughout India. Satheesh et al. [48,49] reported that aerosol layers above 3 km were found over peninsular India coastal regions and central India. Using high-altitude balloon observations, Padmakumari et al. [50] have reported higher aerosol layers up to 4 km altitude in the Himalayan foothills during CAIPEEX 2009. Furthermore, the highest extinction was recorded over Kanpur in April–June at altitudes between 2 km and 4 km, according to Mishra et al. [51]. Komppula et al. [52] noted that a high extinction coefficient at the surface might be linked to high BC mass concentrations during the winter and postmonsoon.

The vertical distribution of extinction coefficient and their frequency distributions in different altitude ranges over the selected region for all the instantaneous data is shown in Fig. 8.6. The extinction histogram peaked between 0 km and 1 km in the winter and postmonsoon, and extinction values in the high-altitude ranges (2.0–5 km) peaks in the summer and monsoon (Fig. 8.6B). The accumulated frequency between 0 km and 2.5 km for winter and postmonsoon was about 78% and 68%, respectively. The summer and monsoon frequency distribution is wide, with most extinction values in the range of 2–5 km, accounting for more than 60% of total daily values. During the winter, summer, monsoon, and postmonsoon, the extinction coefficient is normally between 2.5 km to 5 km, contributing to around 22%, 51%, 48%, 32% of total occurrence, respectively. During the summer and monsoon, the frequency distribution indicated a significant percentage (>36%) of the above 3 km range, confirming the transport of dust particles from northwestern winds. Approximately 50% of data are below 2 km are above 2 km during the summer and monsoon season, showing a mixture of fine and coarse mode particles.

8.3.5 Types of air masses and their effects on aerosol characteristics

The HYSPLIT model has been used to determine the 5-day air-mass backward trajectory at the height of 500 m above ground level in order to analyze seasonal variability

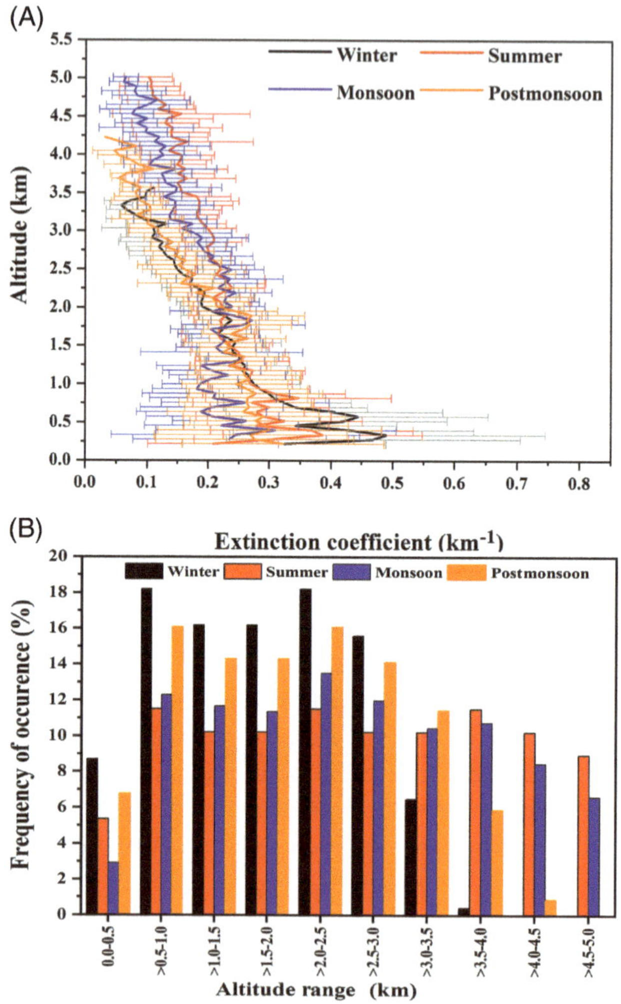

Fig. 8.6 (A) Vertical distribution of Extinction coefficient over Anantapur. (B) Frequency distributions of vertical extinction coefficient at the wavelength of 532 nm in different seasons.

in aerosol sources for the whole period. In addition, multidimensional clustering analysis was utilized to identify various aerosol types in several places [45,53,54]. Fig. 8.7 displays the cluster-mean trajectories for each season and their relative contributions to the total. The three separate cluster-mean trajectories are labelled with separate colors (red-C1, blue-C2, and green-C3) and the percentage contribution of each color to the overall total. As shown in Fig. 8.7A, cluster C1 contains approximately 50% of the back trajectories that originated from the northeast part of India during the winter and are responsible for the dominance of polluted continental and smoke aerosols. Similarly, 19% of the trajectories in Cluster C3 have origins nearly close to the measurement site and

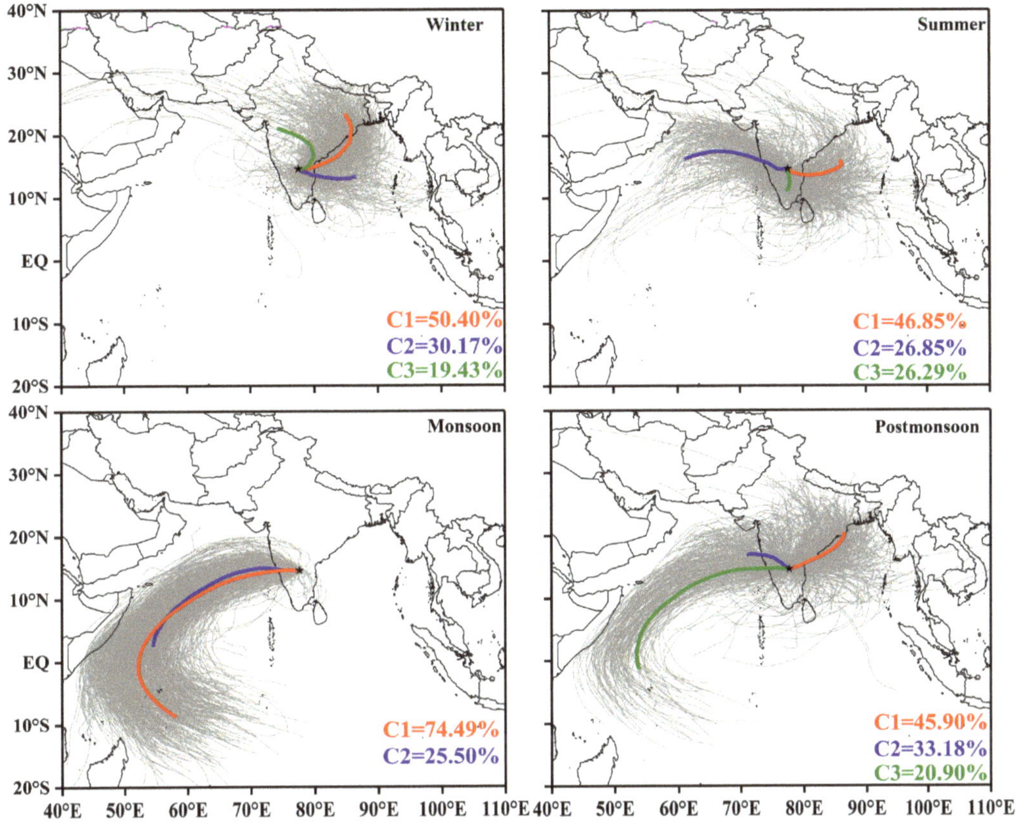

Fig. 8.7 ***Cluster analysis for the 5-day air mass backward trajectories derived from the HYSPLIT model at an altitude of 500 m above the ground level arriving at the Anantapur for different seasons.***

somewhat to the northwest (which may be pollution movement from nearby regions) and local burning activity. The cluster analysis during the summer clearly shows three primary paths. Long-range transport of dust particles originated from border countries are associated with the dominance of polluted dust aerosols in the trajectory group clusters C1 and C2 (72%) (Fig. 8.7B). During the monsoon season, air masses shifted from the northwestern to the southwestern part of India and the cluster groups C1 (74%) and C2 (26%) originate from the Arabian Sea towards the observed region, contributing to the prevalence of dust particles (Fig. 8.7C). The cluster C1 (45%) originates from IGP regions during the postmonsoon season, resulting in a considerable preponderance of fine mode aerosols over Anantapur (Fig. 8.7D). However, cluster C2, which accounted for 33% of all trajectories, travelled short distances, indicating the presence of localized emissions from anthropogenic activities.

8.3.6 Fire count and MERRA-2 winds for source identification

Fig. 8.8 shows the MODIS-derived fire counts merged with MERRA-2 wind fields for each season during the whole period. During the winter, air masses generally originate from the northeasterly. A potential source field of absorbing aerosol supports the high BC mass concentration over the measurement location [9]. Air masses occasionally pass over the Bay of Bengal before arriving at the observation site (Fig. 8.8A). During the summer, red spots detected are likely sources of aerosol absorption in central India, identified by fire count analysis, considerable biomass burning activity, and extensive

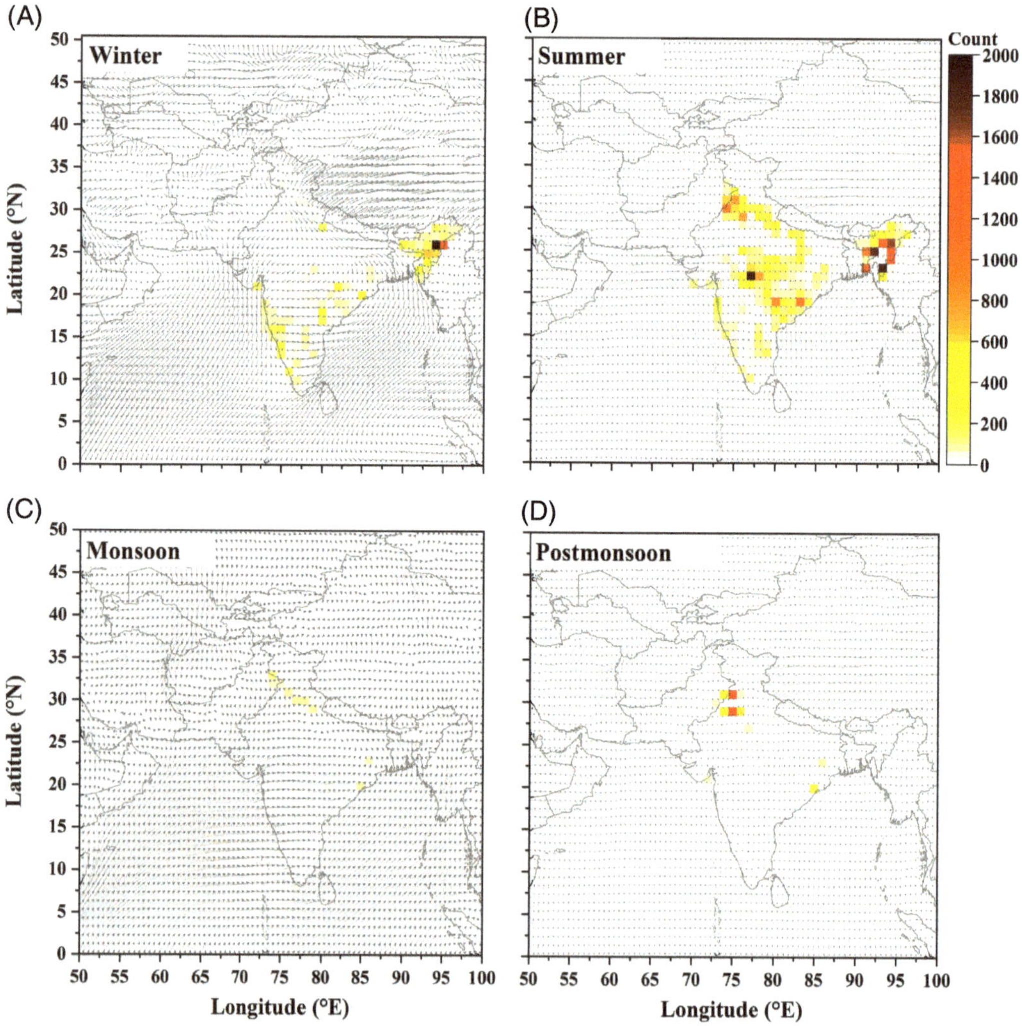

Fig. 8.8 ***Spatial distribution of MODIS cloud corrected fire counts and MERRA-2 winds during the study period over India.***

agricultural areas (Fig. 8.8B). The approach of a north-westerly air mass enhances the transport of mixed aerosols to the observational site, which corresponds to the path indicated by 5 days air mass back trajectories throughout the summer. During the monsoon season, winds mainly originate from the Arabian Sea, and absorbing aerosols over the measuring site is due to local sources (Fig. 8.8C). Synoptic wind patterns originate from central IGP during the postmonsoon, providing an additional absorbing aerosol over the measurement location (Fig. 8.8D).

8.4 Conclusions

Anantapur presents a unique opportunity to research aerosol characterization. The present study analyzed the seasonal variation of different aerosol types and their percentage contributions from December 2008 to November 2017 over Anantapur. The primary conclusions derived from this study can be summarized as follows:

1. The CALIPSO identified that the polluted dust was predominant in summer (69%) and monsoon (53.3%), whereas polluted continental (63.6%) was dominant during the winter season.
2. During winter and postmonsoon, approximately 80%of polluted continental aerosols are trapped below 2.4 km, mainly due to traffic, industrial emissions. Meanwhile, more than 60% of polluted dust aerosol layers were observed above 2 km during summer, mainly due to strong northwest winds transporting dust particles from the Thar desert and absorbing aerosols originated from central India.
3. Due to the local anthropogenic activities and long-range transport, more than 60% of extinction coefficient distributions are above 2km during the summer and southwest monsoon; however, during the winter and postmonsoon, the majority of accumulated frequencies between 0 km and 2.5 km is about 78% and 68%, respectively.
4. During the winter and postmonsoon, air masses originate from IGP, which provides additional sources of absorbing aerosols over the measurement site.

References

[1] S.S. Babu, M.R. Manoj, K.K. Moorthy, M.M. Gogoi, V.S. Nair, S.K. Kompalli, S.K. Satheesh, K. Niranjan, K. Ramagopal, P.K. Bhuyan, D. Singh, Trends in aerosol optical depth over Indian region: Potential causes and impact indicators, J. Geophys. Res. 118 (2013) 11794–11806.

[2] IPCC, 2021: Climate Change 2021: The Physical Science Basis. Contribution of Working Group I to the Sixth Assessment Report of the Intergovernmental Panel on Climate Change. Masson-Delmotte, V. P. Zhai, A. Pirani, S.L. Connors, C. Péan, S. Berger, N. Caud, Y. Chen, L. Goldfarb, M.I. Gomis, M. Huang, K. Leitzell, E. Lonnoy, J.B.R. Matthews, T.K. Maycock, T. Waterfield, O. Yelekçi, R. Yu, and B. Zhou (eds.). Cambridge: Cambridge University Press.

[3] S. Ramachandran, R. Srivastava, S. Kedia, T.A. Rajesh, Contribution of natural and anthropogenic aerosols to optical properties and radiative effects over an urban location, Environ. Res. Lett. 7 (2011) 034028.

[4] V. Ramanathan, G. Carmichael, Global and regional climate changes due to black carbon, Nat. Geosci 1 (2008) 221–227. doi:10.1038/ngeo156.

[5] V. Ramanathan, P.J. Crutzen, J.T Kiehl, D. Rosenfeld, Aerosol, climate and hydrological cycle, Science 294 (2001) 2119–2124.
[6] U.C. Dumka, D.G. Kaskaoutis, S. Tiwari, P.D. Safai, S.D. Attri, V.K. Soni, N. Singh, N. Mihalopoulos, Assessment of biomass burning and fossil fuel contribution to black carbon concentrations in Delhi during winter, Atmos. Environ. 194 (2018) 93–109 doi:10.1016/j.atmosenv.2018.09.033.
[7] N.S. Hussain, T. Chakradhar Rao, G. Balakrishnaiah, K. Rama Gopal, K. Raja Obul Reddy, N. Siva Kumar Reddy, R. … Ramakrishna Reddy, Investigation of black carbon aerosols and their characteristics over tropical urban and semi-arid rural environments in peninsular India, J. Atmos. Sol. Terr. Phys. 167 (2018) 48–57.
[8] P.R. Sinha, U.C. Dumka, R.K. Manchanda, D.G. Kaskaoutis, S. Sreenivasan, K. Krishna Moorthy, S. Suresh Babu, Contrasting aerosol characteristics and radiative forcing over Hyderabad, India due to seasonal mesoscale and synoptic-scale processes, Q. J. R. Meteorol. Soc 139 (671) (2012) 434–450.
[9] R.O.R. Kalluri, B. Gugamsetty, R.G. Kotalo, L.R. Thotli, C.R. Tandule, A. Bhavyasree, Long-term (2008–2017) analysis of atmospheric composite aerosol and black carbon radiative forcing over a semi-arid region in southern India: Model results and ground measurement, Atmos. Environ. 240 (2020) 117840.
[10] G. Myhre, F. Stordal, M. Johnsrud, Y.J. Kaufman, D. Rosenfeld, T. Storelvmo, I. Isaksen, Aerosol-cloud interaction inferred from MODIS satellite data and global aerosol models, Atmos. Chem. Phys. 7 (2007) 3081–3101.
[11] D. Rosenfeld, Suppression of rain and snow by urban and industrial air pollution, Science 287 (5459) (2000) 1793–1796.
[12] D. Rosenfeld, Y. Rudich, R. Lahav, Desert dust suppressing precipitation: A possible desertification feedback loop, Proc. Natl. Acad. Sci. 98 (11) (2001) 5975–5980, doi:10.1073/pnas.101122798.
[13] R.C. Levy, S. Mattoo, L.A. Munchak, L.A. Remer, A.M. Sayer, F. Patadia, N.C. Hsu, The Collection 6 MODIS aerosol products over land and ocean, Atmos. Meas. Tech. 6 (2013) 2989–3034.
[14] G. Myhre, B.H. Samset, M. Schulz, Y. Balkanski, S. Bauer, T.K. Berntsen, Bian, et al., Radiative forcing of the direct aerosol effect from AeroCom Phase II simulations, Atmos. Chem. Phys. 13 (2012) 1853–1877.
[15] P.B. Russell, M. Kacenelenbogen, J.M. Livingston, O.P. Hasekamp, J. Redemann, S. Ramachandran, B. Holben, R.E.T. Al, J. Geophys. Res. Atmos. (2014) 9838–9863.
[16] R. Boiyo, K.R. Kumar, T. Zhao, Optical, microphysical and radiative properties of aerosols over a tropical rural site in Kenya, East Africa: Source identification, modification and aerosol type discrimination, Atmos. Environ. 177 (2018) 234–252.
[17] Tandul, C.R. e, R.O.R. Kalluri, B. Gugamsetty, R.G. Kotalo, L.R. Thotli, R.R. Rajuru, S. Vaddin, Decadal climatology of the spatial and vertical distributions of tropospheric aerosol over the Arabian Sea based on satellite observations, Int. J. Climatol. 40 (2020)4676–4689.
[18] G. Begam, V.C. Viswanath, Y.N. Ahammed, K. Kumar, S.S. Babu, R.R. Reddy, Measurement and analysis of black carbon aerosols over a tropical semi-arid station Kadapa, India, Atmos. Res. 171 (2016) 77–91.
[19] U.C. Dumka, R.K. Manchanda, P.R. Sinha, S. Sreenivasan, K.K. Moorthy, S.S. Babu, Temporal variability and radiative impact of black carbon aerosol over tropical urban station Hyderabad, J. Atmos. Sol.-Terr. Phys 105–106 (2013) 81–90.
[20] S.K. Satheesh, V. Vinoj, K.K. Moorthy, Radiative effects of aerosols at an urban location in southern India: observations versus model, Atmos. Environ. 44 (2010) 5295–5304.
[21] G. Carabali, H. Raúl, M. Valdés-barrón, R. Bonifaz-alfonzo, D. Riveros-rosas, V.M. Velasco-herrera, F.A. Vázquez-gálvez, Aerosol climatology over the Mexico City basin: Characterization of optical properties, Atmos. Res. 194 (2017) 190–201.
[22] M. Kumar, M.P. Raju, R.K. Singh, A.K. Singh, R.S. Singh, T. Banerjee, Wintertime characteristics of aerosols over middle Indo-Gangetic plain: Vertical profile, transport and radiative forcing, Atmos. Res. 183 (2017) 268–282.
[23] T.A. Rajesh, S. Ramachandran, Black carbon aerosol mass concentration, absorption and single scattering albedo from single and dual spot aethalometers: Radiative implications, J. Aerosol Sci. 119 (2018) 77–90.
[24] K.R.O. Reddy, G. Balakrishnaiah, K. Gopal, N.S.K. Reddy, T. ChakradharRao, T. Lokeswara Reddy, N.S. Hussain, M.V. Reddy, R.R. Reddy, S.K.R. Boreddy, S.S. Babu, Long term (2007–2013)

observations of columnar aerosol optical properties and retrieved size distributions over Anantapur, India using multi wavelength solar radiometer, Atmos. Environ. 142 (2016) 238–250.
[25] S. Srivastava, M. Kumar, R.S. Singh, B.N. Rai, R.K. Mall, T. Banerjee, Long-term observation of black carbon aerosols at an urban location over the central Indo-Gangetic plain, South Asia, Atmósfera 32 (2) (2019) 95–113.
[26] Y. Hu, D. Winker, M. Vaughan, B. Lin, A. Omar, C. Trepte, D. Flittner, P. Yang, S.L. Nasiri, B. Baum, R. Holz, W. Sun, Z. Liu, Z. Wang, S. Young, K. Stamnes, J. Huang, R. Kuehn, CALIPSO/CALIOP cloud phase discrimination algorithm, J. Atmos. Oceanic Technol. 26 (2009) 2293–2309.
[27] D.M. Winker, J. Pelon, J.A. Coakley Jr., S.A. Ackerman, R.J. Charlson, P.R. Colarco, P. Flamant, Q. Fu, R.M. Hoff, C. Kittaka, T.L. Kubar, H. Le Treut, M.P. McCormick, G. Megie, L. Poole, K. Powell, C. Trepte, M.A. Vaughan, B.A. Wielicki, The CALIPSO mission: A global 3D view of aerosols and clouds, Bull. Am. Meteorol. Soc. 91 (2010) 1211–1229.
[28] S.P. Burton, R.A. Ferrare, M.A. Vaughan, A.H. Omar, R.R. Rogers, C.A. Hostetler, J.W. Hair, Aerosol classification from airborne HSRL and comparisons with the CALIPSO vertical feature mask, Atmos. Meas. Tech. 6 (2013) 1397–1412.
[29] J. Kar, M.A. Vaughan, Z. Liua, A.H. Omar, C.R. Trepte, J. Tackett, T.D. Fairlie, R. Kowch, Detection of pollution outflow from Mexico City using CALIPSO lidar measurements, Remot. Sens. Environ. 169 (2015) 205–211.
[30] R.R. Rogers, C.A. Hostetler, J.W. Hair, R.A. Ferrare, Z. Liu, M.D. Obland, D.B. Harper, A.L. Cook, K.A. Powell, M.A. Vaughan, D.M. Winker, Assessment of the CALIPSO lidar 532 nm Attenuated backscatter calibration using the NASA LaRC airborne high spectral resolution lidar, Atmos. Chem. Phys. 11 (2011) 1295–1311.
[31] M. Ottaviani, B. Cairns, J. Chowdhary, B. Van Diedenhoven, K. Knobelspiesse, C. Hostetler, R. Ferrare, S. Burton, J. Hair, M.D. Obland, R. Rogers, Polarimetric retrievals of surface and cirrus clouds properties in the region affected by the Deep water Horizon oil spill, Remot. Sens. Environ. 121 (2012) 389–403.
[32] Y.H. Yang, C.Y. Chan, J. Tao, M. Lin, G. Engling, Z.S. Zhang, T. Zhang, L. Su, Observation of elevated fungal tracers due to biomass burning in the Sichuan Basin at Chengdu City, China, Sci. Total Environ. 431 (2012) 68–77.
[33] S. Ramachandran, S Kedia, R. Srivastava, Aerosol optical depth trends over different regions of India, Atmos. Environ. 49 (2012) 338–347.
[34] Z. Liu, M.A. Vaughan, D.M. Winker, C.A. Hostetler, L.R. Poole, D. Hlavka, W. Hart, M. McGill, Use of probability distribution functions for discriminating between cloud and aerosol in lidar backscatter data, J. Geophys. Res. 109 (2004) D15202, doi:10.1029/2004JD004732.
[35] Z. Liu, M. Vaughan, D. Winker, C. Kittaka, B. Getzewich, R. Kuehn, A. Omar, K. Powell, C. Trepte, C. Hostetler, The CALIPSO lidar cloud and aerosol discrimination: Version 2 algorithm and initial assessment of performance, J. Atmos. Ocean. Tech. 26 (2009) 1198–1213.
[36] A.S. Young, A.M. Vaughan, A. Garnier, L.J. Tackett, D.J. Lambeth, A.K. Powell, Extinction and optical depth retrievals for CALIPSO's Version 4 data release, Atmos. Meas. Tech. 11 (2018) 5701–5727.
[37] C. Kittaka, D.M. Winker, M.A. Vaughan, A. Omar, L.A. Remer, Inter-comparison of column aerosol optical depths from CALIPSO and MODIS-Aqua, Atmos. Meas. Tech. 4 (2011) 131–141.
[38] M.A. Vaughan, K.A. Powell, R.E. Kuehn, S.A. Young, D.M. Winker, C.A. Hostetler, W.H. Hunt, Z.Y. Liu, M.J. McGill, B.J. Getzewich, Fully automated detection of cloud and aerosol layers in the CALIPSO lidar measurements, J. Atmos. Ocean. Tech. 26 (2009) 2034–2050.
[39] Draxler, R.R., Rolph, G., 2003. Hysplit (Hybrid Single-Particle Lagrangian Integrated Trajectory) Model Access via NoaaArl Ready Website (http://www.arl.noaa.gov/ready/hysplit4.html) NOAA Air Resources Laboratory, Silver Spring.
[40] R.R. Draxler, G. Hess, An overview of the Hysplit-4 modelling system for trajectories, dispersion, and deposition, Aust. Meteorol. Oceanogr. J. 47 (1998) 295–308.
[41] F. Wang, D. Chen, S. Cheng, J. Li, M. Li, Z. Ren, Identification of regional atmospheric PM10 transport pathways using HYSPLIT, MM5-CMAO and synoptic pressure pattern analysis, Environ. Model. Softw. 25 (2010) 927–934.
[42] U.C. Dumka, K.K. Moorthy, R. Kumar, P. Hegde, R. Sagar, P. Pant, N. Singh, S.S. Babu, Characteristics of aerosol black carbon mass concentration over a high altitude location in the central Himalayas from multi-year measurements, Atmos. Res. 96 (2010) 510–521.

[43] G.D. Rolph, Real-Time Environmental Applications and Display System, (READY) Website, NOAA Air Resources Laboratory, College Park, MD, 2016, http://www.ready.noaa.gov.
[44] K.R. Gopal, S. Arafath, G. Balakrishnaiah, K. Raja Obul Reddy, N. Siva Kumar Reddy, A.P. Lingaswamy, S. PavanKumari, K. Uma Devi, R.R. Reddy, S.S. Babu, Columnar-integrated aerosol optical properties and classification of different aerosol types over the semi-arid region Anantapur, Sci. Total Environ. 527-528 (2015) 507–519.
[45] R.O.R. Kalluri, G. Balakrishnaiah, K. Rama Gopal, K.S.N. Reddy, T. ChakradharRao, T. Lokeswara Reddy, S. NazeerHussain, V.M. Reddy, R.R. Reddy, S.S. Babu, Seasonal variation of near surface black carbon and satellite derived vertical distribution of aerosols over a semi-arid station in India, Atmos. Res. 184 (2017) 77–87 doi:10.1016/j.atmosres.2016.09.003.
[46] A. Bridhikitti, Atmospheric aerosol layers over Bangkok Metropolitan Region from CALIPSO observations, Atmos. Res. 127 (2013) 1–7.
[47] K.K. Moorthy, S.K. Satheesh, S.S. Babu, C.B.S. Dutt, Integrated campaign for aerosols, gases and radiation budget (ICARB): An overview, J. Earth Syst. Sci. 117 (S1) (2008) 243–262.
[48] S.K. Satheesh, K. Krishna Moorthy, S. Suresh Babu, V. Vinoj, C.B.S. Dutt, Climate implications of large warming by elevated aerosol over India, Geophys. Res. Lett. 35 (2008), doi:10.1029/2008GL034944.
[49] S.K. Satheesh, K.K. Moorthy, S.S. Babu, V. Vinoj, V.S. Nair, S.N. Beegum, C.B.S. Dutt, D.P. Alappattu, P.K. Kunhikrishnan, Vertical structure and horizontal gradients of aerosol extinction coefficients over coastal India inferred from airborne lidar measurements during the integrated campaign for aerosol, gases and radiation budget (ICARB) field campaign, J. Geophys. Res. 114 (2009) D05204, doi:10.1029/2008JD011033.
[50] B. Padmakumari, R.S. Maheskumar, S.B. Morwal, G. Harikishan, M. Konwar, J.R. Kulkarni, B.N. Goswami, Aircraft observations of elevated pollution layers near the foothills of the Himalayas during CAIPEEX-2009, Q. J. R. Meteorol. Soc. 139 (2013) 625–638.
[51] A. Mishra, S.N. Tripathi, D.S. Kaul, E. Welton, Study of MPLNET derived aerosol climatology over Kanpur, India, and validation of CALIPSO Level 2 Version 3 Backscatter and Extinction products, J. Atmos. Oceanic Technol. 29 (2012) 1285–1294, doi:10.1175/JTECH-D-11- 00162.1.
[52] M. Komppula, T. Mielonen, A. Arola, K. Korhonen, H. Lihavainen, A.P. Hyvärinen, H. Baars, R. Engelmann, D. Althausen, A. Ansmann, D. Müller, T.S. Panwar, R.K. Hooda, V.P. Sharma, V.M. Kerminen, K.E.J. Lehtinen, Y. Viisanen, One year of Raman-lidar measurements in Gual Pahari EUCAARI site close to New Delhi in India: Seasonal characteristics of the aerosol vertical structure, Atmos. Chem. Phys. 12 (2012) 4513–4524.
[53] K.R. Kumar, N. Kang, V. Sivakumar, D. Griffith, Temporal characteristics of columnar aerosol optical properties and radiative forcing (2011–2015) measured at AERONET's Pretoria_CSIR_DPSS site in South Africa, Atmos. Environ. 165 (2017) 274–289.
[54] H.B. Menon, S. Shirodkar, S. Kedia, R. S., S. Babu, K.K. Moorthy, Temporal variation of aerosol optical depth and associated shortwave radiative forcing over a coastal site along the west coast of India, Sci. Total Environ. 468–469 (2014) 83–92.
[55] M.R. Manoj, S.K. Satheesh, K.K. Moorthy, M.M. Gogoi, S.S. Babu, Decreasing trend in black carbon aerosols over the Indian region, Geophys. Res. Lett. (2019) 0–3.
[56] D.M. Winker, J. Pelon, J.A. Coakley Jr., S.A. Ackerman, R.J. Charlson, P.R. Colarco, P. Flamant, Q. Fu, R.M. Hoff, C. Kittaka, T.L. Kubar, H. Le Treut, M.P. McCormick, G. Megie, L. Poole, K. Powell, C. Trepte, M.A. Vaughan, B.A. Wielicki, The CALIPSO mission: A global 3D view of aerosols and clouds, Bull. Am. Meteorol. Soc. 91 (2010) 1211–1229.

CHAPTER 9

Remote Sensing of Cloudiness: Challenges and Way Forward

Soumi Dutta[a], Sagnik Dey[a], Larry Di Girolamo[b]
[a]Centre for Atmospheric Sciences, Indian Institute of Technology, Delhi, India
[b]Department of Atmospheric Sciences, University of Illinois at Urbana-Champaign, United States

9.1 Introduction

Clouds influence earth's climate by producing a cooling effect in the shortwave and a warming effect in the longwave. The net cloud radiative forcing depends on cloud macrophysical and microphysical characteristics. In the global hydrological cycle, clouds also play a crucial role. The radiative effect of clouds on climate, in turn, causes changes in clouds, creating a complicated system of climate feedback. Cloudiness or cloud fraction (CF) is the fraction of the area covered by clouds and is declared as one of the essential climate variables (ECVs) by WMO [1]. Net cloud radiative forcing is strongly dependent on CF [2,3]. Slingo [4] estimated that a 15–20% increase in low-level CF can balance the warming effect due to the doubling of CO_2. Ohring et al. [5] pointed out that CF needs to be measured with an error within ~1% to predict future climate accurately. Therefore, robust estimation of CF or cloud cover is of very much importance.

Discrepancies among climate models in simulating CF have been a relevant subject of research in climate science [6–8]. Models rely on observational datasets for evaluation and improvement. Clouds can be detected from the surface, aircraft, and satellites. Aircraft observations cannot provide global coverage of cloudiness. Surface observations are not directly comparable with models because of their discrete scale unit (Okta) and scarcity of ground-based stations in remote land and oceanic regions. Satellites are the only observational data sources that provide continuous global coverage. Multiple satellite CF products are currently available for use in climate science (Tables 9.1 and 9.2). These data products are routinely used for various applications, but a comprehensive intercomparison exercise by the Global Water and Energy Cycle (GEWEX) cloud assessment group [9,10] revealed that the global CF varied in the range 0.56–0.73 across various sensors, with larger discrepancies in CFs for low-level, mid-level and high-level clouds. In the absence of any "true" CF datasets, estimating cloud radiative feedback remains one of the most challenging tasks in climate science.

Atmospheric Remote Sensing
DOI: https://doi.org/10.1016/B978-0-323-99262-6.00018-3

Table 9.1 Examples of various state-of-the-art satellites and their sensors detecting clouds and providing CF as a routine product.

Type of orbit	Satellites	Period of observation	Sensors	Time of observation	Observation area	References
Polar	Aqua	2002–present	MODIS CERES	01:30 am and pm	Global	[35]
	Terra	2000–present	MISR MODIS ASTER CERES	10:30 am and pm	Global	[36]
	CALIPSO	2006–present	CALIOP	01:30 am and pm	Global	[37]
	CloudSat	2006–2016	CPR	01:30 am and pm	Global	[38]
Geostationary	INSAT-3D	2013–present	Imager	Half hourly temporal resolution	Centred at 82°E	[39]
	Kalpana-1	2008–2017	VHRR	Half hourly temporal frequency	Centred at 74°E	[33]
	GOES-R	2016–present	ABI	15-minute temporal frequency	Centred at 75°W and 137°W	[40]
	Himawari-8	2015–present	AHI	10-minute temporal frequency	Centred at 140.7°E	[41]
	MeteoSat-8	2017–present	SEVIRI	20-minute temporal frequency	Centred at 41.5°E	[42]

Table 9.2 Examples of state-of-the-art CF datasets that combine multiple satellite data products.

Datasets	Data period	Type of sensors	Resolution	References
ISCCP (International Satellite Cloud Climatology Project)	1983–2017	Passive	3-hourly global coverage with 1° × 1° spatial resolution	[22]
AVHRR PATMOS-x (The Advanced Very High Resolution Radiometer Pathfinder Atmospheres–Extended)	1979–present	Passive	6-hourly global coverage with 0.1° × 0.1° spatial resolution	[34]
CloudSat-CALIPSO combined CF	2006–2011	Active	Day and night global coverage with 2 × 2 degrees horizontal resolution	[43]

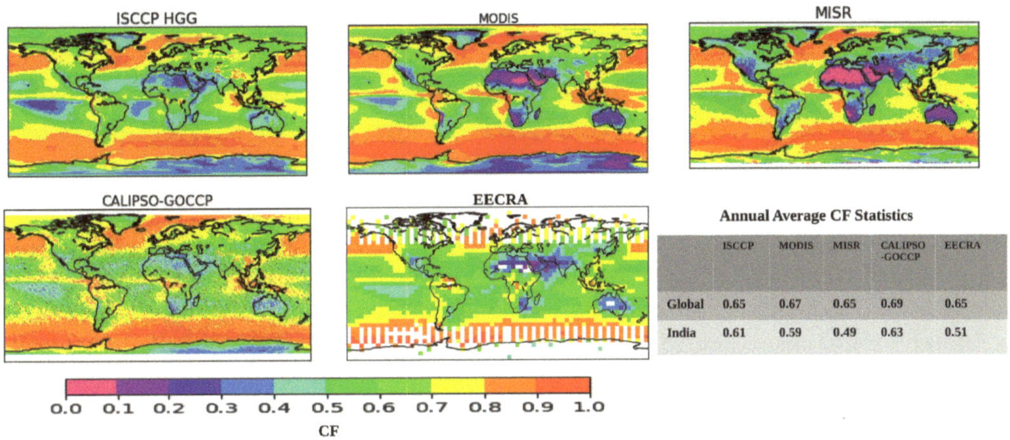

	ISCCP	MODIS	MISR	CALIPSO-GOCCP	EECRA
Global	0.65	0.67	0.65	0.69	0.65
India	0.61	0.59	0.49	0.63	0.51

Fig. 9.1 ***Global distribution of CF from four popular satellite datasets and ground-based observations compiled in Warren Atlas.*** Mean CF values for the global and the Indian subcontinent region are shown in the table.

9.2 Comparative assessment of CF datasets

Fig. 9.1 shows the spatial patterns of CF climatology from four state-of-the-art satellite datasets and one ground-based dataset. Multiangle Imaging SpectroRadiometer (MISR), Moderate Resolution Imaging Spectroradiometer (MODIS), and International Satellite Cloud Climatology Project (ISCCP) are chosen among passive remote sensing datasets, while CALIPSO-GOCCP (GCM-Oriented Cloud-Aerosol Lidar and Infrared Pathfinder Satellite Observation) is active remote sensing data set. Along with these, ground-based observation is chosen from EECRA (Extended Edited Synoptic Cloud Reports Archive) data archive. The overall patterns of cloudiness are similar across the datasets, but there are large discrepancies on the regional scale. For example, the average global CF ranges from 0.65 to 0.69 for these datasets, but over the Indian subcontinent, the values range from 0.49 to 0.63.

The GEWEX assessment pointed out four major reasons for these observed discrepancies among the remote sensing observations: (1) "resolution effect," (2) "view angle effect," (3) "instrument sensitivity and differences in cloud detection techniques," and (4) "diurnal cycle effect." These factors are discussed in detail in the following subsections of this chapter.

9.2.1 Resolution effect

CF is computed as the ratio of the number of cloudy pixels to the total number of cloudy and clear pixels in a scene [11]. Depending on the cloud detection algorithm of the respective sensor, a pixel is declared as "clear" or "cloudy" (this process is referred to as "cloud masking"). No current sensor detects clouds at its original pixel resolution. For example, MODIS has a pixel resolution of 250 m, but it detects clouds at 1 km. Similarly, MISR detects clouds at 1.1 km while its original pixel resolution is 275 m. Therefore, each sensor has to define the entire grid (at which cloud masking is done) as either "clear" or "cloudy," even if not all original pixels are cloudy. In the case of any sub-pixel-level cloud, defining the grid as "clear" underestimates CF. On the other hand, if the entire grid is masked as "cloudy," CF is overestimated. This overestimation in CF is known as the "resolution effect." In the case of MISR, it uses the "clear-conservative cloud masking" where it detects a 1.1 km pixel as "cloudy" if it contains any amount of sub-pixel-level cloud [11]. In the case of MODIS, the cloud detection algorithm chooses a threshold for sub-pixels and declares the larger 1 km grid as "cloudy" or "clear" based on the sub-pixel threshold.

Zhao and Di Girolamo [12] demonstrated that either way, CF is overestimated due to the "resolution effect." Using two different Advanced Spaceborne Thermal Emission and Reflection Radiometer (ASTER) scenes of very high (15 m) pixel resolution (Fig. 9.2), they further showed that CF in the two ASTER scenes was almost the same (0.09 and 0.08). However, due to the resolution effect, CF was estimated as 0.81 (for cloud detection at 1 km) and 0.83 (at 1.1 km) for the left scene (A), while CF was estimated as 0.30 (cloud detection at 1 km) and 0.32 (at 1.1 km) for the right scene (B). The analysis suggested that the spatial patterns of clouds matter most when it comes to the resolution effect. Dey et al. [13], using hundreds of ASTER scenes, showed a

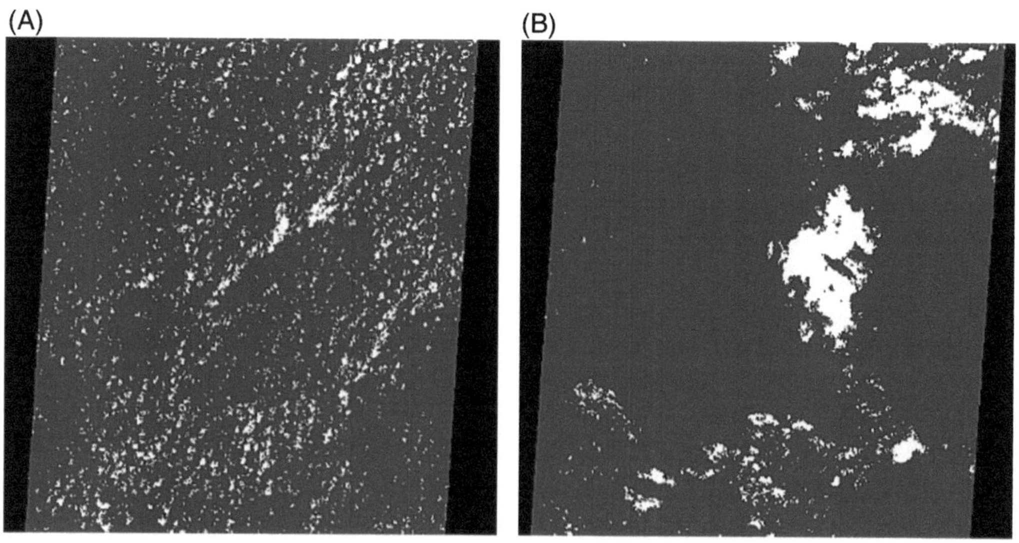

Fig. 9.2 ***Spatial patterns in cumulus clouds over the Atlantic Ocean captured by ASTER.*** The figure is adopted from Zhao and Di Girolamo [12].

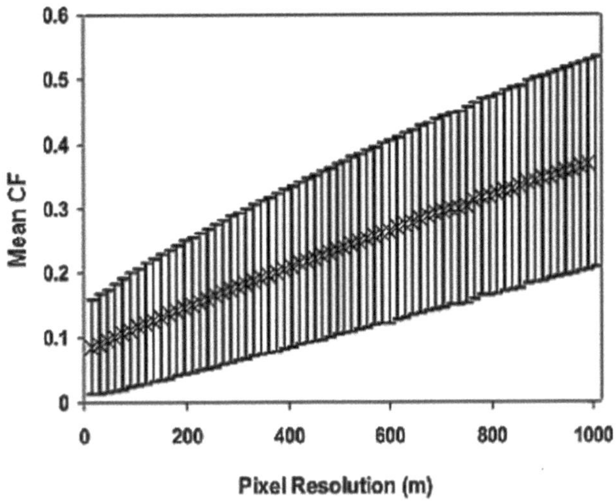

Fig. 9.3 ***Variation of the mean and corresponding standard deviations of CF with degraded pixel resolution following a clear conservative cloud mask strategy.*** The figure is adopted from Dey et al. [13].

continuous increase in CF as the resolution degrades from 15 m to 1 km following a clear conservative cloud masking strategy (Fig. 9.3).

Di Girolamo and Davies [11] proposed several techniques to correct the CF bias due to the resolution effect. One of those is a machine learning pattern recognition technique that determines the necessary correction based on scene-specific information. This technique resulted in unbiased estimates of actual CF relative to a training data set, but it could not have been applied globally due to the absence of globally represented high-resolution training datasets. Jones et al. [14] developed this training set over the ocean from the high-resolution ASTER images. They applied the pattern recognition technique to correct CF estimates from MISR's Radiometric Camera-by-Camera Cloud Mask. A detailed explanation of the pattern recognition technique is available in Jones et al. [14]. They found this technique to reduce average MISR CF from 0.50 to 0.20 over the tropical Western Atlantic (10–20°N and 50–60°W). Subsequently, this technique has been included in the standard production of the MISR CF (Pattern Recognition Corrected CF) since 2014.

Dutta et al. [15] carried out an extensive analysis to evaluate the new resolution-corrected MISR CF product. A near-global (50°N to 50°S) cloud climatology of an 18-year period (March 2000 to February 2018) using this new resolution-corrected CF product was developed. A large reduction in CF (by 0.18 averaged over 50N-50S) was found after resolution-correction across various regions with reductions >0.4 in regions dominated by shallow cumulus clouds (Fig. 9.4). The new MISR resolution-corrected cloud climatology was evaluated against ASTER scenes randomly

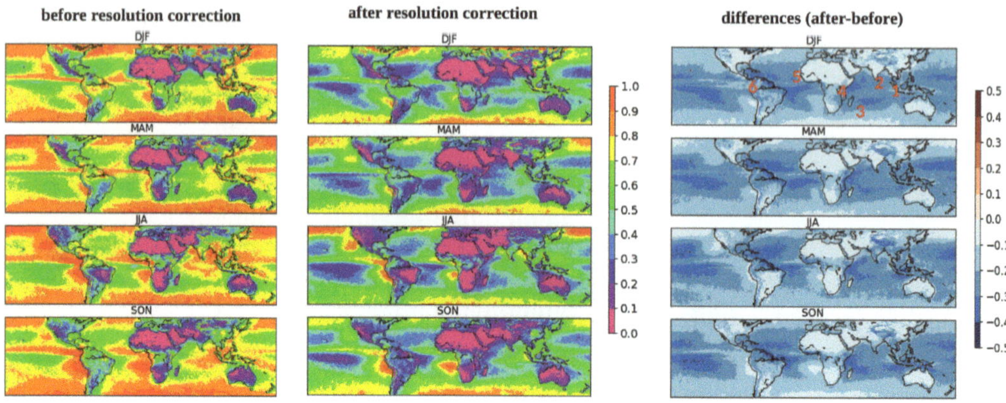

Fig. 9.4 ***Spatial distribution of CF from MISR (before and after resolution-correction) and their differences.*** Locations of six regions (1 to 6) selected for ASTER data analysis are pointed in the top plot of the right column. This figure is adopted from Dutta et al. [15].

collected from six different regions across the globe. The regions were chosen based on the criteria: (1) large (>0.20) reduction in CF post resolution-effect correction, (2) regions where boundary layer clouds often occur, (3) spatially uniform surface to better facilitate ASTER cloud masking method, and (4) regions where ASTER routinely collects data. The new MISR resolution-corrected CF was found to be within ±0.05–0.08 of ASTER CF, which is considered as "true" CF because of its' very high pixel resolution. Currently, the resolution effect has been addressed only in MISR data. Once all other passive sensors address the issue, the cloud climatology will converge further.

9.2.2 View-angle effect

Effects of instrument viewing geometry on cloud properties are undesirable but inevitable when using passive remote sensing satellite data with a very wide swath like MODIS, which has a swath width of 2330 km compared to MISR's much narrower swath of 360 km. MODIS CFs have been reported to have a large view angle effect, with CFs increasing substantially toward the swath edge [16]. For example, Maddux et al. [16] showed that between 35°N and 35°S, the MODIS CF increases from 0.57 to 0.71 between nadir and edge-of-scan (~67°) observations. More oblique views cause an increase in CF mainly for three significant reasons: (1) have a larger instantaneous field of view, (2) observe more cloud sides, and (3) are more sensitive to detecting thin clouds.

The impact of view angle on global cloud statistics must be considered when conducting research using cloud datasets. Dutta et al. [15] have corrected the "view angle effect" for MOD35 (collection 6.1) cloud mask of MODIS of resolution 1-km × 1-km

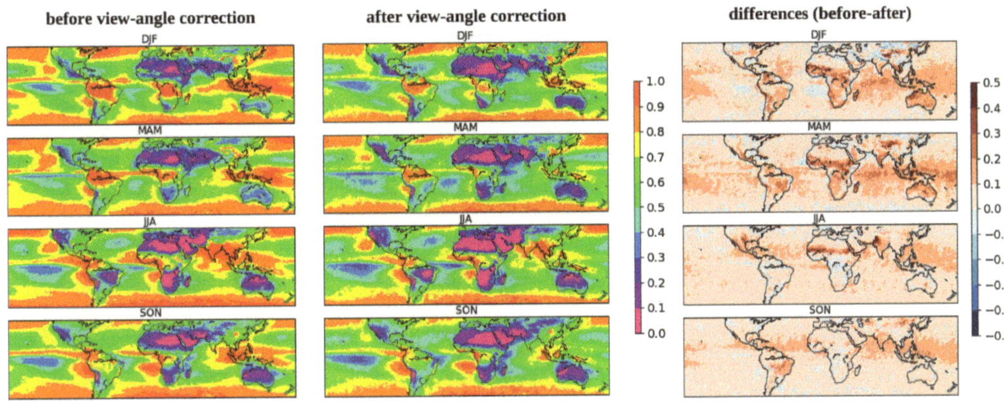

Fig. 9.5 ***Spatial distribution of CF from MODIS (before and after view-angle correction) and their differences.*** This figure is adopted from Dutta et al. [15].

pixel size [17]. Compared to the wider swath of MODIS, the narrower MISR swath is considered as "view-angle effect" free. So, to correct the MOD35 cloud mask from the "view angle" biases, they selected only descending nodes of MODIS granules to match with MISR daytime collection and only MODIS data that fall within the MISR swath (viewing zenith angles <18°). After the view angle correction, a reduction in MODIS CF happens between 50°S and 50°N for all seasons of 0.07 (from 0.66 to 0.59), with regional changes as high as 0.4 (Fig. 9.5). Cumulus and thin cirrus-dominated regions and seasons experience the largest changes. Comparison of view angle corrected MODIS CF with ASTER revealed [14] that MODIS CF overestimates "true" ASTER CF in a range of 0.4–0.5 even after view-angle correction, which points towards the large impact of resolution-effect in MODIS CF.

Surface observations also suffer from the view angle effect in the measurement of CF. The surface observer reports the amount of an imaginary sky-dome obscured by clouds. According to the international codes, this amount reports sky cover (S), which is made in oktas (eighths). S can be larger than the actual earth view (E), which is the vertical nadir view. Henderson-sellers and McGuffie [18] proposed a cubic relationship that can predict E from S.

$$E = -0.001 + 0.082S + 0.269S^2 - 0.019S^3$$

Using this equation, the view-angle correction has been applied to the ground-based observation data [19]. Here, S is the original EECRA data, and E is the view-angle corrected one. Like MODIS, CF is reduced after view-angle correction from 0.61 to 0.57 on an annual scale, with a reduction up to 0.4 on a regional scale. Spatial distribution of CF from ground-based observation (before and after view-angle correction) and their differences are shown in Fig. 9.6. Table 9.3 contains the average statistics of

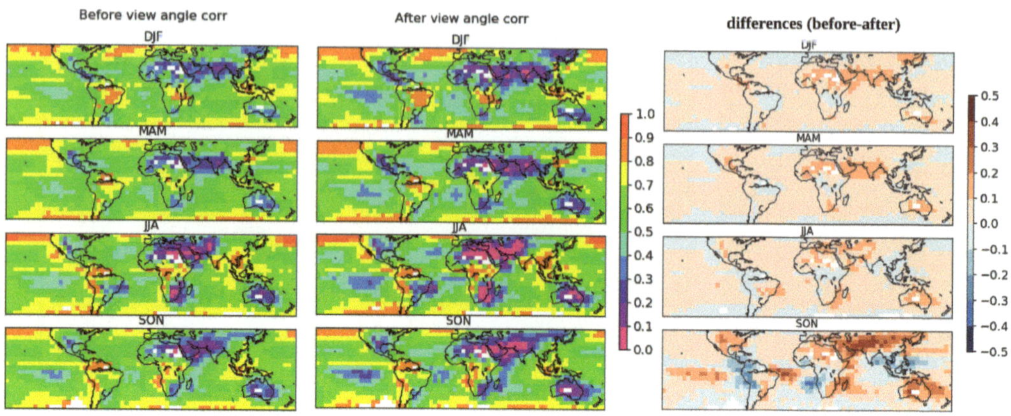

***Fig. 9.6** Spatial distribution of CF from ground-based observation (before and after view-angle correction) and their differences.*

MISR, MODIS and ground-based (EECRA) observed CF after applied resolution and view-angle effect corrections respectively. Like MODIS and ground-based observations, a view-angle correction needs to be applied to the other datasets as well, which suffer from view-angle biases.

9.2.3 Climatology of height-stratified CF

Knowledge about height-stratified CF and cloud vertical structure is critical to an improved understanding of atmospheric circulation processes. However, explicit and accurate knowledge about the vertical distribution of clouds seems elusive because of discrepancies among sensors in detecting low, mid, and high-level

Table 9.3 Comparative statistics of near-global CF (between 50°N and 50°S) derived from the MISR, MODIS, and EECRA before and after the respective corrections.

	DJF	MAM	JJA	SON	Annual
MISR before resolution effect correction	0.66	0.65	0.65	0.65	0.65
MISR after resolution effect correction	0.48	0.46	0.46	0.47	0.47
MODIS before view angle effect correction	0.70	0.64	0.64	0.64	0.66
MODIS after view angle effect correction	0.61	0.58	0.58	0.58	0.59
EECRA before view angle effect correction	0.61	0.61	0.60	0.60	0.61
EECRA after view angle effect correction	0.58	0.58	0.57	0.56	0.57

clouds. According to the GEWEX report [9], uncertainties and biases in CF stem from the error in cloud detection when low-level and mid-level clouds are present underneath optically thick high-level or optically-thin cirrus, single-layer low-level clouds are present that are smaller than the grid size of cloud mask, and single-layer thin cirrus that is not detected by visible bands. Type of instruments (passive or active), sensor wavelength, its sensitivity to the threshold of cloud detection, and the spatiotemporal sampling contribute to the discrepancies in height-stratified CF climatology [20]. Passive remote sensing instruments with visible and infrared (IR) bands can detect cloud types mainly in the presence of single-layer low and high-level clouds. On the contrary, active remote sensing instruments having lidar and radar can penetrate through the atmosphere and are capable of detecting multilayer clouds.

To improve the understanding of the height-stratified distribution of clouds, it is essential to look into the strengths and weaknesses of various satellite instruments and ground-based observations. Currently, resolution-corrected MISR CFbA (Cloud fraction by Altitude [21]), the new ISCCP [22] H-series product, CALIPSO-GOCCP [23], and combined CloudSat-CALIPSO [24] data products provide height-stratified cloud information. MODIS also provides cloud top height [25], which can be used to map distributions of low, mid, and high-level clouds. Along with these satellite products, the ground-based EECRA dataset [19] reports height-stratified CF across the globe. The strengths and weaknesses of these state-of-the-art passive, active, and ground-based remote sensing measurements are summarized in Table 9.4.

It is clear that no single dataset can be treated as the "true" source of height-stratified CF. Several attempts have been made to merge active and passive remote sensing datasets for various parameters like aerosol optical depth, precipitation, and clouds [26,27]. The only way forward to develop a unified climatology of height-stratified CF is the statistical merging of active and passive remote sensing datasets keeping in mind the instrument sensitivities and the strengths of the respective cloud detection methodologies. The synergy of active and passive remote sensing instruments has been attempted by [28] to reconstruct 3-D macro and microphysical properties of clouds, but more consolidated efforts are required in the coming years to address this issue.

9.2.4 Diurnal cycle of CF

Cloud diurnal variation is critical to understanding cloud dynamics, which observations from sun-synchronous or polar-orbiting satellites fail to capture. Geostationary satellites are capable of providing information at very high temporal resolution. ISCCP datasets have been the most versatile source of observations at a higher temporal scale (3-hourly) since 1983 on a global scale. Over the years, numerous geostationary satellites were launched by various countries to continue regional observations of clouds (for weather

Table 9.4 Summary of active, passive, and ground-based remote sensors in detecting multilayer clouds.

Dataset	Time period	Strength and weakness
MISR CFbA (New MISR CFbA is derived from resolution-corrected MISR CF product). Horizontal resolution is 0.5° × 0.5°. Vertical resolution is 500 m.	2000–present	MISR with its visible bands and stereo technique, is more capable of detecting low level clouds in single-layer as well as in the presence of thin cirrus above it. But thin cirrus clouds are not detected with MISR.
ISCCP H-series (Improved version of the ISCCP D-series algorithm). Horizontal resolution 1° × 1°. Low-, mid- and high-level clouds are available.	1983–2017	In multilayer cloud condition ISCCP is capable of detecting thin cirrus clouds with its IR band. Single layer low level clouds are detected with its visible band. But semi-transparent cirrus clouds are misidentified as mid level clouds during night.
CALIPSO-GOCCP (GCM-Oriented CALIPSO Cloud Product (CALIPSO-GOCCP), is aimed at a single goal: evaluating GCM (General Circulation Model) prediction of cloudiness. Horizontal resolution 1° × 1°. Low-, mid-, and high-level clouds are available.	2007–2017	CALIPSO-GOCCP data being produced from active lidar sensor, is always sensible to thin cirrus clouds of upto optical depth 5. But it can not detect low or mid level clouds in multilayer condition as lidar signal gets attenuated.
CloudSat-CALIPSO (Combined radar and lidar data to have multi-layer cloud structure). Horizontal resolution is 2° × 2°. Vertical resolution is 480 m.	2006–2011	CloudSat radar signal can penetrate through multilayer clouds and thus detects low level clouds, while in case of CALIPSO, the lidar is good at detecting thin cirrus. Due to very narrow swaths the combined data product has poor spatio-temporal frequency.
NDP-026E (Ground based data based on EECRA).	1954–1997	Low level clouds are observed from the ground stations, while mid- and high-level clouds can not be observed when the sky is obscured with low level clouds. Also, there are possibilities of human artifacts in ground-based observations.

applications) at a high temporal resolution. This is complemented by the ground-based observations compiled based on station measurements over land and ship-borne measurements over oceans [29].

Although the Indian National Satellite System (INSAT) was launched in 1982 for monitoring clouds over the south Asian monsoon region, the entire record of the dataset was not publicly available [30]. European Meteosat-5 was relocated at 63°E to have observations (spatial and temporal subsampling of 30 km and three hours) over the Indian subcontinent and was included in the ISCCP datasets in 1998. In a recent study by Kottayil et al. [31], the diurnal cycle of deep convective clouds (DCC) over the Indian summer monsoon region has been observed using ISCCP B1 3-hourly dataset. Clear evidence of diurnal scale signature in CF is found over this region from the anomaly in deep convective clouds (Fig. 9.7). Over the Indian land regions, a systematic increase in the diurnal amplitude has been observed from winter to monsoon, while oceanic regions do not show a unique diurnal peak from the study. In 2002, the Kalpana-1 satellite was launched as the first dedicated Indian Meteorological Satellite to study the changes in weather and is positioned at an orbital slot of 74°E [32]. Because of its' half-hourly temporal scale resolution, it can observe the formation and progress of any dynamic cloud system over the Indian subcontinent. Verma et al. [33] showed a close agreement between CF derived from Kalpana-1, INSAT-3D, and MODIS with a two-year record of retrieval. Such datasets allow an opportunity to carry out more in-depth analysis in the future for a better understanding of the cloud dynamics in the south Asian monsoon region.

The detailed discussion on the current uncertainties in CF estimation (Sections 9.2.1, 9.2.2, 9.2.3, and 9.2.4) reveals the fact that significant challenges in satellite remote sensing stem from "resolution effect," "view-angle effect," "instrument sensitivity, and differences in cloud detection methodology," and "cloud diurnal variability." It

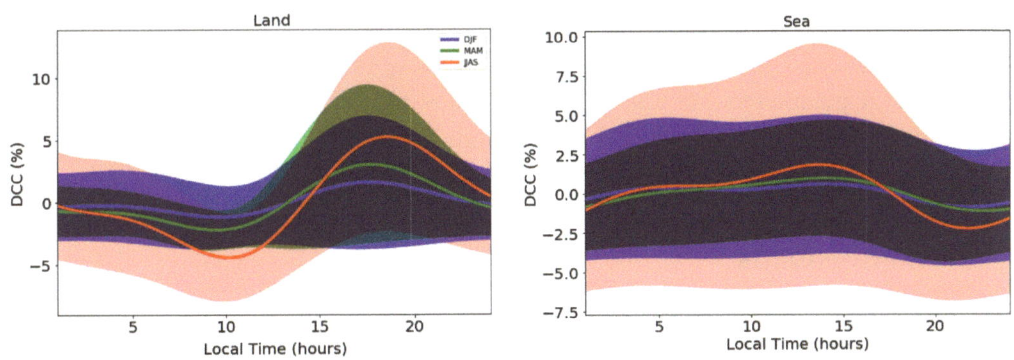

Fig. 9.7 ***Diurnal cycle anomaly observed over land and ocean of Indian summer monsoon region for DJF, MAM, and JJAS.*** The shading represents one sigma standard deviation. This figure is adopted from Kottayil et al. [31].

is clear from the discussion that global cloudiness is highly overestimated due to the resolution-effect and view-angle effect of passive sensors. Climatology of cloud types is biased due to the different cloud detection methodologies and instrument sensitivity. It is necessary to look into the diurnal scale variability of cloudiness to have a wholesome idea of cloud dynamics. Considering all the uncertainties in satellite cloud remote sensing, there is a dire need for "true" cloudiness data on a continuous scale across the globe to improve models and climate prediction and robust estimation of Earth's energy budget.

9.3 Summary

Satellites are the primary and indispensable observational data sources for cloud information on a global scale. Currently, a large discrepancy in CF statistics exists among various popular data products, which hinders in deriving a "true" global CF climatology and reducing the uncertainties in cloud radiative feedback estimation. In the absence of a "true" global cloud dataset, the only way forward is to address the primary causes of uncertainties (resolution-effect, view-angle effect, ability of the sensors to detect low, mid, and high-level clouds, and the differences in sampling frequency) in the respective datasets. Once the CF climatology from various datasets converges after correcting these uncertainties, it will be more prudent to fuse multiple datasets according to their strengths and derive a unified global height-stratified cloud climatology.

The global cloud community has started addressing these critical issues. The finite resolution scale effect has been corrected in the MISR standard CF data product. View-angle correction has been applied to MODIS and ground-based observations. Efforts are ongoing to fuse passive and active remote sensing cloud products and understand the ability of the sensors in identifying individual cloud types based on their morphological, physical, and optical characteristics. It is recommended that measurements from sun-synchronous and geostationary satellites need to be synergized, and cloud detection methods need to be standardized. Without improvement in global cloud climatology, it is difficult to reduce the uncertainty in climate forcing. In addition to CF, various cloud microphysical parameters are also routinely retrieved and reported by satellites. This chapter is beyond the scope to highlight the challenges there and requires a separate discussion.

Acknowledgements

The first author acknowledges the Institute Fellowship from IIT Delhi for doctoral study. The corresponding authors acknowledge IIT Delhi for the Institute Chair fellowship. The authors acknowledge the computing support from the DST-FIST program (SR/FST/ESII-016/2014).

References

[33] S. Bojinski, M. Verstraete, T.C. Peterson, C. Richter, A. Simmons, M. Zemp, The concept of essential climate variables in support of climate research, applications, and policy, Bull. Am. Meteorol. Soc. 95 (9) (2014) 1431–1443.

[1] X. Dong, B. Xi, K. Crosby, C.N. Long, R. Stone, M. Shupe, A 10-yr climatology of Arctic cloud fraction and radiative forcing at Barrow, Alaska, J. Geophys. Res. 115 (2010) D12124.

[2] T. Saud, S. Dey, S. Das, S. Dutta, A satellite-based 13-year climatology of net cloud radiative forcing over the Indian monsoon region, Atmos. Res. 182 (2016) 76–86.

[3] A. Slingo, Sensitivity of the earth's radiation budget to changes in low clouds, Nature 343 (1990) 49–51.

[4] G. Ohring, B. Wielicki, R. Spencer, B. Emery, R. Datla, Satellite instrument calibration for measuring global climate change: Report of a workshop, Bull. Am. Meteorol. Soc. 86 (9) (2005) 1303–1314.

[5] G. Cesana, H. Chepfer, How well do climate models simulate cloud vertical structure? A comparison between CALIPSO-GOCCP satellite observations and CMIP5 models, Geophys. Res. Lett. 39 (20) (2012) L20803.

[6] W.L. Gates, J.S. Boyle, C. Covey, C.G. Dease, C.M. Doutriaux, R.S. Drach, M. Fiorino, Taylor R., E. K., D.N Williams, An overview of the results of the atmospheric model intercomparison project (AMIP I), Bull. Am. Meteorol. Soc. 80 (1) (1999) 29–56.

[7] P.P. Vignesh, J.H. Jiang, P. Kishore, H. Su, T. Smay, N. Brighton, I. Velicogna, Assessment of CMIP6 cloud fraction and comparison with satellite observations, Earth Space Sci. 7 (2020) e2019EA00097.

[8] Stubenrauch, C. J., Rossow, W. B., and Kinne, S. (2012). Assessment of global cloud datasets from satellites: A project of the World Climate Research Programme Global Energy and Water Cycle Experiment (GEWEX) Radiation Panel. WCRP Rep. 23/2012, (p. 176).

[9] C.J. Stubenrauch, W.B. Rossow, S. Kinne, S. Ackerman, G. Cesana, H. Chepfer, et al., Assessment of global cloud datasets from satellites: Project and database initiated by the GEWEX radiation panel, Bull. Am. Meteorol. Soc. 94 (7) (2013) 1031–1049.

[10] L. Di Girolamo, R. Davies, Cloud fraction errors caused by finite resolution measurements, J. Geophys. Res. 102 (D2) (1997) 1739–1756.

[11] G. Zhao, L. Di Girolamo, Cloud fraction errors for trade wind cumuli from EOS-Terra instruments, Geophys. Res. Lett. 33 (2006) L20802.

[12] S. Dey, L. Di Girolamo, G. Zhao, Scale effect on statistics of the macrophysical properties of trade wind cumuli over the tropical western Atlantic during RICO, J. Geophys. Res. 113 (2008) D24214.

[13] A.L. Jones, L. Di Girolamo, G. Zhao, Reducing the resolution bias in cloud fraction from satellite derived clear-conservative cloud masks, J. Geophys. Res. 117 (2012) D12201.

[14] S. Dutta, L. Di Girolamo, S. Dey, Y. Zhan, C.M. Moroney, G. Zhao, The reduction in near-global cloud cover after correcting for biases caused by finite resolution measurements, Geophys. Res. Lett. 47 (2020) e2020GL090313.

[15] B.C. Maddux, S.A. Ackerman, S. Platnick, Viewing geometry dependencies in MODIS cloud products, J. Atmos. Oceanic Technol. 27 (9) (2010) 1519–1528.

[16] Ackerman, S. A., and Frey, R. (2015). MODIS atmosphere L2 cloud mask product. NASA MODIS Adaptive Processing System, Goddard Space Flight Center, USA.

[17] A. Henderson-Sellers, K. McGuffie, Are cloud amounts estimated from satellite sensor and conventional surface-based observations related? Int. J. Remote Sens. 11 (3) (1990) 543–550.

[18] C.J. Hahn, S.G. Warren, NDP-026E, Carbon Dioxide Information Analysis Center, Oak Ridge National Laboratory, Oak Ridge, TN, 2007.

[19] H. Chepfer, G. Cesana, D. Winker, B. Getzewich, M Vaughan, Comparison of two different cloud climatologies derived from CALIOP Level 1 observations: The CALIPSO-ST and the CALIPSO-GOCCP, J. Atmos. Oceanic Technol. 30 (2012) 725–744.

[20] L. Di Girolamo, A. Menzies, G. Zhao, K. Mueller, C. Moroney, D.J. Diner, JPL Publ. D-62358, Jet Propulsion Laboratory, Pasadena, CA, 2010, pp. 1–24.

[21] A.H. Young, K.R. Knapp, A. Inamdar, W. Hankins, W.B. Rossow, The International Satellite Cloud Climatology Project H-Series climate data record product, Earth Syst. Sci. Data 10 (2018) 583–593.

[22] H. Chepfer, S. Bony, D. Winker, G. Cesana, J.L. Dufresne, P. Minnis, C.J. Stubenrauch, S. Zeng, The GCM-oriented CALIPSO cloud product (CALIPSO-GOCCP), J. Geophys. Res. 115 (2010) D00H16.

[23] K. Sassen, Z. Wang, D. Liu, Global distribution of cirrus clouds from CloudSat/Cloud-Aerosol Lidar and Infrared Pathfinder Satellite Observations (CALIPSO) measurements, J. Geophys. Res. 113 (2008) D00A12.

[24] A. Mitra, L. Di Girolamo, Y. Hong, Y. Zhan, K.J. Mueller, Assessment and error analysis of Terra-MODIS and MISR cloud-top heights through comparison with ISS-CATS lidar, J. Geophys. Res. Atmos. 126 (2021) e2020JD034281.

[25] L. Sun, Y. Fu, A new merged dataset for analyzing clouds, precipitation and atmospheric parameters based on ERA5 reanalysis data and the measurements of the Tropical Rainfall Measuring Mission (TRMM) precipitation radar and visible and infrared scanner, Earth Syst. Sci. Data 13 (2021) 2293–2306.

[26] X. Xia, B. Zhao, T. Zhang, et al., Satellite-derived aerosol optical depth fusion combining active and passive remote sensing based on Bayesian maximum entropy, IEEE Trans. Geosci. Remote Sens. 60 (2022) 1–13.

[27] L. Hoppler, F. Godde, M. Gutleben, T. Kolling, B. Mayer, T. Zinner, Synergy of active- and passive remote sensing: An approach to reconstruct three-dimensional cloud macro- and microphysics, Atmos. Meas. Tech. Discuss. (2020) [preprint].

[28] R. Eastman, S.G. Warren, Diurnal cycles of cumulus, cumulonimbus, stratus, stratocumulus, and fog from surface observations over land and ocean, J. Clim. 27 (2014) 2386–2404.

[29] M.M. Wonsick, R.T. Pinker, Y. Govaerts, Cloud variability over the Indian monsoon region as observed from satellites, J. Appl. Meteorol. Climatol. 48 (9) (2009) 1803–1821.

[30] A. Kottayil, K. Satheesan, V. John, O., R Antony, Diurnal variation of deep convective clouds over Indian monsoon region and its association with rainfall, Atmos. Res. 255 (2021) 105540.

[31] M.V. Shukla, P.K. Thapliyal, J.H. Bisht, K.N. Mankad, P.K. Pal, R.R. Navalgund, Inter-satellite calibration of Kalpana thermal infrared channel using AIRS hyperspectral observations, IEEE-GRSL 9 (2012) 687–689.

[32] S. Verma, P.V.N. Rao, H.B.K. Shaeb, M.V.R. Seshasai, B. PadmaKumari, Cloud fraction retrieval using data from Indian geostationary satellites and validation, Int. J. Remote Sens. 39 (22) (2018) 7965–7977.

[34] A.K. Heidinger, M.J. Foster, A. Walther, X. Zhao, The Pathfinder Atmospheres-Extended AVHRR climate dataset, Bull. Am. Meteorol. Soc. 95 (2014) 909–922.

[35] C.L. Parkinson, Aqua: An earth-observing satellite mission to examine water and other climate variables, IEEE Trans. Geosci. Remote Sens. 41 (2) (2003) 173–183.

[36] Y.J. Kaufman, D.D. Herring, K.J. Ranson, G.J. Collatz, Earth observing system AM1 mission to earth, IEEE Trans. Geosci. Remote Sens. 36 (1998) 1045–1055.

[37] D.M. Winker, M.A. Vaughan, A. Omar, Y. Hu, K. Powell, Z. Liu, W.H. Hunt, S.A. Young, Overview of the CALIPSO mission and CALIOP data processing algorithms, J. Atmos. Oceanic Technol. 26 (11) (2009) 2310–2323.

[38] G.L. Stephens, D.G. Vane, R.J. Boain, G.G. Mace, K. Sassen, Z. Wang, et al., The Cloudsat mission and the A–train, Bull. Am. Meteorol. Soc. 83 (2002) 1771–1790.

[39] C.B. Lima, S.S. Prijith, M.V.R. Sesha Sai, P.V.N. Rao, K. Niranjan, M.V. Ramana, Retrieval and validation of cloud top temperature from the geostationary satellite INSAT-3D, Remote Sens. 11 (2019) 2811.

[40] T.J. Schmit, P. Griffith, M.M. Gunshor, J.M. Daniels, S.J. Goodman, W.J. Lebair, A closer look at the ABI on the GOES-R series, Bull. Am. Meteorol. Soc. 98 (4) (2017) 681–698.

[41] H. Shang, H. Letu, T.Y. Nakajima, Z. Wang, R. Ma, T. Wang, Y. Lei, D. Ji, S. Li, J. Shi, Diurnal cycle and seasonal variation of cloud cover over the Tibetan plateau as determined from Himawari-8 new-generation geostationary satellite data, Sci. Rep. 8 (2018) 1105.

[42] M.T. Bushair, S.I. Rani, B.P. Jangid, et al., Evaluation of the benefits of assimilation of Meteosat-8 observations in an NWP system over the Indian Ocean region, Meteorol. Atmos. Phys. 133 (2021) 1555–1576.

[43] Kay, J. & National Center for Atmospheric Research Staff (Eds). The Climate Data Guide: Combined CloudSat spaceborne radar and CALIPSO spaceborne lidar cloud fraction dataset. https://climatedataguide.ucar.edu/climate-data/combined-cloudsat-spaceborne-radar-and-calipso-spaceborne-lidar-cloud-fraction-dataset. (Last modified 21 April 2014).

CHAPTER 10

Overview of aerosol–cloud interactions over Indian summer monsoon region using remote sensing observations

Vijay P. Kanawade[a], N. Arun[b], Chandan Sarangi[b], Abin Thomas[a]
[a]Centre for Earth, Ocean and Atmospheric Sciences, University of Hyderabad, Hyderabad, India
[b]Department of Civil Engineering, Indian Institute of Technology Madras, Chennai, India

10.1 Climatic effects of atmospheric aerosols

Gases and particulates form the major constituents of the earth's atmospheric composition. The particles suspended in air are distinguished based on their size, chemical composition, and water content into aerosols, clouds droplets, and hydrometeors. Atmospheric aerosols are suspended solid or liquid particles in the atmosphere. They have a wide size range spectrum with four orders of magnitude difference between the smallest and the largest particle. The size, mass, and composition of the aerosol particles determine the effect of aerosols on radiation and climate [1]. They are generated from natural (such as volcanoes, sea spray, etc.) and anthropogenic (biomass burning, vehicular pollution, etc.) sources through mainly two major processes such as direct suspension of bulk matter in the atmosphere or nucleation of co-emitted gaseous vapors to form new particles (via gas-to-particle conversion) in the atmosphere [2]. Their lifetime in the atmosphere varies widely and mainly depends on their size, location, and atmospheric circulation. Across the globe, several high aerosol loading hotspots exist, such as over southeast Asian countries like China and India, central-eastern Atlantic, and over the Amazons. While extremely low concentrations exist in pristine areas like Antarctica [3].

Aerosols are an important climate forcing as they can exert a cooling and warming on the earth's climate by scattering and absorption, respectively, of solar and terrestrial radiation. This effect is referred to as the aerosols direct effect [4,5]. The absorbing aerosols like soot or black carbon (BC) particles warm the air surrounding them, leading to the evaporation of cloud droplets and thereby decreasing the cloud cover and suppressing precipitation [6]. The warming effect of absorbing aerosols at an altitude can stabilize the atmosphere and weaken convective updraft. This rapid adjustment in the physical and radiative properties of clouds due to the presence of absorbing aerosols is referred to as the semi-direct effect [7,8]. However, significant uncertainties exist in this effect due to the vertical distribution of absorbing aerosols [9]. For instance, cloud thickening in the presence of absorbing aerosols above the cloud has been reported in several places

Atmospheric Remote Sensing
DOI: https://doi.org/10.1016/B978-0-323-99262-6.00017-1

[10–13]. A recent climate model intercomparison study found that a tenfold increase in BC emissions resulted in a strong positive direct effect that was partially offset by a negative semi-direct effect [14]. In the last two decades, extensive observations and modelling efforts led to improved estimates of aerosol direct radiative forcing [15]. The climate models expanded from including sulfate aerosols for direct radiative forcing [16] to multi aerosol microphysical schemes with multi-model ensembles [17–19]. The accurate estimation of aerosol direct radiative forcing is hampered mainly by the complexities in aerosol mixing, chemical nature, and vertical distribution of aerosols [20,21]. The development in aerosol and physics modules in global models was complemented by the measured aerosol properties. Observational campaigns [21–23] and long-term remote sensing from the ground [24–26] over the last more than two decades have offered crucial datasets to advance our understanding of aerosols. The Aerosol Robotic Network (AERONET) is a network of ground-based CIMEL Sun photometers that provides ground-truth calibration data for NASA's and other international satellite mission [24]. AERONET has been providing daily data of various optical and radiative properties of aerosols over the last 25 years [27]. AERONET provides continuous cloud-screened and quality-controlled observations data from CIMEL Sun photometers which are globally distributed [24]. The satellite retrieved AOD measurements have also been validated extensively using ground-based AERONET AOD measurements [28,29].

The fraction of aerosols acting as cloud condensation nuclei (CCN) and ice nuclei (IN) that affect clouds and thus rain formation processes indirectly, thereby modifying cloud radiative forcing and precipitation patterns, referred to as aerosol indirect effect (AIE). The high CCN concentrations, particularly over continental regions, yield a high number of smaller-sized cloud droplets in moisture deficit conditions. The high concentrations of smaller cloud droplets lead to an increase in cloud albedo and thereby cause a negative cloud radiative forcing known as the Twomey effect [30] or first indirect effect [31] or cloud albedo effect [32]. This resulted in the increased lifetime and decreased precipitation efficiency of clouds under moisture deficient conditions, thus enhancing the clouds cooling effect known as the Albrecht effect [33] or the second indirect effect. Despite extensive studies over the last three decades, aerosol's impact on clouds and precipitation patterns is the least understood in the climate system. The aerosol-induced changes in cloud macrophysical (cloud thickness, liquid water path, etc.) and cloud microphysical (cloud effective radius, cloud droplet number concentrations, etc.) properties remain poorly understood to estimate the precise radiative forcing from aerosol–cloud interactions [34,35]. While numerous studies have made vast progress in understanding the processes involved in warm marine clouds, the aerosol–cloud interactions in the mixed-phase clouds remain highly uncertain due to the complexity of microphysical pathways [36]. IPCC [15] also reveals that aerosol-modulated changes in cloud radiative forcing hampers by the largest uncertainty in estimating the climate forcing for future climate projections. The ambiguities in the aerosol–cloud interactions stem from the uncertainties in meteorological influence on clouds, surface feedbacks, changes in microphysical feedbacks, and spatio-temporal heterogeneity in aerosols [37]. The uncertainties in parameterizations of aerosol and cloud processes and biases in the satellite observations further make it difficult to accurately model aerosol indirect effects [38,39]. Therefore, several studies conducted in-situ ground-based and airborne measurements to quantify the aerosol indirect effect [40–42]. The satellite-based studies of aerosol–cloud interaction tend to underestimate compared to modelling studies in conjunction with in-situ measurements due to uncertainties in the satellite retrieval. The aerosols grow hygroscopically in the presence of clouds, and AOD cannot be retrieved in cloudy pixels [42]. Therefore, the absence of collocated aerosol and cloud makes it challenging to study the association between aerosol, CCN and clouds [39,43]. Recent studies attempted to accurately estimate the CCN ingested by the clouds by calculating cloud droplet number concentration from satellite retrieved cloud effective radius and cloud optical thickness [44,45]. Such efforts have helped to isolate the aerosol and meteorological impact on low marine clouds [46].

The complexity in the aerosol–cloud interaction arises from variations in dominant phase changes and microphysical and dynamical processes associated with different types of clouds. The region experiences tropical and subtropical climates, which causes extreme temperatures, rainfall, and moisture transport [47]. Concurrent measurements of aerosol size distribution, composition, cloud properties, microphysical parameters as well as the development of physical process scale studies based on observations over a varying space and times-scales, and translating them to climate models are essential to improve our understanding of the role of aerosols in modifying weather, hydrological cycle, and climate over India. Previous studies imply that ground-based remote sensing is suited to study aerosol–cloud microphysical interactions [48–51]. The changes in the atmospheric burden of aerosols and trace gases led to significant changes in temperature patterns, aerosol–cloud interactions, and thus radiative forcing. Therefore, it is crucial to improve the representation of critical processes in climate models (aerosol–radiation, aerosol–cloud interactions, and their feedback mechanisms) to understand future climate changes.

The Indo-Gangetic Plain (IGP) is a highly polluted and populated region in the Indian landmass and is considered a global hotspot of aerosol loading. This is due to its geographical location, regional meteorology, the influx of long-transported natural dust aerosols, and high emission rates of aerosols. Several researchers have shown the increasing trend in aerosol loading over the Indian region [25,52–54], with more focus given to the IGP. Moreover, air pollution is no more an urban problem in India, and it is now just as harmful in rural areas as in urban areas [55]. Therefore, it is interesting to identify the pattern and magnitude of the increase in aerosol loading all over the Indian region. The dry winter season is known for its calm meteorological conditions with shallow boundary layer height. The variation in planetary boundary layer dynamics and associated micro-meteorological conditions can prevent the dispersion of aerosols and increase their concentration near the surface and associated feedbacks, thereby affecting the quality of the air near the surface. Observing the trend in aerosol loading during the winter season can show the influence of local sources. Several studies have shown that the rate of increase of AOD is significantly higher during the winter season [52,56]. The changes in the aerosol loading can change the direct and indirect radiative forcing over the region. A variation in aerosol indirect effect was observed with the change in liquid water path (LWP) over the outflow regions of the Indian peninsula, that is, the cloud effective radius increased with aerosol loading when the clouds had low LWP and vice versa during the winter season [57]. The CCN efficiency of aerosols in the northern Indian Ocean during winter was size-dependent and increased with supersaturation [58]. Further, the high aerosol loading is limited to continental regions, but it can also be transported to long distances onto adjoining seas such as the Bay of Bengal and the Arabian Sea [54]. Thus, the aerosols transported onto outflow marine regions can impact aerosol loading and associated changes in the direct radiative effect over marine regions.

Unlike the calm winter season, the strong westerly winds during the premonsoon and monsoon season bring mineral dust to the Indian landmass [59,60]. The complex mixture of natural and anthropogenic aerosols further aids the complexity in understanding radiative properties over continental and marine regions to derive climatic impacts. The presence of dust aerosols transported to the Bay of Bengal from the Indian landmass was observed during the premonsoon season, and the long-term reduction in mineral dust transport over the Bay of Bengal was also reported [61]. The mineral dust and other absorbing aerosols are known to influence the Indian summer monsoon circulation [62,63]. A strong latitudinal and longitudinal gradient in aerosol properties and radiative forcing was observed over the marine regions around India during the premonsoon season using data from ICARB, satellites, and model estimates [64]. Therefore, it is crucial to understand the contribution of anthropogenic aerosol influence on the direct radiative effect over the Bay of Bengal region during the premonsoon season. This chapter highlights the current understanding, challenges, and knowledge gaps of aerosol–cloud interactions over the Indian summer monsoon region.

10.2 Aerosol–cloud interaction studies over the Indian subcontinent

Numerous field observations have been conducted over India and adjoining seas to investigate the impact of aerosol pollution on clouds over continents and outflow regions over the marine region. The long record of aerosols data from Moderate Resolution Imaging Spectroradiometer (MODIS) and Multiangle Imaging Spectroradiometer (MISR) showed that the higher aerosol loading overlay the industrialized, highly populated, and urbanized regions, that is, the South East Asian region [65]. The Indian subcontinent is also one of the hotspots of high aerosol loading and has received considerable attention since the 1960s. Mani and colleagues studied solar radiation attenuation due to wind-blown dust aerosols during the premonsoon season over continental India [66,67]. The Indian Ocean Experiment (INDOEX) was conducted in the northern Indian Ocean to investigate the influence of anthropogenic aerosols onto the outflow regions [68,138]. Several field campaigns were also conducted over the continental Indian region by the Geosphere-Biosphere Program of the Indian Space Research Organisation (ISRO) to study aerosol physical, optical, chemical and radiative properties [23,69–71]. Aerosols vertical distribution and radiative properties were also studied using a multi-platform approach during the Integrated Campaign on Aerosol and Radiation Budget (ICARB) [72–74]. ISRO-GBP initiated a long-term observation network of aerosols in 1985 as a part of the Aerosol Radiative Forcing over India (ARFI) project to improve our understanding of regional aerosol characterization, incorporating the heterogeneity in space, time, and spectral domains, to estimate aerosol radiative forcing [52,75]. The regional aerosol warming experiment (RAWEX) was also conducted as a part of ARFI to quantify regional atmospheric warming by elevated aerosols over the Indian region and assess their radiative impact. On the regional climate [76–78]. Extensive field observations to study aerosols, clouds, and aerosol-mediated changes in clouds over the last two decades have improved our understanding of the complexity of the aerosol associated feedback processes on clouds and their sensitivity to aerosol loading changes over India and adjoining marine regions. Here, we attempt to synthesize our understanding of aerosol-mediated changes in clouds and associated feedback processes over Indian summer monsoon regions, particularly those studied based on remote sensing datasets.

10.3 Indian summer monsoon season

The Indian summer monsoon (ISM) is the world's largest monsoon system. ISM is the most striking manifestation of the seasonal variation in the planetary-scale atmospheric overturning in response to the seasonal variation in the incoming solar radiation [79]. ISM is characterized by widespread rainfall over Indian regions from June to September, and changes in wind patterns over the Indian Ocean, with a strong easterly jet referred to as tropical easterly jet (TEJ) near the tropopause extending from south-east Asia to west Africa. The ISM produces about 80% of total water to the country in just four months, from the beginning of June to September, referred to as the southwest monsoon. The earliest results from remote sensing satellites revealed the northward propagation of convective cloud systems associated with the Inter-Tropical Convergence Zone (ITCZ) towards the Indian landmass [80]. Since then, the development of remote sensing technologies has enabled us to understand the intra- and inter-annual variation of rainfall and associated cloud micro- and macro-physical properties in both space and time. The long-term observations from CLOUDSAT showed the presence of vertically deep convective clouds over the north Bay of Bengal, the north Arabian sea, the south equatorial Indian Ocean, and along the west coast of India [81]. Bhat and Kumar [82] found the largest frequency of occurrence of cumulonimbus towers over the foothills of the Himalayas and plains of northern India and minimum over the Arabian Sea and the equatorial Indian Ocean. There is a consensus that deep convective systems are primarily found over the land than oceans [82,83]. The intra-seasonal variability of the ISM was found to have distinct changes to cloud morphology. While an increase in cloud water content vertically is seen over the Bay of Bengal, along the west coast of India, and the northeast Arabian sea during the active spell, the break spell periods are characterized by low cloud water content along the west coast of India and the Bay of Bengal [81]. The break spell over the ISM region is associated with the formation of shallow clouds [81].

10.4 Cloud radiative forcing over the Indian summer monsoon region

Rajeevan and Srinivasan [84] estimated cloud radiative forcing over the Indian region using Earth Radiation Budget Experiment (ERBE). They found a negative net cloud radiative forcing (NETCRF) of about 30 W m^{-2} during the monsoon season. While the previous study showed a zero net cloud radiative forcing over the tropical convective region due to the near cancellation of shortwave and longwave forcings [85]. The imbalance between shortwave and longwave cloud radiative over the Indian region was attributed to the presence of high clouds with large optical depth [84]. Later, Thampi and Roca [86] also found the high NETCRF over the Indian summer monsoon region using Clouds and Earth's Radiant Energy System (CERES) data and concluded that the changes in the atmospheric water vapor strongly influence the NETCRF over this region. The analyses using CloudSat retrievals of the vertical distribution of cloud optical depth and cloud water content also corroborated the finding of high NETCRF over the Indian Summer Monsoon region [87]. The long-term (13-years) CERES and MODIS observations showed that the variability in NETCRF can be linked to the changes in the cloud properties like cloud fraction, cloud optical depth, and cloud top pressure [88]. Further, the high aerosol loading over India was also found to deepen the cloud systems during the Indian summer monsoon season, leading to aerosol-induced increased surface accumulated rainfall and enhanced cooling of clouds [89]. Previous study also showed that the cloud radiative forcing changes with the cloud types, and that the low clouds have a net cooling effect [90]. While during the Indian summer monsoon, taller clouds lead to a negative cloud radiative forcing [84,88]. There is an increasing trend in the mean cloud cover during the Indian summer monsoon (0.01 per year) [88]. A recent study contradicted this by showing an opposite trend in cloud cover and cloud top height within the monsoon season [91]. They observed a decreasing trend in cloud cover and cloud top height in June and July but an increasing trend in August and September. This was attributed to the changes in upper tropospheric easterlies and vertical winds [91]. Sathiyamoorthy et al. [92] found that the low clouds dominate the Arabian Sea compared to the Bay of Bengal due to lower tropospheric thermal inversion over the Arabian Sea. The tropospheric stability and free tropospheric relative humidity influence the strength of aerosol–cloud radiative forcing in low marine clouds [93]. Overall, it is critical to quantify the meteorological influence on aerosol–cloud interactions over both the Arabian Sea and the Bay of Bengal, where the atmospheric and aerosol conditions are significantly distinct during the Indian summer monsoon season.

Over the past two decades, the importance of aerosol radiative effects on the monsoonal circulation and rainfall has been widely discussed with the help of satellite remote sensing and general circulation models [94–96]. The wind-blown dust aerosols over the North African and Arabian Peninsula can alter the monsoonal influx of moisture onto the Indian region by heating the atmosphere [60]. The localized elevated warming due to the dust aerosols over the ISM region can increase land-sea temperature contrast causing strong convergence of moisture from the surrounding seas and thereby initially strengthening the summer monsoon season [97]. Despite the initial increase in the land-sea temperature contrast and increase in the rainfall, these elevated layers eventually increase the lower tropospheric stability over the ISM and surrounding regions [98,99]. This will profoundly affect the boundary layer dynamics and convective activity over the area. Moreover, the increased cloudiness and wet scavenging of absorbing aerosols can cause prominent solar dimming at the surface, reducing the monsoon precipitation later into the season [100]. The anthropogenic aerosol forcing over the South Asian monsoon region was found to slow down the overturning tropical circulation and weaken the monsoon system [101]. Anthropogenic aerosol forcing was also found to suppress short-term intra-seasonal rainfall resulting in elongated break spells [102]. Further, the difference in the cloud formation mechanisms is reflected in the raindrop size distribution observed at the surface [103–105]. The rainfall can be identified as stratiform and convective based on the raindrop size distribution properties and rain rate changes from the surface [106]. The response of cloud microphysical properties and meteorological parameters to the stratiform and convective rainfall regimes can differ as the underlying cloud formation mechanisms differ. Understanding such distinct responses in different rain regimes can help comprehend cloud feedback in the different cloud or rain types.

10.5 Physical mechanisms of aerosol-induced changes in cloud radiative effects

Several mechanisms are proposed to delineate aerosol-mediated adjustments to clouds, but almost all of them are poorly understood and often strongly modulated by the environmental conditions in which the aerosol–cloud interaction occurs. The aerosol–cloud interaction is a multi-dimensional and complex interaction owing to a multitude of classifications of cloud regimes (warm, cold, mixed, deep convective, glaciated) [37]. A wide spectrum of cloud regimes forms over the Indian region during each season.

This is primarily driven by the seasonal changes in the dynamic and thermodynamic structure of the atmosphere. To understand cloud adjustments associated with aerosols, space-borne instruments use columnar AOD, which is a measure of the extinction induced by aerosol loading on the atmospheric column. However, for assessing aerosol–cloud–precipitation–radiation interactions, columnar AOD is not an ideal aerosol quantity because it may partly involve the contribution of extinction of aerosol particles that may not have been involved in cloud formation and thereby precipitation and radiation changes. This is the fundamental problem wherein isolating the effect of different clouds, thermodynamical, and meteorological conditions are extremely difficult. The ideal solution would be probing the atmosphere multitude and multi-dimensionally approach to address the classic aerosol–cloud–precipitation–radiation puzzle. While aerosol–cloud interactions can be studied systematically by decomposing clouds, the effect of meteorology remains unclear [37,107,108].

CCNs are the small fractions of aerosols that enable the condensation of water vapor and growth into cloud droplets under the right ambient saturation conditions. The activation of an aerosol to CCN mainly depends on the particle size, the chemical composition, and its mixing state [109–111]. The Kohler theory lays the foundational basis for explaining the hygroscopic growth of aerosols into CCNs. Sotiropoulou et al. [112] found that anthropogenic emissions around the world have the highest uncertainty for the autoconversion processes in GCM. This is probably due to the large spatial variability in the aerosol concentrations and their intrinsic chemical composition. Furthermore, studies have also found considerable variability in the CCN properties at spatially and temporally [40]. A study conducted over a semi-rural area in India also concluded the drastic changes in the shape of the particle number size distribution between the pre-monsoon and monsoon season [113]. Interestingly, they also found that the number of concentrations and geometric sizes depended on the air mass source. Andreae [40] found that $CCN_{0.4}$ (SS at 0.4%) concentrations and aerosol optical thickness can be expressed in terms of the power law. However, it has been found that AOD is poorly correlated with CCN, especially at very concentrations. Stier [114] found that ~71% of the global $CCN_{0.2}$ (SS at 0.2%) and AOD have correlation coefficients less than 0.5. This poor translation of AOD into CCN is primarily attributed to regional meteorology. The uncertainties associated with using AOD as a proxy for CCN in studying aerosol–cloud radiation interactions can be overcome by using cloud droplet number concentration (CDNC) estimates [45].

Panicker et al. [115] demonstrated an inverse association between AIE and Indian summer monsoon rainfall. They reported positive (negative) AIE values during drought (excess monsoon) years between 2001 and 2004. The quantitative analyses revealed an aerosol effect on the dispersion of cloud drop size distribution in warm cumuli [116]. Anil [117] found that AIE calculated based on relative changes in cloud droplet number concentrations is overestimated as compared to AIE calculated based on relative changes in cloud droplet effective radius. INDOEX campaign observations revealed that CCN concentrations increased by order of magnitude along the north-south transect of the Inter-tropical convergence zone [118]. INDOEX observations further revealed that cloud droplet concentration increases with aerosol number concentrations over the marine environment [138]. Ministry of Earth Sciences (MoES), Government of India, initiated Cloud-Aerosol Interaction Precipitation Enhancement Experiment (CAIPEEX) in 2009 to improve our understanding of the physics and dynamics of aerosol–cloud interactions and to formulated scientific basis for rain enhancement using cloud seeding experiments over Indian region. CAIPEEX has conducted phase-wise observations of aerosols, cloud properties, CCN concentrations and environmental conditions using airborne and ground-based platforms [41,119]. CAIPEEX observations also showed the more number of cloud droplets activated over the continent under high aerosol loading, particularly over the Ganges valley [119]. While an increase in aerosol loading was found to significantly increase the cloud droplet number concentrations [119], the high aerosol loading causes a narrowing of the cloud droplet spectrum resulting in suppression of collision coalescence and warm rain formation at an elevated layer [137]. Prabha et al. [41] showed that the mixed-phase clouds have a large amount of super-cooled liquid water ($>3\ gm^{-3}$) under high aerosol loading conditions.

The initial warm rain suppression due to aerosol indirect effect can then trigger feedback mechanisms to form taller invigorated convective clouds [120,121]. The decrease in the size of cloud droplets increases their mobility, and they may get carried aloft due to strong updrafts in the convective clouds resulting in deeper clouds [121–123]. This may lead to increased cloud cover, precipitation, and cloud cooling effect at the surface. The ice formation above the freezing level results in the release of latent heat, which provides fuel to the cloud to invigorate further in the vertical. This is referred to as aerosol-induced cloud invigoration mechanism, which was initially observed over North Atlantic Ocean and the Amazon basin [124]. The taller clouds when hits the tropopause, may extend horizontally due to the tropopause thermal capping or the presence of tropical easterly jet, which can form thicker stratiform anvil cloud, containing smaller ice hydrometeors. Such thicker stratiform anvil clouds can reflect incoming solar radiation more effectively masking the warming effect due to longwave emission by cloud tops, leading to an enhanced cloud cooling effect at the surface [125]. Sarangi et al. [89] observed a similar positive association between aerosol loading and precipitation during the Indian summer monsoon season, indicating the critical role of aerosols in deepening the monsoon cloud system. Fig. 10.1 illustrates the case of such aerosol-induced cloud invigoration effect under a polluted scenario, which is widely reported over marine regions [121,123,125] and rarely over continental regions [89,124,126]. Sarangi et al. [126] shown high aerosol concentrations in the atmosphere can change macrophysical (number of cloud layers, cloud top, cloud thickness, etc.), microphysical (cloud droplet number, cloud droplet size, etc.) and radiative properties of clouds which in turn leads to heavy rainfall events and a cooling effect of nearly 1°C during the Indian summer monsoon.

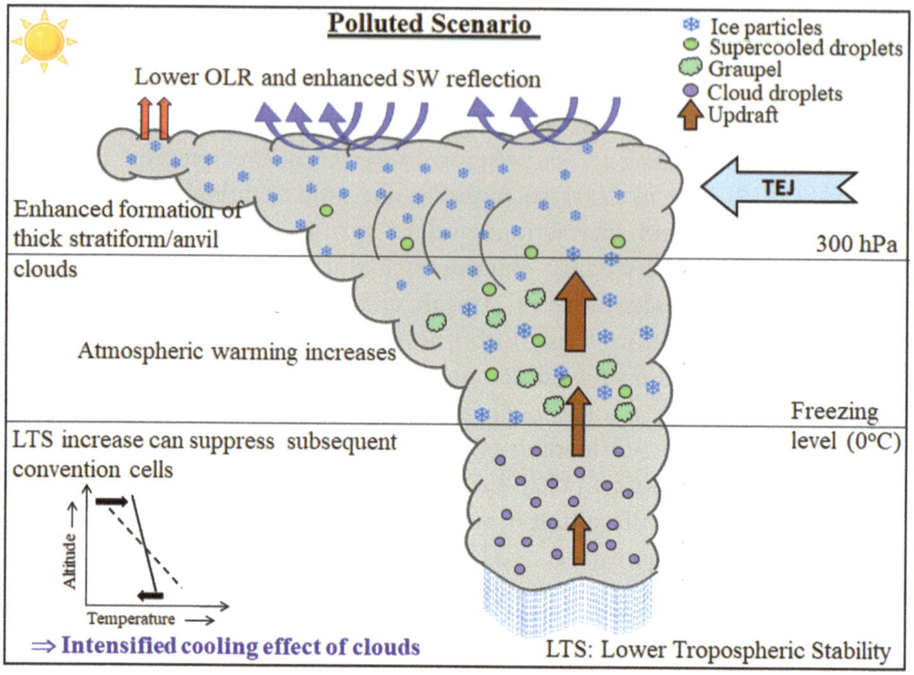

Fig. 10.1 ***Schematic showing processing leads to clouds' intensified cooling effect under polluted scenario over the India summer monsoon region, adopted and modified from [126].***

Two major hypotheses are proposed to explain the aerosol–cloud-climate effect for convective clouds, such as thermodynamic invigoration through latent heating and convective updraft [120,127] and cloud microphysical alterations [128,129]. The increase in aerosol number concentrations may lead to increase latent heat release and, accordingly, increased buoyancy and updraft, which can promote the vertical deepening of the cloud. This can also lead to a decrease in the ice particle sizes in stratiform anvil clouds and therefore decreases their fall velocities, which leads to increased stratiform anvil cloud cover and lifetime [128,126]. Under polluted scenarios, aerosol-mediated invigoration of clouds causes more water mass aloft across the freezing level due to higher buoyancy and thereby enhances the formation of ice particles in growing convective clouds. These clouds essentially grow up to the tropopause, where they expand horizontally into thick stratiform anvil clouds due to the tropopause thermal capping. The large ice particles fall off as precipitation at the surface, and the smaller ice particles in stratiform anvil clouds can remain in the atmosphere for a longer period of time, increasing the cloud radiative cooling effect. The presence of tropical easterly jet stream over India during the summer monsoon season may also produce thick stratiform anvil clouds over downwind regions. A variety of environmental factors may also affect the invigoration of clouds. For instance, wind shear has a decisive role in determining whether high aerosol concentrations may invigorate clouds. Several studies have shown cloud invigoration [121,130–132] under weak wind conditions and suppression of convection under strong wind shear [133–135]. The atmospheric humidity also has a positive effect under more humid conditions [132,136].

10.6 Critical knowledge gaps and recommendations

To summaries, we identify the critical knowledge gaps in aerosol–cloud–radiation interactions research over India and adjoining seas: (1) spatio-temporal heterogeneity in aerosol sources leading to varying magnitude of aerosol numbers, composition, and size from one location to another, (2) aerosol-related changes in different cloud regimes (warm, cold, mixed, continental, maritime) and thus hydrological cycle and climate, (3) rain formation processes under varying thermodynamical conditions, and (4) quantifying the aerosol impact on the initiation of warm and cold rain processes. Despite extensive research using multiple approaches (ground, aircraft, satellite, and modeling) over more than two decades in India and elsewhere, the current understanding of aerosol-related changes to the cloud, radiation, rainfall, and climate remains elusive. Thus, it is crucial to develop a superior process-level understanding of aerosol processes, interactions, and feedback mechanisms to accurately estimate the mean radiative forcing over India and adjoining Seas. Mixed-phase clouds contribute a significant part of Indian summer monsoon clouds, which are not understood completely and need further focused process studies are required. Moreover, a fundamental question also arises about the significance of aerosol indirect effect at the climatic scale, that is, are the aerosol–cloud interactions substantial enough to manifest climate scale modifications to our climate (for example, cloud cover trends, extreme precipitation trends or surface temperature trends), and if yes, how to quantify the associations. We propose a high spatial and temporal resolution network for ground-based measurements of aerosol, CCN, INP, cloud properties and environmental factors from different techniques to obtain the vertical distribution of all these parameters and regional modeling simulation to better understand the aerosol–cloud-climate processes and feedbacks over India.

References

[1] W.C. Hinds, Aerosol Technology, John Wiley & Sons, New York, 1982.

[2] R. Jaenicke, Natural aerosols, Ann. N. Y. Acad. Sci. 338 (1980) 317–329. doi: 10.1111/j.1749-6632.1980.tb17129.x.

[3] L.A. Remer, R.G. Kleidman, R.C. Levy, Y.J. Kaufman, D. Tanré, S. Mattoo, J.V. Martins, C. Ichoku, I. Koren, H. Yu, B.N. Holben, Global aerosol climatology from the MODIS satellite sensors, J. Geophys. Res. 113 (2008), doi:10.1029/2007jd009661.

[4] R.J. Charlson, S.E. Schwartz, J.M. Hales, R.D. Cess, J.A. Coakley, J.E. Hansen, D.J. Hofmann, Climate forcing by anthropogenic aerosols, Science 255 (1992) 423–430. doi:10.1126/science.255.5043.423.

[5] P. Forster, V. Ramaswamy, P. Artaxo, T. Berntsen, R. Betts, D.W. Fahey, J. Haywood, J. Lean, D.C. Lowe, G. Myhre, J. Nganga, R. Prinn, G. Raga, M. Schulz, R. Van Dorland, Changes in Atmospheric Constituents and in Radiative Forcing, Cambridge University Press, Cambridge, 2007.
[6] O. Boucher, D. Randall, P. Artaxo, C. Bretherton, G. Feingold, P. Forster, V.-M. Kerminen, Y. Kondo, H. Liao, U. Lohmann, P. Rasch, S.K. Satheesh, S. Sherwood, B. Stevens, X.Y. Zhang, Clouds and aerosols. in: T.F. Stocker, D.Q., G.-K. Plattner, M. Tignor, S.K. Allen, J. Doschung, A. Nauels, Y. Xia, V. Bex, P.M. Midgley (Eds.), Climate Change 2013: The Physical Science Basis. Contribution of Working Group I to the Fifth Assessment Report of the Intergovernmental Panel on Climate Change, Cambridge University Pres, 2013, pp. 571–657.
[7] A.S. Ackerman, O.B. Toon, D.E. Stevens, A.J. Heymsfield, V. Ramanathan, E.J. Welton, Reduction of tropical cloudiness by soot, J. Sci. 288 (2000) 1042–1047. doi:10.1126/science.288.5468.1042 %.
[8] J. Hansen, M. Sato, R. Ruedy, Radiative forcing and climate response. J. Geophys. Res., 102 (1997) 6831–6864. doi:10.1029/96JD03436.
[9] T.C. Bond, E. Bhardwaj, R. Dong, R. Jogani, S. Jung, C. Roden, D.G. Streets, N.M. Trautmann, Historical emissions of black and organic carbon aerosol from energy-related combustion, 1850–2000, Global Biogeochem. Cycles 21 (GB2018) (2007). doi:10.1029/2006GB002840.
[10] K. Ding, X. Huang, A. Ding, M. Wang, H. Su, V.-M. Kerminen, T. Petäjä, Z. Tan, Z. Wang, D. Zhou, J. Sun, H. Liao, H. Wang, K. Carslaw, R. Wood, P. Zuidema, D. Rosenfeld, M. Kulmala, C. Fu, U. Pöschl, Y. Cheng, M.O. Andreae, Aerosol-boundary-layer-monsoon interactions amplify semi-direct effect of biomass smoke on low cloud formation in southeast Asia, Nat. Commun. 12 (2021) 6416. doi:10.1038/s41467-021-26728-4.
[11] B.T. Johnson, K.P. Shine, P.M. Forster, The semi-direct aerosol effect: Impact of absorbing aerosols on marine stratocumulus, Q. J. R. Meteorol. Soc. 130 (2004) 1407–1422. doi:10.1256/qj.03.61.
[12] E.M. Wilcox, Stratocumulus cloud thickening beneath layers of absorbing smoke aerosol, Atmos. Chem. Phys. 10 (2010) 11769–11777. doi:10.5194/acp-10-11769-2010.
[13] E.M. Wilcox, R.M. Thomas, P.S. Praveen, K. Pistone, F.A.M. Bender, V. Ramanathan, Black carbon solar absorption suppresses turbulence in the atmospheric boundary layer, Proc. Natl Acad. Sci. 113 (2016) 11794. doi:10.1073/pnas.1525746113.
[14] C.W. Stjern, B.H. Samset, G. Myhre, P.M. Forster, Ø. Hodnebrog, T. Andrews, O. Boucher, G. Faluvegi, T. Iversen, M. Kasoar, V. Kharin, A. Kirkevåg, J.-F. Lamarque, D. Olivié, T. Richardson, D. Shawki, D. Shindell, C.J. Smith, T. Takemura, A. Voulgarakis, Rapid adjustments cause weak surface temperature response to increased black carbon concentrations, J. Geophys. Res. Atmos. 122 (11) (2017) 462–481. doi:10.1002/2017JD027326.
[15] IPCC, Climate Change 2021: The Physical Science Basis. Contribution of Working Group I to the Sixth Assessment Report of the Intergovernmental Panel on Climate Change, Cambridge University Press, Cambridge, 2021.
[16] R.J. Charlson, J. Langner, H. Rodhe, C.B. Leovy, S.G. Warren, Perturbation of the northern hemisphere radiative balance by backscattering from anthropogenic sulfate aerosols, Tellus A: Dyn. Meteorol. Oceanogr. 43 (1991) 152–163. doi:10.3402/tellusa.v43i4.11944.
[17] J. Gliß, A. Mortier, M. Schulz, E. Andrews, Y. Balkanski, S.E. Bauer, A.M.K. Benedictow, H. Bian, R. Checa-Garcia, M. Chin, P. Ginoux, J.J. Griesfeller, A. Heckel, Z. Kipling, A. Kirkevåg, H. Kokkola, P. Laj, P. Le Sager, M.T. Lund, C. Lund Myhre, H. Matsui, G. Myhre, D. Neubauer, T. van Noije, P. North, D.J.L. Olivié, S. Rémy, L. Sogacheva, T. Takemura, K. Tsigaridis, S.G. Tsyro, AEROCOM phase III multi-model evaluation of the aerosol life cycle and optical properties using ground- and space-based remote sensing as well as surface in situ observations, Atmos. Chem. Phys. 21 (2021) 87–128. doi:10.5194/acp-21-87-2021.
[18] G. Myhre, W. Aas, R. Cherian, W. Collins, G. Faluvegi, M. Flanner, P. Forster, Ø. Hodnebrog, Z. Klimont, M.T. Lund, J. Mülmenstädt, C. Lund Myhre, D. Olivié, M. Prather, J. Quaas, B.H. Samset, J.L. Schnell, M. Schulz, D. Shindell, R.B. Skeie, T. Takemura, S. Tsyro, Multi-model simulations of aerosol and ozone radiative forcing due to anthropogenic emission changes during the period 1990–2015, Atmos. Chem. Phys. 17 (2017) 2709–2720. doi:10.5194/acp-17-2709-2017.
[19] P. Xian, J.S. Reid, E.J. Hyer, C.R. Sampson, J.I. Rubin, M. Ades, N. Asencio, S. Basart, A. Benedetti, P.S. Bhattacharjee, M.E. Brooks, P.R. Colarco, A.M. da Silva, T.F. Eck, J. Guth, O. Jorba, R. Kouznetsov, Z. Kipling, M. Sofiev, C. Perez Garcia-Pando, Y. Pradhan, T. Tanaka, J. Wang, D.L.

Westphal, K. Yumimoto, Current state of the global operational aerosol multi-model ensemble: An update from the international cooperative for aerosol prediction (ICAP), Q J R Meteorol Soc 145 (2019) 176–209, doi:10.1002/qj.3497.

[20] V. Buchard, A.M. da Silva, C.A. Randles, P. Colarco, R. Ferrare, J. Hair, C. Hostetler, J. Tackett, D. Winker, Evaluation of the surface pm 2.5 in version 1 of the NASA MERRA aerosol reanalysis over the United States, Atmos. Environ. 125 (2016) 100–111. doi:10.1016/j.atmosenv.2015.11.004.

[21] P.K. Quinn, T.S. Bates, Regional aerosol properties: Comparisons of boundary layer measurements from Ace 1, Ace 2, Aerosols99, INDOEX, Ace ASIA, TARFOX, and NEAQS. 110 (2005). doi:10.1029/2004JD004755.

[22] K.K. Moorthy, S. Satheesh, Evolution of aerosol research in India and the RAWEX-GVAX: An overview, Current Science 111 (2016) 53–75.

[23] S.N. Tripathi, V. Tare, N. Chinnam, A.K. Srivastava, S. Dey, A. Agarwal, S. Kishore, R.B. Lal, M. Manar, V.P. Kanawade, S.S.S. Chauhan, M. Sharma, R.R. Reddy, K.R. Gopal, K. Narasimhulu, L.S.S. Reddy, S. Gupta, S. Lal, Measurements of atmospheric parameters during Indian Space Research Organization Geosphere Biosphere Programme land campaign II at a typical location in the Ganga Basin: 1, Phys. Opt. Properties 111 (D23209) (2006). doi:10.1029/2006JD007278.

[24] B.N. Holben, T.F. Eck, I. Slutsker, D. Tanré, J.P. Buis, A. Setzer, E. Vermote, J.A. Reagan, Y.J. Kaufman, T. Nakajima, F. Lavenu, I. Jankowiak, A. Smirnov, Aeronet—A federated instrument network and data archive for aerosol characterization, Remote Sens. Environ. 66 (1998) 1–16. doi:10.1016/S0034-4257(98)00031-5.

[25] K.K. Moorthy, S.S. Babu, M.R. Manoj, S.K. Satheesh, Buildup of aerosols over the Indian region, Geophys. Res. Lett. 40 (2013) 1011–1014, doi:10.1002/grl.50165.

[26] T. Takamura, T. Nakajima, Overview of Skynet and its activities, Ópti. Pura Y Apl. 37 (2004) 3303–3308.

[27] D.M. Giles, A. Sinyuk, M.G. Sorokin, J.S. Schafer, A. Smirnov, I. Slutsker, T.F. Eck, B.N. Holben, J.R. Lewis, J.R. Campbell, E.J. Welton, S.V. Korkin, A.I. Lyapustin, Advancements in the aerosol robotic network (Aeronet) version 3 database: Automated near-real-time quality control algorithm with improved cloud screening for sun photometer aerosol optical depth (AOD) measurements, Atmos. Meas. Tech. 12 (2019) 169–209. doi:10.5194/amt-12-169-2019.

[28] M. Petrenko, C. Ichoku, Coherent uncertainty analysis of aerosol measurements from multiple satellite sensors, Atmos. Chem. Phys. 13 (2013) 6777–6805. doi:10.5194/acp-13-6777-2013.

[29] L.A. Remer, Y.J. Kaufman, D. Tanré, S. Mattoo, D.A. Chu, J.V. Martins, R.-R. Li, C. Ichoku, R.C. Levy, R.G. Kleidman, T.F. Eck, E. Vermote, B.N. Holben, The MODIS aerosol algorithm, products, and validation, J. Atmos. Sci. 62 (2005) 947–973. doi:10.1175/JAS3385.1.

[30] S. Twomey, The influence of pollution on the shortwave albedo of clouds, J. Atmos. Sci. 34 (1977) 1149–1152. doi:10.1175/1520-0469(1977)034<1149:Tiopot>2.0.Co;2.

[31] V. Ramaswamy, O. Boucher, J. Haigh, D. Hauglustaine, J. Haywood, G. Myhre, T. Nakajima, G. Shi, S. Solomon, Radiative Forcing of Climate Change IPCC, (2001) 349–416.

[32] U. Lohmann, J. Feichter, Global indirect aerosol effects: A review, Atmos. Chem. Phys. 5 (2005) 715–737. doi:10.5194/acp-5-715-2005.

[33] B.A. Albrecht, Aerosols, cloud microphysics, and fractional cloudiness, Science 245 (1989) 1227–1230. doi:10.1126/science.245.4923.1227.

[34] N. Bellouin, J. Quaas, E. Gryspeerdt, S. Kinne, P. Stier, D. Watson-Parris, O. Boucher, K.S. Carslaw, M. Christensen, A.-L. Daniau, J.-L. Dufresne, G. Feingold, S. Fiedler, P. Forster, A. Gettelman, J.M. Haywood, U. Lohmann, F. Malavelle, T. Mauritsen, D.T. McCoy, G. Myhre, J. Mülmenstädt, D. Neubauer, A. Possner, M. Rugenstein, Y. Sato, M. Schulz, S.E. Schwartz, O. Sourdeval, T. Storelvmo, V. Toll, D. Winker, B. Stevens, Bounding global aerosol radiative forcing of climate change, Reviews of Geophysics 58 (2020) e2019RG000660, doi:10.1029/2019rg000660.

[35] E. Gryspeerdt, J. Mülmenstädt, A. Gettelman, F.F. Malavelle, H. Morrison, D. Neubauer, D.G. Partridge, P. Stier, T. Takemura, H. Wang, M. Wang, K. Zhang, Surprising similarities in model and observational aerosol radiative forcing estimates, Atmos. Chem. Phys. 20 (2020) 613–623. doi:10.5194/acp-20-613-2020.

[36] U. Lohmann, Anthropogenic aerosol influences on mixed-phase clouds, Curr. Clim. Change Rep. 3 (2017) 32–44. doi:10.1007/s40641-017-0059-9.

[37] J. Fan, Y. Wang, D. Rosenfeld, X. Liu, Review of aerosol–cloud interactions: Mechanisms, significance, and challenges, J. Atmos. Sci. 73 (2016) 4221–4252. doi:10.1175/jas-d-16-0037.1.
[38] M.W. Christensen, D. Neubauer, C.A. Poulsen, G.E. Thomas, G.R. McGarragh, A.C. Povey, S.R. Proud, R.G. Grainger, Unveiling aerosol–cloud interactions – part 1: Cloud contamination in satellite products enhances the aerosol indirect forcing estimate, Atmos. Chem. Phys. 17 (2017) 13151–13164. doi:10.5194/acp-17-13151-2017.
[39] E. Gryspeerdt, T. Goren, O. Sourdeval, J. Quaas, J. Mülmenstädt, S. Dipu, C. Unglaub, A. Gettelman, M. Christensen, Constraining the aerosol influence on cloud liquid water path, Atmos. Chem. Phys. 19 (2019) 5331–5347. doi:10.5194/acp-19-5331-2019.
[40] M.O. Andreae, Correlation between cloud condensation nuclei concentration and aerosol optical thickness in remote and polluted regions, Atmos. Chem. Phys. 9 (2009) 543–556. doi:10.5194/acp-9-543-2009.
[41] T.V. Prabha, A. Khain, R.S. Maheshkumar, G. Pandithurai, J.R. Kulkarni, M. Konwar, B.N. Goswami, Microphysics of premonsoon and monsoon clouds as seen from in situ measurements during the cloud aerosol interaction and precipitation enhancement experiment (CAIPEEX), J. Atmos. Sci. 68 (2011) 1882–1901. doi:10.1175/2011jas3707.1.
[42] D. Rosenfeld, M.O. Andreae, A. Asmi, M. Chin, G. de Leeuw, D.P. Donovan, R. Kahn, S. Kinne, N. Kivekäs, M. Kulmala, W. Lau, K.S. Schmidt, T. Suni, T. Wagner, M. Wild, J. Quaas. Global observations of aerosol-cloud-precipitation-climate interactions. Rev. Geophys., 52 (2014), 750–808. doi:10.1002/2013RG000441.
[43] E. Gryspeerdt, J. Quaas, N. Bellouin, Constraining the aerosol influence on cloud fraction, Journal of Geophysical Research 121 (2016) 3566–3583, doi:10.1002/2015JD023744.
[44] R. Bennartz, J. Rausch, Global and regional estimates of warm cloud droplet number concentration based on 13 years of Aqua-Modis observations, Atmos. Chem. Phys. 17 (2017) 9815–9836. doi:10.5194/acp-17-9815-2017.
[45] Y. Zhu, D. Rosenfeld, Z. Li, Under what conditions can we trust retrieved cloud drop concentrations in broken marine stratocumulus? Journal of Geophysical Research: Atmospheres,123 (2018) 8754–8767, doi:10.1029/2017jd028083.
[46] D. Rosenfeld, Y. Zhu, M. Wang, Y. Zheng, T. Goren, S. Yu, Aerosol-driven droplet concentrations dominate coverage and water of oceanic low-level clouds, J. Sci. 363 (2019) eaav0566. doi:10.1126/science.aav0566%.
[47] S. Ramachandran, R. Cherian, Regional and seasonal variations in aerosol optical characteristics and their frequency distributions over India during 2001–2005, J. Geophys. Res. Atmos 113 (D08207) (2008). doi: 10.1029/2007JD008560.
[48] A.S. Frisch, G. Feingold, C.W. Fairall, T. Uttal, J.B. Snider, On cloud radar and microwave radiometer measurements of stratus cloud liquid water profiles, J. Geophys. Res. Atmos. 103 (1998) 23195–23197. doi:10.1029/98JD01827.
[49] B.L. Madhavan, Y. He, Y. Wu, B. Gross, F. Moshary, S. Ahmed, Development of a ground based remote sensing approach for direct evaluation of aerosol-cloud interaction, Atmosphere 3 (2012) 468–494.
[50] J.H. Seinfeld, C. Bretherton, K.S. Carslaw, H. Coe, P.J. DeMott, E.J. Dunlea, G. Feingold, S. Ghan, A.B. Guenther, R. Kahn, I. Kraucunas, S.M. Kreidenweis, M.J. Molina, A. Nenes, J.E. Penner, K.A. Prather, V. Ramanathan, V. Ramaswamy, P.J. Rasch, A.R. Ravishankara, D. Rosenfeld, G. Stephens, R. Wood, Improving our fundamental understanding of the role of aerosol−cloud interactions in the climate system, Proc. Natl. Acad. Sci. 113 (2016) 5781–5790. doi:10.1073/pnas.1514043113.
[51] E.T. Sena, A. McComiskey, G. Feingold, A long-term study of aerosol–cloud interactions and their radiative effect at the southern great plains using ground-based measurements, Atmos. Chem. Phys. 16 (2016) 11301–11318. doi:10.5194/acp-16-11301-2016.
[52] S.S. Babu, M.R. Manoj, K.K. Moorthy, M.M. Gogoi, V.S. Nair, S.K. Kompalli, S.K. Satheesh, K. Niranjan, K. Ramagopal, P.K. Bhuyan, D. Singh, Trends in aerosol optical depth over Indian region: Potential causes and impact indicators. Journal of Geophysical Research-Atmospheres, 118 (2013) 11, 794–711. doi:10.1002/2013JD020507.
[53] R. Srivastava, Trends in aerosol optical properties over south Asia, Int. J. Climatol. 37 (2017) 371–380. doi:10.1002/joc.4710.
[54] A. Thomas, C. Sarangi, V.P. Kanawade, Recent increase in winter hazy days over central India and the Arabian sea, Sci. Rep. 9 (2019) 17406. doi:10.1038/s41598-019-53630-3.

[55] A.R. Ravishankara, L.M. David, J.R. Pierce, C. Venkataraman, Outdoor air pollution in India is not only an urban problem, J Proc. Natl. Acad. Sci. 117 (2020) 28640–28644. doi:10.1073/pnas.2007236117.
[56] S. Dey, L. Di Girolamo, A decade of change in aerosol properties over the Indian subcontinent, Geophys. Res. Lett. 38 (L14811) (2011). doi:10.1029/2011GL048153.
[57] S. Jose, V.S. Nair, S.S. Babu, Anthropogenic emissions from south Asia reverses the aerosol indirect effect over the northern Indian ocean, Sci. Rep. 10 (2020) 18360. doi:10.1038/s41598-020-74897-x.
[58] V.S. Nair, V.N. Jayachandran, S.K. Kompalli, M.M. Gogoi, S.S. Babu, Cloud condensation nuclei properties of south Asian outflow over the northern Indian ocean during winter, Atmos. Chem. Phys. 20 (2020) 3135–3149. doi:10.5194/acp-20-3135-2020.
[59] S.K. Satheesh, J. Srinivasan, Enhanced aerosol loading over arabian sea during the pre-monsoon season: Natural or anthropogenic? Geophys. Res. Lett. 29 (2002) 21–24. doi:10.1029/2002GL015687.
[60] V. Vinoj, P.J. Rasch, H. Wang, J.-H. Yoon, P.-L. Ma, K. Landu, B. Singh, Short-term modulation of Indian summer monsoon rainfall by west Asian dust, Nat. Geosci. 7 (2014) 308–313. doi:10.1038/ngeo2107.
[61] N.B. Lakshmi, V.S. Nair, S. Suresh Babu, Vertical structure of aerosols and mineral dust over the Bay of Bengal from multisatellite observations. Journal of Geophysical Research: Atmospheres, 122 (2017) 12,845–812,861. doi:10.1002/2017JD027643.
[62] S.K. Pandey, V. Vinoj, K. Landu, S.S. Babu, Declining pre-monsoon dust loading over south Asia: Signature of a changing regional climate, Sci. Rep. 7 (2017) 16062. doi:10.1038/s41598-017-16338-w.
[63] S.K. Satheesh, K.K. Moorthy, S.S. Babu, V. Vinoj, C.B.S. Dutt, Climate implications of large warming by elevated aerosol over India, Geophys. Res. Lett. 35 (2008) L19809, doi:10.1029/2008GL034944.
[64] V.S. Nair, S.S. Babu, K.K. Moorthy, S.S. Prijith, Spatial gradients in aerosol-induced atmospheric heating and surface dimming over the oceanic regions around India: Anthropogenic or natural? J. Clim. 26 (2013) 7611–7621. doi:10.1175/JCLI-D-12-00616.1.
[65] M. Mehta, R. Singh, A. Singh, N. Singh, Anshumali, Recent global aerosol optical depth variations and trends: A comparative study using MODIS and MISR level 3 datasets, Remote Sens. Environ. 181 (2016) 137–150. doi:10.1016/j.rse.2016.04.004.
[66] A. Mani, O. Chacko, S. Hariharan, A study of Ångström's turbidity parameters from solar radiation measurements in India, Tellus 21 (1969) 829–843. doi:10.3402/tellusa.v21i6.10149.
[67] A. Mani, O. Chacko, N.V. Iyer, Atmospheric turbidity over India from solar radiation measurements, Sol. Energy 14 (1973) 185–195. doi:10.1016/0038-092X(73)90033-9.
[68] K.K. Moorthy, A. Saha, B.S.N. Prasad, K. Niranjan, D. Jhurry, P.S. Pillai, Aerosol optical depths over peninsular india and adjoining oceans during the INDOEX campaigns: Spatial, temporal, and spectral characteristics, J. Geophys. Res. 106 (2001) 28539–28554, doi:10.1029/2001JD900169.
[69] K.K. Moorthy, S.S. Babu, S.K. Satheesh, Aerosol characteristics and radiative impacts over the Arabian sea during the intermonsoon season: Results from ARMEX field campaign, J. Atmos. Sci. 62 (2005) 192–206. doi:10.1175/jas-3378.1.
[70] V.S. Nair, K.K. Moorthy, D.P. Alappattu, P.K. Kunhikrishnan, S. George, P.R. Nair, S.S. Babu, B. Abish, S.K. Satheesh, S.N. Tripathi, K. Niranjan, B.L. Madhavan, V. Srikant, C.B.S. Dutt, K.V.S. Badarinath, R.R. Reddy, J. Geophys. Res. 112 (2007), doi:10.1029/2006jd008099.
[71] V. Tare, S.N. Tripathi, N. Chinnam, A.K. Srivastava, S. Dey, M. Manar, V.P. Kanawade, A. Agarwal, S. Kishore, R.B. Lal, M. Sharma, Measurements of atmospheric parameters during Indian Space Research Organization Geosphere Biosphere Program land campaign II at a typical location in the ganga basin: 2. Chemical properties, J. Geophys. Res. Atmos. 111 (D23210) (2006). doi:10.1029/2006jd007279.
[72] S.N. Beegum, K.K. Moorthy, V.S. Nair, S.S. Babu, S.K. Satheesh, V. Vinoj, R.R. Reddy, K.R. Gopal, K.V.S. Badarinath, K. Niranjan, S.K. Pandey, M. Behera, A. Jeyaram, P.K. Bhuyan, M.M. Gogoi, S. Singh, P. Pant, U.C. Dumka, Y. Kant, J.C. Kuniyal, D. Singh, Characteristics of spectral aerosol optical depths over India during ICARB, J. Earth Syst. Sci. 117 (2008) 303–313. doi:10.1007/s12040-008-0033-y.

[73] K.K. Moorthy, V.S. Nair, S.S. Babu, S.K. Satheesh, Spatial and vertical heterogeneities in aerosol properties over oceanic regions around India: Implications for radiative forcing, Q.J.R. Meteorol. Soc. 135 (2009) 2131–2145, doi:10.1002/qj.525.

[74] K.K. Moorthy, S.K. Satheesh, S.S. Babu, C.B.S. Dutt, Integrated campaign for aerosols, gases and radiation budget (ICARB): An overview, J. Earth Syst. Sci. 117 (2008) 243–262. doi:10.1007/s12040-008-0029-7.

[75] M.R. Manoj, S.K. Satheesh, K.K. Moorthy, M.M. Gogoi, S.S. Babu, Decreasing trend in black carbon aerosols over the Indian region, Geophys. Res. Lett. 46 (2019) 2903–2910. doi:10.1029/2018GL081666.

[76] S.S. Babu, V.S. Nair, M.M. Gogoi, K.K. Moorthy, Seasonal variation of vertical distribution of aerosol single scattering albedo over Indian sub-continent: RAWEX aircraft observations, Atmos. Environ. 125 (2016) 312–323. doi:10.1016/j.atmosenv.2015.09.041.

[77] M.M. Gogoi, V.N. Jayachandran, A. Vaishya, S.N.S. Babu, S.K. Satheesh, K.K. Moorthy, Airborne in situ measurements of aerosol size distributions and black carbon across the Indo-Gangetic plain during SWAAMI–RAWEX, Atmos. Chem. Phys. 20 (2020) 8593–8610. doi:10.5194/acp-20-8593-2020.

[78] M.M. Gogoi, N.B. Lakshmi, V.S. Nair, S.K. Kompalli, K.K. Moorthy, S.S. Babu, Seasonal contrast in the vertical profiles of aerosol number concentrations and size distributions over India: Implications from RAWEX aircraft campaign, J. Earth Syst. Sci. 128 (2019) 225. doi:10.1007/s12040-019-1246-y.

[79] K.E. Trenberth, D.P. Stepaniak, J.M. Caron, The global monsoon as seen through the divergent atmospheric circulation, J. Clim. 13 (2000) 3969–3993. doi:10.1175/1520-0442(2000)013<3969:Tgmast>2.0.Co;2.

[80] D.R. Sikka, S. Gadgil, On the maximum cloud zone and the ITCZ over Indian longitudes during the southwest monsoon, Monthly Weather Rev. 108 (1980) 1840–1853. doi:10.1175/1520-0493(1980)108<1840:OTMCZA>2.0.CO;2.

[81] M. Rajeevan, P. Rohini, K. Niranjan Kumar, J. Srinivasan, C.K. Unnikrishnan, A study of vertical cloud structure of the Indian summer monsoon using Cloudsat data, Clim. Dyn. 40 (2013) 637–650. doi:10.1007/s00382-012-1374-4.

[82] G.S. Bhat, S. Kumar, Vertical structure of cumulonimbus towers and intense convective clouds over the south Asian region during the summer monsoon season, J. Geophys. Res. Atmos. 120 (2015) 1710–1722. doi:10.1002/2014JD022552.

[83] K. Saikranthi, T. Narayana Rao, B. Radhakrishna, S.V.B. Rao, Morphology of the vertical structure of precipitation over India and adjoining oceans based on long-term measurements of TRMM pr, J. Geophys. Res. Atmos. 119 (2014) 8433–8449. doi:10.1002/2014JD021774.

[84] M. Rajeevan, J. Srinivasan, Net cloud radiative forcing at the top of the atmosphere in the Asian monsoon region, J. Clim. 13 (2000) 650–657. doi:10.1175/1520-0442(2000)013<0650:Ncrfat>2.0.Co;2.

[85] J.T. Kiehl, V. Ramanathan, Comparison of cloud forcing derived from the earth radiation budget experiment with that simulated by the NCAR community climate model, J. Geophys. Res. 95 (1990) 11679–11698, doi:10.1029/JD095iD08p11679.

[86] B.V. Thampi, R. Roca, Investigation of negative cloud radiative forcing over the Indian subcontinent and adjacent oceans during the summer monsoon season, Atmos. Chem. Phys. 14 (2014) 6739–6758. doi:10.5194/acp-14-6739-2014.

[87] V. Ravi Kiran, M. Rajeevan, H. Gadhavi, S.V.B. Rao, A. Jayaraman, Role of vertical structure of cloud microphysical properties on cloud radiative forcing over the Asian monsoon region, Clim. Dyn. 45 (2015) 3331–3345. doi:10.1007/s00382-015-2542-0.

[88] T. Saud, S. Dey, S. Das, S. Dutta, A satellite-based 13-year climatology of net cloud radiative forcing over the Indian monsoon region, Atmos. Res. 182 (2016) 76–86. doi:10.1016/j.atmosres.2016.07.017.

[89] C. Sarangi, S.N. Tripathi, V.P. Kanawade, I. Koren, D.S. Pai, Investigation of the aerosol–cloud–rainfall association over the Indian summer monsoon region, Atmos. Chem. Phys. 17 (2017) 5185–5204. doi:10.5194/acp-17-5185-2017.

[90] A. Slingo, Sensitivity of the earth's radiation budget to changes in low clouds, Nature 343 (1990) 49–51. doi:10.1038/343049a0.

[91] S.S. Prijith, C.B. Lima, M.V. Ramana, M.V.R.S. Sai, Intra-seasonal contrasting trends in clouds due to warming induced circulation changes, Sci. Rep. 11 (2021) 16985. doi:10.1038/s41598-021-96246-2.

[92] V. Sathiyamoorthy, C. Mahesh, K. Gopalan, S. Prakash, B.P. Shukla, Characteristics of low clouds over the Arabian sea, J. Geophys. Res. Atmos. 118 (2013) 489–503, doi:10.1002/2013JD020553.

[93] Y.-C. Chen, M.W. Christensen, G.L. Stephens, J.H. Seinfeld, Satellite-based estimate of global aerosol–cloud radiative forcing by marine warm clouds, Nat. Geosci. 7 (2014) 643–646. doi:10.1038/ngeo2214.

[94] J.C. Collier, G.J. Zhang, Aerosol direct forcing of the summer Indian monsoon as simulated by the NCAR cam3, Clim. Dyn. 32 (2009) 313–332. doi:10.1007/s00382-008-0464-9.

[95] G.A. Meehl, J.M. Arblaster, W.D. Collins, Effects of black carbon aerosols on the Indian monsoon, J. Clim. 21 (2008) 2869–2882. doi:10.1175/2007JCLI1777.1.

[96] S. Menon, J. Hansen, L. Nazarenko, Y. Luo, Climate effects of black carbon aerosols in China and India, J. Sci. 297 (2002) 2250–2253. doi:10.1126/science.1075159 %.

[97] K.M. Lau, K.M. Kim, Observational relationships between aerosol and Asian monsoon rainfall, and circulation, Geophys. Res. Lett. 33 (L21810) (2006), doi:10.1029/2006gl027546.

[98] V. Ramanathan, C. Chung, D. Kim, T. Bettge, L. Buja, J.T. Kiehl, W.M. Washington, Q. Fu, D.R. Sikka, M. Wild, Atmospheric brown clouds: Impacts on south Asian climate and hydrological cycle, J. Proc. Natl. Acad. Sci. U. S. A. 102 (2005) 5326–5333. doi:10.1073/pnas.0500656102.

[99] C. Sarangi, S.N. Tripathi, A.K. Mishra, A. Goel, E.J. Welton, Elevated aerosol layers and their radiative impact over Kanpur during monsoon onset period, J. Geophys. Res. Atmos. 121 (2016) 7936–7957, doi:10.1002/2015JD024711.

[100] W.K.M. Lau, K.-M. Kim, Fingerprinting the impacts of aerosols on long-term trends of the Indian summer monsoon regional rainfall, Geophys. Res. Lett. 37 (2010). doi:10.1029/2010GL043255.

[101] M.A. Bollasina, Y. Ming, V. Ramaswamy, Anthropogenic aerosols and the weakening of the south Asian summer monsoon, J. Sci. 334 (2011) 502–505. doi:10.1126/science.1204994.

[102] P. Dave, M. Bhushan, C. Venkataraman, Aerosols cause intraseasonal short-term suppression of Indian monsoon rainfall, Sci. Rep. 7 (2017) 17347. doi:10.1038/s41598-017-17599-1.

[103] E.N. Anagnostou, C. Kummerow, Stratiform and convective classification of rainfall using SSM/I 85-ghz brightness temperature observations. Journal of Atmospheric and Oceanic Technology, 14 (1997) 570–575. doi:10.1175/1520-0426.

[104] D. Rosenfeld, C.W. Ulbrich, Cloud microphysical properties, processes, and rainfall estimation opportunities, in: R.M. Wakimoto, R. Srivastava (Eds.), Cloud microphysical properties, processes, and rainfall estimation opportunities, Radar and Atmospheric Science: A Collection of Essays in Honor of David Atlas (2003) 237–258. doi:10.1007/978-1-878220-36-3_10.

[105] A. Thomas, V.P. Kanawade, K. Chakravarty, A.K. Srivastava, Characterization of raindrop size distributions and its response to cloud microphysical properties, Atmos. Res. 249 (2021) 105292. doi:10.1016/j.atmosres.2020.105292.

[106] V.N. Bringi, V. Chandrasekar, J. Hubbert, E. Gorgucci, W.L. Randeu, M. Schoenhuber, Raindrop size distribution in different climatic regimes from disdrometer and dual-polarized radar analysis. Journal of the Atmospheric Sciences, 60 (2003) 354–365. doi:10.1175/1520-0469(2003)060<0354:Rsdidc>2.0.Co;2.

[107] L. Oreopoulos, N. Cho, D. Lee, Using MODIS cloud regimes to sort diagnostic signals of aerosol-cloud-precipitation interactions, J Geophys Res Atmos. 122 (2017) 5416–5440, doi:10.1002/2016JD026120.

[108] L. Oreopoulos, N. Cho, D. Lee, A global survey of apparent aerosol-cloud interaction signals, Journal of Geophysical Research: Atmospheres 125 (2020) e2019JD031287, doi:10.1029/2019JD031287.

[109] D. Bhattu, S. Tripathi, Ccn closure study: Effects of aerosol chemical composition and mixing state, J. Geophys. Res. Atmos. 120 (2015) 766–783.

[110] U. Dusek, G.P. Frank, L. Hildebrandt, J. Curtius, J. Schneider, S. Walter, D. Chand, F. Drewnick, S. Hings, D. Jung, S. Borrmann, M.O. Andreae, Size matters more than chemistry for cloud-nucleating ability of aerosol particles, Science 312 (2006) 1375–1378. doi:10.1126/science.1125261.

[111] G. McFiggans, P. Artaxo, U. Baltensperger, H. Coe, M.C. Facchini, G. Feingold, S. Fuzzi, M. Gysel, A. Laaksonen, U. Lohmann, T.F. Mentel, D.M. Murphy, C.D. O'Dowd, J.R. Snider, E. Weingartner, The effect of physical and chemical aerosol properties on warm cloud droplet activation, Atmos. Chem. Phys. 6 (2006) 2593–2649. doi:10.5194/acp-6-2593-2006.

[112] R.-E.P. Sotiropoulou, A. Nenes, P.J. Adams, J.H. Seinfeld, Cloud condensation nuclei prediction error from application of Köhler theory: Importance for the aerosol indirect effect, J. Geophys. Res. Atmos. 112 (D12202) (2007). doi:10.1029/2006JD007834.

[113] S. Shika, H. Gadhavi, M.N.S. Suman, R. Ravikrishna, S.S. Gunthe, Atmospheric aerosol properties at a semi-rural location in southern India: Particle size distributions and implications for cloud droplet formation, SN Appl. Sci. 2 (2020) 1007. doi:10.1007/s42452-020-2804-2.

[114] P. Stier, Limitations of passive remote sensing to constrain global cloud condensation nuclei, Atmos. Chem. Phys. 16 (2016) 6595–6607. doi:10.5194/acp-16-6595-2016.

[115] A.S. Panicker, G. Pandithurai, S. Dipu, Aerosol indirect effect during successive contrasting monsoon seasons over Indian subcontinent using Modis data, Atmos. Environ. 44 (2010) 1937–1943. doi:10.1016/j.atmosenv.2010.02.015.

[116] G. Pandithurai, S. Dipu, T.V. Prabha, R.S. Maheskumar, J.R. Kulkarni, B.N. Goswami, Aerosol effect on droplet spectral dispersion in warm continental cumuli, J. Geophys. Res. 117 (2012), doi:10.1029/2011JD016532.

[117] V. Anil Kumar, G. Pandithurai, P.P. Leena, K.K. Dani, P. Murugavel, S.M. Sonbawne, R.D. Patil, R.S. Maheskumar, Investigation of aerosol indirect effects on monsoon clouds using ground-based measurements over a high-altitude site in western ghats, Atmos. Chem. Phys. 16 (2016) 8423–8430. doi:10.5194/acp-16-8423-2016.

[118] W. Cantrell, G. Shaw, C. Leck, L. Granat, H. Cachier, Relationships between cloud condensation nuclei spectra and aerosol particles on a south-north transect of the Indian ocean, J. Geophys. Res., 105 (2000) 15313–15320, doi:10.1029/2000JD900219.

[119] J. Kulkarni, R. Maheskumar, S. Morwal, B. Kumari, M. Kumari, C. Konwar, R. Deshpande, R. Joshi, G. Bhalwankar, P. Pandithurai, S. Safai, K. Narkhedkar, A. Dani, S. Nath, V. Nair, P. Sapre, S. Puranik, V. Kandalgaonkar, R. Mujumdar, B.N. Goswami, The cloud aerosol interaction and precipitation enhancement experiment (CAIPEEX): Overview and preliminary results, Curr. Sci. 102 (2012) 413–425.

[120] M.O. Andreae, D. Rosenfeld, P. Artaxo, A.A. Costa, G.P. Frank, K.M. Longo, M.A.F. Silva-Dias, Smoking rain clouds over the amazon, J. Sci. 303 (2004) 1337–1342. doi:10.1126/science.1092779.

[121] I. Koren, Y. Kaufman, D. Rosenfeld, L. Remer, Y. Rudich, Aerosol invigoration and restructuring of Atlantic convective clouds, Geophys. Res. Lett. 32 (L14828) (2005). doi:10.1029/2005GL023187.

[122] J. Fan, T. Yuan, J. Comstock, S. Ghan, A. Khain, L. Leung, Z. Li, V. Martins, M. Ovchinnikov, Dominant role by vertical wind shear in regulating aerosol effects on deep convective clouds, J. Geophys. Res. 114 (D22206) (2009). doi:10.1029/2009JD012352.

[123] I. Koren, G. Feingold, L.A. Remer, The invigoration of deep convective clouds over the Atlantic: Aerosol effect, meteorology or retrieval artifact? Atmos. Chem. Phys. 10 (2010) 8855–8872. doi:10.5194/acp-10-8855-2010.

[124] I. Koren, J.V. Martins, L.A. Remer, H. Afargan, Smoke invigoration versus inhibition of clouds over the amazon, Science 321 (2008) 946–949. doi:10.1126/science.1159185.

[125] I. Koren, G. Dagan, O. Altaratz, From aerosol-limited to invigoration of warm convective clouds, J Science 344 (2014) 1143–1146. doi:10.1126/science.1252595.

[126] C. Sarangi, V.P. Kanawade, S.N. Tripathi, A. Thomas, D. Ganguly, Aerosol-induced intensification of cooling effect of clouds during Indian summer monsoon, Nat. Commun. 9 (2018) 3754. doi:10.1038/s41467-018-06015-5.

[127] W.-K. Tao, J.-P. Chen, Z. Li, C. Wang, C. Zhang, Impact of aerosols on convective clouds and precipitation, Atmospheric Research 50 (2012) 142–157, doi:10.1029/2011rg000369.

[128] J. Fan, L.R. Leung, D. Rosenfeld, Q. Chen, Z. Li, J. Zhang, H. Yan, Microphysical effects determine macrophysical response for aerosol impacts on deep convective clouds, J. Proc. Natl. Acad. Sci. 110 (2013) E4581–E4590. doi:10.1073/pnas.1316830110.

[129] Z.A. Kanji, L.A. Ladino, H. Wex, Y. Boose, M. Burkert-Kohn, D.J. Cziczo, M. Krämer, Overview of ice nucleating particles J. Meteorol. Monogr. 58 (2017) 1.1–1.33. doi:10.1175/amsmonographs-d-16-0006.1.

[130] A. Khain, D. Rosenfeld, A. Pokrovsky, Aerosol impact on the dynamics and microphysics of deep convective clouds, Q.J.R. Meteorol. Soc.,131 (2005) 2639–2663, doi:10.1256/qj.04.62.

[131] A.P. Khain, K.D. Beheng, A. Heymsfield, A. Korolev, S.O. Krichak, Z. Levin, M. Pinsky, V. Phillips, T. Prabhakaran, A. Teller, S.C. van den Heever, J.-I. Yano, Representation of microphysical processes in

cloud-resolving models: Spectral (bin) microphysics versus bulk parameterization, Rev. Geophys. 53 (2015) 247–322, doi:10.1002/2014RG000468.
[132] W.-K. Tao, X. Li, A. Khain, T. Matsui, S. Lang, J. Simpson, Role of atmospheric aerosol concentration on deep convective precipitation: Cloud-resolving model simulations, J. Geophys. Res.,112 (2007), doi:10.1029/2007JD008728.
[133] J. Fan, M. Ovtchinnikov, J.M. Comstock, S.A. McFarlane, A. Khain, Ice formation in arctic mixed-phase clouds: Insights from a 3-D cloud-resolving model with size-resolved aerosol and cloud microphysics, J. Geophys. Res. 114 (2009), doi:10.1029/2008JD010782.
[134] A.P. Khain, N. BenMoshe, A. Pokrovsky, Factors determining the impact of aerosols on surface precipitation from clouds: An attempt at classification, J. Atmos. Sci. 65 (2008) 1721–1748. doi:10.1175/2007jas2515.1.
[135] S.S. Lee, L.J. Donner, V.T.J. Phillips, Y. Ming, Examination of aerosol effects on precipitation in deep convective clouds during the 1997 arm summer experiment, Q.J.R. Meteorol. Soc. 134 (2008) 1201–1220, doi:10.1002/qj.287.
[136] J. Fan, R. Zhang, G. Li, W.-K. Tao, Effects of aerosols and relative humidity on cumulus clouds, J. Geophys. Res. 112 (2007) D14204, doi:10.1029/2006JD008136.
[137] M. Konwar, S.K. Das, S.M. Deshpande, K. Chakravarty, B.N. Goswami, Microphysics of clouds and rain over the western ghat, J. Geophys. Res. Atmos. 119 (2014) 6140–6159, doi:10.1002/2014jd021606.
[138] V. Ramanathan, P.J. Crutzen, J. Lelieveld, A.P. Mitra, D. Althausen, J. Anderson, M.O. Andreae, W. Cantrell, G.R. Cass, C.E. Chung, A.D. Clarke, J.A. Coakley, W.D. Collins, W.C. Conant, F. Dulac, J. Heintzenberg, A.J. Heymsfield, B. Holben, S. Howell, J. Hudson, A. Jayaraman, J.T. Kiehl, T.N. Krishnamurti, D. Lubin, G. McFarquhar, T. Novakov, J.A. Ogren, I.A. Podgorny, K. Prather, K. Priestley, J.M. Prospero, P.K. Quinn, K. Rajeev, P. Rasch, S. Rupert, R. Sadourny, S.K. Satheesh, G.E. Shaw, P. Sheridan, F.P.J. Valero, Indian ocean experiment: An integrated analysis of the climate forcing and effects of the great Indo-Asian haze, J. Geophys. Res. 106 (2001) 28371–28398, doi:10.1029/2001jd900133.

CHAPTER 11

Aerosol loading over the Northern Indian Ocean using space-borne measurements

Harshbardhan Kumar[a,b], **Shani Tiwari**[a,b]
[a]Academy of Scientific and Innovative Research (AcSIR), Ghaziabad, India
[b]CSIR – National Institute of Oceanography, DonaPaula, Goa, India

11.1 Introduction

The rapid increase in population growth and industrialization during the last few decades (mainly over south and east Asian countries) causes a large burden of atmospheric pollutants which directly or indirectly affects the air quality, monsoonal circulation, hydrological cycle, and climate change [1–3]. Apart from the permanent and trace gases, earth's atmosphere contains a variety of solids or liquid particles dispersed into the air (as a fluid medium). The mixture of solid or liquid particles that are suspended into the atmosphere is popularly known as aerosols, having a typical size range from a few nanometers to several tens of micrometers. They are emitted into the atmosphere via several natural (such as wind-driven marine sea-salt spray, forest fire emission, dust storm, volcanic emission, etc.) and anthropogenic sources (like vehicular exhaust, industrial emission, coal combustion, biomass burning emission, etc.). Depending on the prevailing circulation patterns, they get distributed vertically or horizontally across the atmospheric column [4]. On a global scale, natural aerosols have a higher abundance as compared to anthropogenic, but on a regional scale, anthropogenic aerosols have a significant contribution and play a crucial role in the earth's radiation budget [5]. Atmospheric aerosols play a vital role in global radiation budget, atmospheric stability, cloud formation, hence precipitation and hydrological cycle also [6,7]. They are associated with large uncertainty in global radiative forcing due to different emission sources, chemical compositions, physicochemical and radiative properties [8]. But at the same time, they are highly uncertain species within the Earth's climate system, because of their dual nature, that is, cooling as well as warming [8–10]. Some aerosols being brighter than the land and ocean surface, reflect the incoming shortwave radiation and thus cool the planet, while others darker aerosols absorb shortwave radiation which heats the atmosphere and alter atmospheric dynamics [11]. The magnitude of light reflected or absorbed by the aerosol particle varies as a function of their size, type, abundance and especially the wavelength of interacting radiation [12]. Aerosol also affect the global

Atmospheric Remote Sensing
DOI: https://doi.org/10.1016/B978-0-323-99262-6.00010-9

climate indirectly by providing a feasible surface for cloud droplet nucleation acting as cloud condensation nuclei (CCN). Some hygroscopic finer fractions of atmospheric aerosols actively participate in cloud formation process and influence the microphysical and radiative properties of the interacting clouds and thus influence the global radiative forcing [13,14]. All aerosol particles do not exhibit the CCN activity which is largely determined by particle size, mixing state, chemical speciation and ambient atmospheric conditions [15]. Though, some CCN's suppress the cloud formation, while some others invigorate the extent of cloud coverage by interfering with the process of convection and dynamics [16]. Thus, the interaction between aerosol and cloud causes one of the largest uncertainties in global radiative forcing [15].

The Indian Ocean is third-largest oceanic division of the world's ocean bodies, extending up to ~25°N, thereafter almost landlocked by South-Asian sub-continental landmasses. The Northern Hemispheric portion of the Indian Ocean is popularly termed as Northern Indian Ocean (NIO), which comprises three marginal sea bodies namely: Arabian Sea (AS), Bay of Bengal (BoB), and Equatorial Indian Ocean (EIO) and extends roughly between 10°S and 25°N. Although, AS and BoB oceanic masses lie on the same latitudinal extent but are physically separated apart by the southern peninsular Indian mainland (Fig. 11.1). During the last few decades, due to the significant economic

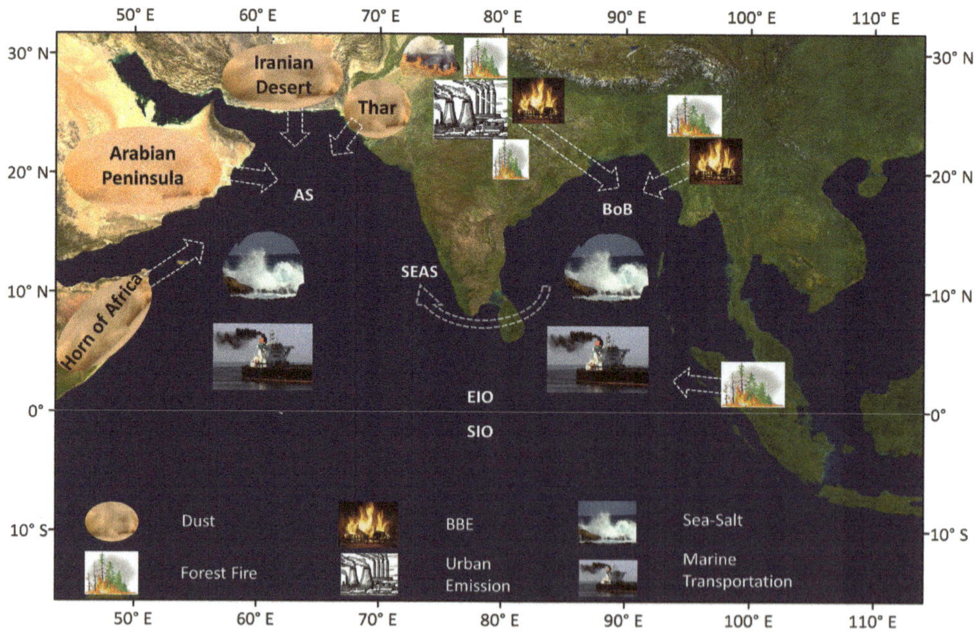

Fig. 11.1 ***Geographical location of Indian Ocean bounded with continental landmasses of aerosol hotspot.*** EIO: Equatorial Indian Ocean; SIO: Southern Indian Ocean.

growth and related energy demands, the anthropogenic emission in the neighboring country of the Indian Ocean is increased and thus aerosol burden into the atmosphere [17,18]. This region also experiences a unique tropical and sub-tropical seasonal wind reversal system which causes large variability in aerosol spatio-temporal distribution and their compositions [19–22]. For example; during premonsoon and monsoon season, AS is mostly dominated by aeolian dust [19,23,24] which is mainly due to the strong winds carrying abundant aeolian dust from north-eastern Africa, Arabian Peninsula, and Thar Desert [20,25,26]. On the other hand, during the postmonsoon (October–November) and winter season (December–February), BoB experienced higher anthropogenic aerosol than AS and attributed to the continental outflow, particularly from heavily aerosol loading regions like Indo-Gangetic Basin (IGB) [21,23,27,28]. In another study, Banerjee et al. [29] also found that during south-west monsoon season (June–September), dust over northern BoB is largely (more than 50% of total dust loading) contributed by regional emission sources (like Thar Desert and adjoining arid regions) while over southern BoB dust transported from the remote-source regions dominate throughout the year. The recent changing emission pattern with the anthropogenic development over the south-east Asian region and prolonger prevailing drought conditions over the arid/semi-arid middle east region, make more susceptible to the NIO marine environment with higher aerosol loading. Thus, to assess aerosol characteristics and their impacts on climate change through different aspects over the Indian Ocean, several campaign-based experiments like Indian Ocean Experiment: INDOEX, Integrated Campaign for Aerosols, Gases and Radiation Budget: ICARB, etc. have been performed. INDOEX experiment primarily documented the great Indo-Asian Haze events that usually occur during dry winter months and causes anthropogenic haze (brown smoke cloud) over the NIO under low-level north-easterly winds. The contribution of anthropogenic aerosols over NIO during the expedition was estimated at nearly 80% of the column aerosol optical depth (AOD) and mean AOD (at 500 nm) over the region was observed ~0.3 [30]. Another extensive campaign ICARB was conducted with the main goal to characterize the physio-chemical and radiative properties of marine boundary layer aerosol. During the ICARB expedition, AODs over the AS region were observed consistently low varying in the range of 0.06–0.3 (at 500 nm), except for two high AOD regions. The first one is western coastal locations that are directly under the confluence of southern Indian mainland observed moderately higher AOD ~ 0.85. Another one is center AS (18°N, 68°E) with AOD (at 500 nm) value ~0.3–0.4 also known as "Western Asian High" region [31]. Similar occurrence of high AODs around the same location has been reported during INDOEX expedition and well explained due to proximity to the western Asian region by Moorthy and Saha [32]. The reported results were very contrasting between the basins, in general higher AODs prevailed over the BoB region compared to AS. Above the 15°N and between 84°E and 92°E in BoB, the measured AOD at 500 nm was notably higher in the range of 0.5–1.0

[31]. The chemical characterization of the filter-based aerosol concentration within the marine boundary layer during the same expedition explains the significant contribution of crustal minerals to the marine environment. Kumar et al. [33], also provide a signature of anthropogenic constituents to the sampled aerosol concentration that explains the out flux from the nearby populated mainland. Their findings also noted a large spatial variability in mineral dust concentration across both the basins and mainly attributed to the high dry deposition fluxes during the inter-monsoon period. Thus, NIO provides a unique platform to investigate aerosol characteristics under different meteorological conditions and their mutual interactions.

11.2 Space-borne observation of aerosols and their transportation

Due to the recent technological development and different constraints in ground-based measurement, space-borne observation of aerosols and other atmospheric pollutants gain enormous attention. It has the ability to provide three-dimension information about the aerosol loading over the region with a high spatial-temporal resolution on a global scale along with a vertical distribution profile. Through space-borne measurement, different atmospheric events like dust storms, cyclones, etc. and their transport pathways can be easily identified. The observation of aerosol from space is quite challenging and associated with different uncertainties due to surface reflectance from the clouds, atmospheric molecules interactions along with sensor calibration and accuracy. Further details about the different sensors onboard different satellites are discussed in the next sections.

11.2.1 Moderate resolution imaging spectro-radiometer (MODIS)

MODIS, as the name of the sensor itself describe, is an imaging spectro-radiometer onboard two satellite platforms, that is, Terra and Aqua launched by NASA in December 1999 and July 2002 respectively; with equator overpass at 10:30 and 13:30 local time. As a passive sensor in space, it measures the intensity of solar reflected radiation in 36 spectral bands ranging from 0.415 μm to 14.235 μm, having a wider swath of ~2330 km provides daily global coverage in a one/two day (Table 11.1) [34]. Considering highly variable earth's surface albedo, AOD is retrieved using different algorithms: Dark Target

Table 11.1 Technical details of some of the major sensors used in space borne aerosol studies.

	Spatial resolution	Aerosol parameters	Swath (km)	Spectral bands	Operational period
MODIS	10 km, 3 km, 1 km	AOD, AE	~2330	36	2000–present
OMI	13 km x 24 km	AOD, SSA, AAOD, UVAI	~2600	3	2006–present
CALIPSO	1 km x 1 km	AOD, βa, σ_a, aerosol height	64 km x 64 km	3	2006–present

(DT) over vegetated/dark land and water bodies, Deep Blue (DB) over a brighter surface such as desert/arid land [35,36]. Further combining the merits of DT and DB algorithms, a merged algorithm Dark Target-Deep Blue (DTB) introduced that significantly increase the spatial coverage with better AOD retrieval [37]. Principally all these retrieval algorithms work based on the Normalized Difference Vegetation Index (NDVI) threshold. For instance, if NDVI <0.2 then DB applied, else NDVI >0.3 DT works well, and for the NDVI falling between 0.2 and 0.3, merged DTB algorithm operated to make better retrieval [38].The global validation exercise of MODIS retrieval against ground based AERONET observation has found an uncertainty envelope of $\pm$ (0.05+0.15 τ) and $\pm$ (0.03+0.05 τ) over land and ocean respectively [37,39]. Apart from this some regional validation studies also recommend a similar margin of uncertainty as global one [40,41,42]. The latest version of MODIS aerosol and cloud data products is collection 6 (C6.1) that has been more optimized and uncertainty reduced than the earlier version.

11.2.2 Ozone monitoring instrument (OMI)

OMI is a nadir viewing UV/Visible imaging spectrometer onboard NASA's Aura satellite since 2004. It has a wide swath with twenty hyperspectral imaging spectrometers resulting in global coverage with high spatial resolution, that is, 13 km × 24 km (Table 11.1) [43]. OMI measures solar backscatter irradiation in ultraviolet and visible spectrum ranging between 270 nm and 500 nm with a higher spectral resolution of 0.5 nm [44]. For OMI aerosols optical properties retrieval, two independent operational algorithms are used. First, OMI Near-UV Aerosol (OMAERUV) retrieval algorithm that typically uses spectral band of 354 nm and 388 nm, while the second, multi-wavelength aerosol retrieval algorithm (OMAERO) uses up to 19 spectral channels across the range of 330–500 nm to retrieve total aerosol optical depth [45]. From the OMAERUV algorithm AOD, single scattering albedo (SSA), absorbing aerosol optical depth (AAOD), and ultraviolet aerosol index (AI) are retrieved. The AI is a function of aerosol optical depth and is defined as the difference between measured (including aerosol effects) spectral contrast at 331 and 360 nm wavelength radiances and the contrast calculated from the radiative transfer theory for a pure molecular (Rayleigh particles) atmosphere [46]. It is highly sensitive to mineral dust, carbonaceous aerosol, and aerosol layer height also. The positive value of AI indicates the presence of UV absorbing aerosols, like dust, smoke whereas the negative value indicates the presence of nonabsorbing aerosols like sulphate [23,47,48]. Recent upgradation and optimization to the algorithm related to the retrieval of the UVAI, and associated uncertainties are explained in Torres et al. [49]. The products from the OMAERUV algorithm are available in daily (OMAERUVG 003, 0.25° × 0.25°) and monthly (OMTO3d 003, 1° × 1°) in spatially averaged grid data format.

11.2.3 Cloud-Aerosol Lidar Infrared Pathfinder Satellite Observation (CALIPSO)

As most of the passive sensors provide columnar integrated loading of aerosol, while a key piece of information missed out in columnar aggregation, that is, the altitudinal distribution of aerosol in the atmosphere. Aerosol near to the surface or within the boundary layer is likely to be quickly removed via dispersal-mixing, deposition and precipitation processes. But the aerosols that are elevated above the boundary layer have more likelihood to travel over a long distance, even across the continental boundary and shed out impacts there. The estimation of vertical distribution and subtype speciation of atmospheric aerosols are crucial information for the understanding of transboundary transportation and determination of its impacts on atmospheric radiation budgetary. There was a great need of active spaceborne sensors to routinely measure the vertical distribution and identification of clouds and aerosols at a global scale. To bridge this observation gap, CALIPSO contains a suite of three different sensors: a dual-wavelength lidar, a three-channel IR radiometer and a single-channel imager launched into the earth's orbit on April 28, 2006. The main objective of this science mission was to provide the capability to well distinguish the aerosol and cloud particles and their optical and physical properties along with their relative vertical location [50]. The prominent sensor, Cloud Aerosol Lidar with Orthogonal Polarization (CALIOP) onboard CALIPSO, is a two-wavelength (532 nm and 1064 nm) polarization sensitive lidar system that efficiently measures the attenuated backscatter signals and depolarization phase [50]. It has the capability to retrieve aerosol and cloud profiles at a vertical resolution of 30 m (0–8.2 km), 60 m (8.2–20.2 km) with a horizontal resolution of 333 m [51]. In aerosol parameters, it provides AOD, aerosol backscatter coefficient (βa) and aerosol extension coefficient (σa,) along with the aerosol vertical height. CALIOP orbital track covers the entire globe in 16 days and it does not provide homogenous global measurements due to its smaller footprints compared to wider swath sensors. The estimated detection sensitivity of CALIOP is better at night time due to the absence of background noise of solar illumination. The detection sensitivity also varies with altitude as the molecular scattering decreases with altitude. There are three basic standard products: Layer, Profile and Vertical Feature Mask (VFM) products for both aerosol and cloud. These are derived from the CALIOP measurement, applying three different selective algorithms that are: Selective Iterative Boundary Locator (SIBYL), Scene Classification Algorithm (SCA) and Hybrid Extinction Retrieval Algorithm (HERA) [52,53]. Attenuated backscatter signals ($km^{-1}sr^{-1}$), vertical layer information and VFM aerosol subtypes are most commonly used for profiling and subtyping of aerosols and clouds. Detailed descriptions on specific data products, their retrieval algorithm, subtyping scheme and lidar ratio selection algorithms are discussed elsewhere [52,54]. Validation and accuracy estimation of CALIOP data retrieval against ground based observation has been done across the globe [55–57].

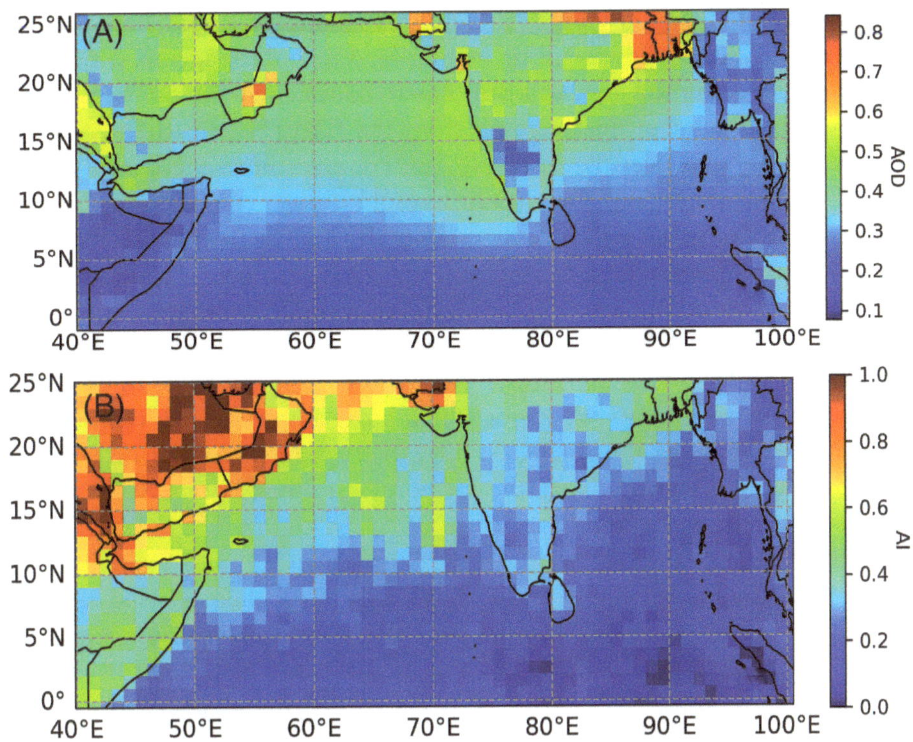

Fig. 11.2 Spatial distribution of mean (A) AOD and (B) AI, retrieved from MODIS/Aqua and OMI satellite respectively, for the period of 2011–2020 over the NIO.

11.3 Spatio-temporal variation of aerosol over the Northern Indian Ocean

As discussed above, due to its unique geographical location and synoptic wind circulation pattern NIO experienced a significant amount of aerosols throughout the year, emitted from both natural and anthropogenic sources [23,29]. The spatial variation of AOD and AI over NIO is shown in Fig. 11.2 which suggest a large spatial variability in aerosol loading over the region. Relatively higher AOD (>0.5) is observed near the coastal region of the Indian subcontinent with a lower spatial gradient of mean AOD over AS than BoB (Fig. 11.2A). A remarkable spatial gradient in AOD value can also be seen in (Fig. 11.2A) where AOD gradually decreases with an increase in distance from the continental land to the ocean region. The spatial variability in aerosol loading is mainly attributed to the different emission sources, meteorological conditions and transport patterns [58]. Typically, during premonsoon and monsoon season, mineral dust aerosols concentrations over AS are largely transported from the south-east Arabian Peninsula by cyclonic wind in lower troposphere and from dry desert region of Iran,

Afghanistan, and Pakistan by anticyclonic wind in middle troposphere [25,59,60]. On the other hand, during postmonsoon/winter season, anthropogenic aerosols are dominant aerosol type over the northern BoB due to continental outflow wind which brings a humongous amount of finer aerosols from the IGB [19,28,61,62]. Relatively lower AOD (≤0.3) is found over the open NIO in comparison to the coastal NIO. A quite similar spatial variability over the NIO in aerosol loading is also reported by earlier studies [63]. Similar to AOD, AI also shows a large spatial variation with relatively higher over the AS (>0.6) compared to BoB (mostly ≤0.4) (Fig. 11.1B), suggesting, relatively higher value of absorbing aerosol (mainly dust) over AS. Relatively higher value of AI over AS is also reported by earlier studies [23,63]. The presence of dust aerosol concentration over NIO shows a good seasonality with the highest value in monsoon season driven by the strong monsoonal winds which bring ample amount of dust from nearby arid and desert regions [24,25]. Thus, for a better understanding of seasonal aerosol loading and their association with the meteorological conditions, AOD variation on a seasonal basis is shown in Fig. 11.3.

Fig. 11.3 reflects a strong seasonal variability in AOD with maximum value over AS (>0.8) during monsoon season which is largely controlled by wind and causes relatively higher aerosol loading in central BoB during monsoon season (i.e., AOD<0.6).

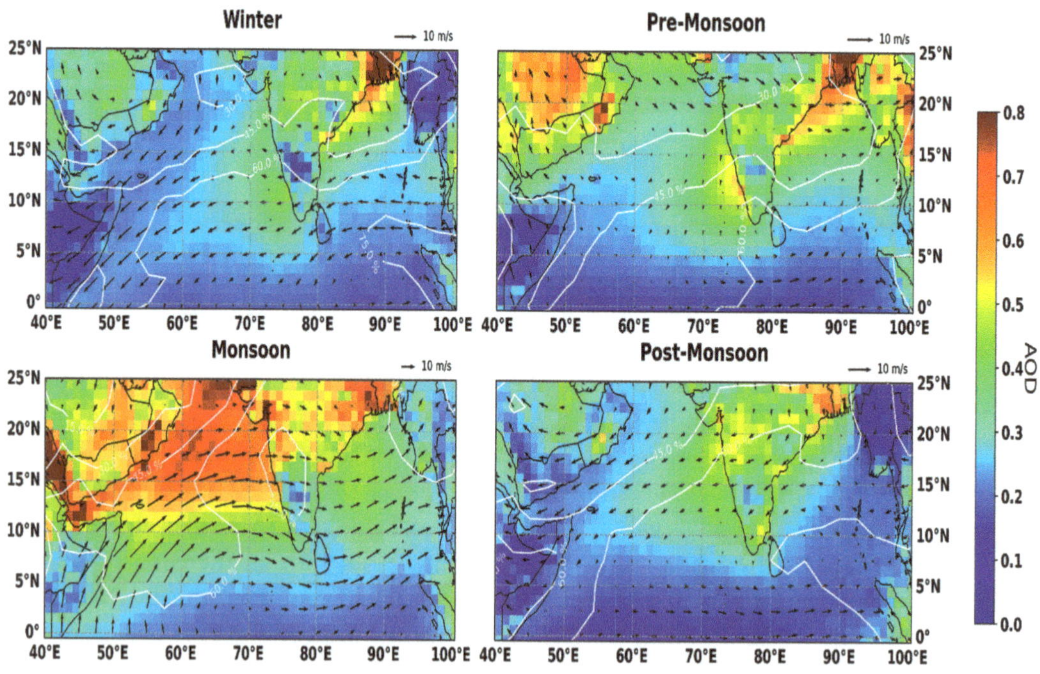

Fig. 11.3 ***Seasonal mean AOD at 550 nm from MODIS/Aqua, overlaid with wind circulation and RH at 850hpa (NCAR Reanalysis) for the period of 2011–2020 over NIO.*** Winds shown with arrows represent the direction and magnitude w.r.t. 10 ms^{-1} scale, while white contour represents RH (%).

A strong wind circulation pattern during summer monsoon season is also reported by Rajeev et al. [64], where they found an enhancement in wind speed from 10 ms^{-1} to 20 ms^{-1}. These strong winds are not only capable to bring dust from Arabian Peninsula [25,26,65,88] but also act as a conduit for moisture from the oceans as well as a source of sea-salt aerosols [66]. Relatively lower AOD over the BoB during monsoon season is possibly due to efficient washout/rainout of elevated aerosol over the transported pathways. Similar aerosol loading is also reported by Tiwari et al. [23] where they found a one-month lag between AS and BoB in maximum aerosol loading during monsoon season. On the other hand, northern BoB experienced maximum aerosol loading (i.e., AOD>0.7) during the postmonsoon/winter season which is largely influenced by continental outflow due to the reversal of wind from south-westerly to north-easterly [62]. Crop residue burning practices during the postmonsoon season over Northern IGB season is quite common and causes a large enhancement in aerosol loading over the region [20,67] which deposited over the BoB due to the continental outflow [68]. Kishcha et al. [69] also found that air pollution over IGB during postmonsoon season significantly affect aerosol loading and its trend over the BoB (particularly north-western BoB) where anthropogenic aerosols such as sulphate and carbonaceous aerosols are dominated.

11.4 Vertical distribution and subtypes of atmospheric aerosol over the Northern Indian Ocean

The precise information of aerosol vertical distribution is one of the crucial aspects in determination of the aerosol climate effect and associated radiative forcing [34]. Accurate representation of this vital piece of information in aerosol–climate interaction process is one of the major uncertain components in both regional and global scale climate models [70]. The usual practice of surface-based measurement of mass and number concentration of aerosol species and subsequent conversion to optical depth using popular *Mie scattering theory* with assumption of vertical profile, is not a robust and flawless technique of estimation of aerosol profile. The surface concentration and associated properties are quite different than the distributed distinct layer across the total column, thus estimation of vertical aerosol profile based on such measurement usually leads to results with large errors [71]. Active remote sensors like lidar systems are proved to be one of the promising observational techniques to acquire the vertical distribution of aerosol layers. However, this technique has a certain degree of uncertainties in retrieval of this precise information and it arises due to the large variation in collocated atmospheric constituents such as presence of thicker clouds or background noise sources as solar illumination [72]. Within the boundary layer, vertically elevated layers of aerosol induce thermal convection that further facilitates higher upliftment and long-range transportation of the aerosols to the free troposphere and hence significantly affect atmospheric boundary layer dynamics [73,74]. Few long-term spaceborne lidar measurement based studies have been carried out over the Indian mainland and adjoining oceanic region to understand the climatological pattern of vertical distribution of aerosols [23,60,75–78]. Apart from these, some short-term in situ measurements such as the Indian Ocean Experiment (INDOEX) which includes lidar based measurements, have been also performed to reveal the aerosol vertical distribution and their long-range transportation over Indian oceanic region [79–83]. The vertical distribution of aerosol subtypes over the Indian ocean depends on several factors like prevailing synoptical meteorological conditions, emission sources, complex air-sea interaction processes, active transportation patterns and chemical speciation during course of transportation [75,84]. Although significant progress has been made in terms of measurement and for better understanding of columnar aerosol loading, a well-defined pattern of height resolved aerosol distribution and subtypes are still rather poor on an extended period of observation, especially over the NIO region.

As earlier discussed in Section 11.3, CALIPSO as a spaceborne lidar system plays promising role in distinguishing cloud and aerosol species along with their relative altitudinal location based on the spectral behavior of the backscattered signal. Utilizing the CALIPSO retrieved backscatter signals for vertical profiles and vertical feature mask (VFM) for subtypes, we have presented a brief overview of vertical distribution of aerosol and their subtypes over the NIO. Fig. 11.4 shows the CALIPSO retrieved backscatter-coefficient profiles at 532 nm from surface to atmospheric height of 5 km over AS and BoB region for the duration of 2006–2012. The monthly mean backscatter profiles clearly reflect a seasonal periodicity in the altitudinal range of aerosol loading over both basins. The maxima occur during summer monsoon months (most prominent signal in June) with vertical distribution lifting up to 3–4 km, while minima observe during dry winter months (October–November) where vertical distribution is confined within height of 2 km over both the basins. Comparatively backscatter signals during the monsoon months are stronger and more prominent over the AS than BoB. This can be attributed to strong monsoonal winds that advect continental arid dust and generate marine aerosols over the entire AS stretch. While during dry winter months, vertical distribution of aerosol layer is relatively higher over BoB than AS, which is mainly attributed to the north-easterly driven continental outflow. Prijith et al. [77] also reported that strong south-westerly monsoonal winds supply a significant fraction of marine aerosols from the AS region to the central Indian main landmass within altitudes range below 2 km. In another study, Tandule et al. [60] also found that vertical distribution of aerosol layers over northern and central AS are mainly confined within the surface to 3 km with maximum elevation during summer months.

Another informative feature retrieval from the CALIPSO observation is Vertical Feature Mask (VFM) products that is capable to describe the vertical and horizontal

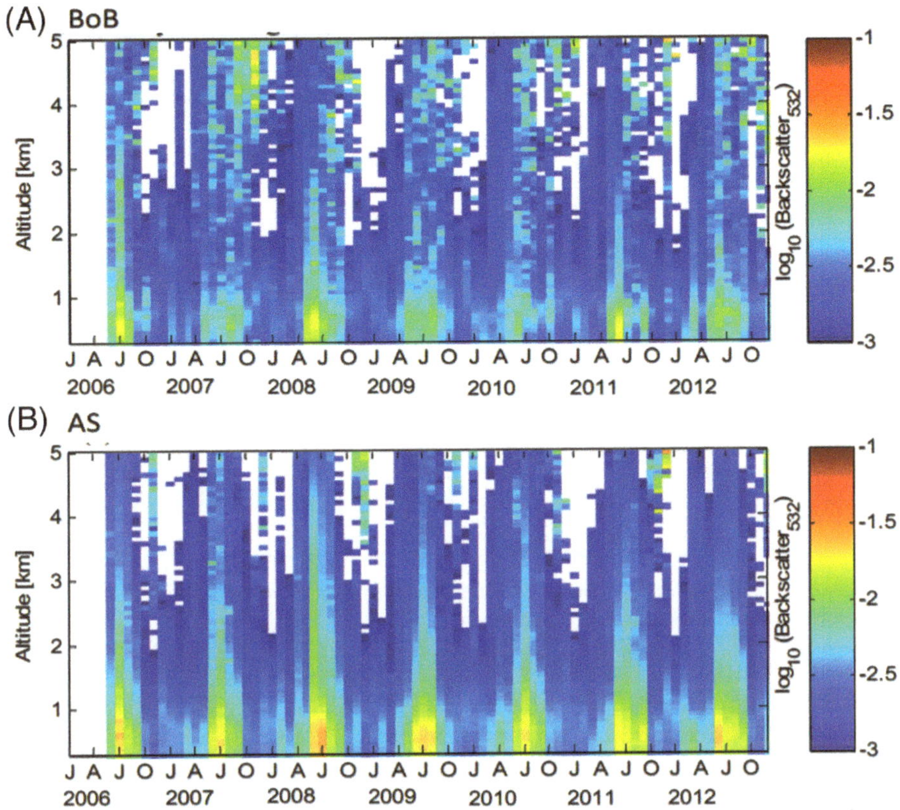

Fig. 11.4 Monthly mean profiles of CALIPSO backscatter-coefficient at 532 nm (night-time) during 2006–2012 over (A) BoB and (B) AS, shown in logarithmic scale (adopted from [22]).

distribution of aerosol sub-types in its overpass. A typical VFM profile of June 29, 2020 for the vertical distribution of aerosol sub-types over AS is shown in Fig. 11.5. The retrieved subtypes profiles clearly confirm the dominance of clean marine and dusty marine aerosol in lower atmospheric layer (up to ~2 km). While a detached elevated layer of continental dust can be easily traced at an altitude of 3–4 km between a latitudinal range of 3.39°N to ~14°N that gradually settled down to 2 km and intermixed with marine aerosols along with horizontal spread of ~14°N to 19°N. The isolated lofted dust layer and its gradual spread within marine environment indicate contribution of long-range transportation from the proximal arid lands. Beyond 21°N the presence of homogenous distribution of dust concentration from surface to up to 5 km indicates the onset of land track observation by the sensor. Since the observation was made during the onset monsoon months (i.e., June), the distinction between the marine aerosol and continental dust layer can be easily made. While in the later months with gradual gain in strength of prevailing monsoonal wind make it ambiguous to identify the separate layer

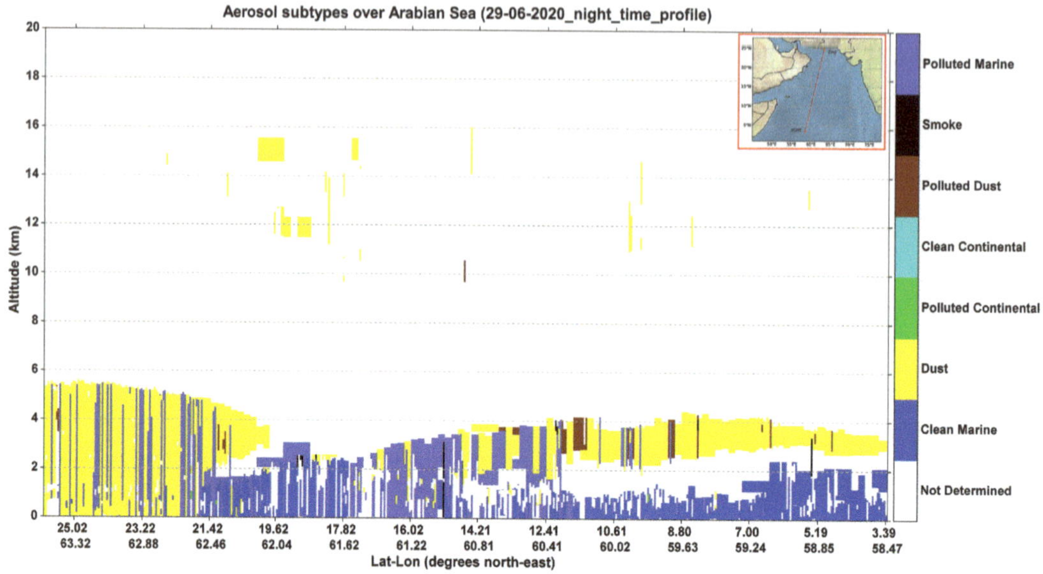

Fig. 11.5 ***CALIPSO VFM profile to differentiate aerosol subtypes over the AS.*** Observation track path over AS region has been shown in red bounded box.

rather than an intermixed dusty marine layer across the profile. So far, it has been well known that AS experiences stronger south-westerly wind action compared to the other stretches of the Indian Ocean which vigorously disturb the sea surface and squeeze out a significant amount of marine sea salt. On the other hand, swirling pattern of south-westerly wind over the Horn of Africa and middle east arid land advect large chunks of dust concentration towards the AS. Thus, CALIPSO retrieved vital information on the vertical extent and subtypes species of aerosol is used to understand the aerosol-induced changes to atmospheric state and cloud processes.

11.5 Long-range transportation of atmospheric aerosol over the Northern Indian Ocean

Long-range transportation and deposition of continental aerosols over different stretches of the Indian Ocean has been well presented in several studies [23,60,75,77,84]. In particular, a significant amount of dust aerosols are transported over AS, north western Indian Ocean mainly from Middle East, and nearby arid regions including horn of Africa [20,26,85,86]. Their findings clearly outlined the pattern of upper tropospheric transportation and deposition mechanisms of continental aerosol to marine environment. The associated spatio-temporal variability in continual process of transportation and deposition is mainly observed due to changes in prevailing wind circulation and their interaction between upper troposphere and lower boundary layer dynamics. In general, chemical characterization of aerosol species using different techniques infers the probable source of emission in terms of whether its locally generated or long-range transported [33]. Airmass back-trajectories are another important proxy to investigate the possible emission source region. Airmass back trajectory analysis use interpolated measured or modelled meteorological fields to estimate the most likely central path over the geographical area and follows a parcel of air backward in hourly steps for a specified period. Hybrid-Single Particle Lagrangian Integrated Trajectory (HYSPLIT) model, developed by National Oceanic and Atmospheric Administration (NOAA) Air Resources Laboratory [87], is one of the air mass back trajectories models that provides simultaneously four-dimensional information (i.e., latitude, longitude, altitude, and time) about the air mass transportation. In this model, a hybrid approach between Lagrangian and Eulerian is used and advection and diffusion calculations are made in a Lagrangian framework following the transport of the air parcel, while pollutant concentrations are calculated on a fixed grid. A typical example of aerosol long-range transportation over AS basin is depicted in Fig. 11.6 with relative percentage contribution of air masses from different regions at different altitude. Fig. 11.6 clearly suggest that within the lower troposphere (at 500 m), aerosol over Northern AS are mainly transported from the sub-continental mainland of Iran, Iraq, along the Persian Gulf by the north-westerly and from the south-western Arabian Peninsula and the Horn of Africa by the south-westerly (Fig. 11.6A). While over central AS aerosols are mainly transported from Arabian Peninsula and the Horn of Africa (Fig. 11.6B). On the other hand, aerosols are transported over south AS mainly from Horn of Africa only (Fig. 11.6C). Apart from this, in mid-troposphere (at altitude 5000 m) aerosols are mainly transported from Iraq, Iran, Afghanistan, Pakistan, and Northwest India except southern AS where significant fraction of aerosols are transported from Southeast Arabian Peninsula. A similar long-range transportation pattern of dust aerosols over AS is also reported by earlier studies [25,85] (Tiwari et al., [23] and reference therein).

11.6 Summary

This chapter provides a brief overview of aerosol loading over the NIO, their emission sources, spatial-temporal variation, transport pattern and role of meteorological conditions in aerosol burden over the region. In addition to this, spaceborne measurement techniques, their retrieval algorithms and uncertainties are also briefly discussed. A large spatial variation in aerosol loading is found over NIO showing a strong seasonal/annual variability. For example, during winter/postmonsoon season northern BoB experiences maximum aerosol loading (AOD>0.7) while central AS experience experienced maximum aerosol loading during summer monsoon season (AOD>0.8). Significant heterogeneity in aerosol type is also found which is largely influenced by regional meteorological conditions. Relatively higher aerosol emission from IGB and surrounding nearby

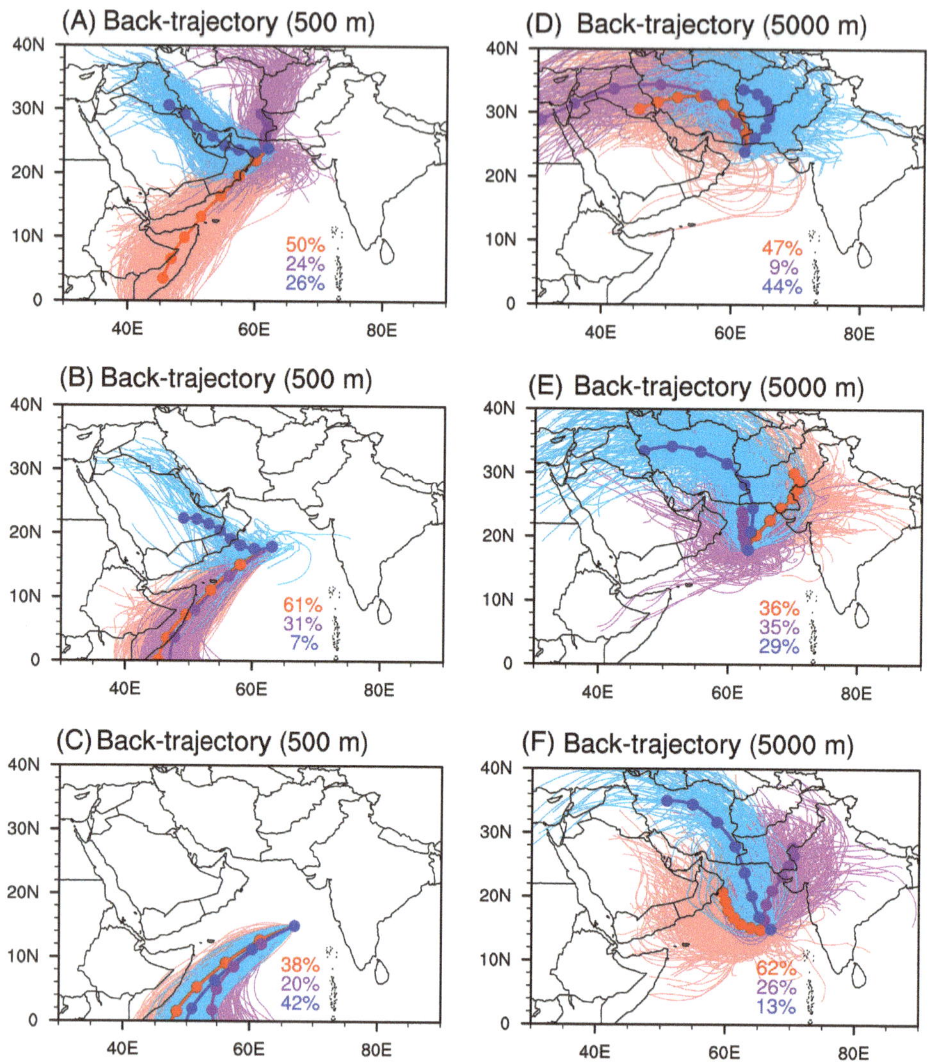

Fig. 11.6 ***84 hours airmass back trajectories over the Arabian Sea from northwest to southeast at altitude 500 m and 5000 m.*** The left and right columns represent back trajectories at the altitude of 500 m and 5000 m, respectively (adopted from [25]).

landmass region is the major sources of aerosol loading over the BoB. On the other hand, strong north-westerly winds coming from the Arabian Peninsula, arid region of Pakistan and Horn of Africa is the largest contributor of aerosol over the Arabian Sea. These strong winds over the NIO uplift atmospheric aerosol, resulting in elevated aerosol layers (up to 5 km) over the region which is more pronounced over AS (at lower altitude, that is, <2 km) during monsoon and over BoB (at higher altitude, that is, >5 km) during postmonsoon season.

Moreover, for a better understanding of future climate change prediction, there is a strong need to study the physio-chemical characteristics of aerosol, their source apportionment and interactions with synoptical meteorological conditions, cloud as well as climate that could be only possible with synergistic approaches of in situ measurement, remote sensing, and model dataset. The accurate information of atmospheric aerosol can be further helpful to policymakers for the development of a sustainable environment.

Acknowledgement

We acknowledge the respective science teams and associated organization for providing useful aerosol data products from MODIS, OMI, and CALIPSO sensors. Mr. Harshbardhan Kumar is thankful to University Grant Commission (UGC) for providing financial support as Junior Research Fellowship.

References

[1] J. Jung, A.H. Souri, D.C. Wong, S. Lee, W. Jeon, J. Kim, Y. Choi, The impact of the direct effect of aerosols on meteorology and air quality using aerosol optical depth assimilation during the KORUS-AQ campaign, J. Geophys. Res. Atmos. 124 (14) (2019) 8303–8319.

[2] D. Sharma, R.L. Miller, Revisiting the observed correlation between weekly averaged Indian monsoon precipitation and Arabian Sea aerosol optical depth, Geophys. Res. Lett. 44 (19) (2017) 10006–10016.

[3] S. Yan, B. Zhu, Y. Huang, J. Zhu, H. Kang, C. Lu, T. Zhu, To what extents dourbanization and air pollution affect fog? Atmos. Chem. Phys. 20 (2020) 5559–5572.

[4] M.V. Ratnam, S.A. Raj, B.L. Madhavan, J.P. Vernier, V.R. Kiran, C.D. Jain, G. Basha, N. Nagendra, B.S. Kumar, A.K. Pandit, B.K. Murthy, Vertically resolved black carbon measurements and associated heating rates obtained using in situ balloon platform, Atmos. Environ. 232 (2020) 117541.

[5] C. Textor, M. Schulz, S. Guibert, S. Kinne, Y. Balkanski, S. Bauer, T. Berntsen, T. Berglen, O. Boucher, M. Chin, F. Dentener, Analysis and quantification of the diversities of aerosol life cycles within AeroCom, Atmos. Chem. Phys. 6 (7) (2006) 1777–1813.

[6] J.C. Kuniyal, R.P. Guleria, The current state of aerosol-radiation interactions: A mini review, J. Aerosol Sci. 130 (2019) 45–54.

[7] Z. Li, W.M. Lau, V. Ramanathan, G. Wu, Y. Ding, M.G. Manoj, J. Liu, Y. Qian, J. Li, T. Zhou, J. Fan, Aerosol and monsoon climate interactions over Asia, Rev. Geophys. 54 (4) (2016) 866–929.

[8] O. Boucher, D. Randall, P. Artaxo, C. Bretherton, G. Feingold, P. Forster, V.-M. Kerminen, Y. Kondo, H. Liao, U. Lohmann, P. Rasch, S.K. Satheesh, S. Sherwood, B. Stevens, X.Y. Zhang, Clouds and aerosols, in: T.F. Stocker, D. Qin, G.-K. Plattner, M. Tignor, S.K. Allen, J. Doschung, A. Nauels, Y. Xia, V. Bex, P.M. Midgley (Eds.), Climate Change 2013: The Physical Science Basis. Contribution of Working Group I to the Fifth Assessment Report of the Intergovernmental Panel on Climate Change, Cambridge University Press, Cambridge, 2013, pp. 571–657.

[9] S. Tiwari, A.K. Srivastava, A.K. Singh, S. Singh, Identification of aerosol types over Indo-Gangetic Basin: Implications to optical properties and associated radiative forcing, Environ. Sci. Pollut. Res. 22 (16) (2015) 12246–12260.

[10] C. Zhao, Y. Yang, H. Fan, J. Huang, Y. Fu, X. Zhang, S. Kang, Z. Cong, H. Letu, M. Menenti, Aerosol characteristics and impacts on weather and climate over the Tibetan Plateau, Natl. Sci. Rev. 7 (3) (2020) 492–495.

[11] S. Ramachandran, Aerosols: Radiative transfer and effects. In: Atmospheric Aerosols, CRC Press, Boca Raton, 2018, pp. 143–190.

[12] R. Ceolato, M.J. Berg, Aerosol light extinction and backscattering: A review with a lidar perspective, J. Quant. Spectrosc. Radiat. Transfer 262 (2021) 107492.

[13] B.A. Albrecht, Aerosols, cloud microphysics, and fractional cloudiness, Science 245 (4923) (1989) 1227–1230.

[14] I. Koren, J.V. Martins, L.A. Remer, H. Afargan, Smoke invigoration versus inhibition of clouds over the Amazon, Science 321 (5891) (2008) 946–949.

[15] J.E. Penner, L. Authors, M. Andreae, H. Annegarn, L. Barrie, J. Feichter, D. Hegg, A. Jayaraman, R. Leaitch, D. Murphy, J. Nganga, G. Pitari, C. Authors, A. Ackerman, P. Adams, P. Austin, R. Boers, O. Boucher, M. Chin, C. Chuang, B. Collins, W. Cooke, P. Demott, Y. Feng, H. Fischer, I. Fung, S. Ghan, P. Ginoux, S.-L. Gong, A. Guenther, M. Herzog, A. Higurashi, Y. Kaufman, A. Kettle, J. Kiehl, D. Koch, G. Lammel, C. Land, U. Lohmann, S. Madronich, E. Mancini, M. Mishchenko, T. Nakajima, P. Quinn, P. Rasch, D.L. Roberts, D. Savoie, S. Schwartz, J. Seinfeld, B. Soden, D. Tanre, K. Taylor, I. Tegen, X. Tie, G. Vali, R. Dingenen, M. Van Van Weele, Y. Zhang, Aerosols, their direct and indirect effects. In: Climate Change 2011: The Physical Science Basis. Contribution of Working Group 1 to the Third Assessment Report of the Intergovernmental Panel on Climate Change, Cambridge University Press, (2001) 291–336.

[16] O. Altaratz, I. Koren, L.A. Remer, E. Hirsch, Cloud invigoration by aerosols-coupling between microphysics and dynamics, Atmos. Res. 140 (2014) 38–60.

[17] A. Mhawish, M. Sorek-Hamer, R. Chatfield, T. Banerjee, M. Bilal, M. Kumar, C. Sarangi, M. Franklin, K. Chau, M. Garay, O. Kalashnikova, Aerosol characteristics from earth observation systems: A comprehensive investigation over South Asia (2000–2019), Remote Sens. Environ. 259 (2021) 112410.

[18] K.K. Moorthy, S. Suresh Babu, M.R. Manoj, S.K. Satheesh, Buildup of aerosols over the Indian region, Geophys. Res. Lett. 40 (5) (2013) 1011–1014.

[19] A.R. Aswini, P. Hegde, S. Aryasree, I.A. Girach, P.R. Nair, Continental outflow of anthropogenic aerosols over Arabian Sea and Indian Ocean during wintertime: ICARB-2018 campaign, Sci. Total Environ. 712 (2020) 135214.

[20] D.G. Kaskaoutis, A. Rashki, E.E. Houssos, D. Goto, P.T. Nastos, Extremely high aerosol loading over Arabian Sea during June 2008: The specific role of the atmospheric dynamics and Sistan dust storms, Atmos. Environ. 94 (2014) 374–384.

[21] D.G. Kaskaoutis, S.K. Kharol, P.R. Sinha, R.P. Singh, K.V.S. Badarinath, W. Mehdi, M. Sharma, Contrasting aerosol trends over South Asia during the last decade based on MODIS observations, Atmos. Meas. Tech. Discuss. 4 (4) (2011) 5275–5323.

[22] A. Thomas, C. Sarangi, V.P. Kanawade, Recent increase in winter hazy days over Central India and the Arabian Sea, Sci. Rep. 9 (1) (2019) 1–10.

[23] S. Tiwari, A.K. Mishra, A.K. Singh, Aerosol climatology over the Bay of Bengal and Arabian Sea inferred from space-borne radiometers and lidar observations, Aerosol Air Qual. Res. 16 (11) (2016) 2855–2868.

[24] V. Vinoj, P.J. Rasch, H. Wang, J.H. Yoon, P.L. Ma, K. Landu, B. Singh, Short-term modulation of Indian summer monsoon rainfall by West Asian dust, Nat. Geosci. 7 (4) (2014) 308–313.

[25] Q. Jin, J. Wei, B. Pu, Z.L. Yang, S.P. Parajuli, High summertime aerosol loadings over the Arabian Sea and their transport pathways, J. Geophys. Res. Atmos. 123 (18) (2018) 10–568.

[26] S. Tiwari, A. Kumar, V. Pratap, A.K. Singh, Assessment of two intense dust storm characteristics over Indo-Gangetic basin and their radiative impacts: a case study, Atmos. Res. 228 (2019) 23–40.

[27] B. Srinivas, M.M. Sarin, Atmospheric pathways of phosphorous to the Bay of Bengal: Contribution from anthropogenic sources and mineral dust, Tellus B: Chem. Phys. Meteorol. 64 (1) (2012) 17174.

[28] G. Nayak, A. Kumar, S. Bikkina, S. Tiwari, S.S. Sheteye and A.K. Sudheer. Carbonaceous aerosols and their light absorption properties over the Bay of Bengal during continental outflow. Environ. Sci.: Process. Impacts, 24(1) (2022) 72–88.

[29] P. Banerjee, S.K. Satheesh, K.K. Moorthy, R.S. Nanjundiah and V.S. Nair. Long-range transport of mineral dust to the northeast Indian Ocean: Regional versus remote sources and the implications. J. Clim., 32(5) (2019) 1525–1549.

[30] V. Ramanathan, P.J. Crutzen, A.P. Mitra, D. Sikka, The Indian Ocean experiment and the Asian brown cloud, Curr. Sci. 83 (8) (2002) 947–955.

[31] K.K. Moorthy, S.K. Satheesh, S.S. Babu, C.B.S. Dutt, Integrated campaign for aerosols, gases and radiation budget (ICARB): An overview, J. Earth System Sci. 117 (1) (2008) 243–262.

[32] K.K. Moorthy, A. Saha, Aerosol study during INDOEX: Observation of enhanced aerosol activity over the mid Arabian Sea during the northern winter, J. Atmos. Sol. Terr. Phys. 62 (1) (2000) 65–72.

[33] A. Kumar, A.K. Sudheer and M.M. Sarin. Chemical characteristics of aerosols in MABL of Bay of Bengal and Arabian Sea during spring inter-monsoon: a comparative study. J. Earth System science, 117(1) (2008) 325–332.

[34] Y.J. Kaufman, D. Tanré, L.A. Remer, E.F. Vermote, A. Chu, B.N. Holben, Operational remote sensing of tropospheric aerosol over land from EOS moderate resolution imaging spectroradiometer, J. Geophys. Res. Atmos. 102 (D14) (1997) 17051–17067.

[35] L.A. Remer, Y.J. Kaufman, D. Tanré, S. Mattoo, D.A. Chu, J.V. Martins, R.R. Li, C. Ichoku, R.C. Levy, R.G. Kleidman, T.F. Eck, The MODIS aerosol algorithm, products, and validation, J. Atmos. Sci. 62 (4) (2005) 947–973.

[36] N.C. Hsu, M.J. Jeong, C. Bettenhausen, A.M. Sayer, R. Hansell, C.S. Seftor, J. Huang, S.C. Tsay, Enhanced Deep Blue aerosol retrieval algorithm: The second generation, J. Geophys. Res. Atmos. 118 (16) (2013) 9296–9315.

[37] R.C. Levy, S. Mattoo, L.A. Munchak, L.A. Remer, A.M. Sayer, F. Patadia, N.C. Hsu, The Collection 6 MODIS aerosol products over land and ocean, Atmos. Measur. Tech. 6 (11) (2013) 2989–3034.

[38] S. Ibrahim, L. Halounova, Statistical study of MODIS algorithms in estimating aerosol optical depth over the Czech, Civil Eng. J. (4) (2019) 523–531.

[39] L.A. Remer, R.G. Kleidman, R.C. Levy, Y.J. Kaufman, D. Tanré, S. Mattoo, J.V. Martins, C. Ichoku, I. Koren, H. Yu, B.N. Holben, Global aerosol climatology from the MODIS satellite sensors, J. Geophys. Res. Atmos. (D14) (2008) 113.

[40] M. Tao, J. Wang, R. Li, L. Wang, L. Wang, Z. Wang, J. Tao, H. Che, L. Chen, Performance of MODIS high-resolution MAIAC aerosol algorithm in China: Characterization and limitation, Atmos. Environ. 213 (2019) 159–169.

[41] J. Wei, Z. Li, Y. Peng, L. Sun, MODIS Collection 6.1 aerosol optical depth products over land and ocean: validation and comparison, Atmos. Environ. 201 (2019) 428–440.

[42] A. Mhawish, T. Banerjee, M. Sorek-Hamer, A. Lyapustin, D.M. Broday, R. Chatfield, Comparison and evaluation of MODIS Multi-angle Implementation of Atmospheric Correction (MAIAC) aerosol product over South Asia, Remote Sens. Environ. 224 (2019) 12–28.

[43] P.F. Levelt, E. Hilsenrath, G.W. Leppelmeier, G.H. van den Oord, P.K. Bhartia, J. Tamminen, J.F. de Haan, J.P. Veefkind, Science objectives of the ozone monitoring instrument, IEEE Trans. Geosci. Remote Sens. 44 (5) (2006) 1199–1208.

[44] M. Dobber, R. Voors, R. Dirksen, Q. Kleipool, P. Levelt, The high-resolution solar reference spectrum between 250 and 550 nm and its application to measurements with the Ozone Monitoring Instrument, Sol. Phys. 249 (2) (2008) 281–291.

[45] O. Torres, A. Tanskanen, B. Veihelmann, C. Ahn, R. Braak, P.K. Bhartia, P. Veefkind, P. Levelt, Aerosols and surface UV products from Ozone Monitoring Instrument observations: An overview, J. Geophys. Res. Atmos. 112 (D24) (2007).

[46] R.L. Curier, J.P. Veefkind, R. Braak, B. Veihelmann, O. Torres, G. de Leeuw, Retrieval of aerosol optical properties from OMI radiances using a multiwavelength algorithm: Application to western Europe, J. Geophys. Res. Atmos. 113 (D17) (2008).

[47] D.G. Kaskaoutis, M.C.R. Kalapureddy, K. Krishna Moorthy, P.C.S. Devara, P.T. Nastos, P.G. Kosmopoulos, H.D. Kambezidis, Heterogeneity in pre-monsoon aerosol types over the Arabian Sea deduced from ship-borne measurements of spectral AODs, Atmos. Chem. Phys. 10 (10) (2010) 4893–4908.

[48] A. Singh, S. Tiwari, D. Sharma, D. Singh, S. Tiwari, A.K. Srivastava, N. Rastogi, A.K. Singh, Characterization and radiative impact of dust aerosols over northwestern part of India: A case study during a severe dust storm, Meteorol. Atmos. Phys. 128 (6) (2016) 779–792.

[49] O. Torres, C. Ahn, Z. Chen, Improvements to the OMI near-UV aerosol algorithm using A-train CALIOP and AIRS observations, Atmos. Measur. Tech. 6 (11) (2013) 3257–3270.

[50] D.M. Winker, J. Pelon, J.A. Coakley Jr, S.A. Ackerman, R.J. Charlson, P.R. Colarco, P. Flamant, Q. Fu, R.M. Hoff, C. Kittaka, T.L. Kubar, The CALIPSO mission: A global 3D view of aerosols and clouds, Bull. Am. Meteorol. Soc. 91 (9) (2010) 1211–1230.

[51] G. Gupta, M.V. Ratnam, B.L. Madhavan, P. Prasad, C.S. Narayanamurthy, Vertical and spatial distribution of elevated aerosol layers obtained using long-term ground-based and space-borne lidar observations, Atmos. Environ. 246 (2021) 118172.
[52] D.M. Winker, M.A. Vaughan, A. Omar, Y. Hu, K.A. Powell, Z. Liu, W.H. Hunt, S.A. Young, Overview of the CALIPSO mission and CALIOP data processing algorithms, J. Atmos. Oceanic Technol. 26 (11) (2009) 2310–2323.
[53] S.A. Young, M.A. Vaughan, The retrieval of profiles of particulate extinction from Cloud-Aerosol Lidar Infrared Pathfinder Satellite Observations (CALIPSO) data: Algorithm description, J. Atmos. Oceanic Technol. 26 (6) (2009) 1105–1119.
[54] M.H. Kim, A.H. Omar, J.L. Tackett, M.A. Vaughan, D.M. Winker, C.R. Trepte, Y. Hu, Z. Liu, L.R. Poole, M.C. Pitts, J. Kar, The CALIPSO version 4 automated aerosol classification and lidar ratio selection algorithm, Atmos. Meas. Tech. 11 (11) (2018) 6107–6135.
[55] M. Kacenelenbogen, M.A. Vaughan, J. Redemann, R.M. Hoff, R.R. Rogers, R.A. Ferrare, P.B. Russell, C.A. Hostetler, J.W. Hair, B.N. Holben, An accuracy assessment of the CALIOP/CALIPSO version 2/version 3 daytime aerosol extinction product based on a detailed multi-sensor, multi-platform case study, Atmos. Chem. Phys. 11 (8) (2011) 3981–4000.
[56] A. Misra, S.N. Tripathi, D.S. Kaul, E.J. Welton, Study of MPLNET-derived aerosol climatology over Kanpur, India, and validation of CALIPSO level 2 version 3 backscatter and extinction products, J. Atmos. Oceanic Technol. 29 (9) (2012) 1285–1294.
[57] G. Pappalardo, U. Wandinger, L. Mona, A. Hiebsch, I. Mattis, A. Amodeo, A. Ansmann, P. Seifert, H. Linné, A. Apituley, L.A. Arboledas, EARLINET correlative measurements for CALIPSO: First intercomparison results, J. Geophys. Res. Atmos. 115 (D4) (2010).
[58] M.M. Gogoi, C.R. Tandule, V. Jayachandran, S.K. Kompalli, V.S. Nair, K.R. Gopal, S.S. Babu, Spatial gradient of aerosol mass concentrations and size distributions over southeastern Arabian Sea and equatorial Indian Ocean during ICARB-2018, Atmos. Environ. 213 (2019) 727–738.
[59] Q. Jin, J. Wei, Z.-L. Yang, Positive response of Indian summer rainfall to Middle East dust, Geophys. Res. Lett. 41 (2014) 4068–4074.
[60] C.R. Tandule, R.O.R. Kalluri, B. Gugamsetty, R.G. Kotalo, L.R. Thotli, R.R. Rajuru, S. Vaddin, Decadal climatology of the spatial and vertical distributions of tropospheric aerosol over the Arabian Sea based on satellite observations, Int. J. Climatol. 40 (10) (2020) 4676–4689.
[61] A.K. Sudheer, M.M. Sarin, Carbonaceous aerosols in MABL of Bay of Bengal: influence of continental outflow, Atmos. Environ. 42 (18) (2008) 4089–4100.
[62] K. Budhavant, S. Bikkina, A. Andersson, E. Asmi, J. Backman, J. Kesti, H. Zahid, S.K. Satheesh, Ö. Gustafsson, Anthropogenic fine aerosols dominate the wintertime regime over the northern Indian Ocean, Tellus B: Chem. Phys. Meteorol. 70 (1) (2018) 1–15.
[63] S.S. Prijith, K. Rajeev, B.V. Thampi, S.K. Nair, M. Mohan, Multi-year observations of the spatial and vertical distribution of aerosols and the genesis of abnormal variations in aerosol loading over the Arabian Sea during Asian summer monsoon season, J. Atmos. Sol. Terr. Phys. 105 (2013) 142–151.
[64] K. Rajeev, S.K. Nair, K. Parameswaran, C.S. Raju, Satellite observations of the regional aerosol distribution and transport over the Arabian Sea, Bay of Bengal and Indian Ocean, Indian J. Mar. Sci. 33 (1) (2004) 11–29.
[65] V. Vinoj, S.K. Satheesh, Measurements of aerosol optical depth over Arabian Sea during summer monsoon season, Geophys. Res. Lett. 30 (5) (2003) 1263.
[66] S.K. Satheesh, J. Srinivasan, Enhanced aerosol loading over Arabian Sea during the pre-monsoon season: Natural or anthropogenic, Geophys. Res. Lett. 29 (18) (2002) 21-1–21-4.
[67] S. Sarkar, R.P. Singh, A. Chauhan, Crop residue burning in northern India: Increasing threat to Greater India, J. Geophys. Res. Atmos. 123 (13) (2018) 6920–6934.
[68] B. Srinivas, M.M. Sarin and V.V.S.S. Sarma, 2011. Atmospheric dry deposition of inorganic and organic nitrogen to the Bay of Bengal: Impact of continental outflow. Marine Chemistry, 127(1-4), pp.170-179.
[69] P. Kishcha, A.M. Da Silva, B. Starobinets and P. Alpert, 2014. Air pollution over the Ganges basin and northwest Bay of Bengal in the early postmonsoon season based on NASA MERRAero data. J. Geophys. Res. Atmos., 119(3), 1555–1570.
[70] B. Koffi, M. Schulz, F.M. Bréon, F. Dentener, B.M. Steensen, J. Griesfeller, D. Winker, Y. Balkanski, S.E. Bauer, N. Bellouin, T. Berntsen, Evaluation of the aerosol vertical distribution in global aerosol

models through comparison against CALIOP measurements: AeroCom phase II results, J. Geophys. Res. Atmos. 121 (12) (2016) 7254–7283.

[71] S.K. Satheesh, V. Vinoj, K.K. Moorthy, Vertical distribution of aerosols over an urban continental site in India inferred using a micro pulse lidar, Geophys. Res. Lett. 33 (20) (2006).

[72] A.C. Povey, R.G. Grainger, D.M. Peters, J.L. Agnew, Retrieval of aerosol backscatter, extinction, and lidar ratio from Raman lidar with optimal estimation, Atmos. Measur. Tech. 7 (3) (2014) 757–776.

[73] R.B. Stull, An Introduction to Boundary Layer Meteorology, Springer, New York, 1999, pp. 620.

[74] S. Ramachandran, T.A. Rajesh and R. Cherian, 2021. Black carbon aerosols over source vs. background region: Atmospheric boundary layer influence, potential source regions, and model comparison. Atmospheric Research, 256, 105573.

[75] C.A. Babu, P. Sivaprasad, Variability and mechanisms of vertical distribution of aerosols over the Indian region, Int. J. Remote Sens. 35 (22) (2014) 7691–7705.

[76] M. Mehta, R. Khushboo, R. Raj, N. Singh, Spaceborne observations of aerosol vertical distribution over Indian mainland (2009–2018), Atmos. Environ. 244 (2021) 117902.

[77] S.S. Prijith, S.S. Babu, N.B. Lakshmi, S.K. Satheesh, K.K. Moorthy, Meridional gradients in aerosol vertical distribution over Indian Mainland: Observations and model simulations, Atmos. Environ. 125 (2016) 337–345.

[78] M.V. Ratnam, P. Prasad, S.T.A. Raj, M.R. Raman, G. Basha, Changing patterns in aerosol vertical distribution over South and East Asia, Sci. Rep. 11 (2021) 308.

[79] D. Müller, K. Franke, F. Wagner, D. Althausen, A. Ansmann, J. Heintzenberg, G. Verver, Vertical profiling of optical and physical particle properties over the tropical Indian Ocean with six-wavelength lidar: 1. Seasonal cycle, J. Geophys. Res. Atmos. 106 (D22) (2001) 28567–28575.

[80] D. Müller, K. Franke, F. Wagner, D. Althausen, A. Ansmann, J. Heintzenberg, G. Verver, Vertical profiling of optical and physical particle properties over the tropical Indian Ocean with six-wavelength lidar: 2. Case studies, J. Geophys. Res. Atmos. 106 (D22) (2001) 28577–28595.

[81] J. Pelon, C. Flamant, P. Chazette, J.F. Leon, D. Tanré, M. Sicard, S.K. Satheesh, Characterization of aerosol spatial distribution and optical properties over the Indian Ocean from airborne LIDAR and radiometry during INDOEX'99, J. Geophys. Res. Atmos. 107 (D19) (2002) INX2–IN28.

[82] S.K. Satheesh, V. Ramanathan, X. Li-Jones, J.M. Lobert, I.A. Podgorny, J.M. Prospero, B.N. Holben, N.G. Loeb, A model for the natural and anthropogenic aerosols over the tropical Indian Ocean derived from Indian Ocean Experiment data, J. Geophys. Res. Atmos. 104 (D22) (1999) 27421–27440.

[83] E.J. Welton, K.J. Voss, P.K. Quinn, P.J. Flatau, K. Markowicz, J.R. Campbell, J.D. Spinhirne, H.R. Gordon, J.E. Johnson, Measurements of aerosol vertical profiles and optical properties during INDOEX 1999 using micropulse lidars, J. Geophys. Res. Atmos. 107 (D19) (2002) INX2–IN18.

[84] N.K. Kala, N.S. Anand, M.R. Manoj, H.S. Pathak, K.K. Moorthy, S.K. Satheesh, Zonal variations of the vertical distribution of atmospheric aerosols over the Indian region and the consequent radiative effects, Atmos. Chem. Phys. Discuss. (2021) 1–32.

[85] D.G. Kaskaoutis, A. Rashki, P. Francois, U.C. Dumka, E.E. Houssos, M. Legrand, Meteorological regimes modulating dust outbreaks in Southwest Asia: The role of pressure anomaly and Inter-Tropical Convergence Zone on the 1–3 July 2014 case, Aeolian Res. 18 (2015) 83–97.

[86] U.C. Dumka, D.G. Kaskaoutis, D. Francis, J-P. Chaboureau, A. Rashki, S. Tiwari, S. Singh, E. Liakakou, and N. Mihalopoulos. "The role of the Intertropical Discontinuity region and the heat low in dust emission and transport over the Thar desert, India: A Premonsoon case study." J. Geophys. Res.: Atmos. 124(23) (2019): 13197–13219.

[87] R.R. Draxler, G.D. Rolph, HYSPLIT (HYbrid Single-Particle Lagrangian Integrated Trajectory). NOAA Air Resources Laboratory, College Park, MD. 2003.

[88] Banerjee, P., Satheesh, S.K. and Moorthy, K.K., 2021. The unusual severe dust storm of May 2018 over northern India: Genesis, propagation, and associated conditions. J. Geophys. Res.: Atmos., 126(7), p. e2020JD032369.

CHAPTER 12

Balloon-Based Remote Sensing of the Atmosphere

Sarvan Kumar
Department of Earth and Planetary Sciences, Veer Bahadur Singh Purvanchal University, Jaunpur, India

12.1 Introduction

Atmospheric science is a field that depends on observations and measurements including chemistry, physics, and meteorology. The chemical and physical reactions occurring in the real atmosphere are not fully possible to reproduce in a confined laboratory. This requires a rigorous observation at different time and size scales. The temporal scales can range from a few minutes to tens of years. The size or spatial scales can range from kilometers to thousands of kilometers, indeed, to the entire world (e.g., satellite measurements). Therefore, a diverse set of measurement tools is required for observing the atmosphere. These tools take the form of different sorts of instrumentation. We must be able to operate a diverse set of instruments from different locations. Depending on the components of the atmosphere which we want to observe, we operate the instruments. For this, the measurements can be made from space, balloons, aircraft, and ground platforms. Thus, we use space-based or satellite platforms, balloon platforms, aircraft platforms, and ground-based platforms for the atmospheric measurements. The traditional aircraft can hardly fly up to the near-space, and the satellites in space are not easy to detect down the near-space [1]. Therefore, near space is a new research hotspot that worth more observation and detection. A high-altitude balloon is one of the near-space exploration vehicles, and it has many advantages. Firstly, a high-altitude balloon can achieve low-cost sustained flight for months or even years. Compared with other air vehicles, the high-altitude balloon has a long endurance time, which can achieve sustained and wider coverage for regional observation and detection [2,3]. Secondly, compared with a satellite, a high-altitude balloon can take more and heavier loads, and increase the observation accuracy and range, for example, the spatial resolution and the signal strength of the optical observation load in near space can be increased greatly [4]. Thirdly, the payloads in the near space with high altitude balloons can be recycled with low cost and low risk, which is difficult to achieve by satellites [5]. All these platforms have their own advantages and disadvantages with a unique set of measurements, contributing toward a better understanding of the earth's atmosphere.

Atmospheric Remote Sensing
DOI: https://doi.org/10.1016/B978-0-323-99262-6.00020-1

For observing the atmospheric parameters there are two basic approaches. The first approach is remote sensing and the second is in-situ. Remote sensing exists in two types, that is, passive and active. Remote sensing may but does not necessarily require that the observer be far removed from the location of the parameter. Satellites and ground-based instruments are far removed from stratospheric ozone, whereas high flying balloons and aircraft are actually in the stratosphere. In-situ measurements involve direct sampling of the atmosphere. A sample of the atmosphere (air) is brought into the instrument and is analyzed for its properties and/or its relative amount of various chemical components. These techniques can be employed either from the ground, aircraft, or balloon platforms (http://www.ccpo.odu.edu/). Many worldwide studies have been done using different types and sizes of balloons for air quality, trace gases and atmospheric parameters, some recent studies in Asia and mainly in India are discussed in ([6,7] and references therein). Some of these case studies are also discussed in upcoming sections.

12.2 Balloon platforms for atmospheric measurements

Balloons are being use for a long time using the two concepts, both flown for the first time in 1783 in France: hot air by the Montgolfier brothers and gas by Charles and Robert [8]. These balloons were used primarily in celebrations during festivals or events, mostly being manned, used little for the science purposes. The first scientific observations were done by the Robertson in 1803 and of Gay-Lussac and Biot in 1804. They demonstrated the decrease of temperature, pressure, and moisture with height and the constant composition of air up to an altitude of 7 km. But progress was slow, limited by the maximum altitude acceptable to the pilots at that time. During the nineteenth century only few ascents of scientific interest were performed. A very detailed historical development of scientific ballooning in metrological and air quality measurement is given by Pommereau [8].

12.2.1 Types of balloon sounding

Basic components of balloon sounding system may consist of balloon, receiver, transmitter, probing instruments, and ground control. A complete radiosonde system consists of a balloon-borne radiosonde instrument package, a radio receiver, a tracking unit, and a recorder. Balloons are used for two types of study; one is vertical measurement of meteorological parameters (weather balloons) and the other is for vertical measurement of air pollutants and trace gases (air quality balloons).

A weather balloon, also known as a sounding balloon, is a balloon (specifically a type of high-altitude balloon) that carries instruments aloft to send back information on atmospheric parameters. The meteorological information (pressure, temperature, humidity, and wind speed) is sent by means of a small, expendable measuring device called a radiosonde. The wind data can be obtained by tracking the balloons with radar, radio direction finding, or navigation systems (such as the satellite-based Global Positioning System (GPS)). The sensors in the radiosonde communicate with ground receiving stations that constantly receive data streams with recorded information. Weather balloons that do not carry an instrument pack are used to determine upper-level winds and the height of cloud layers. For such balloons, a theodolite or total station is used to track the balloon's azimuth and elevation, which are then converted to estimated wind speed and direction and/or cloud height, as applicable. Sounding type meteorological balloons are a very important tool in upper-air observations. For weather forecasting and computer model data and to diagnose current conditions weather balloons are launched around the world. Between 900 and 1300 locations around the globe do routine releases, two or four times daily.

High altitude balloons (HABs) are a convenient and low-cost alternative to measuring high altitude air quality nowadays. HABs are derived from weather balloons that were invented in the 1800s by French meteorologist Leon Teisserenc de Bort who then discovered tropopause and stratosphere. The balloon that he used was called an instrumented balloon filled with hydrogen gas to observe the air quality around 8–17 km of height. With the development of off-the-shelf electronic components, HAB has now become a popular means to conduct near-space experiments, including measuring air quality, or certain types of pollutants. One of the most economical ways for researchers is to use latex balloons; with payloads these balloons can easily reach near-space at a low cost. But one of a disadvantage of this method is that the weight of the payload is restricted due to the limited balloon loading capacity.

The balloon expands as it rises, usually reaching an altitude of at least 90,000 ft (27,400 m) before it bursts. A small parachute lowers the instruments to the ground. Teardrop-shaped balloons are also used for the horizontal sounding of the atmosphere. Atmospheric pressure, temperature, and humidity information may be sent by radio from a balloon; monitoring of its movement provides information about winds at its flight level. Techniques also have been developed whereby many horizontal sounding balloons can be monitored by earth-orbiting satellites that relay information to earth-based stations.

12.3 Different scientific balloon systems and their use

Based on the user preferences, a variety of balloon systems are available to the scientist wanting to study the earth's, or another planet's atmosphere, the characteristics, and performances of which are summarized in Table 12.1. Different balloon systems are described from the point of view of the user, with their application.

Table 12.1 Performances of scientific balloons (Pommereau, 2015 [8], Copyright permission 2015 Elsevier).

Balloon	Volume (m^3)	Payload (kg)	Altitude (km)	Duration
Zero-pressure	5000–2,000,000	50–2000	25–40	Hours–days
Pressurized sphere	50–600	2–20	12–19	Weeks–months
Pressurized tracers	3–8	2–3	1–5	Weeks
Pressurized pumpkin	1,000,000	1000	35	(Goal 100 days)
IR Montgolfier	45,000	60	27 (day) 18 (night)	Weeks–months

12.3.1 Open zero-pressure balloons

The balloon most commonly used for scientific purposes is the zero-pressure balloon (Figs. 12.1 and 12.2). Its volume varies from 50 Km^3 to 2 Mm^3 or more, which can carry between 100 kg and 2000 kg between 25 km and 40 km. In most applications, the flight is limited to within the telemetry range from the control station of 300–400 km, that is for a duration of a few hours, though it can be extended by using a down-range receiving station [8].

The size of the balloon may be up to 300 m high at liftoff, and the high weight of the payload, requires the use of a mobile crane or else auxiliary balloons to keep the payloads off the ground during the launch operation. The payloads descents for recovery after the flight is terminated using one or several parachutes. Nowadays the ascent or descent speed and altitude of the balloon can also be adjusted by alternately valving gas and dropping ballast during greater scientific measurements [7,8].

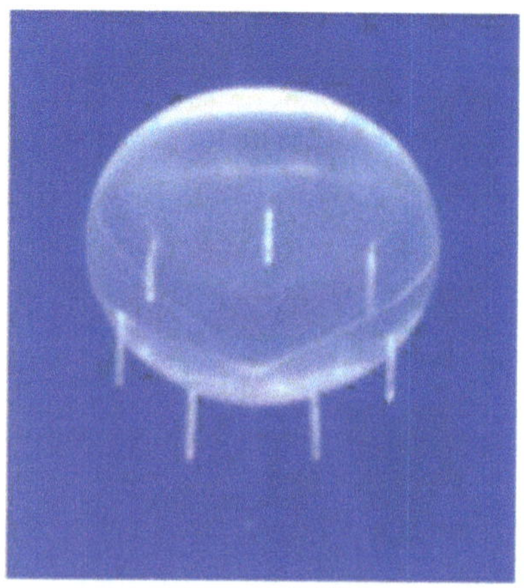

Fig. 12.1 ***Zero-pressure balloon.*** The excess gas is vented by seven open ducts at the bottom of the envelope (Pommereau, 2015 [8], Copyright permission 2015 Elsevier).

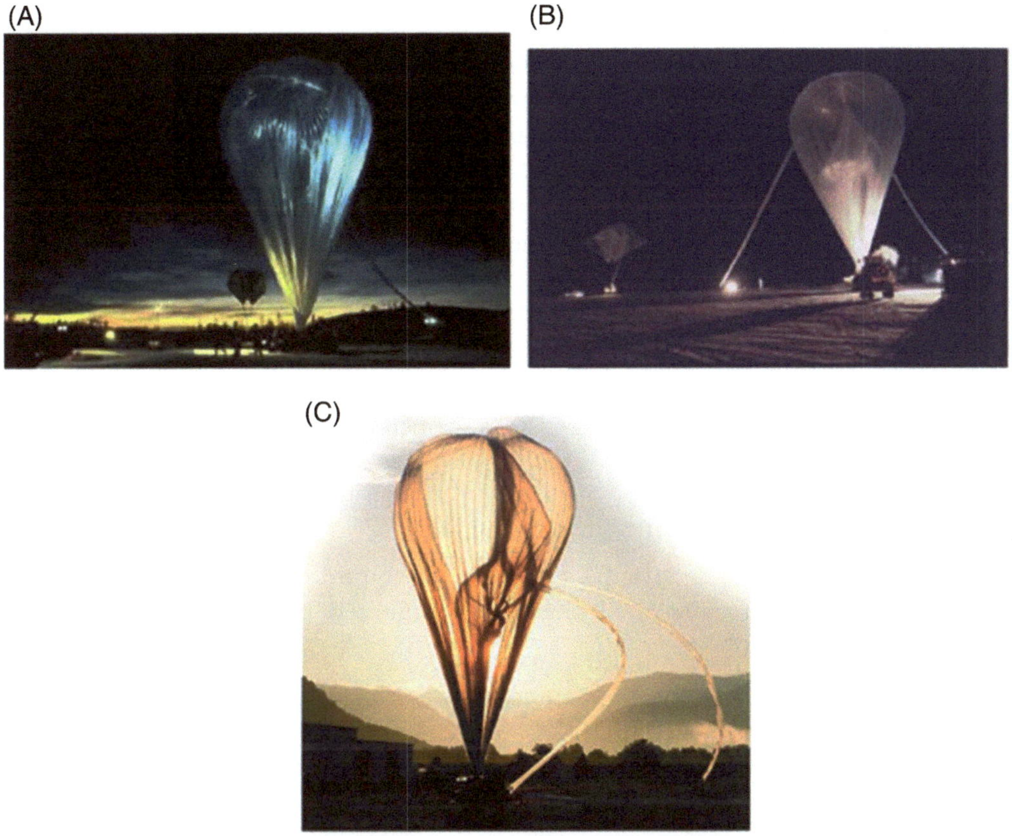

Fig. 12.2 (A) and (B) Inflation of a 100,000 m^3 zero-pressure balloon in the northern Sweden in January 2000 for studying stratospheric clouds and ozone depletion. ©CNRS. (C) Early-morning inflation of a 100,000 m^3 zero-pressure balloon at Gap in the Alps. ©CNES (Pommereau, 2015 [8], Copyright permission 2015 Elsevier).

12.3.2 Light, small, zero-pressure balloons

The large zero-pressure balloons are extremely sensitive to surface wind, and hence cannot be launched frequently, they are difficult to use in atmospheric science although being powerful tools. Recording of atmospheric variability requires frequent observations but being costly and their operation needs heavy equipment and large facilities, it is not recommended financially. An alternative approach, explored during recent years by CNES, is the reduction of the size of the balloons. This is achieved by using lighter material, 16 μm and more recently 12 μm thick, but reinforced at the gores assembly, and also by reducing dramatically the weight of the payload by applying miniaturization techniques to scientific instruments as well as to operational subsystems. Finally, launch techniques have also been revised. The payloads are designed to be dragged, perhaps on a sledge, on the launch pad, thus not requiring auxiliary balloons or a crane [8].

12.3.3 Tethered balloons

Open balloons that are only a little pressurized are called tethered balloons. Though the concept is simple, tethered balloons have been used for the study of the lowermost boundary layer. The use of tethered balloons is limited to the lowermost 1–2 km levels in light wind and in areas with little air traffic. However, a promising system for studying the sea–air interface in remote oceanic regions is the drifting balloon attached to a floating guide rope as tested in 2000 by the CNES in the tropical Atlantic [8].

12.3.4 Super-pressure balloons

The super-pressure balloons (Fig. 12.3) were developed originally in the USA at NCAR for studying the meteorology of the upper troposphere. They were spherical balloons of 43 μm thick polyester and 2.70 m diameter, able to carry a 2.5 kg payload at 200 hPa (12 km) [8].

12.3.5 Ultra-long-duration balloons (ULDB)

Unless a new, high-strength material appears in the future, the size and therefore the altitude of spherical super-pressure balloons will be limited. The only way to overcome

Fig. 12.3 ***Launch of a 10 m diameter super-pressure constant-level balloon at Kiruna, northern Sweden, in January 1999 for studying the dynamics of the stratosphere.*** ©CNES (Pommereau, 2015 [8], Copyright permission 2015 Elsevier).

the limit is to change completely the architecture of the balloon and adopt a pumpkin shape. A large pumpkin balloon of 58 m diameter for carrying 1.6 tons at 33.5 km for 100 days is under development at NASA [8].

12.3.6 Infra-red montgolfier (MIR)

Because of the smaller lift of hot air compared with hydrogen or helium, the MIR requires a larger volume than gas balloons for the same payload. But, in turn, it is a zero-pressure balloon and so the stress applied to the material is weak. The MIR is thus a robust balloon whose duration is limited only by the capacity to fly over very low-emitting, high-altitude clouds [8].

12.4 Balloon-borne campaigns to study atmosphere

Balloon platforms are used for different types of atmospheric studies worldwide; results from some important and recent studies over Asian especially Indian region are discussed in coming sections.

a. Transport of tropospheric and stratospheric ozone over India: balloon-borne observations and modeling:

A high altitude (troposphere and stratosphere) balloon campaign experiments was done by Sinha et al. [6], in his study they described the spatio-temporal variation of vertical profiles of ozone (O_3) measured by balloon-borne ozonesondes over two tropical sites of Trivandrum (TVM) and Hyderabad (HYD) in India from January 2009 to December 2010. They found that in the lower troposphere, the mixing ratios of O_3 over both the cities were similar around (18–65 ppbv). In the free troposphere, the O_3 mixing ratios over HYD were higher than those over TVM throughout the year. They also observed a strong seasonal variation of O_3 in the lower stratosphere between 18 km and 24 km over TVM, however, it is not pronounced for HYD. they also observed the transport of air masses from the biomass burning region of central Africa, Southeast Asia, and the Indo Gangetic Plains (IGP) influenced and led to enhancements of lower and mid-tropospheric O_3 over HYD and TVM while the isentropic (325 K) potential vorticity (PV) at 100 hPa showed transport of O_3-rich air from the lower stratosphere to the upper troposphere during winter and spring months over both sites. They found that the free tropospheric O_3 mixing ratios contribute substantially to the tropospheric column O_3 (TCO) with an annual average fraction of 30% and reveal similar seasonal variations over HYD and TVM. They compared the vertical profiles of O_3 obtained from the Monitoring Atmospheric Composition and Climate-Interim Implementation (MACC-II) reanalysis and the Meteorological Research Institute-Chemistry Climate Model version 2 (MRI-CCM2) with the ozonesonde data over both sites. The good agreement with measurements in the stratosphere while significant differences in the tropospheric columns in the simulated magnitude, phase, and vertical gradient of O_3 from both MRI-CCM2 and MACC-II were found in this study by Sinha et al. [6].

Similar many more balloon-born aerosol and air quality measurement campaigns were done by TIFR balloon facility Hyderabad as described in Sinha et al. [6] and references therein.

a. Asian tropopause aerosol layer measurement campaign using balloon sounding

(I) The Balloon Measurement Campaigns of the Asian Tropopause Aerosol Layer (BATAL)

Between 2014 and 2017 a series of American and Indian space agencies (NASA-ISRO) sponsored balloon-born campaigns were conducted in Saudi Arabia and India to find the nature, formation, and transportation of atmospheric pollutants in the upper troposphere and lower stratosphere (UTLS) region during the Asian summer monsoon (ASM) by Vernier et al. [7].

Observations from the satellites have shown the presence of pollutants (e.g., carbon monoxide (CO), hydrogen cyanide (HCN), and methane (CH_4)) in the UTLS region during the ASM [9,15]. Model simulations (e.g., [10]), also revealed the same thing, these studies suggest that deep convection can serve as a conduit for the transport of gas-phase boundary layer pollutants to the UTLS. Under the Asian anticyclone flow conditions that characterize the UTLS between the eastern Mediterranean Sea and China during the ASM, and which encompasses northern India the pollutants can mount up in this region of the atmosphere. Fig. 12.4 shows

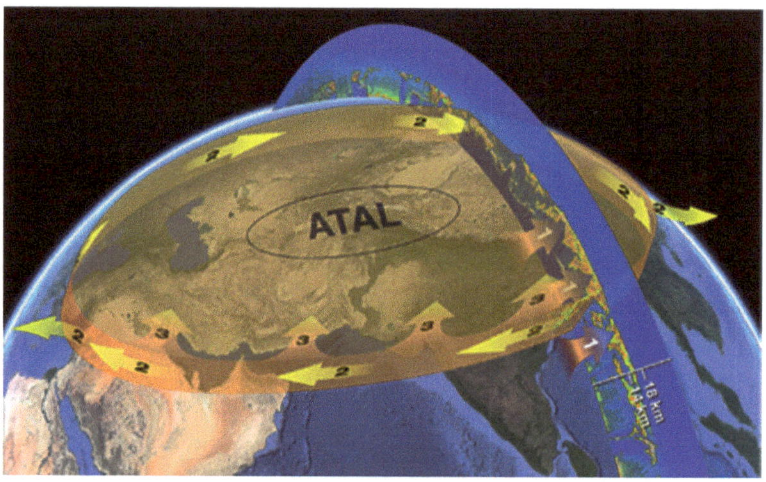

Fig. 12.4 ***Deep convection from the ASM transports air masses from the boundary layer to the UTLS (feature 1).*** An overpass of the CALIPSO lidar at 1930 UTC August 12, 2017 shows the vertical extension of convection through backscatter measurements. The monsoonal outflow is transported through tropical easterlies prevailing in the UTLS along the southern branch of the Asian anticyclone (feature 2). Air in the tropical upper troposphere along the southern edge of the Asian anticyclone can be uplifted into the lower stratosphere via diabatic ascent (feature 3). The ATAL is a pool of aerosols filling the Asian anticyclone shown as a light yellowish color. (Vernier et al., 2018 [7], ©American Meteorological Society. Used with permission).

a sketch of the transport mechanisms associated with the ASM. An overpass of the Cloud-Aerosol Lidar and Infrared Pathfinder Satellite Observations (CALIPSO) satellite for 12 August 2017 shows the vertical extension of convective storms reaching the UTLS over India. Air transported via the ASM anticyclone to the tropical upper troposphere can reach the lower stratosphere via a relatively slow adiabatic ascent [11]. In addition, strong mesoscale convective systems (MCSs) in India can directly penetrate the lower stratosphere.

During the ASM aerosol concentrations are increased in the UTLS along with trace gases. observations from the CALIPSO satellite have shown a recurrent aerosol layer in clear-sky summertime backscatter profiles near 13–18 km from the eastern Mediterranean Sea to western China and India [12], which is known as the Asian tropopause aerosol layer (ATAL). The ATAL was confirmed through solar occultation observations by Stratospheric Aerosol and Gas Experiment (SAGE) II [13] after improving the cloud–aerosol separation approach using the ratio between aerosol extinction coefficients retrieved at two wavelengths. Analysis of long-term satellite measurements of UTLS aerosols suggests that ATAL's aerosol optical depth has increased by 2–3 times since the late 1990s, pointing out its possible connection with Asian pollution growth [12]. They have chosen three balloon sites in India to sample air masses directly influenced by convection and to capture the horizontal gradient of aerosol in the UTLS region from the edge to the center of the ASM anticyclone. The location in Saudi Arabia was selected for measurements downwind from the monsoon region and to study the evolution of air masses along with their transport.

12.4.1 Description of the BATAL campaigns

The BATAL campaigns have comprised a total of ~68 balloon flights between 2014 and 2017 from India and Saudi Arabia, with payload weights ranging from 1.5 to 50 kg [7]. The effort has involved a number of institutes from the United States (Science Systems and Applications, Inc., National Aeronautics and Space Administration (NASA) Langley Research Center, and the University of Wyoming), France (Laboratoire de Physique et Chimie de l'Environnement et de l'Espace (LPC2E) Centre National de la Recherche Scientifique (CNRS)/University of Orléans), Switzerland (Swiss Federal Institute of Technology), Saudi Arabia (King Abdullah University of Science and Technology (KAUST)), and India (National Atmospheric Research Laboratory (NARL), Tata Institute of Fundamental Research (TIFR), Banaras Hindu University (BHU), and Physical Research Laboratory (PRL)). Fig. 12.5 shows the position of the launch sites (Gadanki, Hyderabad, and Varanasi, India; and Thuwal, Saudi Arabia) and the trajectories of the balloon flights conducted during the campaigns. The balloons were drifted westward in strong easterly winds of up to 40 m/s near the cold-point tropopause, mostly. Most of the payloads launched from India were recovered after landing within 300 km from the launch site, while most of the balloon flights from Thuwal landed in the Red Sea.

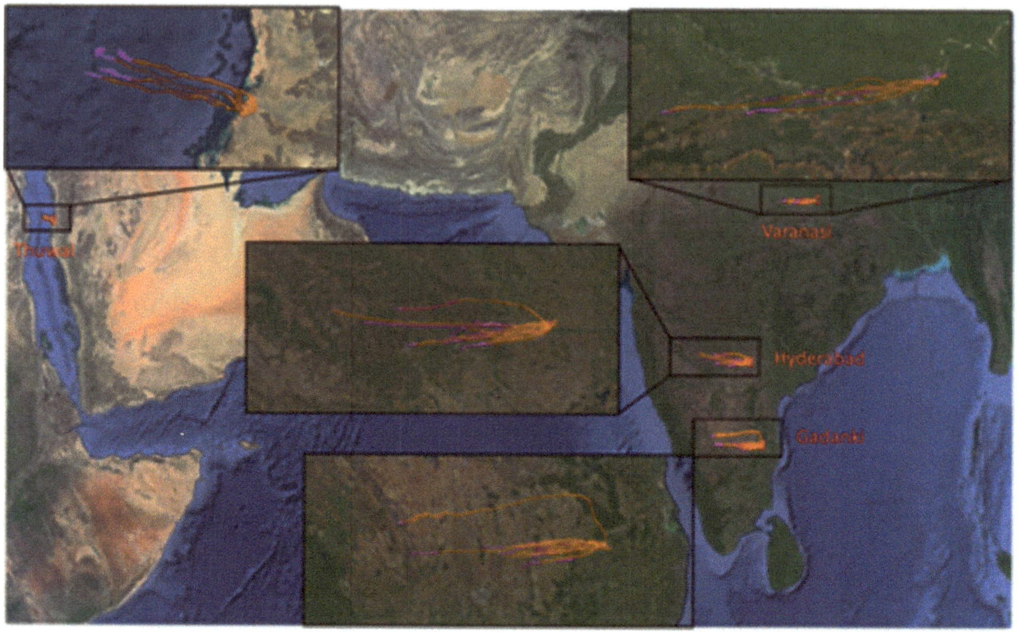

Fig. 12.5 ***Balloon trajectories from Gadanki (13.46°N, 79.17°E), Hyderabad (17.47°N, 78.58°E), Varanasi (25.27°N, 82.99°E), and Thuwal (22.32°N, 39.10°E) during the BATAL campaigns between 2014 and 2017.*** The ascending and descending balloon profiles are shown in orange and magenta, respectively (Vernier et al., 2018 [7], ©American Meteorological Society. Used with permission).

12.4.2 Types of balloon flight use in BATAL campaign

Different balloon platforms (Table 12.2 and Fig. 12.6) were used over the 4-yr BATAL effort to address the science objectives described in the earlier section:

Heavy balloon flight (HF): The optical and physical properties of the ATAL were addressed with aerosol payloads onboard HFs with a total payload weight greater than 30 kg using 3000 m^3 polyethylene balloons. The HFs flights took place from the TIFR balloon facility, Hyderabad, in August 2015 and 2017. During those flights [7] also studied the behavior of the ozone and water vapors near the ATAL.

Zero-pressure flight (ZF): Using ZFs floating near the tropopause for several hours, the chemical composition of the ATAL was studied using an aerosol impactor onboard balloon flight. The float duration was critical to collect a sufficient concentration of aerosols on the filters mounted on the aerosol impactor. Filters were sent postflight for ion chromatography (IC) analysis. The maximum floating duration above 14 km of 2 hour and 50 minutes was achieved during ZF3 [7].

Light flight/medium flight (LF/MF): [7] have studied the optical properties of the ATAL using a backscatter sonde, called COBALD, flown under a meteorological sonde for LFs with 1200-g latex balloons and a total payload weight of less than 1.3 kg. The MFs included, in addition to LF payloads, ozone and water vapor sensors and sometimes

Table 12.2 Balloon flight types during BATAL (Vernier et al., 2018 [7], ©American Meteorological Society. Used with permission).

Flight type	Approx payload weight	Balloon type	Measurements
HF	>30 kg	H2 filled, 3000 m^3, polyethylene	Particle size distribution, volatility, ozone, water vapor, particle backscatter, meteorological parameters
ZF	>10 kg	H2 filled, 300 m^3, polyethylene	Extended aerosol measurements near the tropopause, particle backscatter, aerosol composition and size distribution
MF	1.3 < MF <10 kg	H2 filled, 2000/3000 g, latex	Combined particle backscatter, ozone, water vapor, particle size distribution
LF	<1.3 kg	H2 filled, 1200 g, latex	Particle backscatter, meteorological parameters
BF	<10 kg	H2/He filled, tandem balloons, 2000/1200 g	Particle sampling using impactor with controlled ascent rate

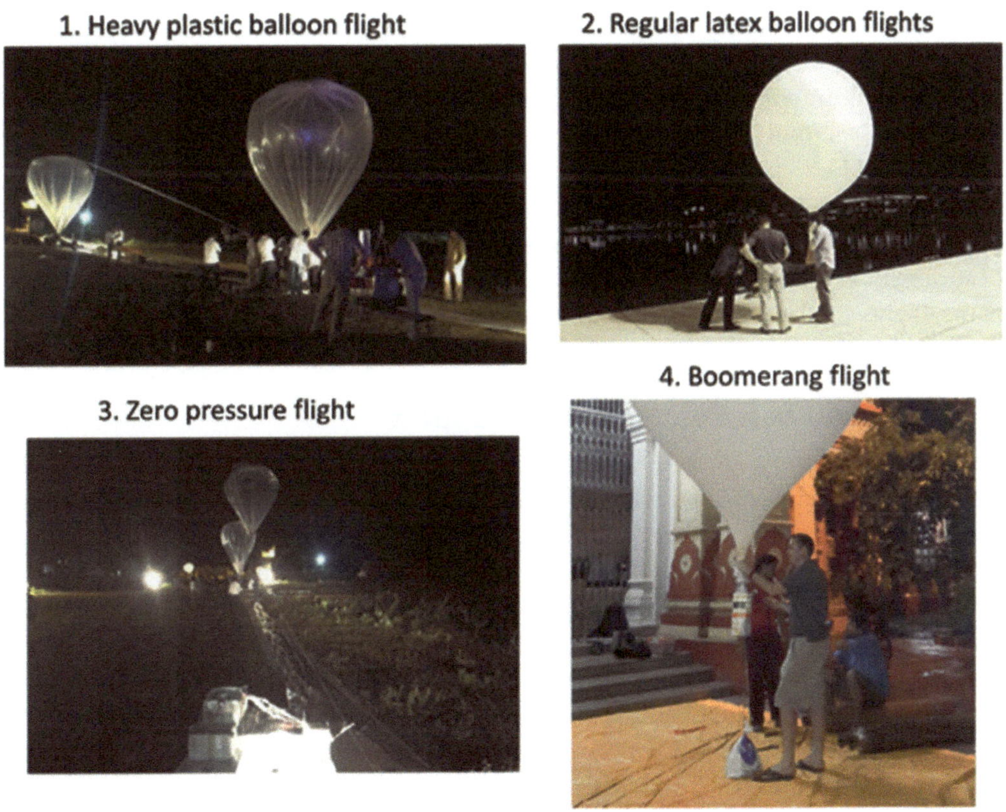

Fig. 12.6 ***Ballooning during BATAL.*** Preparation of (top left) HF from Hyderabad using plastic balloons, (top right) regular MF/LF with latex balloons, (bottom left) ZF using plastic balloon, and (bottom right) controlled flight using the Boomerang system attached to the neck of the balloon (Vernier et al., 2018 [7], ©American Meteorological Society. Used with permission).

a lightweight aerosol counter to study the size distribution of the ATAL and the chemical composition of the UTLS.

Boomerang flight (BF): To study the influence of convection on the ATAL, Vernier et al. [7] used a controlled balloon system, developed by Smith and Williamson, Inc., called Boomerang. The rationale was to have a flexible and simple system to launch payloads of less than 10 kg in weather conditions that would not allow for ZF or HF. The goal was to extend the measurement duration in the UTLS region and attempt to fly in convective storms. Boomerang was used with a 3000-g balloon and a secondary balloon (1200 g) to which it was attached with a pressure-sensitive cutdown device. The two Boomerang flights conducted from Varanasi in August 2016 included an aerosol impactor and COBALD.

Payloads: Vernier et al. [7] have launched payloads specifically designed for aerosol measurements; To characterize the optical, physical, and chemical properties of the ATAL, these included an optically based instrument to measure aerosol backscatter properties (COBALD) and optical particle counters (OPCs) from the University of Wyoming (Wyo), NASA Langley Research Center (Lan), and LPC2E for aerosol size distribution and volatility. In addition, they have launched an aerosol impactor for composition assessment. A further detailed description of payloads used in the campaign can be found in Table 12.3. Ozone and water vapor concentration measurements accompanied aerosol observations on several occasions to study the behavior of the ATAL and UTLS chemistry, using an ozonesonde and a CFH, respectively. They used iMet radiosondes on board most of the balloon flights, and occasionally a Meisei radiosonde (RS-11G) to transmit data, follow the balloon trajectory, and obtain the landing location. SPOT and global spectral model (GSM)-based locators was also used for most of the flights to determine the exact location of the landing and improve the chance of recovery [7].

12.4.3 Ground-based and modeling support

Throughout the campaign, at the National Atmospheric Research Laboratory (Gadanki) and the King Abdulah University for Science and Technology (Thuwal), observations were done from additional instrumentations. These include ground-based lidar observations that can be compared with COBALD in situ backscatter measurements from the boundary layer to the middle stratosphere and can be used to validate CALIPSO satellite aerosol retrievals. Throughout the BATAL campaign in India, weather forecasts (http://forecast.narl.gov.in/weather) provided by NARL and near-real-time Doppler weather radar images obtained from the India Meteorological Department (www.imd.gov.in/section/dwr/dynamic/dwr.htm) supported flight planning for balloon launches.

Starting with the 2017 campaign, regional and global chemical transport simulations have been carried out using the Weather Research and Forecasting (WRF) Model coupled with chemistry (WRF-Chem), version 3.6.1, and the NASA Goddard Earth

Table 12.3 Information on the payloads flown during the BATAL campaigns. The aerosol size is expressed in radii (Vernier et al., 2018 [7], ©American Meteorological Society. Used with permission).

BATAL payloads	Laboratory/company	Measured parameters	Measurement technique
iMet radiosonde	iMet	Temperature, pressure, relative humidity, wind speed/direction	Piezoresistive, thermistor, capacitive
Ozonesonde	Droplet Measurement Technologies, United States	Ozone concentration	Electrochemical
CFH	Droplet Measurement Technologies, United States	Water vapor concentration (UTLS)	Chilled-mirror principle
COBALD backscatter sonde	Eidgenössische Technische Hochschule (ETH), Switzerland	Scattering ratio at 455 and 940 nm	Light scattering with diodes
Lan OPC	NASA Langley Research Center, United States	Aerosol concentration (8 channels, 0.15–5 μm)	Laser counting
Wyo OPC 1	University of Wyoming, United States	Aerosol concentration (8 channels, 0.094–17 μm)	Laser counting
Wyo OPCH	University of Wyoming, United States	Aerosol concentration (8 channels, 0.094–17 μm)	Laser counting
Wyo CN counter	University of Wyoming, United States	Aerosol concentration (*r* greater than a few nanometers)	Saturation chamber/optical counting
LOAC	LPC2E, France	Aerosol concentration (19 channels, 0.1–50 μm)	Laser counting
BAI	California Measurements–NASA Langley Research Center, United States	Chemical composition	Aerodynamical impaction
Boulder counter	LightHouse–NASA Langley Research Center, United States	Ice measurements (6 channels, 5–100 μm)	Laser counting
Mixing cloud particle counter	Bretchel–NASA Langley Research Center, United States	Aerosol concentration (*r* greater than a few nanometers)	Saturation chamber/laser counting

Observing System (GEOS), version 5 (https://gmao.gsfc.nasa.gov/forecasts/), with chemical and aerosol forecasts over South Asia.

12.4.4 Findings of BATAL campaign

This section provides an overview of the balloon-borne activities from India and Saudi Arabia between 2014 and 2017 during the BATAL campaigns. With 68 balloon flights and payloads dedicated to study UTLS aerosols, together with water vapor and ozone sensors [7] have showed the first set of advanced in situ measurements of the ATAL with COBALD backscatter sonde, OPC size distributions, and offline chemical analysis. Authors have found that the ATAL is composed of mostly small (r <0.25 μm) liquid (~80–95%) aerosols with relatively low SR at 532 nm (SR <1.2), consistent with CALIOP satellite observations. Chemical analysis of particles collected near the tropopause indicates the dominant presence of nitrate aerosol with a concentration of about 100 ng m^{-3}. Balloon and satellite observations consistently showed a drop in aerosol concentration. In addition, enhanced water vapor and ice particles have been observed during the BATAL campaigns at and above the cold-point tropopause. These indicate that convection can potentially influence aerosol loadings well above the tropopause level.

(I) Balloon-borne measurements of temperature, water vapor, ozone, and aerosol backscatter on the southern slopes of the Himalayas [StratoClim 2016–2017] [14]

The Asian summer monsoon anticyclone (ASMA) is a major meteorological system of the UTLS region during boreal summer. Due to rapid lifting from the boundary layer by deep convection and subsequent horizontal confinement UTLS contain enhanced tropospheric trace gases and aerosols [14]. Given its dynamical structure, the ASMA represents an efficient pathway for the transport of pollutants to the global stratosphere. A detailed understanding of the thermal structure and processes in the ASMA requires accurate in situ measurements. The StratoClim project was performed state-of-the-art balloon-borne measurements of temperature, water vapor, ozone, and aerosol backscatter from two stations on the southern slopes of the Himalayas by Brunamonti et al. [14]. A total, 63 balloon soundings were conducted during two extensive monsoon-season campaigns, in August 2016 in Nainital, India (29.4°N, 79.5°E), and in July–August 2017 in Dhulikhel, Nepal (27.6°N, 85.5°E) and one shorter postmonsoon campaign was also carried out in November 2016 in Nainital [14]. These measurements provide unprecedented insights into the UTLS thermal structure, the vertical distributions of water vapor, ozone and aerosols, cirrus cloud properties, and interannual variability in the ASMA. An overview of all the data collected during the three campaign periods, with a focus on the UTLS region and the monsoon season is given in Brunamonti et al. [14]. They have analyzed the vertical structure of the ASMA in terms of significant levels and layers, identified from the temperature and potential temperature lapse rates and Lagrangian backward trajectories, providing a framework for relating the measurements to local thermodynamic properties and the large-scale anticyclonic flow. Both the monsoon-season campaigns show evidence of deep convection and confinement extending up to 1.5–2 km above the cold-point tropopause (CPT), yielding a body of air with high water vapor and low ozone which is prone to be lifted further and mixed into the free stratosphere. Enhanced aerosol backscatter also reveals the signature of the Asian tropopause aerosol layer (ATAL) over the same region of altitudes. The Dhulikhel 2017 campaign was characterized by a 5 K colder CPT on average than in Nainital 2016 and a local water vapor maximum in the confined lower stratosphere, about 1 km above the CPT [14].

12.5 Summary and conclusion

The balloon platform can be used in place of satellite and ground-based platforms for in-situ, localized, and short time duration measurement. The chapter describes detailed information about different types of remote sensing platforms including balloon platforms, their characteristics, and their uses in atmospheric studies. Different types of instruments which can be used by balloon platforms are also discussed in the chapter. The chapter discussed the use ofTropospheric-stratospheric balloons, which is a relatively cheap method for obtaining remotely sensed data. The measurements of atmospheric aerosols, particulate matters, trace gases, meteorological parameters, and the study of monsoon convection by remote sensing balloons are well discussed. High altitude atmospheric trace gases and pollution measurement and influences of UTLS region by loading of pollutants during Asian summer monsoon are also discussed. Three UTLS region balloon-borne measurement campaign studies and their results are discussed in the chapter. Few important results from balloon-based campaigns in India are summarized as,

1. In ozone measurement campaign Sinha et al. [6] have found the that the free tropospheric O_3 mixing ratios contribute substantially to the tropospheric column O_3 (TCO) with an annual average fraction of 30% and reveal similar seasonal variations over HYD and TVM. Authors compared the vertical profiles of O_3 obtained from the Monitoring Atmospheric Composition and Climate -Interim Implementation (MACC-II) reanalysis and the Meteorological Research Institute-Chemistry Climate Model version 2 (MRI-CCM2) with the ozonesonde data over both sites.
2. Enhanced water vapor and ice particles have been observed during the BATAL campaigns at and above the cold-point tropopause. These indicate that convection can potentially influence aerosol loadings well above the tropopause level. A suite of aerosol and ice measurements would contribute to an improved understanding of aerosol-cloud interactions in the UTLS region. Additionally, the results of the BATAL campaigns may be used to formulate future balloon-borne and airborne campaigns in the regions.
3. Balloon campaign over sub-Himalayan region show evidence of deep convection and confinement extending up to 1.5–2 km above the cold-point tropopause (CPT), yielding a body of air with high water vapor and low ozone which is prone to be lifted further and mixed into the free stratosphere. Enhanced aerosol backscatter also reveals the signature of the Asian tropopause aerosol layer (ATAL) over the same region of altitudes. The Dhulikhel 2017 campaign was characterized by a 5 K colder CPT on average than in Nainital 2016 and a local water vapor maximum in the confined lower stratosphere, about 1 km above the CPT [14].

References

[1] P. Wang, L. Liu, J. Wu, G. Tang, Environmental adaptability analysis of near space laser communication optical system, Proceedings of the 2012 International Conference on Optoelectronics and Microelectronics, Changchun, China, 23–25 August 2012, pp. 331–333.

[2] S.H. Alsamhi, S.K. Gupta, N.S. Rajput, Performance evaluation of broadband service delivery via tethered balloon technology, Proceedings of the 2016 11th International Conference on Industrial and Information Systems (ICIIS), Roorkee, India, 3–4 December2016 133–138 .
[3] J.A. Shaw, P.W. Nugent, N.A. Kaufman, N.J. Pust, D. Mikes, C. Knierim, N. Faulconer, R.M. Larimer, A.C. Desjardins, W.B. Knighton, Multispectral imaging systems on tethered balloons for optical remote sensing education and research, J. Appl. Remote. Sens. 6 (2012) 63613.
[4] W. Chen, P. Deng, X. Zhang, H. Wu, Integration of space and ground collaboration based on near space platform, Proceedings of the 2017 8th International Conference on Mechanical and Aerospace Engineering (ICMAE), Prague, Czech Republic 22–25 July, 2017, pp.596–602.
[5] Z. Wang, M. Huang, L. Qian, B. Zhao, G. Wang, High-altitude balloon-based sensor system design and implementation, Sensors 20 (2020). 2080 doi:10.3390/s20072080.
[6] P.R. Sinha, L.K. Sahu, R.K. Manchanda, V. Sheel, M. Deushi, M. Kajino, M.G. Schultz, N. Nagendra, P. Kumar, D.B. Trivedi, S.K. Koli, Transport of tropospheric and stratospheric ozone over India: Balloon-borne observations and modelling analysis, Atmos. Environ. 131 (2016) 228–242. doi:10.1016/j.atmosenv.2016.02.001.
[7] J.-P. Vernier, T.D. Fairlie, T. Deshler, Venkat Ratnam, Gadhavi M., Kumar H., S. B., M. Natarajan, A.K. Pandit, Akhil Raj, T. S., Hemanth Kumar, Jayaraman A., Singh A., K. A., N. Rastogi, P.R. Sinha, S. Kumar, S. Tiwari, T. Wegner, N. Baker, D. Vignelles, G. Stenchikov, I. Shevchenko, J. Smith, K. Bedka, A. Kesarkar, V. Singh, J. Bhate, V. Ravikiran, Durga Rao, Ravindrababu M., Patel S., Vernier A., Wienhold H., G. F., H. Liu, T.N. Knepp, L. Thomason, J. Crawford, L. Ziemba, J. Moore, S. Crumeyrolle, M. Williamson, G. Berthet, F. Jégou, J.-B Renard, BATAL: The Balloon Measurement Campaigns of the Asian Tropopause Aerosol Layer, Bull. Am. Meteorol. Soc., 99(5), (2018) 955–973.
[8] J.-P. Pommereau, Observations platforms: Balloons. In: Encyclopedia of Atmospheric Sciences, second ed., Academic Press, Cambridge, Massachusetts, United States, 2015, pp. 255–263. doi:10.1016/B978-0-12-382225-3.00254-1.
[9] M. Park, W.J. Randel, A. Gettelman, S. Massie, J. Jiang, Transport above the Asian summer monsoon anticyclone inferred from Aura MLS tracers, J. Geophys. Res. 112 (2007) D16309. doi:10.1029/2006JD008294.
[10] M.G. Lawrence, J. Lelieveld, Atmospheric pollutant outflow from southern Asia: A review, Atmos. Chem. Phys. 10 (2010) 11017–11096. doi:10.5194/acp-10-11017-2010.
[11] H. Garny, W.J. Randel, Transport pathways from the Asian monsoon anticyclone to the stratosphere, Atmos. Chem. Phys. 16 (2016) 2703–2718. doi:10.5194/acp-16-2703-2016.
[12] J.-P. Vernier, et al., Increase in upper tropospheric and lower stratospheric aerosol levels and its potential connection with Asian pollution, J. Geophys. Res. Atmos. 120 (2015) 1608–1619. doi:10.1002/2014JD022372.
[13] L.W. Thomason, J.-P. Vernier, Improved SAGE II cloud/aerosol categorization and observations of the Asian tropopause aerosol layer: 1989–2005, Atmos. Chem. Phys. 13 (2013) 4605–4616. doi:10.5194/acp-13-4605-2013.
[14] S. Brunamonti, T. Jorge, P. Oelsner, S. Hanumanthu, B.B. Singh, K.R. Kumar, S. Sonbawne, S. Meier, D. Singh, F.G. Wienhold, B.P. Luo, M. Boettcher, Y. Poltera, H. Jauhiainen, R. Kayastha, J. Karmacharya, R. Dirksen, M. Naja, M. Rex, S. Fadnavis, T. Peter, Balloon-borne measurements of temperature, water vapor, ozone and aerosol backscatter on the southern slopes of the Himalayas during StratoClim 2016–2017, Atmos. Chem. Phys. 18 (2018) 15937–15957. doi:10.5194/acp-18-15937-2018.
[15] Randel, W. J., L. Emmons, D. Kinnison, P. Bernath, K. A. Walker, C. Boone, and H. Pumphrey, 2010: Asian monsoon transport of pollution to the stratosphere. Science, 328, 611–613. doi:10.1126/science.1182274.

CHAPTER 13

Vertical distribution of atmospheric brown clouds using Lidar remote sensing over Indian region

Manish Jangid, Amit Kumar Mishra
School of Environmental Sciences, Jawaharlal Nehru University, New Delhi, India

13.1 Introduction

The term "climate change" is widely used as an umbrella term to capture various adverse effects of global warming on climate which includes, sea level rise, glacier melt, ecosystem damage, loss of agricultural yield, increase in frequency of extreme meteorological events, etc. [1–3]. The science community across the world [4] accept the fact that the greenhouse gases (such as carbon dioxide, nitrous oxide, methane, etc.) emitted via anthropogenic activates are the main culprit for the observed global warming (i.e., rise in average earth's surface temperature). On the other hand, aerosols (tiny liquid/solid particles suspended in air) are supposed to partially counter balance the global warming through extinction of incoming solar radiation reaching to the earth's surface [5,6]. However, the radiative impact of aerosols on climate is not so straight forward. Depending on chemical composition, aerosols may induce warming or cooling at different atmospheric layers and surface [7,8]. For example, scattering aerosols such as sulphate and sea salt are known as cooling agent, whereas absorbing aerosols such as black carbon, brown carbon and dust induce warming. Therefore, the study of aerosol properties and their spatio-temporal variations are needed to estimate the accurate signature of radiative effects of these aerosols.

Aerosols can directly impact the earth's radiation budget by scattering and/or absorbing the radiation and indirectly through modification in cloud macro- and microphysics and thereby changes in radiation fluxes [6–8]. The uncertainties in estimation of aerosol radiative forcing depend on many factors such as space-time dependent changes in aerosol optical and microphysical properties. Aerosol vertical distribution is also one of the key parameters in the estimation of aerosol radiative forcing [9,10]. The inadequate information about vertical positioning of aerosol layers with respect to clouds and vertical distribution of optical properties are considered as important sources of uncertainties in estimation of aerosol radiative effect on climate [7,10]. Therefore, the study of vertical distribution of aerosols and its accurate representation in global climate

Atmospheric Remote Sensing
DOI: https://doi.org/10.1016/B978-0-323-99262-6.00004-3

models are necessary to narrow down the uncertainties associated with radiative forcing calculation.

Since in-situ observations of aerosol vertical distribution using aircraft or unmanned aerial vehicles (commonly termed as UAV) are tedious and expensive methods, the use of active remote sensing techniques like Light detection and ranging (Lidar) instruments mounted on ground and/or satellite are very popular nowadays (e.g., [11–13]). Lidar uses its own energy source (such as laser) to illuminate the distant target and records the reflected irradiance to provide information about the target. It provides range resolved information about the target using simple time-of-flight calculation [14]. The Cloud Aerosol Lidar with Orthogonal Polarization (CALIOP) on board Cloud-Aerosol Lidar and Infrared Pathfinder Satellite Observation (CALIPSO) is one of the successfully operated satellite-borne Lidars, which provides information about the vertical distribution of aerosols and clouds in the atmosphere [15,16]. The space-borne Lidar (such as CALIOP) along with the ground-based Micro-Pulse Lidar Network (MPLNET) are extensively used to study the aerosol and cloud vertical structure in last two decades across the globe [11–13,17–19].

These Lidar techniques complemented with satellite and ground-based passive remote sensors are widely used to characterize the 3-D structures of atmospheric brown clouds (ABCs) over many regions of the world [11,13,18–22]. The word "ABCs" is used to define the regional scale pollution plumes mainly containing the different types of aerosols from both anthropogenic and natural origins such as dust, smoke, and urban-industrial pollution plumes in the atmosphere [23–25]. A bird-eye-view of the earth can identify various ABCs' hotspots over the globe, and Indian subcontinent is one of them. The widespread cover of this regional haze over the Indian region is a matter of great concern due to its impact on air quality, weather conditions, human health, and agriculture [24,25]. Being a developing and second largest populated country, India is one of ABCs' hotspots of the world. Considering the significant impact of aerosols on human population, comprehensive understanding of the spatio-temporal variations of ABCs and its components is required over Indian region.

In the view of importance of information on aerosol vertical distribution as an input for global/regional climate models, this chapter focuses on basic theoretical framework of Lidar techniques and its application to get information on aerosol vertical distribution. Further, the spatiao-temporal variation of ABCs over Indian subcontinent will be discussed as a case study.

13.2 Theoretical background of light detection and ranging

Light detection and ranging (Lidar) technologies have been developed to provide information on vertical profiling of atmosphere and surface since 1960 [26–28]. It supports the advancements in various research areas of earth and environmental sciences namely, atmospheric science, terrestrial ecology, oceanography, cryospheric and hydrological studies, geomorphology, etc. [29]. There are two types of Lidar sensor, one is ground-based and another is space-born Lidar. It is an active remote sensor. The working principal of Lidar technology is fairly simple. It has active rapidly firing laser source of lights. The emitted light travels to target (surface, atmospheric constituents, tree etc.) and reflect to Lidar receiver, which records the information. The distance travelled is calculated by using time (which is called time of flight), then it converted into elevation. Lidar technology is very useful for may application in the field of atmospheric sciences.

In this study we have used space-born CALIOP on board CALIPSO satellite for the estimation of vertical distribution of aerosol. The CALIOP instrument observes vertically-resolved signature of aerosol and cloud distribution in the atmosphere across the globe. CALIPSO was launched in a sun-synchronized polar orbit on April 28, 2006 as a part of "A-train" constellation. It revolves around the planet on polar orbit and provides both day and night measurements of aerosol and cloud properties. The orbit of the CALIPSO system has a repeat cycle of 16 days, producing sub-satellite longitudinal tracks every 172 km at the equator [30]. The CALIOP uses Nd:YAG laser as a transmitter source and provides information about the vertical profiles of atmospheric constituents using Mie-Rayleigh scattering code. Aerosols in the clear sky conditions, over bright surfaces and even beneath thin clouds can be observed by CALIOP [16]. The vertical resolution of lidar profiles is 30 m/60 m at 532 nm/1064 nm for all the altitudes chosen in this study.

CALIOP level 2 version 3.01 datasets give range-resolved approximations of aerosol extinction at 532 nm from the attenuated backscatter coefficients at two wavelength points (532 nm and 1064 nm) measured by CALIOP using the Hybrid Extinction Retrieval Algorithm. This algorithm is a hybrid collection of methods that have been incorporated to form a single, fully automated retrieval system [31]. It yields aerosol data in the form of aerosol extinction coefficients at 532 nm and 1064 nm wavelengths. Particle extinction at 532 nm derived from CALIOP level 2 Ver. 3.01 is used for the analysis of the vertical profiles of aerosols [15,16,31,32] from 2007 to 2019. Only daytime data sets are used in this study.

It is also provided six different types of aerosol profiles which are dust, polluted dust, smoke, marine, clean continental, and polluted continental aerosol. Additional information on the CALIOP aerosol identification algorithm and aerosol products is provided by Omar et al. [33]. In this study, we have analyzed extinction profiles and different types of aerosol profiles in all four seasons from 2007 to 2019 over the northwest India, Indian core monsoon region (ICMR) and south India regions. The number of screened profiles and unscreened profiles of extinction and all six different types of aerosols data is used for analyzed vertical distribution for frequency of occurrence.

Screening methodology using quality flags and other attributes can be found in Mishra et al. [34].

13.3 Vertical distribution of ABCs over India

The present study is carried out over the three different regions of the Indian subcontinent. Fig. 13.1 shows three different regions which cover major parts of Indian subcontinent. Blue, black, and magenta color boxes are representing North-West India [21°–31°N; 68°–80°E], Indian core monsoon region [ICMR: 21°–27°N; 80°–90°E] and South India [9°–21°N; 73°–80°E], respectively. The background of Fig. 13.1 shows the spatial distribution of MODIS-derived aerosol optical depth (AOD) during the premonsoon season (2003–2019). The high AOD (>0.3) prevalence can be seen over all three regions. These regions are highly populated and highly polluted due to multiple sources of anthropogenic activities like biomass burning, agricultural waste, emissions through the thermal power plant, transportation, industries, etc. and natural sources like dust and sea salt transported from the western arid region and nearby seas, respectively through south-west monsoon [20,35]. Fig. 13.1 shows the occurrence of ABCs over Indian region during premonsoon region. This region experiences four different seasons which are premonsoon [March, April, May (MAM)], monsoon [June, July, August, September (JJAS)], postmonsoon [October, November (ON)], and winter [December, January, February (DJF)].

Fig. 13.2 shows monthly average vertical profiles from 0.3 km to 8 km altitude range of total extinction at 532 nm from 2007 to 2019 (total 156 months) over

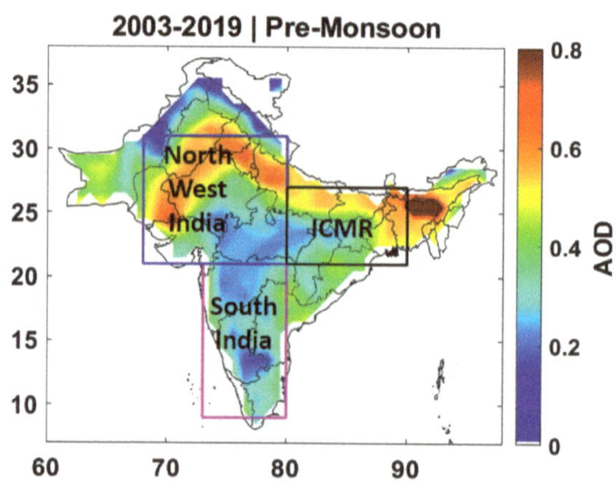

Fig. 13.1 ***Spatial distribution of MODIS-derived aerosol optical depth (AOD) during premonsoon season (2003–2019).*** Blue, black, and magenta color boxes are representing North-West India [21°–31°N; 68°–80°E], Indian Core Monsoon Region [ICMR: 21°–27°N; 80°–90°E] and South India [9°–21°N; 73°–80° E] region respectively which are used in this study.

North-west India, ICMR and South India in Fig. 13.2A–C, respectively. We observed seasonal and interannual variability in total aerosol extinction over all three different locations. We found higher aerosol concentration in North-West (Fig. 13.2A) region followed by ICMR (Fig. 13.2B), and South India (Fig. 13.2C). The thick layer or high concentration of aerosol at low-level altitude from 0.3 km to 1.5 km during both

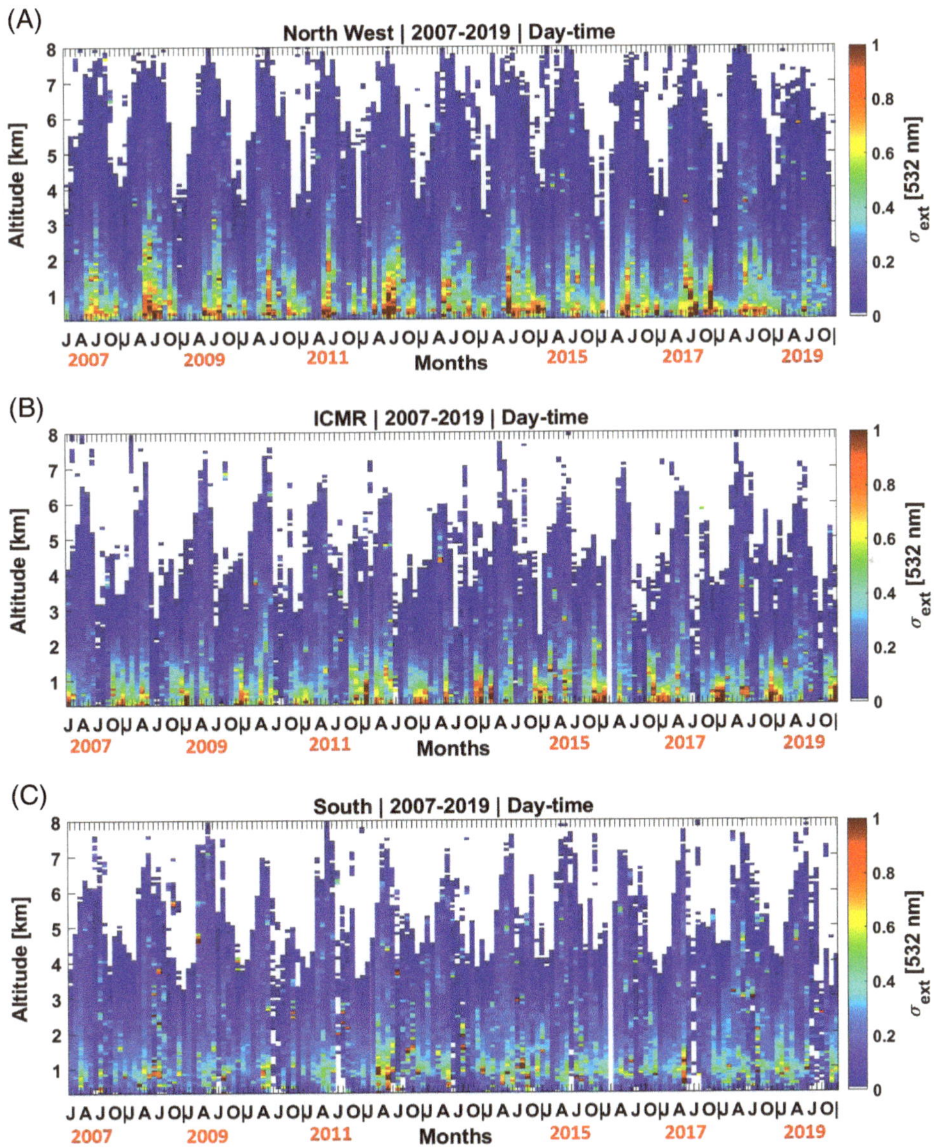

Fig. 13.2 Monthly averaged day time vertical profile of extinction at 532 nm during 2007 to 2019 over (A) North-West India, (B) ICMR, and (C) South India region.

postmonsoon and winter seasons is evident over all three regions. The extinction profiles are vertically extended up to 4–8 km in premonsoon and monsoon seasons over all regions. Generally, premonsoon and monsoon seasons are characterized by elevated aerosol layers (EAL; aerosol layers ~3–5 km) over all three regions with maximum occurrence over north-west part followed by ICMR and south region (Fig. 13.2). The EAL (3–5 km) shows the highest values in May and the lowest in December. The vertical distribution of aerosols varied according to the season which showed a cyclic pattern of increase and decreases over the Indian region. The top of the atmosphere layer reached its maximum during May and then decreased till October, after which it reaches to the maximum in December and then declined again [21]. It was found over the Kanpur region that the vertical distribution of aerosol found to have two zones, a well-mixed layer of aerosol up to 4 km and a layer of a sharp aerosol decrease in concentration at 4.5 km and above [19]. The small-scale clouds that have been created by dust were measured up to ~5 km and aerosol dust particles were perceived up to ~10 km over the Tibetan plateau [30].

Fig. 13.3 shows day time mean vertical profile of extinction in (1) winter, (2) premonsoon, (3) monsoon, and (4) postmonsoon seasons over the North-West region (red line), ICMR (blue line), and South India (black line) during 2007–2019. We observed high values of total extinction during the winter season at lower altitudes (0.3–1.5 km) over the ICMR (Fig. 13.3A). Whereas north-western region experiences aerosol extinction values at lower altitudes in the monsoon season (Fig. 13.3C). Aerosol vertical profile at both north-west and ICMR region shows similar pattern during the postmonsoon season (Fig. 13.3D). The seasonal variability in aerosol vertical profiles can be seen over all three regions. The high aerosol extinction over north-west region as compared to ICMR and south region during monsoon season can be attributed to less rainfall over north-west region. The prevalence of massive agriculture crop residue burning over north India region [36] increases the aerosol extinction over the north-west and ICMR as compared to south.

Fig. 13.4 provides clear information about the frequency of occurrence of different types of aerosols at different locations and in different seasons. It shows the daytime vertical distribution of averaged frequency (%) of occurrence of dust (red line), polluted dust (blue line), smoke (black line), marine (cyan line), clean continental (pink line), and polluted continental (green line) aerosols during 2007–2019 over North-West region (Fig. 13.4A–D), ICMR (Fig. 13.4E–H), and South India region in (Fig. 13.4E–I) in all four seasons. Dust profile occurrence of frequency is highest in premonsoon season at altitude 1–5 km over north-west region followed by ICMR and south India and similar pattern observed for the frequency of occurrence of polluted dust. Smoke aerosols are found higher at altitudes 1–3 km in all seasons. The frequency occurrence of marine aerosols is observed highest during the winter season followed by postmonsoon, premonsoon, and monsoon seasons at altitudes 0.5–1 km over the ICMR region. Clean

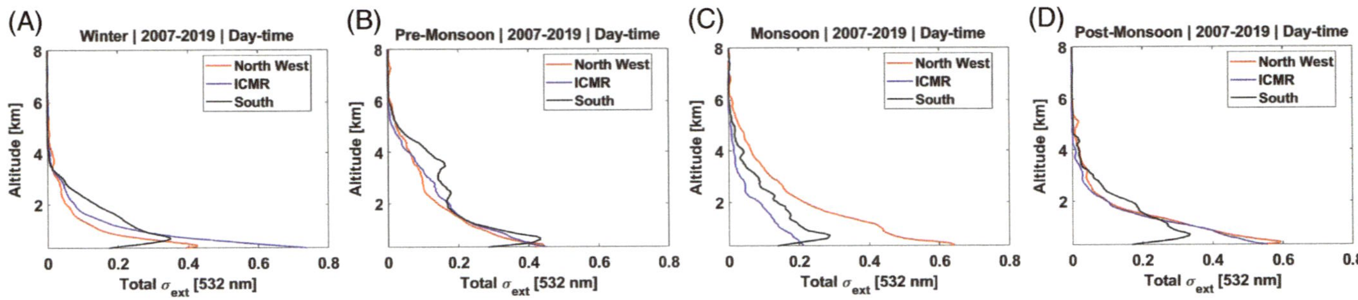

Fig. 13.3 Vertical distribution of averaged day time total extinction coefficient over North-West region (red line), ICMR (blue line), and South India (black line) during 2007–2019 in (A) winter, (B) premonsoon, (C) monsoon, (D) postmonsoon season.

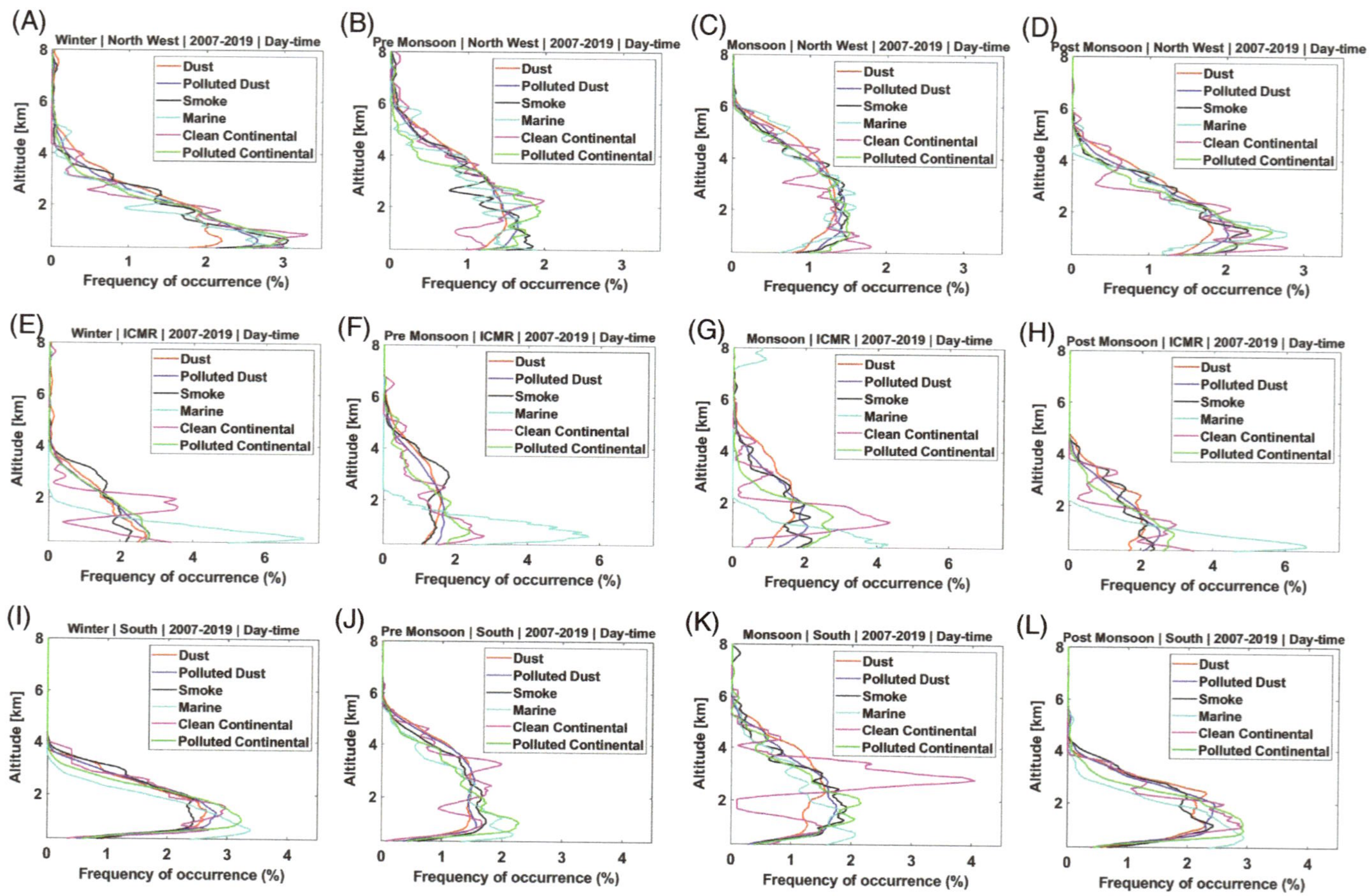

Fig. 13.4 Vertical distribution of averaged frequency of occurrence (%) of dust (red line), polluted dust (blue line), smoke (black line), marine (cyan line), clean continental (pink line) and polluted continental (green line) aerosols during 2007–2019 over North-West region in (A) winter, (B) premonsoon, (C) monsoon, and (D) postmonsoon, ICMR in (E) winter, (F) premonsoon, (G) monsoon, and (H) postmonsoon and South India in (I) winter, (J) premonsoon, (K) monsoon, and (L) postmonsoon season.

continental aerosols occurrence frequency observed highest during monsoon season over south India at altitude level 2–3.5 km (shows in Fig. 13.4K) followed by ICMR (altitude level 1–2 km) and North-west India. Polluted continental aerosols occurrence frequency is highest over ICMR during winter season followed by postmonsoon, premonsoon, and monsoon seasons. Higher concentration of polluted continental aerosol is observed at altitude 1–2 km over south India during winter season followed by postmonsoon, premonsoon, and monsoon seasons.

The vertical profiles of aerosols depend on metrological conditions and local anthropogenic/natural pollution sources and their transport. The biomass burning events produce massive heat at ground level, which could lead to strong convection of pollutants [43] commonly known as pyro-convection [36]. This phenomenon strongly affects the north-western region because this region is dominated by biomass burning activities during the postmonsoon season. North-west India is dust dominant region whereas, ICMR is rarely associated with the crop residue events and dust events but is affected by regional transported of agriculture crop residual burning aerosol and dust aerosol from north-west India region [44,45]. These dust aerosols and crop residue burning aerosols further mix with local pollutants to form the mixed aerosol (polluted-dust) over the region [8]. The monsoon of the Asian region is strongly related to changes in aerosols during premonsoon and monsoon [37–39]. A three-fold increase in aerosol concentrations is observed during the premonsoon season (late 1990–2015) over the Asian region [22,46]. Aerosol vertical profiles are majorly dominated by dust aerosol in the premonsoon region and biomass burning aerosol in the postmonsoon and winter season [19,21,36], which impacts the monsoon system of the region [38,40,41]. Manoj et al. [42] experienced the dust-dominated aerosols from 1 km to 4 km altitude over Indo-Gangetic Plains and these types of aerosols' vertical profiles are commonly observed in premonsoon and monsoon seasons [21,42]. Ratnam et al. [11] have shown a contrasting difference between aerosol vertical distribution over two ABCs hotspots (northern India and eastern China) using multiyear CALIOP observations. They have reported the statistically significant decreasing trend of boundary layer aerosols and increasing trend of free tropospheric aerosols over Indian region in changing climate scenarios, attributed to increase in convective activities over this region.

13.4 Conclusion

In view of importance of aerosol vertical distribution in climate related studies, we have provided a concise note on Lidar measurements and their application to study atmospheric aerosols in this chapter. The spatio-temporal variation of vertical profiles of ABCs have been studied by using CALIOP level 2 version 3.01 derived from CALIPSO satellite over the Indian subcontinent. This study is also provided information about frequency occurrence of different component of ABCs (i.e., type of aerosols) in different seasons. The vertical aerosol profile shows seasonal and interannual variation over the region. The vertically extended aerosol profiles have been found during premonsoon season, mainly due to formation of extended mixing boundary layers caused by surface heating. We found high concentration of dust aerosols at north-west region. The south India experienced high frequency of clean-continental aerosol and ICMR have high concertation of marine aerosol. Maximum frequency of dust is found at the altitude 2–5 km, polluted dust at the altitude 1–3 km, smoke aerosol at the altitude 3.5–4 km, marine aerosol at the altitude 0.5–1 km, clean continental the altitude 2–3.5 km and polluted continental at the altitude 1–2 km. Our study provides the information about vertical distribution of aerosols in all four seasons and shows spatial variation of all aerosol types over Indian region. The output of this study can be useful for climate and weather modelers to narrow down the uncertainties in their calculation/predictions.

References

[1] L.F. Girard, F. Nocca, Climate change and health impacts in urban areas: Towards hybrid evaluation tools for new governance, Atmosphere 11 (12) (2020) 1344.

[2] N. Seddon, A. Smith, P. Smith, I. Key, A. Chausson, C. Girardin, …, B. Turner, Getting the message right on nature-based solutions to climate change, Global Change Biol. 27 (8) (2021) 1518–1546.

[3] B. R., Singh, O. Singh, (2012). Study of impacts of global warming on climate change: Rise in sea level and disaster frequency. In (Ed.), Global Warming - Impacts and Future Perspective. IntechOpen. https://doi.org/10.5772/50464.

[4] S. Manabe, Role of greenhouse gas in climate change, Tellus A: Dyn. Meteorol. Oceanogr. 71 (1) (2019) 1620078.

[5] H. Feng, B. Zou, Satellite-based estimation of the aerosol forcing contribution to the global land surface temperature in the recent decade, Remote Sens. Environ. 232 (2019) 111299.

[6] V. Ramaswamy, W. Collins, J. Haywood, J. Lean, N. Mahowald, G. Myhre, …, T. Storelvmo, Radiative forcing of climate: The historical evolution of the radiative forcing concept, the forcing agents and their quantification, and applications, Meteorol. Monogr. 59 (2019) 14 -1.

[7] O. Boucher, D. Randall, P. Artaxo, C. Bretherton, G. Feingold, P. Forster, V.-M. Kerminen, Y. Kondo, H. Liao, U. Lohmann, P. Rasch, S.K. Satheesh, S. Sherwood, B. Stevens, X.Y. Zhang, Clouds and aerosols, in Climate Change 2013: The Physical Science Basis, in: T.F. Stocker, D. Qin, G.K. Plattner, M. Tignor, S.K. Allen, J. Boschung et al (Eds.), 2013. Contribution of Working Group I to the Fifth Assessment Report of the Intergovernmental Panel on Climate Change (pp. 571–657). Cambridge, UK and New York: Cambridge University Press, doi:10.1017/CBO9781107415324.016.

[8] M. Jangid, A.K. Mishra, I. Koren, C. Sarangi, K. Kumar, S. Singh, S. Tripathi, Observation of aerosol induced 'lower tropospheric cooling' over Indian core monsoon region, Environ. Res. Lett. 16 (12) (2021) 124057.

[9] Fountoulakis, I., Papachristopoulou, K., Proestakis, E., Amiridis, V., Kontoes, C., & Kazadzis, S. (2022). Effect of the aerosol vertical distribution on the modelling of solar radiation.

[10] A.K. Mishra, I. Koren, Y. Rudich, Effect of aerosol vertical distribution on aerosol-radiation interaction: A theoretical prospect, Heliyon 1 (2) (2015) e00036. https://doi.org/10.1016/j.heliyon.2015.e00036.

[11] M.V. Ratnam, P. Prasad, S.T. Raj, M.R. Raman, G. Basha, Changing patterns in aerosol vertical distribution over South and East Asia, Sci. Rep. 11 (1) (2021) 1–11.

[12] J. Shen, N. Cao, Comprehensive observation and analysis of aerosol optical properties and vertical distribution in Nanjing, China, Atmos. Environ. 239 (2020) 117767.

[13] T. Su, Z. Li, C. Li, J. Li, W. Han, C. Shen, …, J. Guo, The significant impact of aerosol vertical structure on lower atmosphere stability and its critical role in aerosol–planetary boundary layer (PBL) interactions, Atmos. Chem. Phys. 20 (6) (2020) 3713–3724.

[14] U. Wandinger, Introduction to lidar. In: Lidar, Springer, New York, 2005, pp. 1–18.
[15] D.M. Winker, J. Pelon, M.P. McCormick, The CALIPSO mission: Space borne lidar for observation of aerosols and clouds, Proc. SPIE 4893, P. Soc. Photo-Opt. Ins. 1 (2003) 1–11. doi:10.1117/12.466539.
[16] D.M. Winker, W.H. Hunt, M.J. McGill, Initial performance assessment of CALIOP, Geophys. Res. Lett. 34 (2007) L19803. doi:10.1029/2007GL030135.
[17] J. Huang, P. Minnis, B. Chen, Z. Huang, Z. Liu, Q. Zhao, ..., J.K. Ayers, Long-range transport and vertical structure of Asian dust from CALIPSO and surface measurements during PACDEX, J. Geophys. Res. Atmos. 113 (D23) (2008).
[18] A. Misra, S.N. Tripathi, D.S. Kaul, E.J. Welton, Study of MPLNET-derived aerosol climatology over Kanpur, India, and validation of CALIPSO level 2 version 3 backscatter and extinction products, J. Atmos. Oceanic Technol. 29 (9) (2012) 1285–1294. DOI: https://doi.org/10.1175/JTECH-D-11-00162.1.
[19] C. Sarangi, S.N. Tripathi, A.K. Mishra, A. Goel, E.J. Welton, Elevated aerosol layers and their radiative impact over Kanpur during monsoon onset period, J. Geophys. Res. Atmos. 121 (13) (2016) 7936–7957. doi:10.1002/2015JD024711.
[20] A. Mhawish, M. Sorek-Hamer, R. Chatfield, T. Banerjee, M. Bilal, M. Kumar, ..., O. Kalashnikova, Aerosol characteristics from earth observation systems: A comprehensive investigation over South Asia (2000–2019), Remote Sens. Environ. 259 (2021) 112410. doi:10.1016/j.rse.2021.112410.
[21] A.K. Mishra, T. Shibata, Climatological aspects of seasonal variation of aerosol vertical distribution over central Indo-Gangetic belt (IGB) inferred by the space-borne lidar CALIOP, Atmos. Environ. 46 (2012) 365–375. doi:10.1016/j.atmosenv.2011.09.052.
[22] J.P. Vernier, T.D. Fairlie, M. Natarajan, F.G. Wienhold, J. Bian, B.G. Martinsson, ..., Bedka, M K., Increase in upper tropospheric and lower stratospheric aerosol levels and its potential connection with Asian pollution, J. Geophys. Res. Atmos. 120 (4) (2015) 1608–1619. doi:10.1002/2014JD022372.
[23] G. Engling, A. Gelencsér, Atmospheric brown clouds: From local air pollution to climate change, Elements 6 (4) (2010) 223–228.
[24] V. Ramanathan, F. Li, M.V. Ramana, P.S. Praveen, D. Kim, C.E. Corrigan, ..., S.C. Yoon, Atmospheric brown clouds: Hemispherical and regional variations in long-range transport, absorption, and radiative forcing, J. Geophys. Res. Atmos. 112 (D22) (2007).
[25] S. Sharma, L. Nunez, V. Ramanathan, Atmospheric brown clouds. In: Oxford Research Encyclopedia of Environmental Science, 2016. https://doi.org/10.1093/acrefore/9780199389414.013.47.
[26] A.G. Kashani, M.J. Olsen, C.E. Parrish, N. Wilson, A review of LiDAR radiometric processing: From ad hoc intensity correction to rigorous radiometric calibration, Sensors 15 (11) (2015) 28099–28128. https://doi.org/10.3390/s151128099.
[27] R.M. Schotland, Some observations of the vertical profile of water vapor by means of a ground based optical radar, in Proceedings of the Fourth Symposium on Remote Sensing of the Environment, vol. 271, Univ. of Mich., Ann Arbor (1966).
[28] C. Weitkamp, Lidar: range-resolved optical remote sensing of the atmosphere. Springer Science & Business. 102 (2006).
[29] Z. Wang, M. Menenti, Challenges and Opportunities in Lidar Remote Sensing, Front. Remote Sens. 2 (2021) 641723. https://doi.org/10.3389/frsen.2021.641723.
[30] Z. Liu, D. Liu, J. Huang, M. Vaughan, I. Uno, N. Sugimoto, ..., D. Winker, Airborne dust distributions over the Tibetan Plateau and surrounding areas derived from the first year of CALIPSO lidar observations, Atmos. Chem. Phys. 8 (16) (2008) 5045–5060. doi:10.5194/acp-8-5045-2008.
[31] A.S., Young, M.D., Winker, A.M., Vaughan, Y., Hu, and E.R., Kuehn, (2008). Cloud-Aerosol lidar Infrared Pathfinder satellite observations CALIOP Algorithm Theoretical Basis Document, Document No: PC-SCI-202 Part 4.
[32] S.A. Young, M.A. Vaughan, The retrieval of profiles of particulate extinction from Cloud-Aerosol Lidar Infrared Pathfinder Satellite Observations (CALIPSO) data: Algorithm description, J. Atmos. Ocean. Tech. 26 (2009) 1105–1119. doi:10.1175/2008JTECHA1221.1.
[33] A.H. Omar, M.D. Winker, M.A. Vaughan, Y. Hu, C.R. Trepte, R.A. Ferrare, K.-P. Lee, C.A. Hostetler, C. Kittaka, R.R. Rogers, R.E. Kuehn, Z. Liu, The CALIPSO automated aerosol classification and lidar ratio selection algorithm, J. Atmos. Ocean. Tech. 26 (2009) 1994–2014. doi:10.1175/2009JTECHA1231.1.

[34] A.K. Mishra, K. Klingmueller, E. Fredj, J. Lelieveld, Y. Rudich, I. Koren, Radiative signature of absorbing aerosol over the eastern Mediterranean Basin, Atmos. Chem. Phys. 14 (14) (2014) 7213–7231. doi:10.5194/acp-14-7213-2014.
[35] C. Sarangi, S.N. Tripathi, S. Tripathi, M.C. Barth, Aerosol-cloud associations over Gangetic Basin during a typical monsoon depression event using WRF-Chem simulation, J. Geophys. Res. Atmos. 120 (2015) 10974–10995. doi:10.1002/2015JD023634.
[36] A.K. Mishra, T. Shibata, Synergistic analyses of optical and microphysical properties of agricultural crop residue burning aerosols over the Indo-Gangetic Basin (IGB), Atmos. Environ. 57 (2012) 205–218.
[37] D. Ganguly, P.J. Rasch, H. Wang, J.H. Yoon, Climate response of the South Asian monsoon system to anthropogenic aerosols, J. Geophys. Res. Atmos. 117 (D13) (2012). https://doi.org/10.1029/2012JD017508.
[38] K.M. Lau, K.M. Kim, Observational relationships between aerosol and Asian monsoon rainfall, and circulation, Geophys. Res. Lett. 33 (21) (2006). doi:10.1007/s00382-006-0114-z.
[39] W. Li, L. Shao, D. Zhang, C.U. Ro, M. Hu, X. Bi, …, J. Chen, A review of single aerosol particle studies in the atmosphere of East Asia: Morphology, mixing state, source, and heterogeneous reactions, J. Clean. Product. 112 (2016). lee. 1330–1349 doi:10.1016/j.jclepro.2015.04.050.
[40] F. Höpner, F.M. Bender, A.M. Ekman, A. Andersson, Ö. Gustafsson, C Leck, Investigation of two optical methods for aerosol-type classification extended to a Northern Indian Ocean Site, J. Geophys. Res. Atmos. 124 (15) (2019) 8743–8763. doi:10.1029/2018JD029685.
[41] K.M. Lau, M.K. Kim, K.M. Kim, Asian summer monsoon anomalies induced by aerosol direct forcing: the role of the Tibetan Plateau, Clim. Dyn. 26 (7-8) (2006) 855–864. doi:10.1029/2006GL027546.
[42] M.R. Manoj, S.K. Satheesh, K.K. Moorthy, J. Trembath, H. Coe, Measurement report: Altitudinal variation of CCN activation across the Indo-Gangetic Plains prior to monsoon onset and during peak monsoon periods: Results from the SWAAMI field campaign, Atmos. Chem. Phys. Discuss. (2021) 8979–97. https://doi.org/10.5194/acp-2020-1233.
[43] J. Trentmann, G. Luderer, T. Winterrath, M.D. Fromm, R. Servranckx, C. Textor, …, M.O. Andreae, Modeling of biomass smoke injection into the lower stratosphere by a large forest fire (Part I): Reference simulation, Atmos. Chem. Phys. 6 (12) (2006) 5247–5260. doi:10.5194/acp-6-5247-2006.
[44] R. Gautam, N.C. Hsu, K.M. Lau, Premonsoon aerosol characterization and radiative effects over the Indo-Gangetic Plains: Implications for regional climate warming, J. Geophys. Res. Atmos. 115 (D17) (2010) https://doi.org/10.1029/2010JD013819.
[45] S.K. Pandey, V. Vinoj, K. Landu, S.S. Babu, Declining pre-monsoon dust loading over South Asia: Signature of a changing regional climate, Sci. Rep. 7 (1) (2017) 1–10. https://doi.org/10.1038/s41598-017-16338-w.
[46] P. Kulkarni, S. Ramachandran, Y. Bhavani Kumar, D. Narayana Rao, M. Krishnaiah, Features of upper troposphere and lower stratosphere aerosols observed by lidar over Gadanki, a tropical Indian station, J. Geophys. Res. Atmos. 13 (2008) D17207. https://doi.org/10.1029/2007JD009411.

CHAPTER 14

Application of remote sensing to study forest fires

Swagata Payra[a], Ajay Sharma[b], Sunita Verma[b]
[a]Department of Remote Sensing, Birla Institute of Technology Mesra, Ranchi, Jharkhand, India
[b]Department of Environment and Sustainable Development, Banaras Hindu University, Varanasi, Uttar Pradesh, India

14.1 Introduction

Forests are the lifeline of our planet and play a vital role in the conservation of flora, fauna, and the environment. Forest cover has recently been susceptible to human pressures as well as natural disasters such as fire, floods, and cyclones. Forest fires are one such disaster that puts forests at risk. Forest fires are caused by natural phenomena as well as anthropogenic activities. It results in severe damage to wildlife, fertile forest floor, timber, human property, and certain rare plant species. As the number of uncontrolled forest fires rises, so does the magnitude of the resulting negative consequences. It is estimated in a study that 90% of the total forest fire events were initiated by anthropogenic activities. Fires are responsible for damaging around 3.73 million hectares of forest land per year, resulting in a loss of Rs. 440 crores. About 54.7% of India's forest is fire-prone, with recurrent forest fires affecting 9.2% and occasional forest fires affecting 45.5% of the forest area [53] (Chandra, 2005F). Such occurrences have brought forth an importance to develop a system that is capable of identifying fire risk areas.

Weather and climate are strongly linked with forest fires and can affect the initiation, frequency, and intensity of forest fires over a region [54]. The prevalence and severity of forest fires are directly influenced by different factors like elevation, aspect, and slope. While aspect and elevation are essential factors in determining slope, that helps in deciding the rate at which fire spreads. Additionally, wind currents are usually uphill on a steep slope and help in the rising of convection heat along the slope. The rate of forest fire spread increases when all of these factors are combined. Due to the dependency of forest fires, it is difficult to determine such spatial parameters. Geographic information system (GIS) techniques have been made it possible to categorize a geographical area based on the degree of fire risk. Further, to improve the accuracy of fire detection algorithms, various parameters and criteria are used. Unfortunately, a forest fire is observed after it has spread over a vast region and is very difficult to control. The results of such events lead to irreparable loss of biodiversity and immense damage to the environment.

Atmospheric Remote Sensing
DOI: https://doi.org/10.1016/B978-0-323-99262-6.00015-8

In addition to irreparable damage to the ecological systems, forest fires can lead to devastating effects which include extinction of various rare species of the fauna and flora, disruption of local weather patterns, and global warming.

Forest fires have released large amounts of air pollution, emitting aerosol, and trace gases in the huge amount that have effects on environment for both the immediate as well as for long term. Along with ~30% of carbon dioxide (CO_2), other gases such as carbon monoxide (CO), oxides of nitrogen (NO_X), methane (CH_4), aerosols, black carbon, other trace substances, and huge amounts of smoke in the atmosphere comes from forest fires. Both upper and lower atmospheres are affected by the pollutant transported from a large distance [55]. Long-range transport of black carbon and its deposition on Himalayan glaciers reduces the snow albedo, traps more solar heat, and hence results in faster melting of glaciers [56–59]. Further, the trapped heat contributes in alteration of energy budget of the Earth's atmosphere system. Thus, forest fires impart significant negative impacts on both regional as well as local systems [60,61].

As a result of the pervasive harmful effects due to forest fires have received increasing recognition in past recent years. The forests are mostly remote unplanned area filled with fry leaves, woods, and are covered with trees and bushes, etc., which acts as fire fuel and make the area more vulnerable. Therefore, a precise mapping and monitoring of the location and temporal distribution of wildfires is important. This will aid in assessing the effects of fire and controlling the spread of future events. Additionally, the maps and tools can be utilized in several ongoing studies, like climate dynamics and land use, land-cover change. The goal is to find the fire quickly, with the exact location and to inform quickly. The invention detects forest fires at an early stage and must be extinguished before they cannot be controlled or repaired, or ensure to cause significant damage. In this case, remote sensing plays an important role. This chapter gives an overview of past and current forest fire events. The remote sensing approach for forest fire detection and mitigation is also covered. Further, the state in the art measurement tools and detection techniques have been discussed along with the basic working principles of the same. Finally, the overall impact of forest fires on all the spheres of the planet has been described.

14.2 Evolution of fire detection techniques

Traditionally, lookout tower was used to detect forest fire. A person was kept for the surveillance at the lookout tower located at the high point in forest. His duty was to look for the smoke or fire coming out from the forest using binocular (NRSC). Osborne fire finder [62] is one of the sophisticated instruments used for forest fire detection. Osborne fire finder consists of disk with a graduated rim with a topographic printer. To determine the forest fire location and direction, a pointer is aimed. Once the forest fire is determined the fire finder instrument send alert to firefighter crew.

These traditional techniques are completely reliable on the human observation which has many drawbacks. To overcome this problem, many automatic video surveillances were developed [63]. Instrument like infrared detectors and charged couple device (CCD) camera were installed at the lookout tower for the surveillance. These cameras and detectors hunt out fire activity and send an alert to the firefighter crew. In the current scenario, many new and advance technologies are developed such as unmanned aerial vehicle (UAV) monitoring system, airborne sensors, and satellite imageries.

Remote sensing is one of the most efficient and cost-effective technique for fire detection and mapping. Wireless sensor network (WSN) and camera surveillance are the two different type of approaches used in the remote sensing. These sensors and cameras are mounted on different platforms like satellite, drone, UAV, planes, and so on to provide images of different areas. These images are further processed in order to recognize and detect forest fires.

14.3 Remote sensing

Remote sensing is a science (art) of acquisition, measurement and interpretation of data, and its information without actually being physical, intimate contact with the object or feature [45].

Remote sensing system required the involvement of energy and radiation. In this technique, sophisticated sensors are used to measure electromagnetic radiation (EMR) that works as a carrier of information from object to sensor [1]. EMR is the waves that move into the space with the speed of light and consists of two fluctuating field (electric field and magnetic field). Both the fields (electrical and magnetic) are right angle to each other moving perpendicular to each other to the direction of propagation [2] (Fig. 14.1). EMR extant wide range of energies, wavelength, or frequency known as electromagnetic spectrum (EMS).

The stored information is carried through intervening medium from object to the sensor. The information are stored in the form of images consisting of digital numbers. Mathematical and statistical algorithms are used to extract the valuable information from the imageries. Remote sensing system require the involvement of energy and radiation.

Remote sensing is not limited to the satellite information but it is a combination of multidisciplinary science like spectroscopy, aerial photography, telecommunication, optics, and so on. Working of remote sensing is based on the interaction of EMR between features and sensors [3]. There are various stages in this process, as shown in Fig. 14.2.

1. EMR is emitted from Sun or self-source.
2. Energy transfer from a source to Earth

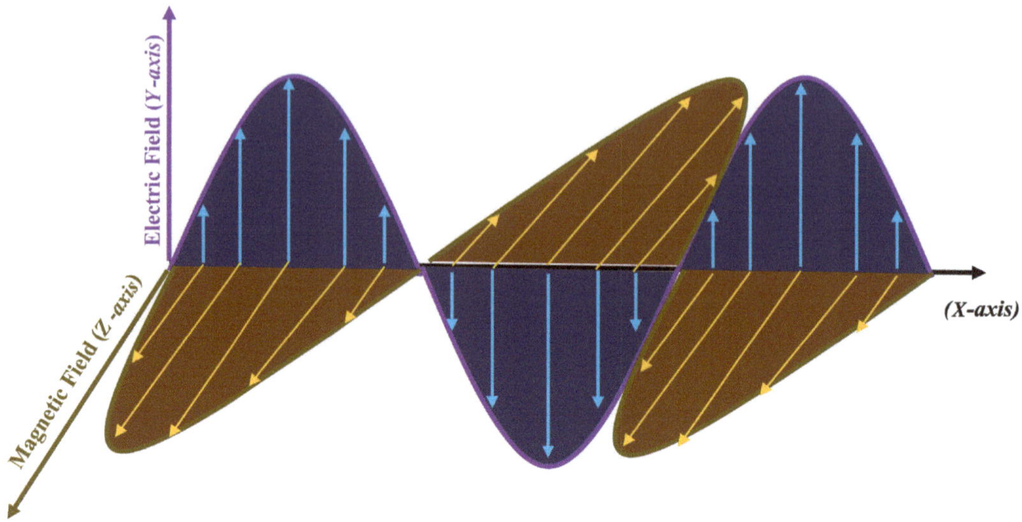

Fig. 14.1 ***Electromagnetic radiation (EMR) waves.***

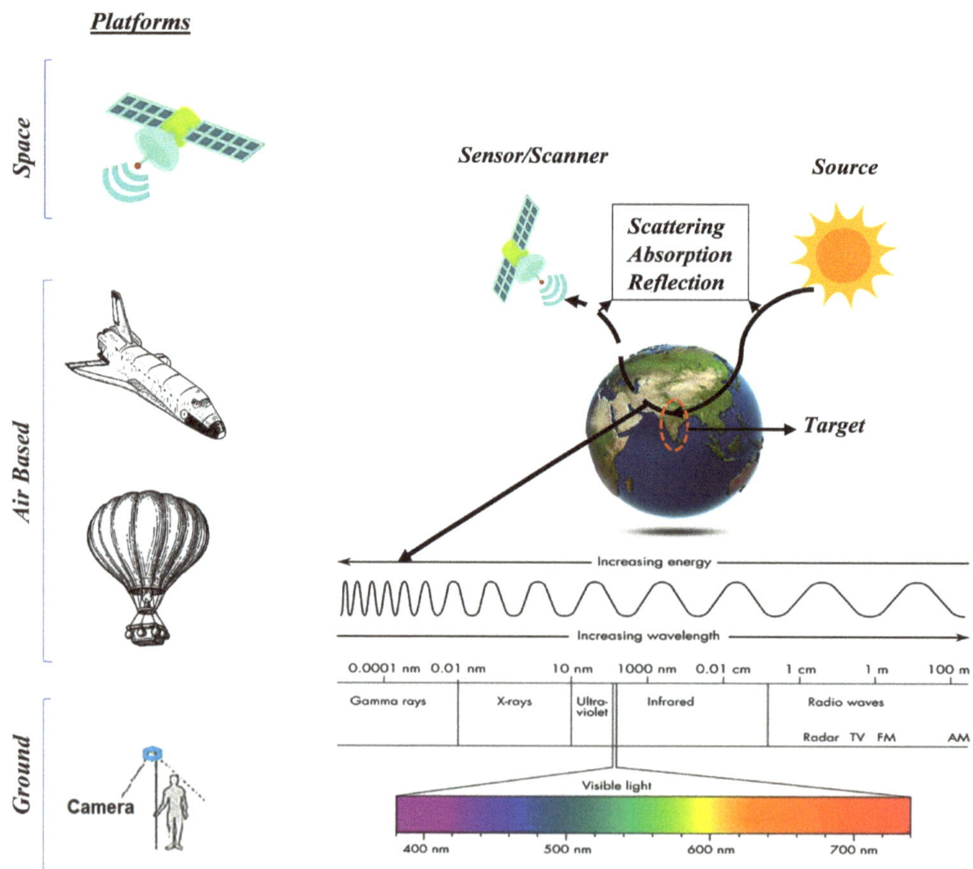

Fig. 14.2 ***Working of remote sensing.***

3. Absorption, scattering
4. EMR interaction with the Earth's surface (reflection)
5. Energy transfer from the surface to the sensor
6. Output generated by sensors
7. Data collection, assimilation and analysis

14.3.1 Thermal remote sensing

Thermal remote sensing is a technique to measure radiation emitting from the ground object of any entity [4]. Forest fire detection is based on the principle of identifying fire pixel according to the temperature difference between the energy emitting from the surface and ambient temperature. The basis for the determination is the ratio of the degree of thermal radiation between temperature and wavelength [5]. This thermal radiation is measured using a thermal band attached in the sensor. Thermal remote sensing plays a vital role in determination of fire pixels. Thermal remote sensing is works on two major laws which are as follows:

1. ***Stefan–Boltzmann law:*** Surface temperature of any object radiate energy and shows specific properties. These properties are calculated by Boltzmann law.

$$\mathrm{M}(\mathrm{T}) = \sigma \mathrm{T}^4 \tag{14.1}$$

where $M(T)$ = Energy emitted from the surface of the blackbody, known as total radiant exitance ($W \cdot m^{-2}$);

σ = Stefan–Boltzmann constant ($5.6697 \times 10^{-8}\ Wm^{-2}K^{-4}$);

T = thermodynamic emission of temperature from [blackbody temperature (K)].

According to the Stefan–Boltzmann law, complete energy released per unit time by a blackbody, from per unit surface area of the blackbody is directly proportional to the fourth power of temperature. Therefore, a precise change in temperature will lead to large change in radiation. This dramatic change in radiation emitted from forests helps to detect forest fire and heat source.

2. ***Wien's displacement law:*** Wien's displacement law explains the relation between temperature and the wavelength of radiation. It states that the wavelength of radiation emitted from a blackbody is inversely proportional to the temperature of the black body.

$$\lambda_{max} = A/T \tag{14.2}$$

where λ_{max} = maximum wavelength of radiation intensity (μm)

A = 2898 μmK;

T = thermodynamic temperature K.

According to Plank's radiation law, middle infrared is highly sensitive to the radiation of high temperature than thermal infrared (TIR) [64]. Therefore, fire detection basically works on the principle to determine the change in the brightness temperature from band ranging from 4 to 11 μm.

14.4 Satellites and sensors for forest fire applications

In general, satellites are used for a variety of tasks (communication, remote sensing of large areas or atmospheric elements, etc.), and it is cost-effective that enables wide use for wildfire detection capability. Many satellite systems and their operations depend on a number of national and international obligation having various rules, regulations, and agreements that may be less appropriate for forest fire monitoring.

In many cases, it is not possible to detect a small fire through satellites that can be stopped in a reasonable amount of time or can eventually lead to a large fire. It is, therefore, important to detect the correct timing of forest fires in order to protect the natural environment and forest resources. In recent years, remote sensing technology has become a powerful tool for precise and rapid control of wildfires.

Various Earth-orbiting satellites and even other aviation vehicles have been used to observe and detect wildfires, as shown in Table 14.1. Advanced very high-resolution radiometer (AVHRR) launched in 1998 and moderate resolution imaging spectroradiometer (MODIS) launched in 1999 and are two satellites sensor that are mostly used for forest fire detection. Unfortunately, these satellites can deliver images in a variety of settings. Images of Earth can be taken every other day, which is a long time to make fire measurements. In addition, the quality of satellite images may be affected by the weather. Visible (VIS), midinfrared (MIR), and thermal bands are used by satellite sensors to detect active forest fire areas during day time and VIS band hunts out fire by detecting light during night time. Detection of forest fire using conventional datasets were done in the early 1980s using various instruments and ground based observation [6]. These methods and ground data have many limitations that are overcame by the using satellite remote sensing data. Various techniques and algorithm like single or multichannel threshold algorithms [7], spatial contextual algorithms [8], radiance base algorithm for fire radiative power [9], weighted fire test algorithm [10] are applied on remote sensing data to detect forest fire by many scholars' in recent time in order to achieve accuracy and good results. Apart from AVHRR and MODIS other fire detection satellite is National Oceanic and Atmospheric Administration (NOAA) geostationary operational environment satellite provides thermal as well as VIS images after every 15 minutes over the United States and every 30 minutes over remaining surface. Optical lines can system of Defense Meteorological Satellite program is also used for fire detection having temporal resolution of 1 day and collects images from VIS bands [11]. Among all AVHRR is the most widely and popularly used satellite product having thermal sensor to detect forest fire having daily overpass time. MOD14A2/MYD14A2, MCD45A1, and MCD64A1 of MODIS products are used to detect forest fire. VIIRS 750 and 375 m provide active fire product VNP14 and VNP14IMG, respectively, sentinel 3 sea and land surface temperature instrument (SLSTR), ESA FIRE CCI, FireCCI51, etc., are some remote sensing forest fire product [12].

Table 14.1 List of sensor used for forest fire detection.

Satellite sensor	MODIS (Terra/Aqua)	Landsat (TM/ETM/ OLI-TIRS)	NOAA (AVHRR)	SNPP (VIIRS)	EO1 (Hyperion)	SPOT (HRV/HVIR)	Sentinel (SLSTR)
Spectral band	Visible NIR SWIR MWIR Thermal IR	Blue Green Red NIR SWIR Thermal	Red NIR MWIR Thermal	22 Bands (VIS, IR, NIR)	220 Bands (NIR, TIR)	Visible NIR SWIR PAN	21 Bands (VIS, SWIR, TIR)
Temporal resolution	Daily	16 Days	Daily	Daily	16 Days	Daily	27 Days
Spatail resolution	250 m to 1 km	30m (multispectral) and 15 m (PAN)	1 km	375 m and 750 m	10–30 m	20 m (multi-spectral) and 10m (PAN)	500 m and 1 km
Swath	2000 km	100 km	2000 km	3000 km	7.7 km	60 km	1.4 km
Purpose	Detection of thermal emission from active fires: burn scars mapping: vegetation classification	Detection and burn scars mapping: vegetation cover mapping	Detection of active fires hot spots region	Detection of forest fire range: burnt area	Hyperspectral fire detection	Detection of active fire: location: burn area: vegetation	Detect active burned area and forest fire range

14.4.1 Landsat

Landsat program is joint mission of NASA and United States Geological Survey (USGS) which is a longest-running enterprise for acquisition of satellite imagery of Earth. USGS is a scientific community of the United States cover all parts of the Earth's surface for surveying through satellites [13]. First Landsat mission was launched in 1972. Further the Landsat legacy continued and recently Landsat 9 was lunched on September 27, 2021. Landsat series of satellite are placed at Sun-synchronous orbit and revisits the same path on every 16 days. Spatial resolution of Landsat MSS (1-5) is 60 m and TM (4-5), ETM (7), and OLI/TRIS (8-9) is 30 m and 15 m for panchromatic band [14]. Landsat TM, ETM, and OLI/TRIS are capable to detect fire as they consist of thermal bands. Landsat 8 TIRS OLI has spatial resolution of 30 m for each band, in addition to TIRS and panchromatic bands [14]. The TIRS band spatial resolution is 100 m, but it is downscaled to 30 m and can be used to identify areas of microburning. The SWIR (Landsat 5 and 7 systems, TM5 and 7) band can be used to calculate normalized burn ratio (NBR) composites and identify the map burn areas. The algorithm developed for TIRS for fire detection appears to be red and SWIR band of the sensor looks as green which is used for monitoring drought condition in forest vegetation. At the time of fire both band will be seen dark.

14.4.2 Sentinel-2

Sentinel is a series of satellite developed by ESA (European Space Agency). Sentinel 2 mission comprise a constellation of two polar orbiting satellite namely sentinel 2A and 2B and were launched in June 2015 and March 2017, respectively. It is a multispectral satellite moving Sun-synchronous orbit. Spatial resolution of sentinel 2 are 10, 20, and 60 m and revisit the same location within 2–5 days. Sentinel 2 adds an option for a well-functioning medium resolution image with the ability to view the Earth's surface in the simple configuration, with the ability to probe the Earth's surface to locate active forest fire with high precision and accuracy [15–17,46]. Most of forest fire algorithms are developed using Sentinel-2 focuses on how to design fireplaces and products [18,19].

14.4.3 Sentinel-3

Sentinel 3 is a joint mission of ESA and EUMETSAT lunched on February 16, 2016. Sentinel-3 is a Sun-synchronous orbit at the NEAR-polar, where the equator at the intersection descends at 10:00 hour local solar time and repeat its cycle after every 27 days (ESA). (1) Ocean and land color instrument, (2) SLSTR, (3) SRAL (SAR radar altimeter), and (4) microwave radiometer are four main instruments on-boarded on sentinel 3.

ATSR is a series sensor installed in ERS-1, ERS-2, and ENVISAT [65]. It provides simultaneous transmission ofVIS by TIR band with moderate spatial resolution at both nadir and high view. Same specification as ATSR is used to build SLSTR, including in

Table 14.2 List of vegetation indices used in remote sensing.

Vegetation index	Name	Formula	References
NDVI	Normalized difference vegetation index	(ρNIR-ρRED)/(ρNIR+ρRED)	[48]
EVI	Enhanced vegetation index	2.5[(ρNIR-ρRED)/(ρNIR+2.4ρRED+1)]	[49]
SAVI	Soil adjusted vegetation index	[(ρNIR- ρRED)/(ρNIR+ρRED+L)](1+L)	[50]
SR	Simple ratio	ρNIR/ρRED	[51]
CTVI	Transformed vegetation index	[(ρNIR-ρRED)/(ρNIR+ρRED)]+0.5	[52]
DVI	Difference vegetation index	ρNIR-ρRED	[51]

the form of observation of high-temperature targets [20]. Active burning area and forest fire range are detected by the algorithm developed by SLSTR, using spectral radiance of VIS, near-infrared (NIR), shortwave infrared (SWIR), MIR, and TIR.

14.4.4 Vegetation-based fire applications

Vegetation is most common feature that widely effected by fire. Thus, studying and mapping vegetation area, health, and stress help us to identify the burned area in a forest. Land use land cover classification is one of the best techniques to demarcate the vegetation extent and different types. Various vegetation indices like NDVI, EVI, SAVI, SR, CTVI, DVI, etc., as shown in Table 14.2, are also used to identify the vegetation health, extent, and verity. To map or identify the vegetation drought, NDWI is used.

14.4.5 MODIS

MODIS is an instrument that is boarded on Tera and Aqua (EOS AM-1 and EOS PM-1, respectively) operational from 2000 to till date. MODIS uses three spatial resolutions 250 m, 500 m, and 1 km with 36 spectral bands (https://modis.gsfc.nasa.gov/about/). MODIS is a polar-orbiting satellite, crosses equator at the same local time (10:30 for Tera and 13:30 for Aqua). MODIS provides data at different temporal resolution which are daily, 8-day, 16-day for short-term data analysis and monthly, quarterly, and yearly for long term data analysis.

MODIS uses 36 spectral band to provide various product like LULC, NDVI, aerosol optical depth, etc. Among all MODIS also provide active fire and burned area product which are encoded as MOD14A1, MYD14A1 [21] at spatial resolution of 1 km and MOD09GHK and MYD09GHK with 500 m resolution [66]. Level 2 Gridded (L2G) surface reflectance data are extracted from three different band at different wavelengths which are band 1 (0.65 μm), 5 (1.2 μm), and 7 (2.1 μm) [22].

14.4.6 Visible infrared imaging radiometer suite (VIIRS)

VIIRS is a key sensor onboarded on Suomi National Polar-orbiting Partnership (Suomi NPP) which is a joint mission of NASA/NOAA lunched on October 28, 2011. Placed at Sun-synchronous orbit with 22 spectral channels having spatial resolution of 375 m of 5 bands and 750 m of 16 spectral bands (https://lpdaac.usgs.gov/data/get-started-data/collection-overview/missions/s-npp-nasa-viirs-overview/). VNP14IMG and VNP14 are the two different active fire products provided by VIIRS with spatial resolution of 375 m and 750 m [23]. It uses VIS, infrared (IR), and NIR band to detect forest fire and burnt areas.

14.4.7 Soil moisture active passive (SMAP)

Operation SMAP was developed and launched by NASA in January 2015 and data collection started in late March 2015. Sensors monitor soil moisture and freezing/thawing conditions twice a day at 6 pm and 6 pm local solar time, approximately polar synchronous solar orbit. It provides a global coverage after every 3 days. Passive microwave of SMAP is currently collecting data since March 31, 2015, till date [24]. SMAP provides various soil moisture data based on the algorithm developed using a passive microwave sensor. In order to identify areas of dry wood, it is important to know the drought conditions. To examine drought condition or evaporative stress, it is important to consider soil moisture that acts as proxy for vegetation moisture.

14.4.8 EO-1 hyperion

Earth-observing (EO) one-Hyperion satellite is a hyperspectral satellite consist of 220 spectral band with spatial resolution of 10–30 m. It is developed by NASA and was lunched on November 21, 2000 till 2017. Spectral band of Hyperion ranges from 357 to 2567 nm. Based on the spectral radiance centered at 2061 and 2429 nm of shortwave infrared of Hyperion a normalized difference index is developed known as hyperspectral fire detection index, helps in demarcation of forest fire region [25].

14.4.9 Advanced very high-resolution radiometer (AVHRR)

AVHRR is a polar-orbiting satellite that was first lunched on 1978 by NASA. AVHRR is carried by NOAA series of satellites which is mainly sent for weather surveillance, sea surface temperature, and to detect wildfires. AVHRR consists of five different channels in which first two were red and NIR (0.6 mm and 0.9 mm respectively), third is located at 3.6 mm and remaining two channels are thermal radiation bands with wavelength 11 mm and 12 mm. Spatial resolution of AVHRR is 1 km with temporal resolution of 1 day. AVHRR uses Fire Identification, Mapping and Monitoring Algorithm consists of two major steps: marking potential fires and removing false fires (Canadian).

14.5 Ground-based fire monitoring system

Forest fire monitoring is different in terms of timing and emphasis. Typically, satellite sensors produce low spatial resolution fire product that depicts the approximate location and size of forest fires. To effectively map small fire hotspots and active fire perimeters, detailed fire suppression monitoring and the use of finer resolution air-based TIR sensors is required. To deploy fire suppression workers and airborne water or retardant drops, higher-resolution fire maps are required. Thus to meet these requirements, some ground base monitoring system are used such as:

a. **Forest fire finder** is a unique, intelligent, and advanced system that relies on analyzing the air rather than detecting forest fires or smoke plumes. These systems learn how to put out forest fires by absorbing Sun rays. The receptor or the sensor depends on the chemical's composition of pollutant and the chemicals in the air [26]. Different mixtures have different receiving properties. Thus, forest fires finder can identify different types of smoke based on sources within a 15 km range, such as wood smoke and industrial smoke. Field-proven applications for rapid detection use this method in the Portuguese forest.
b. **Forest watch** is a sensor and camera based optical system offering a semiautomatic fire detection system developed by enviro-vision solutions of South Africa. During the day, the camera tower monitors the forest at a range of 16–20 km for a column of smoke, and at night, it looks for fire glow.
c. **Fire hawk**—It is a computer-based system developed for demarcation of forest fire locations. Identification of fire location helps in risk management system which consists of three different layers as follow:
   (i) Installing camera or sensor at appropriate location for monitoring.
   (ii) Setup wireless link to provide communication among layers.
   (iii) Fire Hawk integrates both GIS and Fire Watch software to provide the shortest path to the forest fire location to generate machine vision layer. Currently, two different locations in South Africa have installed Fire Hawk.
d. **FireWxNet** is an abbreviation of Fire and Weather Network, which is a portable wireless system used for monitoring wildland fire and weather conditions. The main aim of the network is to investigate behavior rather than detect wildfires. It uses a hybrid camera that detect the forest fire and wireless sensor to provide weather data [27]. The system uses multilayered architecture that starts with a directional antenna at the top of the mountain and ends with a series of sensor network to monitor the necessary environmental variables, which also provides visual data using installed sensor on web cameras. Apart from that a small GPS device that provides location information is also installed to the system.
e. **Wireless Sensor Network (WSN)** is an infrastructure less ground based fire monitoring system which has been explained in other recent publications [28,29–31]. WSN is used for monitoring environmental as well as physical condition like

temperature, pressure, and movement of pollutants [32]. It is a self-configured instrument that consist of hundreds of sensors node [33]. This sensor detects the level of temperature that helps in classification of forest fire. In WSN cameras, fire detectors, and 802.11g (slandered Wi-Fi technology that supports 54 mbps for communication) access points are used [34]. The goal of this network is to identify and examine fires in rural and forest areas, as well as send notifications to central servers. To run WSN, a collaboration between WSN provider and Ethernet, ZigBee, or GPRS communication modules is required [35]. Ref. [36] explains the WSN and solar energy units-based forest fire detection system.

Some other approached used for ground forest fire detection are:

f. Video camera are subjected to sense VIS spectrum that help in detection and recognition of smoke during day and night.

g. Thermal camera is used to sense the heat exchange of the fire.

h. Infrared spectrometer is also used which help in identification of spectral properties of smoke.

i. Light detection and ranging system is capable to detect laser ray emitting from smoke particles.

j. CCD are the device that store amount of photon that strike the surface of CCD in a form of array. CCD consists of image sensor containing photodiodes or capacitors which are sensitive to light.

14.6 Fire assessment

Biomass regeneration is done by radar datasets in the zone of tropical forests [37], to monitor and analyze the magnitude and burn severity of wildfire patches in the boreal forests [38], and to predict ecosystem recovery in Mediterranean climates. AVHRR and other low-resolution sensors operating in VIS and infrared wavelength ranges have proved to be effective for two purposes. The first one is the evaluation of impacts of forest-fire on long-scale land use land cover change [39], and the second is automated wildfire monitoring [40]. The scale-related to influences of evaluating vegetation regeneration postfire events are revealed in multiresolution studies using both AVHRR and Landsat imagery [41]. Once a forest fire event has occurred, a combination of these coarser resolution (AVHRR) and high resolution (SPOT, Landsat, and Radar) images can used to assess the size and impacts of forest fires. All these sensors are capable to detect different class of forest fire, as shown in Table 14.3.

Fuels are the major contributors of forest fire. It is a physical property of living and dead biomass that influence wildfire spread, severity, and intensity. Probability of fire in an area is termed as fire risk. Factors like size (particle diameter), loading (weight per unit area), and bulk density (weight per unit volume) that influence fire initiation and behavior are compiled for the determination of fire risk. An association of fuel elements of typical species, size, forms, continuity, and arrangement have been identified that will

Table 14.3 Types of fires.

Class of information	Type of information
Alpha type	Fire: start and end dates, location, size, and cause
Beta type	Fuels biome classification and fuel loading forest inventory (number), age class, size class
Gamma type	Fire characterization (crown, surface, etc.), fuel consumption and structural involvement (wildland urban interface)
Delta type	Number of fires, areas burnt (by forest type), cause of fires (number)
Epsilon type	Gas and aerosol emission data
Eta type	Total expenditure of fire program, total fire suppression costs, and total direct losses of merchantable timber, structure losses

demonstrate specific fire behavior under specific burning conditions [42]. The physical properties and impact of wildfire vary with time. Typically, fire regimes are described by the frequency, intensity, magnitude, and pattern of fires. A fire regime is a broad description of fire's role in a certain area or ecological environment; it refers to the type of fires that occur over a long period of time.

Forest fire assessment through remote sensing (satellite imageries) are done by calculating fire count. Fire count is defined as the number of hotspot or pixels region demarcated by satellite, sensor, or scanner over a given area at particular time period. For example: state/country/city/district political boundary, spherical, circular, square or regular box, buffer zone from certain feature and so on. Temporal variability can be 1 hour, 24 hours, and 3 days for short period of time, 7 days and 1 year for long term. Thermal anomaly algorithm is used by users to calculate fire count which is either active or in active fir hotspot zone. Fig. 14.3 is showing various stages of estimation of fire product through remote sensing.

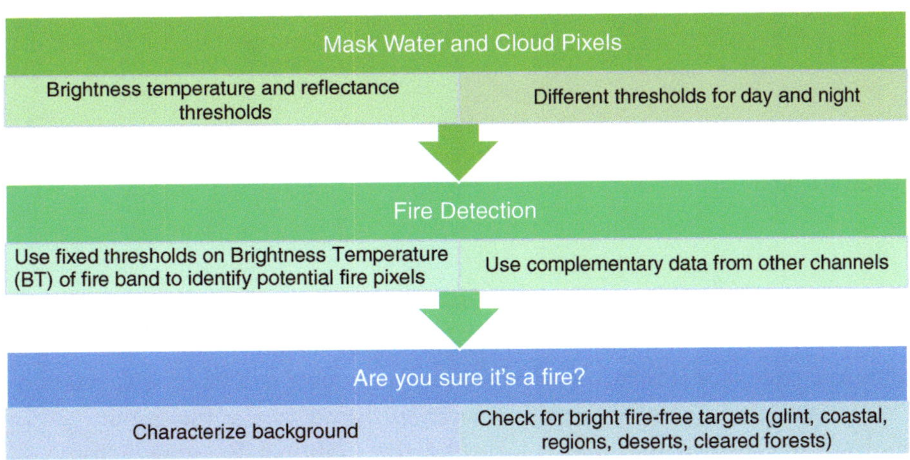

Fig. 14.3 ***Stages of fire count detection.***

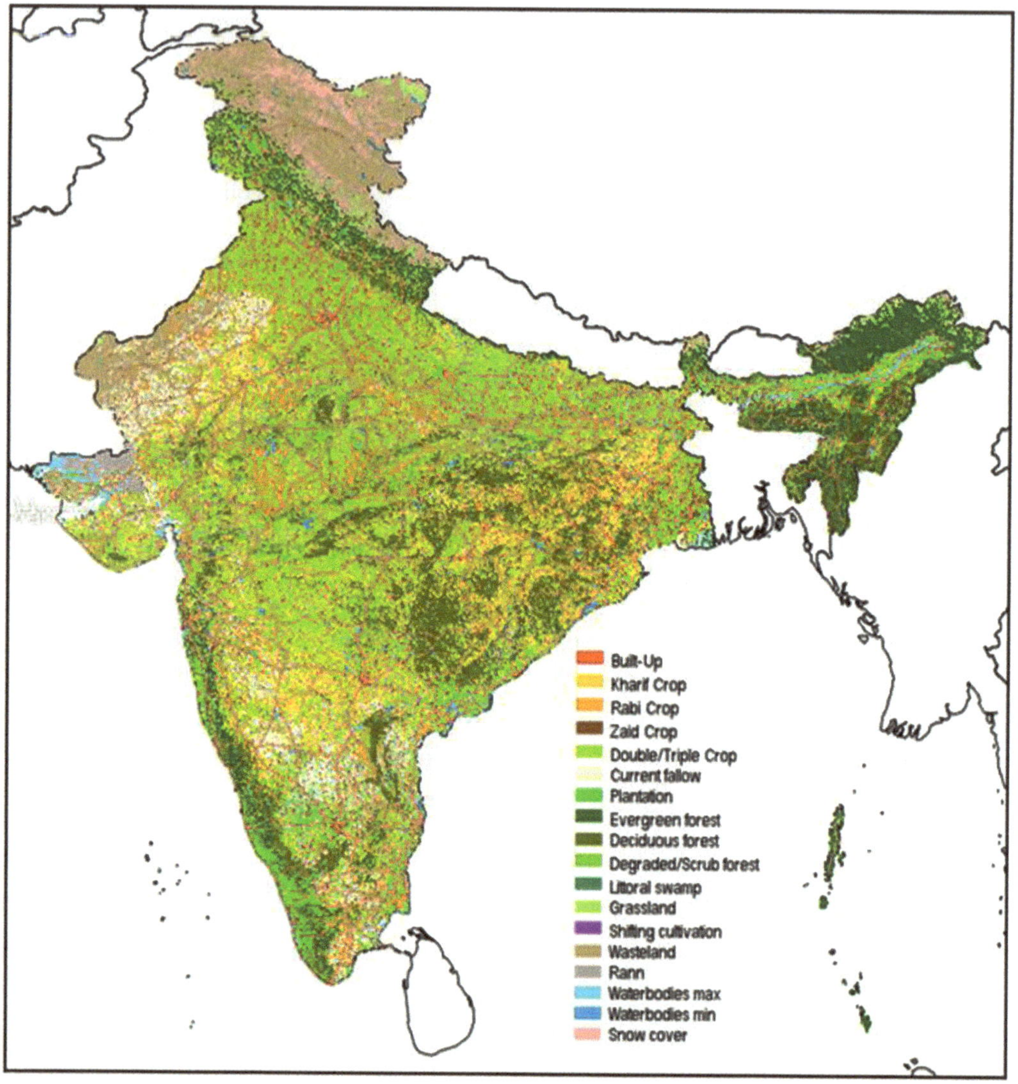

Fig. 14.4 ***Land use land cover (LULC) Map of India.*** Source: NRSC – ISRO, https://www.nrsc.gov.in/LULC_Overview_Img.

Remote sensing data products allow the analysis of indicators such as vegetation health, size, moisture, and density that affect forest fire behavior. To identify fire fuel at the landscape scale, one needs to study land use land cover and various vegetation indices. This information helps for fire risk assessment based on fire fuel input.

Land cover classifications (Fig. 14.4), like the National Remote Sensing Center (NRSC), provides about the information and location of wildfire.

14.7 Burnt area and burn severity mapping (postfire)

Image segmentation and fire identification techniques are commonly used to detect changes and study the burning depth of forest areas [43]. Remote sensing provides best ground to estimate forest fire area using various indices that can be calculated using different spectral bands. These indices are specially designed to discriminate and highlight the forest fire area using spectral band from VIS band (Red), and infrared range (NIR, MIR, and TIR) [44]. Detailed analysis and mapping of burnt area mapping are exclusively based on in situ data (GPS, laser station, etc.) with integration with remotely sensed data. Effects of forest fire through a huge impact on climate change and variability. Understanding fire control at different scale (Fig. 14.5) helps us to know the dominant factor that influences fire.

a. Healthy vegetation versus burned areas: exploiting spectral response curves are shown in Fig. 14.6.

b. Burned area: NBR.

NBR is used to identify burned area with NIR and shortwave infrared spectral band. Demarcation of burnt area and its severity are more prominent using this algorithm

$$NBR = \frac{NIR\text{-}SWIR}{NIR+SWIR}$$

c. Burn severity: differenced normalized burn ratio (dNBR).

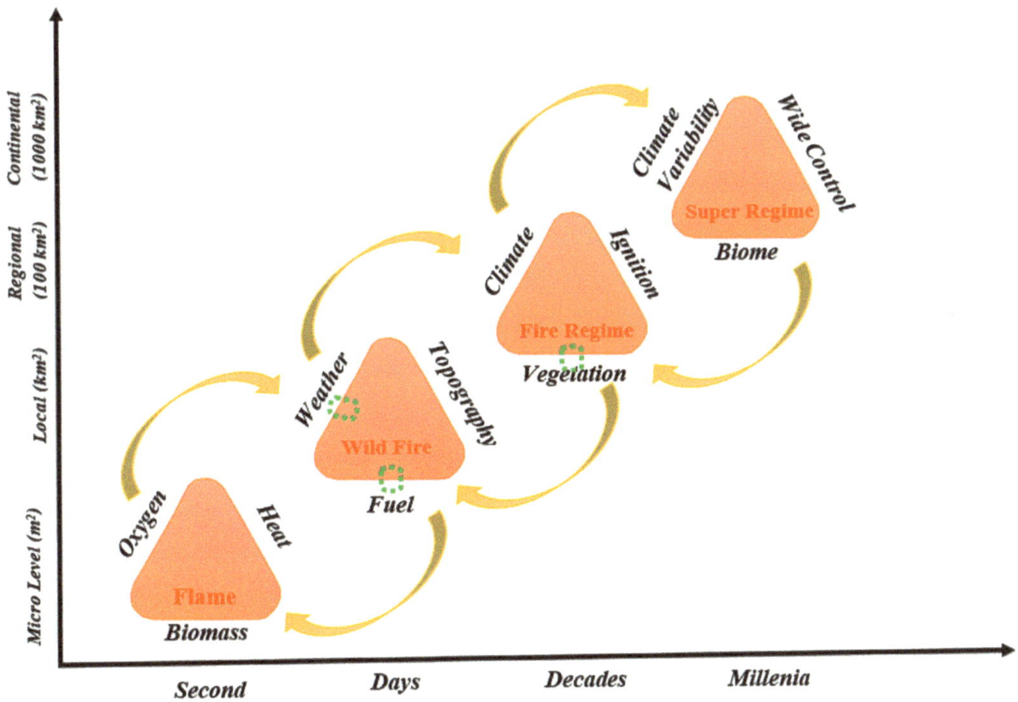

Fig. 14.5 *Controls on fire at different scales.*

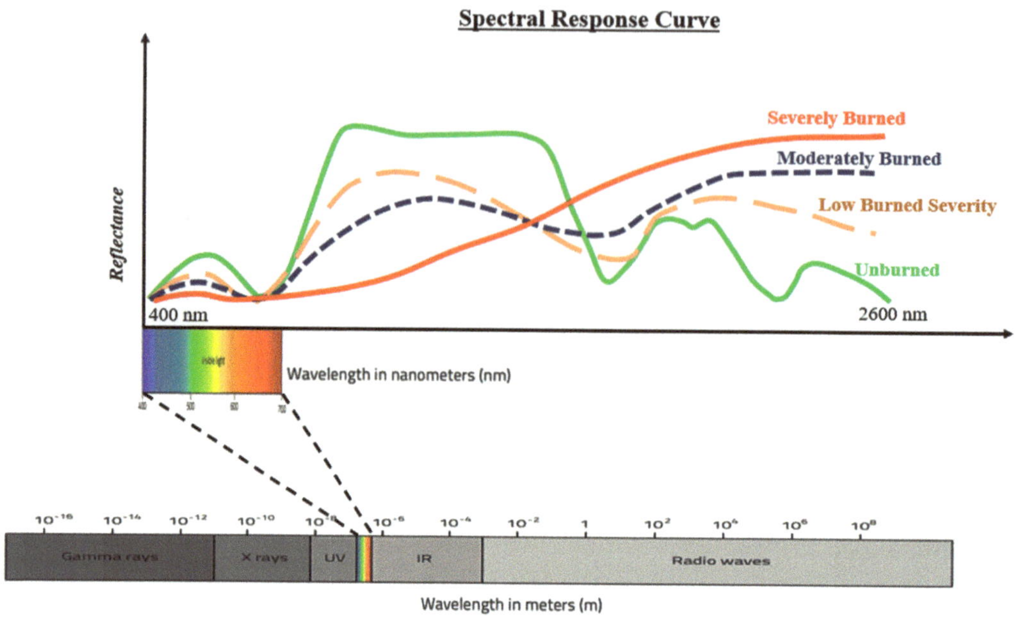

Fig. 14.6 ***Spectral curve of burning severity.***

Difference normalized burn ratio (dNBR) is the modified version of NBR. It is used to estimate the burned area before and after fire event. Determination of severity caused by pre- and post-fire condition can be compared using this index

$$\text{dNBR} = \textit{Prefire}\ \text{NDR} - \textit{Postfire}\ \text{NBR}$$

d. Burn severity: relativized differenced normalized burn ratio (RdNBR)

RdNBR is a variant of dNBR developed by Miller and Thode in 2007. The purpose of this algorithm was to remove the bias that is associated with prefire vegetation. Pixel wise measurement of relative change in vegetation if estimated by using this algorithm.

e. Burn area index (BAI)

BAI uses reflectance unit of red and NIR region of spectrum to locate terrain that are affected by fire. In postfire image charcoal signal is emphasizes using this index which make the pixel look bright. Thus, brighter pixel represents the burned area.

$$\text{BAI} = \frac{1}{(0.1 - \text{Red})^2 + (0.06\text{NIR})}$$

f. Char soil index (CSI)

It is simple ration between NIR and shortwave infrared (SWIR) spectrum. CSI discriminates burned areas well after high burn severity [49].

$$\text{CSI} = \frac{\text{NIR}}{\text{SWIR}}$$

g. Normalized burn ratio thermal (NBRT)

This index is modified version of NBR used thermal band to enhance the burned areas. Before using thermal band, it must be calibrated to brightness temperature. NBRT was initially developed for Landsat data (TM, ETM) but further it is extent to all the multispectral satellite having thermal band.

$$NBR = \frac{NIR\text{-}SWIR \times TIR}{NIR + SWIR \times TIR}$$

h. Midinfrared burned index (MIRBI)

MIRBI is designed to analysis burn severity and vegetation regrowth process. It measures the difference in canopy density of consequence to the forest fire used for evaluation of forest growth recovery. It uses only shortwave infrared channel to estimate the burned area and forest health.

$$MIRBI = 10{*}SWIR - 9.8{*}SWIR + 2$$

14.8 Mapping postfire vegetative growth

Mapping the extent of vegetation growth after a forest fire is key step to estimate the damage and severity caused by the fire that takes several decades to restore (Fig. 14.7). Vegetation regrowth follows a pattern of succession as they vary across the world from different forest type, shrublands, and so on. Monitoring vegetation regrowth and identification as the first dominant succession plant species to change

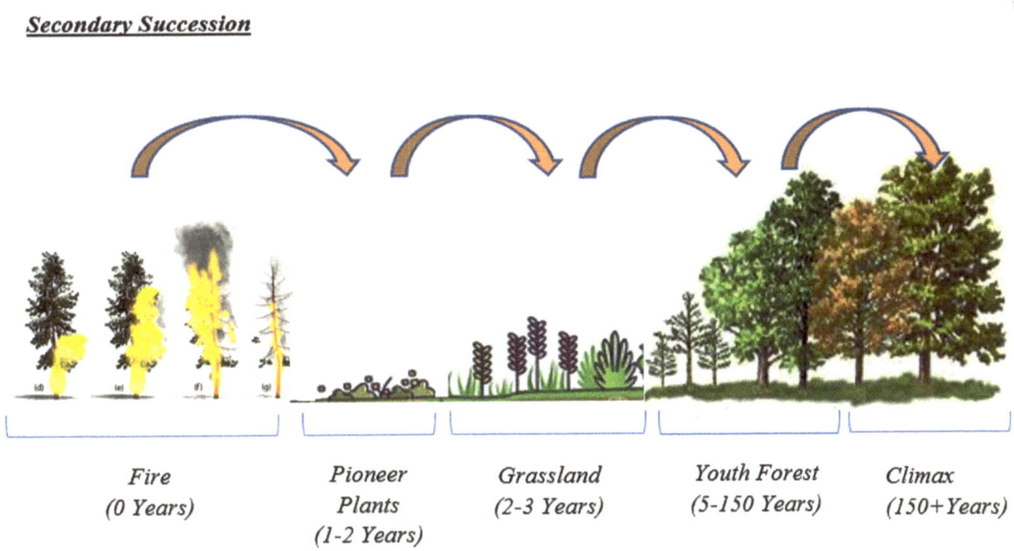

Fig. 14.7 ***Vegetation regrowth after forest fire.***

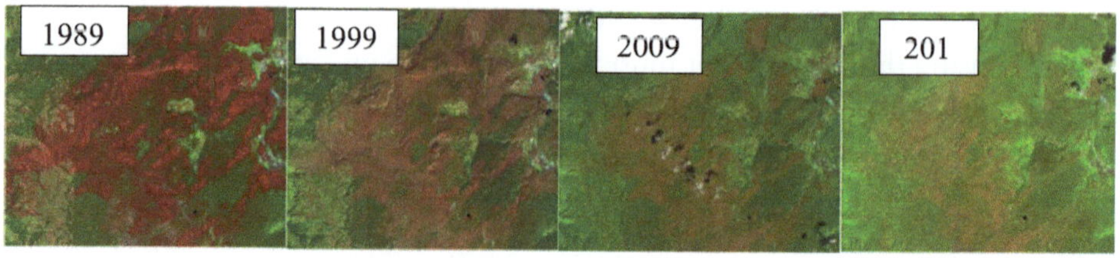

Fig. 14.8 ***Changes in vegetation (1989–2019) at interval of every 10 years.*** Source: Gray, E. T. [[47], November 5], https://svs.gsfc.nasa.gov/13129.

over time. Some of the methods used to map forest regrowth are mentioned below.
a. Vegetation regrowth: NBR.

High NBR results (close to 1) indicate vegetation availability and receptive growth. The NDVI interim network provides solutions for plant restoration after a fire. A time series analysis of vegetation regrowth can observe throughout.
b. Postfire vegetation regrowth: NDVI.

Fire incident at Yellowstone National Park during months of summer season in 1988, natural (lightning) as well as anthropogenic (human activities) source ignited wildfire the extent to a huge stretch. Out of 2,221,800 acres 793,000 acres of the park were burned. Landsat 5 and 8 normalized difference vegetation index (NDVI) is calculated using Landsat TM and OLI/TIRS to estimate the slow regrowth of vegetation over the burned area. The NDVI plot (Fig. 14.8) shows an increase in greenness growth over a period of 30 years, indicating the renewal of vegetation through flames.
c. Postfire vegetation regrowth: land cover change.

Land use land cover classification can help in understanding plant growth over time and provide information about vegetation types that are useful in understanding the next stage.

14.9 Conclusions

This study sheds light on the use of remote sensing to detect and analyze forest fires. Forest fires have become a major problem in recent times. It is difficult to identify a fire in the traditional way, as observers had to assess and interpret the fire. The traditional concept of wildfire has been replaced by the latest technology and advanced satellite data. Satellites like MODIS, VIIRS, AVHRR, Sentinel 3 provide high-quality and reliable fire products. Various EMS like NIR, MIR, VIS, and TIR multispectral satellites are used, which can emphasize high intensity and light intensity.

Improved remote sensing technology has created a broad EMS that helps identify forest fires and monitor their severity. The large area coverage and the frequency of remote sensing data make it possible to identify the large wildfires, making it easier to plan rescue work through easily and fast. Compared to conventional methods, remote detection is relatively inexpensive and efficient.

Remote sensing can store large amounts of data (worldwide) using EMS. Indices such as NBR (normalized burned ratio), BAI, and total NBRT are commonly used to measure the burnt area and its severity. Red, NIR, and SWIR are the EMS that was used to calculate these Indices. Various fire monitoring system such as forest fire finder, fire watch, and fire hawks provides ground dataset that helps in validation that improve the accuracy and precision of remote sensing datasets.

Vegetation-based fire detection and mapping such as NDVI, EVI, and so on are the one of the most reliable methods of detecting forest fires using remote sensing datasets. Pre- and Post-forest fire mapping help us to know the extent and intensity of wildfire. The application of vegetation-based study also helped us to know the forest fire extent by understanding the vegetation regrowth.

Acknowledgments

The authors would like to thank funding support under MoES (MoES/16/18/2017-RDEAS) and ISRO RESPOND (ISRO/RES/3/806/19-20) project.

References

[1] R.S. Dwivedi, An Introduction to Geospatial Technology, Geospatial Technologies for Land Degradation Assessment and Management, Taylor & Francis, 2018, pp. 1–30.

[2] J.F. Nye, Polarization effects in the diffraction of electromagnetic waves: the role of disclinations, Proc. Royal Soc. London. A. Mathematical Physical Sci. 387 (1792) (1983) 105–132.

[3] G.O. Reddy, Satellite remote sensing sensors: principles and applications. In: Geospatial Technologies in Land Resources Mapping, Monitoring and Management, Springer, Cham, 2018, pp. 21–43.

[4] A. Prakash, Thermal remote sensing: concepts, issues and applications, Int. Archives Photogrammetry Remote Sens. 33 (B1; PART 1) (2000) 239–243.

[5] M.F. Gao, Z.H. Qin, S.C. Liu, A study of forest fire detection based on MODIS data, Remote Sens. Land Resources 17 (2) (2005) 60–63.

[6] J.H. Zhang, F.M. Yao, C. Liu, L.M. Yang, V.K. Boken, Detection, emission estimation and risk prediction of forest fires in China using satellite sensors and simulation models in the past three decades—an overview, Int. J. Environ. Res. Public Health 8 (8) (2011) 3156–3178.

[7] S.P. Flasse, P. Ceccato, A contextual algorithm for AVHRR fire detection, International Journal of Remote Sensing 17 (2) (1996) 419–424.

[8] Z. Xie, W. Song, R. Ba, X. Li, L. Xia, A spatiotemporal contextual model for forest fire detection using Himawari-8 satellite data, Remote Sensing 10 (12) (2018) 1992.

[9] L. Giglio, W. Schroeder, C.O. Justice, The collection 6 MODIS active fire detection algorithm and fire products, Remote Sensing Environ. 178 (2016) 31–41.

[10] N. Zhang, L. Sun, Z. Sun, Y. Qu, Detecting low-intensity fires in East Asia using VIIRS data: an improved contextual algorithm, Remote Sensing 13 (21) (2021) 4226.

[11] T.E. Lee, S.D. Miller, F.J. Turk, C. Schueler, R. Julian, S. Deyo, …, S. Wang, The NPOESS VIIRS day/night visible sensor, Bull. Amer. Meteor. 87 (2) (2006) 191–200.

[12] F. Rumiano, E. Wielgus, E. Miguel, S. Chamaillé-Jammes, H. Valls-Fox, D. Cornélis, …, A. Tran, Remote sensing of environmental drivers influencing the movement ecology of sympatric wild and domestic ungulates in semi-arid savannas, a review, Remote Sens. 12 (19) (2020) 3218.

[13] S. Verma, A. Sharma, P.K. Yadava, P. Gupta, J. Singh, S. Payra, Rapid flash flood calamity in Chamoli, Uttarakhand, region during Feb 2021: an analysis based on satellite data, Natural Hazards 112 (2) (2022) 1–15.

[14] T. He, S. Liang, D. Wang, Y. Cao, F. Gao, Y. Yu, M. Feng, Evaluating land surface albedo estimation from Landsat MSS, TM, ETM+, and OLI data based on the unified direct estimation approach, Remote Sensing of Environment 204 (2018) 181–196.
[15] I.A. Csiszar, W. Schroeder, Short-term observations of the temporal development of active fires from consecutive same-day ETM+ and ASTER imagery in the Amazon: implications for active fire product validation, IEEE J. Sel. Top. Appl. Earth Obs. Remote Sens. 1 (4) (2008) 248–253.
[16] S.W. Murphy, C.R. de Souza Filho, R. Wright, G. Sabatino, R.C. Pabon, HOTMAP: Global hot target detection at moderate spatial resolution, Remote Sens. Environ. 177 (2016) 78–88.
[17] W. Schroeder, P. Oliva, L. Giglio, B. Quayle, E. Lorenz, F. Morelli, Active fire detection using Landsat-8/OLI data, Remote Sens. 185 (2016) 210–220.
[18] C. Quintano, A. Fernández-Manso, O. Fernández-Manso, Combination of Landsat and Sentinel-2 MSI data for initial assessing of burn severity, Int. J. Appl. Earth Obs. Geoinf. 64 (2018) 221–225.
[19] E. Roteta, A. Bastarrika, M. Padilla, T. Storm, E. Chuvieco, Development of a Sentinel-2 burned area algorithm: generation of a small fire database for sub-Saharan Africa, Remote Sens. Environ. 222 (2019) 1–17.
[20] M.J. Wooster, W. Xu, T. Nightingale, Sentinel-3 SLSTR active fire detection and FRP product: pre-launch algorithm development and performance evaluation using MODIS and ASTER datasets, Remote Sens. Environ. 120 (2012) 236–254.
[21] C.O. Justice, L. Giglio, S. Korontzi, J. Owens, J.T. Morisette, D. Roy, …, Y. Kaufman, The MODIS fire products, Remote Sens. Environ. 83 (1-2) (2002) 244–262.
[22] L. Giglio, T. Loboda, D.P. Roy, B. Quayle, C.O. Justice, An active-fire based burned area mapping algorithm for the MODIS sensor, Remote Sens. Environ. 113 (2) (2009) 408–420.
[23] W. Schroeder, Visible infrared imaging radiometer suite (VIIRS) 375 m & 750 m active fire detection data sets based on NASA VIIRS land science investigator processing system (SIPS) reprocessed data—Version 1. 2017.
[24] D. Jensen, J.T. Reager, B. Zajic, N. Rousseau, M. Rodell, E. Hinkley, The sensitivity of US wildfire occurrence to pre-season soil moisture conditions across ecosystems, Environ. Res. Lett. 13 (1) (2018) 014021.
[25] C.F. Waigl, A. Prakash, M. Stuefer, D. Verbyla, P. Dennison, Fire detection and temperature retrieval using EO-1 Hyperion data over selected Alaskan boreal forest fires, Int. J. Appl. Earth Obs. Geoinf. 81 (2019) 72–84.
[26] R. Valente de Almeida, P. Vieira, Forest fire finder–DOAS application to long-range forest fire detection, Atmos. Meas. Tech. 10 (6) (2017) 2299–2311.
[27] C. Hartung, R. Han, C. Seielstad, S. Holbrook, FireWxNet: a multi-tiered portable wireless system for monitoring weather conditions in wildland fire environments, Proceedings of the 4th International Conference on Mobile SYSTEMS, applications and Services, 2006, pp. 28–41.
[28] L. Yu, N. Wang, X. Meng, Real-time forest fire detection with wireless sensor networks, Proceedings. 2005 International Conference on Wireless Communications, Networking and Mobile Computing, 2005, 2, IEEE, 2005, pp. 1214–1217.
[29] U. Dampage, L. Bandaranayake, R. Wanasinghe, K. Kottahachchi, B. Jayasanka, Forest fire detection system using wireless sensor networks and machine learning, Scientific Reports 12 (1) (2022) 1–11.
[30] Garg, A. Forest fire using optimized solar powered wireless sensor networks. J. Homepage*:* www.ijrpr.com. ISSN, *2582*, 7421.
[31] R. Vinodhini, C. Gomathy, Fuzzy based unequal clustering and context-aware routing based on glow-worm swarm optimization in wireless sensor networks: forest fire detection, Wireless Personal Commun. 118 (4) (2021) 3501–3522.
[32] Y.E. Aslan, I. Korpeoglu, Ö. Ulusoy, A framework for use of wireless sensor networks in forest fire detection and monitoring, Computers, Environ. Urban Syst. 36 (6) (2012) 614–625.
[33] M.A. Matin, M.M. Islam, Overview of wireless sensor network, Wireless Sensor Networks-Technol. Protocols 1 (3) (2012) 1–22.
[34] J. Lloret, M. Garcia, D. Bri, S. Sendra, A wireless sensor network deployment for rural and forest fire detection and verification, Sens 9 (11) (2009) 8722–8747.
[35] J. Zhang, W. Li, N. Han, J. Kan, Forest fire detection system based on a ZigBee wireless sensor network, Front. Forestry China 3 (3) (2008) 369–374.

[36] U. Ganesh, M. Anand, S. Arun, M. Dinesh, P. Gunaseelan, R. Karthik, Forest fire detection using optimized solar-powered Zigbee wireless sensor networks, Int. J. Sci. Eng. Res. 4 (2013) 586–596.
[37] A. Luckman, J. Baker, T.M. Kuplich, C.D.C.F. Yanasse, A.C. Frery, A study of the relationship between radar backscatter and regenerating tropical forest biomass for spaceborne SAR instruments, Remote Sens. Environ. 60 (1) (1997) 1–13.
[38] E.S. Kasischke, L.L. Bourgeau-Chavez, N.H. French, Observations of variations in ERS-1 SAR image intensity associated with forest fires in Alaska, IEEE Trans. Geosci. Remote Sens. 32 (1) (1994) 206–210.
[39] D. Ehrlich, E.F. Lambin, J.P. Malingreau, Biomass burning and broad-scale land-cover changes in Western Africa, Remote Sens. Environ. 61 (2) (1997) 201–209.
[40] I. Fernández, A. Cabaneiro, T. Carballas, Organic matter changes immediately after a wildfire in an Atlantic forest soil and comparison with laboratory soil heating, Soil Biol. Biochem. 29 (1) (1997) 1–11.
[41] L.T. Steyaert, F.G. Hall, T.R. Loveland, Land cover mapping, fire regeneration, and scaling studies in the Canadian boreal forest with 1 km AVHRR and Landsat TM data, J. Geophys. Res. Atmos. 102 (D24) (1997) 29581–29598.
[42] D.F. Merrill, M.E. Alexander, Glossary of Forest Fire Management Terms, Canadian Committee on Forest Fire Management, National Research Council of Canada, Ottawa, 1987.
[43] B., Kurnaz, C., Bayik, S. Abdikan, Forest Fire Area Detection by Using Landsat-8 and Sentinel-2 Satellite Images: A Case Study in Mugla, Turkey, 11 May 2020, PREPRINT (Version 1) available at Research Square [https://doi.org/10.21203/rs.3.rs-26787/v1].
[44] E. Chuvieco, M.P. Martin, A. Palacios, Assessment of different spectral indices in the red-near-infrared spectral domain for burned land discrimination, Int. J. Remote Sens. 23 (23) (2002) 5103–5110.
[45] G.C. Davenport, Remote sensing applications in forensic investigations, Historical Archaeol. 35 (1) (2001) 87–100.
[46] S.S. Kumar, D.P. Roy, Global operational land imager Landsat-8 reflectance-based active fire detection algorithm, Int. J. Digit. Earth 11 (2) (2018) 154–178.
[47] E.T. Gray, Yellowstone Fire: Then and Now, NASA Visualization Explorer, 2019. https://svs.gsfc.nasa.gov/13129.
[48] J.W. Rouse, R.H. Haas, J.A. Schell, D.W. Deering, Monitoring vegetation systems in the Great Plains with ERTS, NASA Special Publication 351 (1974) (1974) 309.
[49] A. Huete, C. Justice, H. Liu, Development of vegetation and soil indices for MODIS-EOS, Remote Sens. Environ. 49 (3) (1994) 224–234.
[50] Z. Jiang, A.R. Huete, J. Li, J. Qi, Interpretation of the modified soil-adjusted vegetation index isolines in red-NIR reflectance space, J. Appl. Remote Sens. 1 (1) (2007) 013503.
[51] C. F. Jordan, Derivation of leaf-area index from quality of light on the forest floor, Ecol 50 (4) (1969) 663–666.
[52] D.W. Deering, Measuring "forage production" of grazing units from Landsat MSS data, Proceedings of the Tenth International Symposium of Remote Sensing of the Environment, 1975, pp. 1169–1198.
[53] S. Chandra, Application of Remote Sensing and GIS Technology in Forest Fire Risk Modeling and Management of Forest Fires: A Case Study in the Garhwal Himalayan Region, Geo-Information for Disaster Management, Springer, 2005, pp. 1239–1254.
[54] M.D. Flannigan, J.B. Harrington, A study of the relation of meteorological variables to monthly provincial area burned by wildfire in Canada (1953–80), J Appl Meteorol Climatol 27 (4) (1988) 441–452.
[55] R. Damoah, N. Spichtinger, C. Forster, P. James, I. Mattis, U. Wandinger, S. Beirle, T. Wagner, A. Stohl, Around the world in 17 days-hemispheric-scale transport of forest fire smoke from Russia in May 2003, Atmos. Chem. Phys. 4 (5) (2004) 1311–1321.
[56] J. Ming, C. Xiao, H. Cachier, D. Qin, X. Qin, Z. Li, J. Pu, Black Carbon (BC) in the snow of glaciers in west China and its potential effects on albedos, Atmos. Res. 92 (1) (2009) 114–123.
[57] T.J. Yasunari, P. Bonasoni, P. Laj, K. Fujita, E. Vuillermoz, A. Marinoni, P. Cristofanelli, R. Duchi, G. Tartari, K.M. Lau, Estimated impact of black carbon deposition during pre-monsoon season from Nepal Climate Observatory-Pyramid data and snow albedo changes over Himalayan glaciers, Atmos. Chem. Phys. 10 (14) (2010) 6603–6615.

[58] S. Kaspari, S.McKenzie Skiles, I. Delaney, D. Dixon, T.H. Painter, Accelerated glacier melt on Snow Dome, Mount Olympus, Washington, USA, due to deposition of black carbon and mineral dust from wildfire, J Geophys Res 120 (7) (2015) 2793–2807.
[59] K. Bali, A.K. Mishra, S. Singh, Impact of anomalous forest fire on aerosol radiative forcing and snow cover over Himalayan region, Atmos. Environ. 150 (2017) 264–275.
[60] M.D. Flannigan, B.D. Amiro, K.A. Logan, B.J. Stocks, B.M. Wotton, Forest fires and climate change in the 21st century, Mitigation and Adaptation Strategies for, Global Change 11 (4) (2006) 847–859.
[61] A. Bytnerowicz, M.J. Arbaugh, C. Andersen, A.R. Riebau, Integrating research on wildland fires and air quality: needs and recommendations, Developments in Environmental Science 8 (2008) 585–602.
[62] J. Fleming, R.G. Robertson, Fire Management Tech Tips: The Osborne Fire Finder, Technical Report 0351 1311-SDTDC, USDA Forest Service. (2003).
[63] E. Den Breejen, M. Breuers, F. Cremer, R. Kemp, M. Roos, K. Schutte, J.S. De Vries, Autonomous forest fire detection, Proceedings of the 3rd International Conference on Forest Fire Research, 2, ADAI-Associacao para o Desenvolvimento da Aerodinamica, Coimbra, Portugal, 1998, pp. 2003–2012.
[64] S. Wang, H. Zhang, Research Progress of Fire Detection Based on Remote Sensing, International Conference on Computer Application and System Modeling, atlantis-press, 2012, pp. 762–766.
[65] S. Casadio, O. Arino, Monitoring the South Atlantic Anomaly using ATSR instrument series, Adv. Space Res. 48 (6) (2011) 1056–1066.
[66] L. Giglio, T. Loboda, D.P. Roy, B. Quayle, C.O. Justice, An active-fire based burned area mapping algorithm for the MODIS sensor, Remote Sens. Environ. 113 (2) (2009) 408–420.

CHAPTER 15

Study of the atmospheric and ionospheric phenomenon using GPS-based remote sensing technique

Sanjay Kumar, S.S. Rao, Mukulika Mondal, A.K. Singh
Atmospheric Research Laboratory, Department of Physics, Institute of Science, Banaras Hindu University, Varanasi, Uttar Pradesh, India

15.1 Introduction to global positioning system

The navigation has its long chronicle from the ancient time but in 1978, the launching of the first global positioning system (GPS) Satellite by the US Department of Defence gives a new imputes to the society. The GPS is a space-based Global Navigation Satellite System (GNSS) that is applied to ascertain ground position and velocity (Location, speed, and direction) information in all weather and any place on or near the earth. The GPS is comprised of 32 satellites at 20,200 km altitude above the earth. The satellites are monitored in six orbits so that at any time at least four satellites will be in look to users anywhere in the world. GNSS is an enterprise phrase adopted to name all of the world's satellite constellations, operating or proposed. Currently, the four GNSS GPS systems are operational to allow position, velocity and precise time (PVT) solutions globally: (I). Navigation Satellite Timing and Ranging (NAVSTAR)—United States system, (II). Global Navigation Satellite System (GLONASS)—the Russian version and (III). GALILEO—from the European Space Agency and IV. BeiDou from China. Apart from these four mainstream GNSS, the QZSS of Japan and NavIC of India regional systems are also operational.

The GPS can be classified into three sections: (1) space segment, (2) control Segment, and (3) user segment. The space segment contains the list of 32 satellites arrayed in six orbital planes placed at 60° offsets in longitude so there are four satellites in each orbital plane and each of which is comparatively inclined at 55° for the earth equatorial plane at an altitude of about 20,200 km in space. These satellites are Sun-synchronous and each satellite takes 11 hours 58 minutes to orbit the earth, hence on each successive day, the same satellite re-enters 4 minutes early. Each satellite weighs about 908 kg and is about 5.2 m across with the solar panels lengthened. The power transmitter is only 50 watts. The GPS signal includes a "pseudo-random Noise code (PRN) code," ephermeris, and almannc data. Each GPS Satellite is designated by a specific number which named by their PRN code. The PRN number is ranging from 1 to 32. Every satellite generates a navigation message based upon data periodically uploaded from the Control Segment

Atmospheric Remote Sensing
DOI: https://doi.org/10.1016/B978-0-323-99262-6.00019-5

and appends the message to a 1.023 MHz PRN Coarse/Acquisition code sequence. The satellite embeds the code sequence onto a 1575.42 MHz L-band carrier to create a spread spectrum ranging signal and sends it to the users. This broadcast is referred as the Standard Positioning Service (SPS) ranging signal. The GPS satellite also transmits a second-ranging signal at 1227.6 MHz known as L2, which carries precise positioning service user two-frequency corrections. The broadcasted navigation signal contains the relevant data like position of the satellite, health and working status, and current date/time as well as other important data.

The control segment is a combination of ground stations that control and command the GPS satellites. The GPS is managed by the United States Air Force. The Master Control Station; MSC, (on Schriever Air Force Base, Colorado Springs) one Backup Mater Control Station; BMCS (on Gaithersburg, Maryland) and six additional monitoring stations operated by the US Air force (on Hawaii, Ascension Islands, Diego Garcia, Kwajalein, Schriever and Cape Canaveral Florida). Six more monitor stations from National Geospatial-Intelligence Agency were added in 2005. The MCS is responsible for several aspects of constellation command and control which includes monitoring of status of routine satellite bus and payload, satellite maintenance and anomaly resolution, management of SPS performance upload of navigation data operations as needed to maintain performance by accuracy performance standards and prompt detection and response to service failures. The job of monitoring stations is to provide satellite-ranging data in real-time to the MCS and keep near-continuous monitoring. Each monitor stations perceive the pseudo-ranges to all satellites and after smoothing of the data transmitted to MCS. The Mater control station measures the satellite's ephemeris and clock offsets and transferred it to one of the three ground control stations. Communication between the control segment and the GPS satellite is done via an S band uplink from one of three ground stations antennas. The four GPS ground antenna stations are Kwajalein Atoll, Ascension Island, Diego Garcia, and Cape Canaveral. The user needs a GPS receiver and antenna to receive the broadcasted signal from the satellites concurrently with precise measurements of the signal transmission delays to resolve the user position and velocity, as well as synchronize to the GPS time. To measure the position and time coordinates, at least four satellites should be in view angle with reference to receiver which enables GPS for the navigation, precise positioning, and time and frequency dissemination.

15.1.1 Error sources in GPS signals

The GPS positioning accuracy is deteriorated by many errors tainted with GPS signal. The root of these errors can be listed in three categories; the satellite-related errors (satellite clock error, satellite orbit error, and satellite interfrequency bias), the atmospheric-related errors (ionospheric delay error and tropospheric delay error), and receiver-related errors (receiver clock error, receiver interfrequency bias, multipath and receiver noise). The errors in GPS signal in detail are discussed elsewhere [1,2].

15.1.2 Total electron content and water vapor measurements using GPS

Since the ionosphere is a dispersive medium for GPS signal, in order to quantify the propagation effects on radio waves traveling through the ionosphere, the refractive index of the medium must be specified. As the GPS signals travel through the ionosphere carrier experience a phase advanced and code experience a group delay and therefore two types of refractive index are induced, that is, phase refractive index and group refractive index. Both delays are proportional to the total electron content (TEC) of the ionosphere and can be used to measure TEC. Carrier phase observation is the measured phase difference between the received signal and the receiver-generated signal. It can be expressed in terms of cycles or in terms of units of range. The conversion of phase cycles to pseudo-range makes it compatible to code pseudo-range. A GPS receiver measures P code pseudo-range and carrier phase using signal transmitted by the satellite at the two frequencies L1 and L2. TEC of the ionosphere can be measured either by using pseudo-range measurement or by using carrier phase measurement [3].

The change in refractive index for GPS signal due to the presence of atmospheric components which includes dry ($O_{2,}$ N_2) and wet components (water vapor) causes change in phase and group velocity which further change total phase and path length of GPS signal. This information can be used to estimate water vapor and need first to estimate zenith tropospheric delay (ZTD) using GPS observation data. The ZTD is the sum of zenith hydrostatic delay (ZHD) due to O_2, N_2, and zenith wet delay (ZWD) due to water vapor. ZHD is computed by using meteorological data observed at a level of GPS antenna or by using the model. Subtracting ZHD from ZTD one can get ZWD which is directly related to perceptible water vapor (PWV). The steps involved to derive water vapor from GPS data are discussed elsewhere [2].

15.2 Study of ionospheric phenomena using GPS

15.2.1 Response to space weather

Space weather refers to the conditions on the Sun and in the solar wind, magnetosphere, and ionosphere that affect the performance and the reliability of the technological systems on earth and in space. Space weather is the effect experienced by the Earth's ionosphere by the following three phenomena:

1. Coronal mass ejection (CME).
2. Solar flares (SFs).

A disruptive change in the Earth's magnetic field compared to the quiet day trend is identified as the geomagnetic storm. The disturbance in the interplanetary medium may be due to either a solar coronal mass ejection (CME) or a high-speed stream (corotating interaction region, CIR) of the solar wind originating from the Sun's surface. Geomagnetic storms are started when a solar wind shock wave interacts with the Earth's magnetosphere. The increased solar wind pressure compresses the magnetosphere and results in the transfer of large energy into the magnetosphere. Interaction with either the solar wind shock wave or the cloud of magnetic field enhances electric fields inside the magnetosphere, resulting in increased movement of plasma through it, and consequently, the electric current both in the magnetosphere and ionosphere is enhanced. The disturbance storm time index (Dst) or SYM-H index is adopted to describe geomagnetic storms.

A geomagnetic storm is divided into three main phases: (1) an initial phase (2) the main phase, and (3) a recovery phase. The onset of a magnetic storm is often characterized by a global sudden increase in H, which is referred to as the storm's sudden commencement (SSC). After that, the H component remains higher at its average level by 20 to 50 nT for a few hours which is referred as the initial phase. Following the SSC, a large global decrease in H begins. When Dst or SYM-H index decreases, to less than −50 nT, indicates the development of the main phase of the storm. The magnitude of the decrease in H represents the severity of the disturbance. The selection of -50 nT to define a storm is somewhat arbitrary. The minimum value of the Dst/ SYM-H index can go down up to -600 nT during geomagnetic storms. The main phase of the geomagnetic storm is sustained for 2–8 hours. The period when Dst/ SYM-H index changes back from its highest negative value to its quiet time value (or zero level) is known as the recovery phase of the storm. The period of the recovery phase may be as short as 8 hours or as long as 7 days. In terms of changes in the Dst index, the geomagnetic storms are classified as moderate (-50 nT>minimum of Dst> -100 nT), intense (-100 nT> minimumDst> -250 nT) or super-storm (minimum of Dst< -250 nT). The occurrence probability of geomagnetic storms depends upon the phase of the solar cycle. CME-driven storms are more common to occur during the solar maxima, the CIR-driven storms are happened during the solar minima.

The interaction between the solar wind and the magnetosphere under southward interplanetary magnetic field (IMF) conditions during major geomagnetic storms causes a change in the region-1 current. This leads to a sudden increase in the dawn-to-dusk polar cap potential. The region 2 currents are affiliated with the inner magnetospheric electric field and are directed from dusk to dawn. The region-2 current is not able to shield the mid and equatorial latitudes from high latitude electric fields for a longer duration, the storm time electric field instantaneously enters from high latitude to the equatorial ionosphere [4]. The storm time electric fields are important because they drastically modify the plasma dynamics and density distribution on varied time scales, from a few to several hours [5]. The storm time electric fields are known as the prompt penetration (PP) electric fields and the disturbance dynamo (DD) electric fields. The PP fields emerge due to the undershielding (UC)/overshielding conditions (OC), the divergence of the asymmetric ring currents, and enhanced polar cap potential drop [6]. The UC happens during the southerly phase of the interplanetary magnetic field (IMF Bz) and OC occurs during the northward IMF Bz. The undershielding/overshielding fields are eastward/westward during the day and westward/eastward during the night and have typical rise and decay times shorter than about 15 minutes and lifetimes of about an hour [7]. The DD fields are observed due to the modification of neutral wind circulation caused by the enhanced energy deposition at high latitudes and have time scales, from a few to tens of hours [8]. Here the mechanical forcing implies the upwelling of the ionospheric plasma caused by the equator-ward flow of meridional winds that result from the storm time heating at high latitudes due to the large energy deposition [9,10]. The upwelled plasma stays at higher altitudes where recombination rates are lower, therefore the plasma survives longer, and increased plasma density or a positive ionospheric storm results. During the day, the electrodynamical forcing resulting in a positive ionospheric storm arises mainly from the PP of the eastward electric fields that are generated due to the undershielding condition that follows the southward turning of the IMF Bz [11].

The PP field and the ambient zonal field are eastward during the daytime at the equator. Therefore increase the ExB vertical drift that may drastically modify the plasma distribution. As a result, poleward expansion of the EIA takes place [12]. The DD electric fields, on the other hand, are known to be the main responsible factor for the negative ionospheric storms that result in the reduction of the daytime TEC at low latitudes [13] and suppression of EIA in terms of its latitudinal extent [12]. The negative ionospheric storms are caused mainly by the DD electric fields that are generated due to changes in the global wind circulation stemming from the enhanced Joule heating at the high latitudes during the geomagnetic storms [14]. The direction of the DD fields is opposite to the ambient zonal fields, both by day and night, at the equator [15]. Their effect in low latitudes during the daytime could be inferred from the weakened or suppressed equatorial electrojet [12].

For a representative case we have shown response of TEC to the space weather event of September 6–9, 2017 in Fig. 15.1. On September 7, 2017, a significant increase in TEC can be seen by pink curve that is due to the combined effect of both solar flares and storm sudden commencement as well as due to pressure pulses of geomagnetic storm event of September 7, 2017. It has been verified from the Dst index and IMF Bz variation on September 8, 2017, that the severe geomagnetic storm happened on 8 September. The increase in TEC during 00:00–03:00 on September 8, 2017, is attributed to PP electric field. Also increased level of TEC on 8 September (see, the green curve in Fig. 15.1) after 11:00 UT is attributed to PP field because of the substorm condition. The increase in TEC on 9 September is attributed to the solar flares as well as DD electric filed.

The results from the various ionospheric studies during different geomagnetic storms could be summarized as follows:

The geomagnetic storm time TEC enhancement, as well as decrements, depends upon the magnitude and polarity of geomagnetic parameters (Dst/SYM-H index and IMF Bz) and local time at given locations. The increment/reduction in TEC during a geomagnetic storm normally depends upon PP electric field or DD electric field.

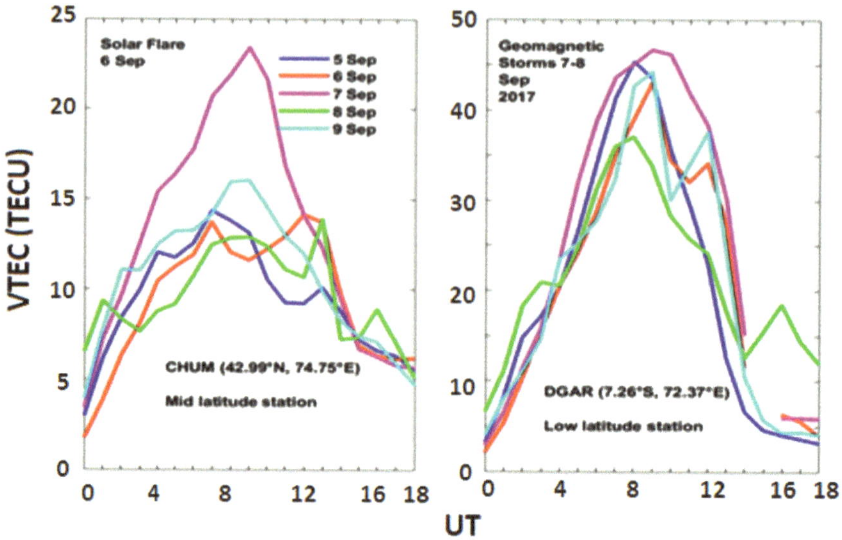

Fig. 15.1 ***Response of ionospheric TEC to the space weather (solar flare and geomagnetic storm) events of September 2017.***

The storm time TEC increases are of the amplitude 10 TECU-35 TECU. In general, the increase or decrease in TEC during and after the geomagnetic storms is associated with electric fields, neutral wind, and local electrodynamics. The ionospheric ionization distribution experiences sea changes during geomagnetic storms. Such large, sudden changes in ionospheric density are detrimental to ground and space positioning and navigation by GPS satellites. In disturbed ionospheric conditions, signal depravities may be so grave as to begin system breakdowns and may appear in huge economic loss.

15.2.2 Solar flares (SFs)

The magnetic field near the sunspots frequently twist, cross and reconnect and proffer a solar flare. The location of the solar active region affects the TEC response during solar flare events. This effect is known as the Central Meridian Distance effect. Thus the potential geo-effectiveness of any solar flare is depended upon where from it emerges. The solar energetic particles originating from the western solar hemisphere obey a Parker spiral trajectory and hence more likely to affect the Earth. However, their nominal trajectories can be deflected by the interactions with the coronal holes emitting higher velocity solar wind [16]. It is known that the soft X-ray does ionization in the E region and EUV does ionization above ~150 km [17]. The location of a solar flare plays an important issue as for as the variation of electron density distribution is concerned. The time delay in ionospheric response during peak and decay of solar flare is considered to be affiliated with solar zenith angle, ionospheric recombination process, or thermospheric-ionospheric coupling process [18,19]. It is still to be established, whether ionospheric response time to solar flare has any connection to the flare activity or is it due to the ionosphere itself?

There are two types of the time lag between solar flare and ionospheric response which have been discussed in the past studies: (1) time lag during peak time of solar flare, (2) time lag during the recovery phase of a solar flare. The delayed ionospheric response to solar flare is owing to the time taken in the charge exchange reaction of O_2^+, O^+, as well as the delay response of neutral density [20]. Also, it has been suggested that the energetic particles released by solar flares at the solar center reach the Earth at a speed slower than the speed of photons [21]. On a geomagnetically disturbed day, solar flares can happen prior and can encompass the main phase and recovery phases of storm. In such cases, the complementation between solar flares and geomagnetic storms produces further complication in understanding of responses of thermosphere–ionosphere system to such events [22]. The complexity is mainly related to filter out the solar flare and storm-induced ionospheric impacts separately. The variation of TEC on September 6, 2017 is shown by red curve in above Fig. 15.1. The blue curve shows the variation of TEC on geomagnetic quiet day when no solar flare or CME encountered to the earth's ionosphere. On 6 September two solar flares namely X2.2 and X9.3 were occurred at ~0857 UT and 1153 UT, respectively. It can be easily seen from red curve that the level of TEC variation is higher than the normal day trend during and after the solar flare occurrence. Also, there was no geomagnetic storm condition was exist on 6 September and thus the observed higher level of TEC was solely due to the solar flare. It is also noted from the figure that the increase in TEC is not an immediate effect but it takes a few minutes to respond. During the decay of solar flare also, the higher level of TEC variation compared to normal day trend sustained for a few hours after the decay of solar flare. Thus, TEC response has time lag to respond solar flare during peak as well as decay of solar flare. The outcome of various studies suggests the following conclusions.

The current view on ionospheric response to solar flare is in three ways. First, it is on the enhanced ionization of the EUV and X-ray flux. Second, it is from the dynamics raised by the first effect, including the possible electrodynamics. The electric field is suggested to be decreased in sunlit hours in the course of a great flare. The third is on the thermosphere as for the thermospheric effect, the solar flare results in great neutral density and temperature changes. The solar zenith angle has a significant controlling effect on the thermospheric response to a solar flare

On a normal quiet day: on a normal day, the daytime ambient electric field at the equator is eastward. This field, in conjunction with the earth's magnetic field which is horizontal at the equator, gives rise to the vertical $\mathbf{E} \times \mathbf{B}$ drift of the ionospheric plasma. In the nighttime, the ambient electric field at the equator is westward. Thus, if the time of occurrence of solar flare falls within sunlit hours then the electron density of the ionosphere is enhanced. On the contrary, if the time of occurrence of solar flare falls within night hours then the electron density of the ionosphere is decreased.

The time lag in the ionospheric response to the X-ray and EUV irradiance during the peak of solar flares depends on the solar zenith angle. The soft X-ray regulates the ionization in the E region, whereas EUV governs the ionization above ~150 km. Therefore, the location of a solar flare alters the thermosphere-ionosphere responses in the TEC. The solar flares' effect on the ionosphere is found to be higher for disk center flare than limb flare. The optical density of the soft X-rays and EUV in the solar atmosphere is the main factor for this variance. Coronal emissions, which govern the soft X-ray spectrum, are usually optically thin in the solar atmosphere.

15.2.3 Response to solar eclipse

A solar eclipse appears when the Moon passes between the Sun and the Earth, and the Moon fully or partially covers the Sun. The solar radiation reduces during a solar eclipse which decreases the ionization. So, the loss rate of plasma on an eclipse day dominates over the production rate. Ionospheric disruption during the solar eclipse can manifest errors in GPS positioning for single-frequency users [23]. The asymmetrical horizontal gradients created by the eclipse convolute phase-ambiguity resolution in precise GPS measurements. GPS allows studying temporal and spatial changes in the composition of the ionosphere during the eclipse. It has been observed that the TEC values decreased during two eclipse events of October 24, 1995 and March 9, 1997 at different latitudes in the Asian region [24]. The values of TEC depression were found to be about 7 TECU and a delay of 0–30 min has been observed for the phase of totality. It is revealed from studies that the TEC decreases after the start of the partial eclipse. TEC reaches a minimum value after ~ 30 minutes to 2 hours from the time of maximum obscuration. This delay is due to the combined effect of the photochemical processes and plasma dynamics in the EIA. The studies reveal the reduction of eclipse day TEC over a large area. Yet, some studies explicate latitudinal and longitudinal effects [25]. It has been reported that the depression of TEC is of an order of a few TECU [26,27], whereas it has been reported to be of the order of about 0.2–0.3 TECU over middle latitudes [28]. The depression in TEC has been reported as more than 40–50% relative to quiet conditions in the region of the equatorial anomaly [29]. In the polar ionosphere, TEC depression reached 1.5–1.7 TECU (about 20% relative to quiet conditions) during the total solar eclipse of November 23, 2003 [30]. The above-cited works imply the fact that the response of the ionosphere to the solar eclipse depends on different factors such as geophysical condition, longitude, latitude, and local time. During a solar eclipse, the obscuration of the Moon reduces the ionizing radiation from the Sun, producing variations in electron concentration and temperature, and neutral compositions and temperature [31].

Ionospheric response to the annular solar eclipse of June 21, 2020 is presented in Fig. 15.2 using TEC obtained from two stations namely tcms and hkws located in the Chinese–Taiwanese region. The percentage of obstruction was 93% at cksv and 85%

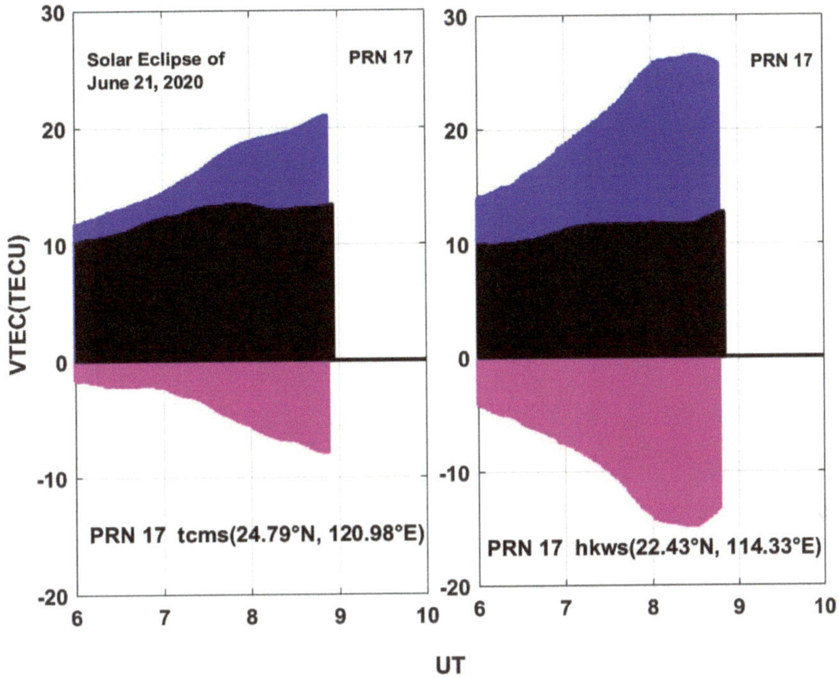

Fig. 15.2 ***The variation of TEC on pre-eclipse day is shown by blue surface plot, on eclipse day of 21 June 2020 is shown by black surface plot and difference of TEC of eclipse day to pre-eclipse day is shown by pink surface plot.*** The path of eclipse could be found at link: https://eclipse.gsfc.nasa.gov/.

at hkws. We have shown TEC variation from satellites designated by Pseudo-Random Noise code (PRN)17. It can be seen from Fig. 15.2 that the TEC was decreased during the solar eclipse by an order of ~ 10±2 TECU. This can be easily verified by pink surface plot shown in Fig. 15.2. It is also noted that the decrease in TEC was observed 15–30 minutes late after the phase of maximum eclipse.

The gravity waves signature in eclipse day TEC is also reported [32]. It is proposed that the sources of gravity waves exist in the thermosphere where molecular oxygen heating starts, the stratosphere where ozone layer heating takes place [33], and the troposphere where water vapor IR absorption and ground cooling happen [34]. The Fourier analysis of the difference in TEC (DTEC) of selected PRN (02 and 06) on eclipse day is carried out at two equatorial Indian stations Hyderabad and Bangalore which reflects the presence of sparse amplitude peaks of 20–50- and 50–90 minutes periods (Fig. 15.3A and B). Thus, observations of TEC response during solar eclipse conclude following results; TEC values are significantly decreased during the solar eclipse event. TEC levels are found to be reduced during the solar eclipse with a 15–30 minutes delay. The presence of gravity waves is noticed on the eclipse day TEC having a periodicity of 20–50 and 50–90 minutes. It is reported that the gravity wave signatures

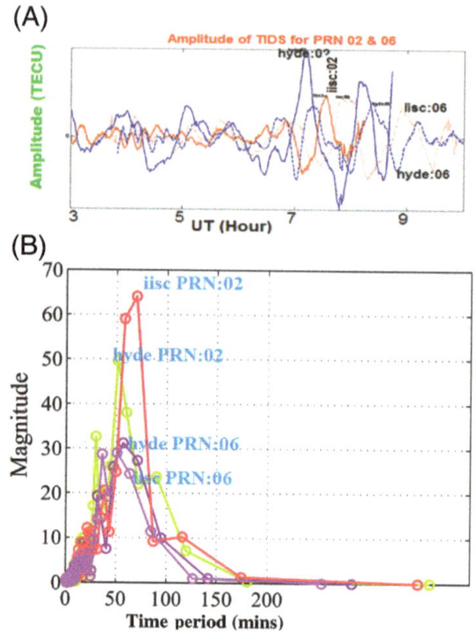

Fig. 15.3 (A) The amplitude of DTEC data obtained by taking the difference of VTEC of given PRN and its best-fit polynomial value of a PRN at Hyderabad (hyde) and Bangalore (iisc) station on solar eclipse day of 21 June 2020. (B) Plot of FFT analysis of DTEC of PRN 02 & 06 on solar eclipse day June 21, 2020 at Hyderabad and Bangalore.

in eclipse day TEC are associated with the eclipse-induced cooling of the stratosphere/mesosphere.

15.2.4 Response to earthquakes

Earthquake is one of the most frightening and destructive phenomena of nature and has terrible after effects. In the course of history, earthquakes have killed many people and have destroyed many communities. During last 100 years, there has been an average about 18 large earthquake(magnitude 7.0–7.9) in each year (http://earthquake.usgs.gov). The objective of prediction of earthquake is important giving warning to possibly harmful areas early enough enabling people to minimize loss of life and their property. Many papers have mentioned electric field disturbances before earthquakes [35–37]. The problem of anomalous electric field penetration from the near-ground surface into the ionosphere is not a generally accepted solution. One of the main hypotheses is ionization of the ground atmospheric gases by radon injected from the cracks of the Earth's crust. This process changes the air conductivity of the boundary layer of the atmosphere and hence modifies the global electric circuit [38].

Nowadays, a special attention is devoted to an investigation of earthquakes precursors observed in the ionosphere as these precursor signals are in short time. It is known that during periods preceding strong seismic events the ionosphere can be disturbed over the epicenter region as well as longer distances from it [39]. Data from ground-based and satellite-based measurements have been used to show perturbations in E- and F-layer density of the ionosphere which may increase or decrease over epicenters before seismic events [40]. The best possible physical mechanism is vertical electromagnetic F2-layer plasma $E \times B$ drift which brought by a seismic genic electric field for the ionospheric anomalies formation [41]. Recent studies have shown that the precursory signals modify the ionization content of the ionosphere, which is reflected in the TEC. Using TEC data from ground-based GPS measurements perturbations in TEC few days before and after the EQs has been reported in previous papers [2]. Using statistical method [e.g., 42], it has been reported preseismic, coseismic, and postseismic ionospheric anomaly in GPS-TEC data for the NDOI-Fiji earthquake of August 19, 2018 [2], as shown in Fig. 15.4A and B. The results showed that the ionospheric perturbation induced from the earthquakes depends on the distance of observation from the epicenter as well as on the direction of observation point with reference to the epicenter. They have also reported perturbation in atmospheric ozone. Thus seismic induced perturbation in ionosphere and atmospheric ozone supports to the lithosphere–atmosphere–ionosphere coupling process.

15.2.5 Response to thunderstorms

Thunderstorms are electric generators, which drive upward currents in the global electric circuit [43,44] and produce peak current up to a few tens of kA [45]. The electric field produced by thunderstorm-driven current may affect $\mathbf{E} \times \mathbf{B}$ plasma drift causing to redistribution of plasma density in the ionosphere and hence perturbation in electron density [46–48]. It has been reported that electric activity associated with even a small thunderstorm affects the D-region by heating electrons and subsequently reducing the electron density [49]. A tripole-charge structured thunderstorm model was simulated and computed upward flowing current into the ionosphere and its effect on electron density distribution [50]. They showed that a source current of 8.3 A in the equatorial region may cause a change of 1–10% in TEC. In addition to electric effects, thunderstorm activity also causes gravity wave generation from overshooting cloud tops of thunderstorm [51]. The gravity wave generated by thunderstorm activity propagates in upward direction and transfer energy into the ionosphere which can be seen in TEC of the ionosphere and in some cases thunderstorm activity can also generate ionospheric plasma bubbles and scintillation activity [40]. Thunderstorm-induced perturbation in electron density distribution of the ionosphere are recently reported by many workers [46,47,51–55]. Fig. 15.5 shows analysis of DTEC data derived from GPS over different stations of Hong Kong for thunderstorm activity of April 1, 2014. The ionospheric perturbation in TEC data has been observed up to a distance around ~500 km and even more from the lightning center. Wavelet analysis of DTEC data shows the presence of gravity waves with period ranging from 10 minutes to 100 minutes (Table 15.1).

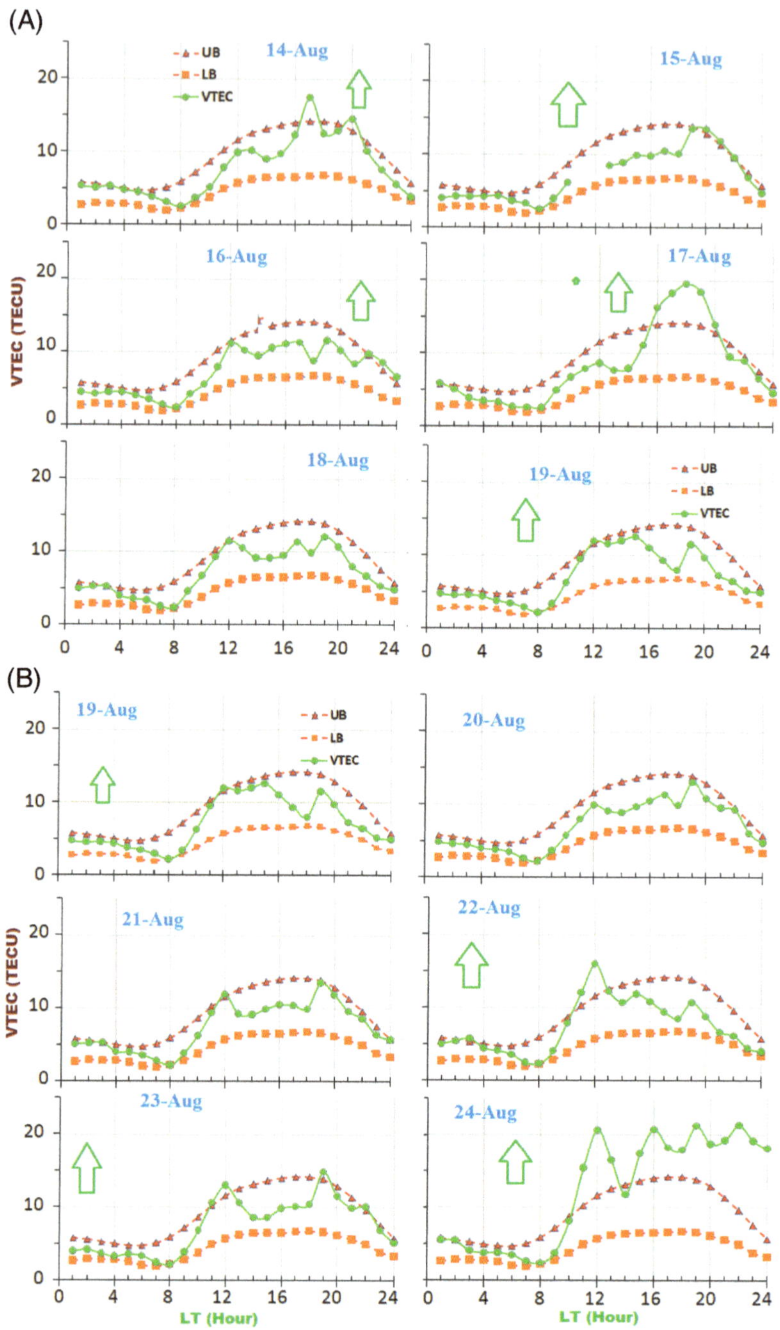

Fig. 15.4 Variation of TEC along with *upper bound (UB)* and *lower bound (LB)* at Maopo'opo, French Polynesia (FTNA), of the earthquake of August 19, 2018 (A) 5-days before from the main shock, (B) 5-days after from the main shock. UB and LB are shown by dotted line.

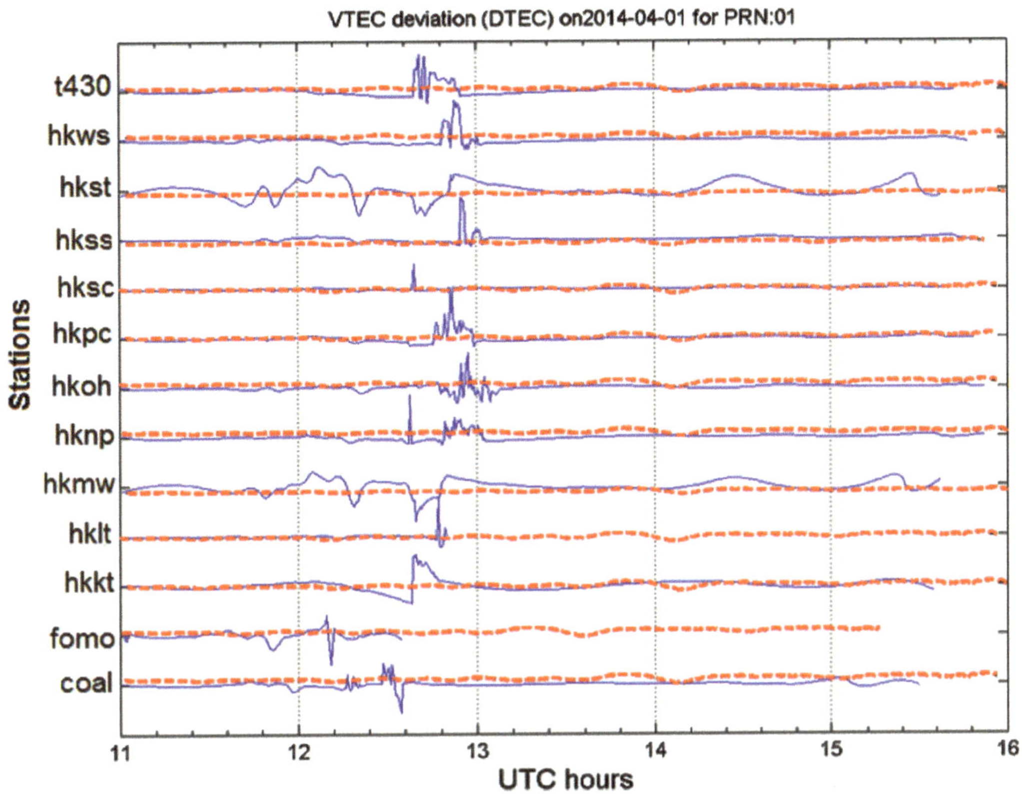

Fig. 15.5 ***Normalized VTEC deviation (DTEC) from best fitted polynomial curve observed with PRN 01 over different GPS stations on April 1, 2014.*** VTEC deviation on nonlightning day is shown by red dotted curve.

Table 15.1 Period of GWS in DTEC data observed from PRN 01 over different GPS stations.

Sr. No.	Name of the HK CORS station with code	Longitude	Latitude	Period (min) in DTEC data
1	COAL	113°,33′ E	22°, 07′ N	69
2	DSMG	113°,34′ E	22°, 09′ N	57
3	FOMO	113°,32′ E	22°, 11′ N	30
4	Kam Tin (HKKT)	114°,18′ E	22°, 22′ N	71
5	Lam Tei (HKLT)	114°,03′ E	22°, 26′ N	48
6	Mui Wo (HKMW)	114°,07′ E	22°, 13′ N	96
7	Ngong Ping (HKNP)	113°,59′ E	22°, 25′ N	58
8	Obelisk Hill (HKOH)	114°,00′ E	22°, 15′ N	79
9	Peng Chau (HKPC)	113°,53′ E	22°, 14′ N	81
10	STNECUTTERS (HKSC)	114°,13′ E	22°, 14′ N	58
11	SIU LANG SHUI (HKSL)	114°,02′ E	22°, 17′ N	59
12	SHAP SZE HEUNG (HKSS)	114°,08′ E	22°, 19′ N	100
13	SHA TIN (HKST)	113°,55′ E	22°, 22′ N	80
14	WONG SHEK (HKWS)	114°,16′ E	22°, 25′ N	62
15	North District Government Offices(T430)	114°,18′ E	22°, 23′ N	76

15.2.6 Response to cyclones

Ionosphere is an upper part of the Earth's atmosphere which is affected by solar and geomagnetic disturbances. Disturbances from upper layer cause the ionospheric perturbation through magnetosphere–ionosphere–thermosphere coupling. In addition to upper layer several phenomenon occurring in the lower atmosphere from meteorological origin can also affect the ionospheric perturbation and is a topic of further investigation. These phenomena include mesoscale convective complexes, deep tropical convection, tropical cyclones (TCs),typhoons, tornadoes, thunderstorms, upper troposphere jet, etc., are some of the disturbances of tropospheric meteorological origin and these are expected to cause ionospheric perturbation. The ionosphere–atmosphere coupling is proposed to be caused by several processes which include atmospheric DC electric field, lightning/thunderstorm-induced electric filed, waves generated in lower atmosphere, neutral particle dynamics originating from large-scale weather system from surface of troposphere [49,56–58]. Among the above disturbances of meteorological origin, TCs are one of the most powerful possible sources which influence the ionosphere through atmosphere-ionosphere coupling.

A TC is a rotation of air mass around low-pressure center originating in equatorial region of troposphere (between ±5° and ±20° latitude) over ocean and has major hazardous effect on both human life and property. It is associated with organized deep convection, thunderstorm activity, and breezy winds at low levels and rainfall [59,60].

TCs can affect the ionosphere from below in different ways which include the vertical large-scale convection of cloudy atmosphere generates upward transport of charged water drops aerosols, generation of spectrum of gravity waves which perturbs the atmosphere–ionosphere electric circuit which further generates an external electric current within a horizontal scale of hundreds of kilometers. This is one of the possible mechanisms which explain interaction between atmospheric and ionospheric components [57,61]. In the previous study, several reports are available that AGWs are generated during low-pressure cyclonic storm which couples with atmosphere–ionosphere system and causes in perturbation in D, E, and F-region of the ionosphere. The critical frequency of F2 layer (foF2) increases and reaches to a maximum before landing of TC [62]. In recent years, ground-based GPS-TEC measurements have been used to show the presences of ionospheric perturbation due to TCs [58,63–65]. The gravity waves generated during the TC may act as seed for the generation of ionospheric irregularities which may further cause radio scintillation which has been reported for TC Tembin, passing near Hong Kong [64]. Fig. 15.6 shows the variations of rate of TEC change index (ROTI) and amplitude scintillation index (S4) observed from GPS over Hong Kong on 26 August 2012. For comparison purpose, the results on 25 August and 27 August 2012 are also shown. Each GPS satellite uses one color to represent its ROTI and S4 values, as indicated in the color bar. It shows that both ROTI and S4 have an evident increase on 26 August. The increase began from 12:00UT (local time = UT + 8 h)

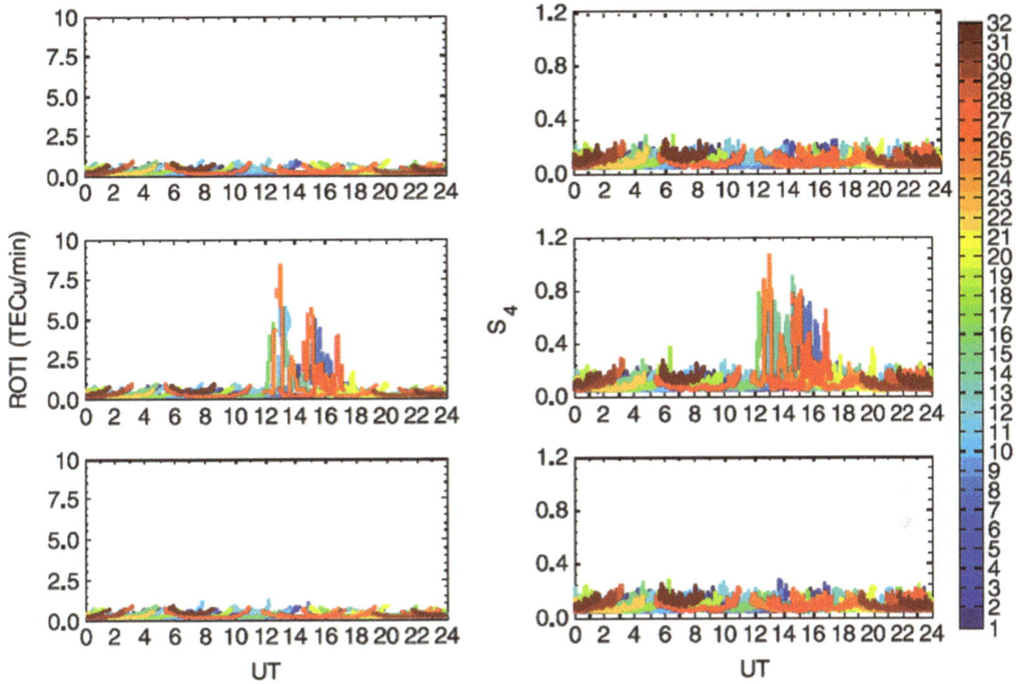

Fig. 15.6 ***The variation of ROTI and amplitude scintillation index (S4) on August 25 (top), (middle row) August 26, and (bottom row) August 27, 2012 over the HokTsui station, Hong Kong.*** Each GPS satellite uses one color to represent its ROTI and S4 values, as indicated in the color bar put on right of the figure [64].

and ended before 18:00UT. It should be noted that the enhancements of ROTI and S4 are agreed well in time. As a close view of the ionospheric responses to Tembin, Fig. 15.5 displays the temporal variations of ROTI and S4 corresponding to the affected GPS satellites (pseudo-random noise (PRN) sequences are 9, 15, 17, 26, and 27) in the nighttime period of August 26, 2012.

15.3 Study of atmospheric phenomena using GPS

GPS is useful to study atmospheric phenomena by measuring water vapor which is sensitive to the surface temperature, onset of monsoon, rainfall, moisture transport by winds and plays very important role in greenhouse effect and climate change [66,67]. Moreover, water vapor is the most abundant greenhouse gas in the atmosphere which has the largest contribution to Earth's greenhouse effect and recognized as important part of the global warming process. In addition to water vapor other atmospheric parameters like pressure, temperature, refractivity, etc., can be measured using GPS-based radio occultation technique in which GPS receiver is fixed on low earth

Table 15.2 Year-wise onset days declared by IMD, days before enhancement in *N* is seen, the date on which wind speed (WS) crossing 8 m/s 21 is seen in the Arabian Sea grid (present study), and the date on which WS crossing 8 m/s 21 seen in the grid [e.g., 73].

Year	Date of monsoon-onset	Days before enhancement in refractivity seen	Date on which WS crossing 8 m/s seen in Arabian sea	Date on which WS crossing 8 m/s seen in 7.5 -20N, 62.5–75 E (Taniguchi and Koike, 2006) grid
2001	23 May	7	21 May	23 May
2002	18 June	11	5 June	5 June
2003	8 June	4	8 June	10 June
2004	18 May	6	18 May	18 June
2005	5 June	10	6 June	8 May
2006	26 May	8	23 May	23 May
2007	28 May	10	2 June	10 June
2008	31 May	12	4 June	30 May
2009	23 May	5	23 May	24 May
2010	31 May	7	6 June	6 June

orbiting satellite [68,69]. Using refractivity data from GPS-RO measurements during 2001–2010, the variation of refractivity during the onset of Indian Summer Monsoon over the east Arabian Sea have been shown [66]. They have reported an enhancement of 5–10 N-units in the refractivity which is observed around 4.8 km (~600 hPa) a few days (9.23 ± 3.6 days) before onset of monsoon over Kerla India (Table 15.2). High precipitation or rainfall affect the human life as well as our environment to a great extent some time can cause flood. Recent papers have shown that high precipitation increases with the surface air temperature at a rate according to Classius–Clapeyron [70]. This suggests that water vapor content in the atmosphere explains the relation between precipitation and surface air temperature.

The increased water vapor in the atmosphere contributes to more warming and enhances the greenhouse effect and often recognized as a part of global warming process (https://climatechangeconnection.org/science/what-about-water-vapour/).

Enhanced value of water vapor data from GPS on July 27, 2006 has been reported to show onset of monsoon over California [71]. Water vapor is also important in numeric weather prediction and monsoon change. Spatial and temporal distribution of water vapor is required for NWP models and hydrological cycles and therefore PWV is considered as an input variable in climatological studies on global as well as local scales and it also affects the energy budget of the atmosphere. Near-real-time water vapor data from GNSS network have been used for weather diagnosis and weather forecasting application. Using network of GNSS station in China (Fig. 15.7A), it has been observed

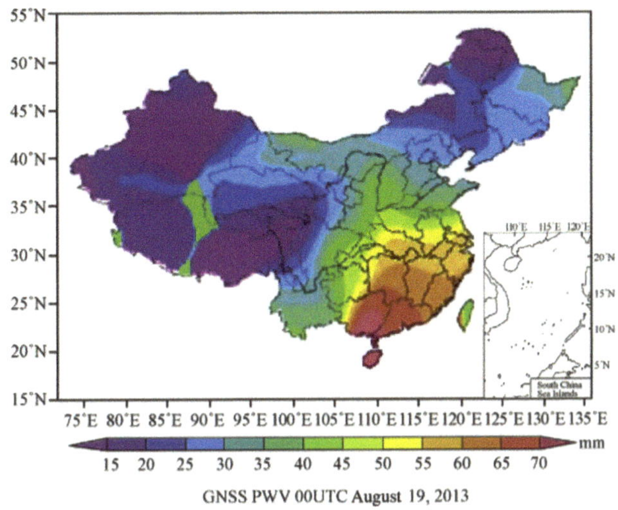

Fig. 15.7A ***Near real-time spatial distribution of PWV over China on August 19, 2013 at 00:00 UTC.***

that the PWV increased by 4–5 mm within 1–3 hours prior to the most cases of precipitation [72].

The network is able to provide real-time PWV which is shown in Fig. 15.7A. This figure shows real-time distribution of PWV for a typical days of August 19, 2913 at 00:00 UTC. An increment of about 5 mm in PWV within 2 hours is used as an index and statistics. They have observed 70% correlation between index and precipitation. Moreover, the index goes to 5 mm within 3–4 hours before precipitation (Fig. 15.7B).

15.4 Summary

GPS is a paradigm for navigation, and the atmosphere including the ionosphere is biggest threats in accurate determination of user's positions. The sources of error arise in the form of the phase and group delay of GPS signals whiles their propagation in the atmosphere/ionosphere. Thus knowledge of these delays is able to provide TEC of the ionosphere and water vapor content of atmosphere which are being used to study several atmospheric and ionospheric phenomenon associated with thunderstorm, cyclones, earthquakes, solar eclipse, monsoon, climate, numeric weather prediction, etc. As for as numeric weather prediction is concerned, ZTD which is related to PWV considered as a key parameters and its tomographic representation at global and regional level is required. Tomographic modeling of atmospheric water vapor required the knowledge of presence of humidity, is a challenging task, and should be addressed in future studies.

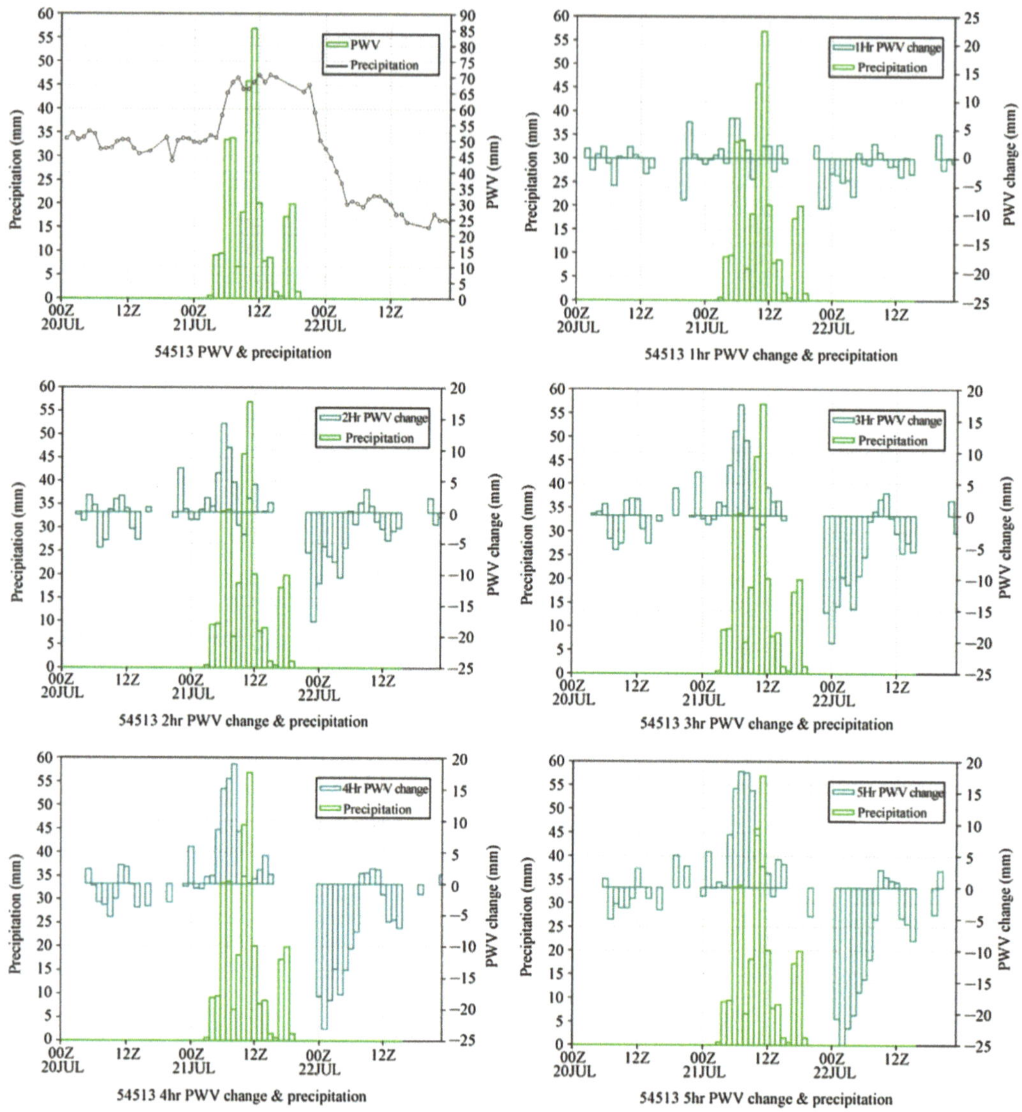

Fig. 15.7B ***Diagram showing relationship between hourly variation in PWV and in precipitation [72].***

Acknowledgments

The work is partially supported by SERB, New Delhi, for the CRG project (File No: CRG/2019/000573) and partially by the Institute of Imminence (IoE) Program (Scheme No: 6031) of BHU, Varanasi. S.K. is thankful to CSIR New Delhi, India, for providing financial assistance under the scientist Pool Scheme [13(9049-A)/2019- POOL]. S.S.R. is thankful to UGC-New Delhi for providing Dr. D.S. Kothari Postdoctoral Fellowship scheme vide sanction No.4-2/2006(BSR)/ES/17-18/0048.

References

[1] M.S. Grewal, L.R. Weill, A.P. Andrews, Global Positioning System Inertial Navigation, and Integration, Second Edition, Wiley-Interscience: A John Wiley & Sons, Inc., Publication, Hoboken, New Jersey, 2007.

[2] S. Kumar, A.K. Singh, R.P. Singh, Probing the tropospheric water vapor using GPS, GPS and GNSS Technology in Geosciences (Science direct), Chapter 6, 1st Edition (2021) pp. 119–134.

[3] G. Blewitt, Carrier phase ambiguity resolution for the global positioning system applied to geodetic baselines up to 2000 km, J. Geophys. Res. 94 (B8) (1989) 10187–10283.

[4] A. Nishida, N. Iwasaki, T. Nagata, The origin of fluctuations in the equatorial electrojet. A new type of geomagnetic variation, Ann. Geophys. 22 (1966) 478–484.

[5] B. Tsurutani, A. Mannucci, B. Iijima, M.A. Abdu, J.H.A. Sobral, A. Humberto, W. Gonzalez, F. Guarnieri, T. Tsuda, A. Saito, K. Yumoto, B. Fejer, T.J. FullerRowell, J. Kozyra, J.C. Foster, A. Coster, V.M. Vasyliunas., Global dayside ionospheric uplift and enhancement associated with interplanetary electric field, J. Geophys. Res. 109 (2004) A08302, doi:10.1029/2003JA010342.

[6] B.G. Fejer, J.W. Jensen, T. Kikuchi, M.A. Abdu, J.L. Chau, Equatorial ionospheric electric fields during the November 2004 magnetic storm, J. Geophys. Res. 112 (2007) A10304, doi:10.1029/2007JA012376.

[7] C.A. Gonzales, M.C. Kelley, B.G. Fejer, J.F. Vickrey, R.F. Woodman, Equatorial electric fields during magnetically disturbed conditions: 2. Implications of simultaneous auroral and equatorial measurements, J. Geophys. Res. 84 (1979) 5803–5812, doi:10.1029/JA084iA10p05803.

[8] T.J. Fuller-Rowell, G.H. Millward, A.D. Richmond, M.V. Codrescu, Storm-time changes in the upper atmosphere at low latitudes, J. Atmos. Sol. Terr. Phys. 64 (2002) 1383–1391, doi:10.1016/S1364-6826(02)00101-3.

[9] G. Kirchengast, K. Hocke, K. Schlegel., The gravity wave TID relationship: Insight via theoretical model EISCAT data comparison, J. Atmos. Terr. Phys. 58 (1996) 233243.

[10] J.H. Sastri, N. Jyoti, V.V. Somayajulu, H. Chandra, C.V. Devasia, Ionospheric storm of early November 1993 in the Indian equatorial region, J. Geophys. Res. 105 (2000) 18,443-18,45.

[11] B.G. Fejer, Low latitude storm time electrodynamics, J. Atmos. Sol. Terr. Phys. 64 (2002) 1401.

[12] S. Sharma, P. Galav, N. Dashora, S. Alex, R.S. Dabas, R. Pandey, Response of low-latitude ionospheric total electron content to the geomagnetic storm of 24 August 2005, J. Geophys. Res. 116 (2011) A05317, doi:10.1029/2010JA016368.

[13] N. Dashora, R. Pandey, GPS data post-processing methods for undifferential receivers located near the crest of EIA, International Beacon Satellite Symposium, Boston, 2007.

[14] M. Blanc, A.D. Richmond, The ionospheric disturbance dynamo, J. Geophys. Res. 85 (1980) 1669–1686, doi:10.1029/JA085iA04p01669.

[15] L. Scherliess, B. Fejer, Storm time dependence of equatorial disturbance dynamo zonal electric fields, J. Geophys. Res. 102 (1997) 24037–24046, doi:10.1029/97JA02165.

[16] N. Gopalswami, P. Makela, H. Xie, CME interactions with coronal holes and their interplanetary consequences, J. Geophys. Res. Atmospheres 114 (A3) (2009) A00A22, doi:10.1029/2008JA013686.

[17] L. Qian, A.G. Burns, P.C. Chamberlin, S.C. Solomon, Flare location on the solar disk: modeling the thermosphere and ionosphere response, J. Geophys. Res. 115 (2010) A09311.

[18] T Basak, S.K. Chakrabarti, Effective recombination coefficient and solar zenith angle effects on low-latitude D-region ionosphere evaluated from VLF signal amplitude and its time delay during X-ray solar flares, Astrophys Space Sci. 348 (315) (2013) 326.

[19] R. Zhang, L. Liu, H Le, Y. Chen., Equatorial ionospheric electrodynamics during solar flares, Geophys. Res. Lett. 44 (2017) 4558–4565.

[20] S.G. Sumod, T.K. Pant, An investigation of solar flare effects on equatorial ionosphere and thermosphere using co-ordinated measurements, Earth, Planets Space, 71 (2019) 125.

[21] B.T. Tsurutani, D.L. Judge, F. Guarnieri, P. Gangopadhyay, A.R. Jones, J. Nuttall, G.A. Zambon, L. Didkovsky, A. Mannucci, B. Iijima, R.R. Meier, T.J. Immel, T. Woods, S. Prasad, L. Floyd, J. Huba, S.C. Solomon, P. Straus, R. Viereck, The October 28, 2003 extreme EUV solar flare and resultant extreme ionospheric effects: comparison to other Halloween events and the Bastille Day event, Geophys. Res. Lett. 32 (3) (2005) L03S09, doi:10.1029/2004GL021475.

[22] S.C. Solomon, L. Qian, Solar extreme-ultraviolet irradiance for general circulation models, J. Geophys. Res. 110 (2005) A10306.

[23] A. Krankowski, I.I. Shagimuratov, L.W. Baran, G.A. Yakimova, The effect of total solar eclipse of October 3, 2005, on the total electron content over Europe, Adv. Space Res. 41 (2008) 628–638.

[24] H.F. Tsai, J.Y. Liu, Ionospheric total electron content response to solar eclipses, J. Geophys. Res. 104 (A6) (1999) 12657–12668.

[25] E.L. Afraimovich, K.S. Palamartchouk, N.P. Perevalova, V.V. Chernukhov, A.V. Lukhnev, V.T. Zalutsky, Ionospheric effects of the solar eclipse of March 9, 1997, as deduced from GPS data, Geophys. Rese. Lett. 25 (4) (1998) 465–468.

[26] L.W. Baran, I. Ephishov, I.I. Shagimuratov, V.P. Ivanov, A.F. Lagovsky, The response of the ionospheric total electron content to the solar eclipse on 1999, Adv. Space Res. 31 (4) (2003) 989–994.

[27] R.A. Bamford, The effect of the 1999 total solar eclipse on the ionosphere, Phys. Chem. Earth (C) 26 (5) (2001) 373–377.

[28] E.L. Afraimovich, O.S. Lesyuta, Ionospheric response to the total solar eclipse of June 21, 2001 as deduced from the data from the African GPS, J. Atmos. Sol. Terr. Phys. 1 (2002) 1933–1941.

[29] J.Y. Liu, H.F. Tsai, M.Q. Chen, Ionospheric total electron content observed during the 24 October 1995 solar eclipse, Adv. Space Res. 24 (11) (1999) 1495–1498.

[30] Z.A.A. Rashid, M.A. Momani, S. Sulaiman, M.A.M. Ali, B. Yatim, G. Fraser, N. Sato., GPS ionospheric TEC measurement during the 23rd November 2003 total solar eclipse at Scott Base Antarctica, J. Atmospheric Solar-Terrestrial Phys. 68 (2006) 1219–1236.

[31] H. Le, L. Liu, X. Yue, W. Wan, The ionospheric responses to the 11 August 1999 solar eclipse: observations and modelling, Ann. Geophys. 26 (2008) 107–116, doi:10.5194/angeo-26-107.

[32] C.S. Zerefos, E. Gerasopoulos, I. Tsagouri, B.E. Psiloglou, A. Belehaki, T. Herekakis, A. Bais, S. Kazadzis, C. Eleftheratos, N. Kalivitis, N. Mihalopoulos, Evidence of gravity waves into the atmosphere during the March 2006 total solar eclipse, Atmos. Chem. Phys. 7 (2007) 4943–4951, doi:10.5194/acp-7-4943-2007.

[33] C.S. Zerefos, D.S. Balis, C. Meleti, A.F. Bais, K. Tourpali, K. Kourtidis, K. Vanicek, F. Cappelani, U. Kaminski, T. Colombo, R. Stubi, L. Manea, P. Formenti, M.O. Andreae, Changes in surface solar UV irradiances and total ozone during the solar eclipse of 11 August, 1999, J. Geophys. Res. 105 (2000) 26463–26473.

[34] G. Chimonas, C.O. Hines, Atmospheric gravity waves induced by a solar eclipse, J. Geophys. Res. 75 (1970) 5545–5551.

[35] G. Kondo, The variation of the atmospheric electric field at the time of earthquake, Memoirs Kakioka Magnetic Observatory, Kakioka, Japan 13 (1968) 17–23.

[36] V.M. Chmyrev, N.V. Isaev, S.V. Bilichenko, G.A. Stanev, Observationby space – borne detectors of electric fields and hydromagnetic wavesin the ionosphere over on earthquake center, Phys. Earth Planet. Int. 57 (1989) 110–114.

[37] X.M. Zhang, H.R. Chen, J. Liu, X.H. Shen, Y.Q. Miao, X.B. Du, J.D. Qian, Ground-based and satellite DC-ULF electric field anomalies around Wenchuan M8.0 earthquake, Adv. Space Res. 50 (1) (2012) 85–95, doi:10.1016/j.asr.2012.03.018.

[38] S.A. Pulinets, Strong earthquake prediction possibility with the help of topside sounding from satellites, Adv. Space Res. 21 (1998) 455–458, https://doi.org/10.1016/S0273-1177(97)00880-6.

[39] S. Priyadarshi, S. Kumar, A.K. Singh, Ionospheric perturbations associated with two recent major earthquakes (M>5.0), Phys. Scr. 84 (2011) 045901.

[40] S Kumar, A.K. Singh, Ionospheric precursors observed in TEC due to earthquake of Tamenglong on 3 January 2016, Current Sci. 113 (4) (2017) 790–795.

[41] Namgaladze, A A., Shagimuratov, II, Zakharenkova, I E., Zolotov, O V., and Martynenko, O V., 2007. Possible mechanism of the TEC enhancements observed before earthquakes. In XXIV IUGG General Assembly, Perugia, Italy 02, 13.

[42] J.Y. Liu, Y.I. Chen, Y.J. Chuo, C.S. Chen, A statistical investigation of pre-earthquake ionospheric anomaly, J. Geophys. Res. (2006), doi:10.1029/2005JA011333.

[43] D. Singh, R.P. Singh, A.K. Singh, S. Kumar, M.N. Kulkarni, AK. Singh, Discharges in the stratosphere and mesosphere, Space Sci. Rev. 169 (2012) 73–121, https://doi.org/10.1007/s11214-012-9906-0.

[44] E.R. Williams, The global electrical circuit: a review, Atmos. Res. 91 (2009) 140–152, https://doi.org/10.1016/j.atmosres.2008.05.018.

[45] E.A. Mareev, S.A. Yashunin, S.S. Davydenko, T.C. Marshall, M. Stolzenburg, C.R. Maggio, On the role of transient currents in the global electric circuit, Geophys. Res. Lett. 35 (15) (2008) L15810, doi: 10.1029/2008GL034554.
[46] C.J. Davis, C.G. Johnson, Lightning-induced intensification of the ionospheric sporadic E layer, Nature 435 (2005) 799–801, https://doi.org/10.1038/nature03638.
[47] C.J. Davis, K.-H. Lo, An enhancement of the ionospheric sporadic-E layer in response to negative polarity cloud-to-ground lightning, Geophysical Res. Letters 35 (2008) (2008) L05815, https://doi.org/10.1029/2007GL031909.
[48] E.H. Lay, X.M. Shao, C.S. Carrano, Variation in total electron content above large thunderstorms, Geophysical Res. Letters 40 (2013) 1945–1949, https://doi.org/10.1002/grl.50499.
[49] X.-M. Shao, E.H. Lay, A.R. Jacobson, Reduction of electron density in the night-time lower ionosphere in response to a thunderstorm, Nat. Geosci. 6 (2013) 29–33.
[50] C.L. Kuo, L.C. Lee, Ionospheric plasma dynamics and instability caused by upward currents above thunderstorms, J. Geophysical Res.: Space Phys. 120 (2015) 3240–3253, https://doi.org/10.1002/2014JA020767.
[51] E.H. Lay, X. Shao, A.K. Kendrick, C.S. Carrano, Ionospheric acoustic and gravity waves associated with midlatitude thunderstorms, J. Geophysical Res. Space Phys. 120 (2015) 6010–6020, https://doi.org/10.1002/2015JA021334.
[52] T.J. Immel, S.B. Mende, M.E. Hagan, P.M. Kintner, S.L. England, Evidence of tropospheric effects on the ionosphere, EOS 90 (2009) 69–70, https://doi.org/10.1029/2009EO090001.
[53] C.G. Johnson, C.J. Davis, The location of lightning affecting the ionospheric sporadic-E layer as evidence for multiple enhancement mechanisms, Geophysical Res. Letters 33 (2006) 1–4, https://doi.org/10.1029/2005GL025294.
[54] Kumar, V.V., Parkinson, M.L., Dyson, P.L., Burns, G.B., 2009a. The effects of thunderstorm-generated atmospheric gravity waves on mid-latitude F-region drifts. J. Atmospheric Solar-Terrestrial Phys. 71, 1904–1915. https://doi.org/10.1016/j.jastp.2009.07.006
[55] S.L. Vadas, DC. Fritts, Thermospheric responses to gravity waves arising from mesoscale convective complexes, J. Atmospheric Solar-Terrestrial Phys. 66 (2004) 781–804, https://doi.org/10.1016/j.jastp.2004.01.025.
[56] J.M. Forbes, S.E. Palo, X. Zhang, Variability of the ionosphere, J. Atmos. Sol. Terr. Phys. 62 (2000) 685–693.
[57] V.M. Sorokin, N.V. Isaev, A.K. Yaschenko, V.M. Chmyrev, M. Hayakawac, Strong DC electric field formation in the low latitude ionosphere over typhoons, J. Atmos. Sol. Terr. Phys. 67 (2005) 1269–1279, doi:10.1016/j.jastp.2005.06.014.
[58] A. Dube, S. Kumar, R. Singh, A.K. Singh, Ionospheric perturbations induced by a Very Severe Cyclonic Storm: a case study of Phailin VSCS, J. Geophys. Res. Space Physics (AGU) 125 (2020) e2019JA027197.
[59] G.J. Holland, Ready Reckoner: Chapter 9, Global guide to tropical cyclone forecasting, WMO/TC-No. 560, Report No. TCP-31. In: World Meteorological Organization, World Meteorological Organization (WMO), Geneva, 1993 Chap. 9.
[60] A.S. Polyakova, N.P. Perevalova, Investigation into impact of tropical cyclones on the ionosphere using GPS sounding andNCEP/NCAR reanalysis data, Adv. Space Res. 48 (2011) 1196–1210, doi:10.1016/j.asr.2011.06.014.
[61] N.V. Isaev, V.M. Sorokin, O.N. Serebryakova, G.A. Stanev, A.K. Yaschenko, E.P. Trushkina, Electric field enhancement and disturbances of plasma density in the ionosphere above the zones of preparation and development of typhoons, Second scientific conference with international participation, Space, Ecology, Nanotechnology, Safety, Varna, Bulgaria (2006).
[62] Y.M. Liu, J.S. Wang, Y.C. Suo, Effects of typhoon on the ionosphere, Adv. Geosci. 29 (2006) 351–360.
[63] T. Mao, J.S. Wang, G.L. Yang, T. Yu, J.S. Ping, Y.C. Suo, Effects of typhoon Matsa on ionospheric TEC, Chin. Sci. Bull. 55 (8) (2010) 712–717, doi:10.1007/s11434-009-0452-4.
[64] Z. Yang, Z. Liu, Observational study of ionospheric irregularities and GPS scintillations associated with the 2012 tropical cyclone Tembin passing Hong Kong, J. Geophys. Res. Space Phys. 121 (2016) 4705–4717.

[65] A. Guha, B. Paul, M. Chakraborty, B.K. De, Tropical cyclone effects on the equatorial ionosphere: First result from the Indian sector, J. Geophysical Res.: Space Phys. 121 (2016) 5764–5777, https://doi.org/10.1002/2016JA022363.

[66] V.V.M.J. Rao, M.V. Ratnam, Y.D. Santhi, M.R. Raman, M. Rajeevan, S.V.B. Rao, On the detection of Onset and Activity of the Indian Summer Monsoon Using GPS RO Refractivity Profiles, Monthly weather Rev. 141 (2012) 2096–2106.

[67] S. Kumar, A.K. Singh, A.K. Prasad, R.P. Singh, Variability of GPS derived water vapor and comparison with MODIS data over the Indo-Gangetic plains, J. Phys., Chem. Earth 55 (2013) 11–18.

[68] C. Rocken, et al., Analysis and validation of GPS/MET data in the neutral atmosphere, J. Geophys. Res. 102 (D25) (1997) 29849–29866.

[69] G.A. Hajj, E.R. Kursinski, L.J. Romans, W.I. Bertiger, S.S. Leroy, A technical description of atmospheric sounding by GPS occultation, J. Atmos. Sol. Terr. Phys. 64 (2002) 451–469.

[70] M. Fujita, T. Sato, Observed behaviours of precipitable water vapour and precipitation intensity in response to upper air profiles estimated from surface air temperature, Sci. Rep. 7 (2017) 4233.

[71] J. Means, D. Cayan, Precipitable water from GPS zenith delays using north American regional reanalysis meteorology, J. Atmos. Oceanic Technol. 30 (3) (2013) 485–495.

[72] Liang, H., 2015. Meteorological applications of precipitable water vapor measurements retrieved by the national GNSS network of China. doi:10.1016/j.geog.2015.03.001.

[73] K. Taniguchi, T. Koike, Comparison of definitions of Indian summer monsoon onset: Better representation of rapid transitions of atmospheric conditions, Geophys. Res. Lett. 33 (2006) L02709, doi: 10.1029/2005GL024526.

CHAPTER 16

Low-latitude upper atmosphere remote sensing using very low frequency (VLF) waves

Ajeet Kumar Maurya[a], Gaurish Tripathi[b], S.B. Singh[b], Rajesh Singh[c], A.K. Singh[b]
[a]Department of Physics, Babasaheb Bhimrao Ambedkar University, Lucknow, Uttar Pradesh, India
[b]Atmospheric Research Laboratory, Department of Physics, Institute of Science, Banaras Hindu University, Varanasi, Uttar Pradesh, India
[c]KSK Geomagnetic Research Laboratory, IIG, Prayagraj (Allahabad), India

16.1 Introduction

Very low frequency (VLF) (3–30 KHz) waves are a powerful tool for remote sensing of the upper atmosphere. The VLF waves have their source in both natural and man-made. The natural sources include lightning discharges from thunderstorms, volcanic eruptions, dust storms, and tornadoes [1]. The newly discovered phenomenon of upper atmospheric lightning discharges, also known as "Transient Luminous Events" (TLEs), which includes sprites, elves, jets, blue starters, etc., are also sources of VLF waves [1]. However, on a global basis, the most significant source of VLF is the lightning discharge from a thunderstorm. The manmade sources of VLF waves include fixed frequency VLF navigational transmitters operated by the Navy of different countries to communicate with a submarine, nuclear explosions, and HF heating ([1] and reference therein).

Based on the frequency range and sources, the VLF study is divided into two categories: (1) broadband VLF waves which involve a complete frequency range (~3–30 kHz), and (2) narrowband VLF waves, which were mainly fixed frequency signals from navigational VLF transmitters are used. The broadband waves comprise *radio atmospherics* [2,3], *Tweeks* [3], *Whistlers* [4–6], and *VLF emissions* [7]. The narrowband data are the fixed frequency VLF signals from various navigational VLF transmitters being operated globally by different countries. The amplitude and phase of these signals have been received for studying various phenomena like D-region ionospheric perturbations caused by phenomena such as solar flares [8], earthquakes (EQs) [9,10], lightning-induced electron precipitations (LEPs) [56], TLEs [11], terrestrial gamma-ray flash [12], the effect of geomagnetic storms [57], etc.

The VLF signals from their sources to receiving stations propagate by multiple reflections through the waveguide formed by the Earth and the lower ionosphere called Earth-ionosphere waveguide (EIWG) and thus form the novel tools for continuous monitoring of the D-region ionosphere [13]. There are two methods by which VLF

Atmospheric Remote Sensing
DOI: https://doi.org/10.1016/B978-0-323-99262-6.00002-X

waves propagate, (1) ray theory or wave-hop theory, (2) waveguide theory. Each theory has its advantage and disadvantage depending on the propagation distance, source and receiver characteristics, etc. The wave-hop theory is limited to the distance range between ~2000 and 3000 km. This is due to the complexity of mathematical calculations as the number of sky waves will be large. The waveguide mode is much useful which represents the propagation mechanism in terms of the waveguide modes where the waveguide is formed by the Earth's surface as the lower part of the boundary and the lower ionosphere as the upper part of the boundary. A comprehensive discussion and required mathematical derivation can be found in a historical paper by Wait [14]. The other important related work can be found by Budden, Wait, Spies, and others [1,15].

Overall, it is well established that VLF waves form undoubtedly a cost-effective tool for continuous monitoring of both ionosphere and magnetosphere. In this chapter, the probing potential of narrowband, as well as broadband VLF waves to study the ionospheric, plasmaspheric, and magnetospheric plasma parameters, will be described mainly over the low latitude region. Thus, the chapter is divided into two parts: first, we shall focus on some of the important case studies of the D-region to establish the usefulness of VLF waves in D-region ionosphere studies categorized as (1) quite time variation, (2) geomagnetic storm, (3) solar flares, (4) solar eclipse, and (5) EQs. In the second part, we shall discuss the broadband VLF waves of (1) Tweeks, (2) whistlers, and (3) VLF emissions observed at low latitudes and their probing potential.

16.2 Remote sensing of D-region of ionosphere using narrowband VLF waves

16.2.1 Quiescent D-region ionosphere

The D-region of the ionosphere with an altitude range of 50–90 km covers the mesosphere and lower thermosphere regions of the Earth's space atmosphere and hence shows several unique features. In this region, motion of electrons and ions is controlled by neutral wind and turbulence, and is responsible for most of the absorption encountered by HF signals. The change in neutral temperature in the mesosphere strongly affects the D-region recombination process and thus results in a variation of electron density and ion composition. This is the most complex part of the ionosphere due to its relatively high pressure, which plays important role in the photochemical process also in this region several different sources of ionization contribute to the ionization.

Fig. 16.1 shows an example of D-region remote sensing with VLF waves on a geomagnetically quiet day. There is a transmitter in Western Australia known as NWC, which operates at 19.8 kHz. The signal propagates in the EIWG and is detected at a receiver at Dehradun, North India. The receiver monitors the amplitude and phase of the transmitter, which are displayed in Fig. 16.1 (upper and lower panel). There is a diurnal variation that is evident, due to the Sun's influence (Solar Lyman-α). As the Sun rises and sets, the intensity of Lyman-α varies as per the change in the solar zenith angle (the

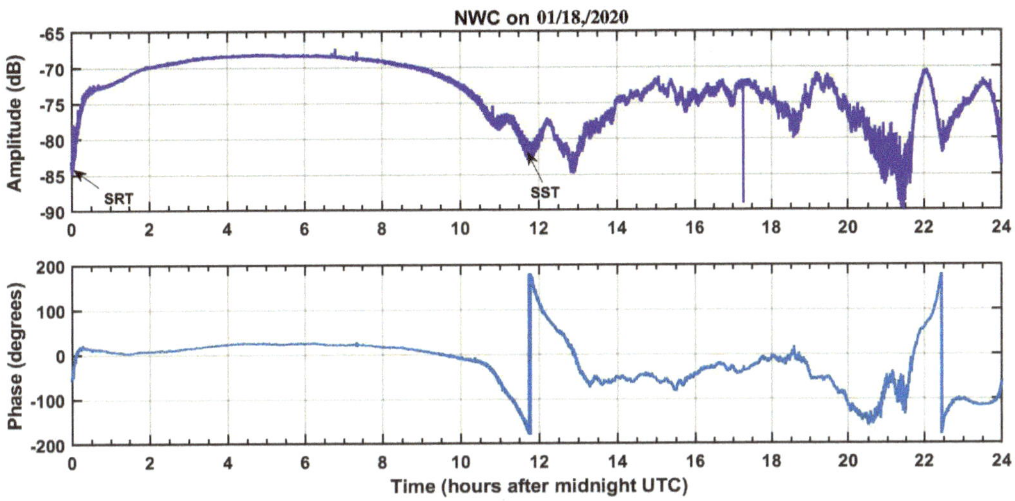

Fig. 16.1a Example plot of normal day NWC (19.2kHz) VLF signal recorded at Dehradun, India.

zenith angle is defined as the angle between the Sun's rays and the vertical direction). The D-region electron density along with the transmitter–receiver path changes, from minimum during sunrise rises to its maximum during local noontime and decreases afterward. The VLF amplitude and phase exactly follow this variation. At night, with the Sun gone down, the electron density is highly variable, which changes VLF propagation conditions rapidly. The diurnal mean variation of the VLF signals shows sunrise terminator (denoted as SRT) and sunset terminators (denoted as SST) (the time at which dayside meets night side) at ~0:10 UT, ~12:00 UT, respectively. The amplitude minimum during sunrise is deeper and clearer than that during sunset. the terminator time is formed as an interference between the ground wave and sky wave [16]. This happens when the phase difference between two waves is 180°, due to reflection from the night side, and thus superposition of two waves results in destructive interference at the terminator and one sees a decrease (minimum) in amplitude. The diurnal variation of three VLF transmitter signals, JJI (22.2 kHz, Ebino Japan, lat = 32.09N; 130.8E), DHO (23.4 kHz, Rhauderfehn, Germany, lat = 53.08N; long =7.60E) and VTX (18.2 kHz, South Vijaynaranyam, India, lat = 8.3N, long = 77.8E) is also shown in Fig. 16.1B for comparison of various TRGCPs. The morning and evening terminators are also shown on each path. One can see the difference in the day/night portion of the signal depending on the TRGCP path length and location.

16.2.2 Geomagnetic storms and D-region ionosphere

The geomagnetic storm, solar flares, and solar eclipse are the major space weather phenomena having origin in the Sun which affect the entire ionosphere, with different effects in the different regions of the ionosphere [17]. The geomagnetic storm is the

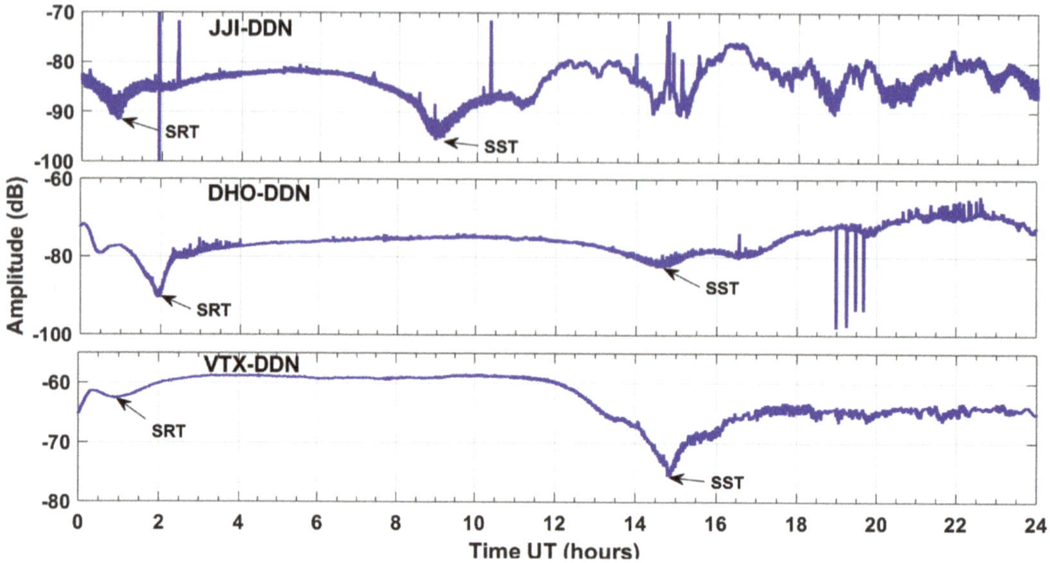

Fig. 16.1b Showing diurnal amplitude variation of three VLF transmitters (JJI, DHO, and VTX) recorded at Dehradun.

result of coronal mass ejection (CME) activity at the surface of the Sun. The CME event is characterized by the release of huge energetic plasma along with a solar magnetic field into the interplanetary space. This modifies the solar wind properties and increases their speed and pressure. The increase in the solar wind pressure initially compresses the near-Earth space environment called the magnetosphere. The solar wind's magnetic field interacts with the Earth's magnetic field and the geomagnetic storm occurs when there is a very efficient exchange of energy between the solar wind and the magnetosphere. This happens when the interplanetary magnetic field (IMF) B_z is in the north direction, causing a reconnection between the Earth's magnetic field and IMF B_z. The intensity of a geomagnetic storm is measured through an index called the disturbance storm time (D_{st}) index. The D_{st} index is derived from the change in the horizontal component of the Earth's magnetic field. The enhancement in the westward ring current around Earth during a geomagnetic storm causes magnetic disturbances on the ground opposite to the Earth's magnetic field. The geomagnetic storm consists of three stages storm sudden commencement (SSC), main stage, and recovery stage based on the variation in the D_{st} index. An example plot of the St. Patrick's day geomagnetic storm of March 17, 2015 is shown in Fig. 16.2(upper panel). The St. Patrick's Day geomagnetic storm began at ~04:45 UT on March 17, 2015, when a double-halo CME hit the Earth's magnetic field. Initially, the IMF B_z went northward reaching ~27 nT for a while giving rise to SSC, then it turned southward ~06:00 UT. The storm reached its peak intensity at ~00:00 UT on March 18, 2015 with a minimum D_{st} ~ -223 nT and thus categorized

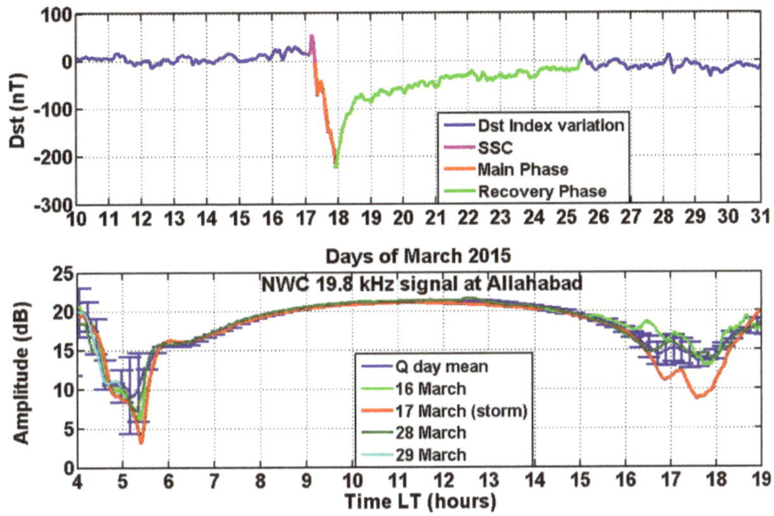

Fig. 16.2 ***Upper panel shows the variation of the Dst index showing various stages of the March 17, 2015, St. Patrick's day geomagnetic storm.*** Lower panel showing the variation of NWC signal before, during, and after the geomagnetic storm.

as a severe geomagnetic storm [18] after that it started recovering and the D_{st} reaches its normal level on 25 March 2015.

The ionospheric effects of a geomagnetic storm are fairly significant and affect the entire ionosphere (~60–1000 km altitude), but the effects are perceived differently in the different regions of the ionosphere. In the D-region, the major disturbances due to geomagnetic storms are mostly seen at the high and mid-latitude which is associated with energetic electron precipitation which causes localized electron density enhancements in the D-region [58]. A less pronounced negative amplitude and phase change during the daytime and a more pronounced effect during the nighttime for the low-mid latitude D-region is observed in the VLF signal for the Australia–Kamchatka transmitter–receiver great circle path. This is also called a negative storm effect [19]. The storm after effect appeared as an important feature of the geomagnetic storm effect on the D-region, which is characterized by a long recovery duration in the VLF signal even after several days of the recovery of the storm. Such a storm affecter effect could last for several days, sometimes up to 10 days [20]. In the mid-latitudes, there is particle loss due to precipitation during the storm recovery and this ionizes the D-region is the major source of the storm affect recovery. While in the low latitude, perturbation in the D region composition results in a slower recovery rate causing a long duration of VLF signal recovery [17].

The geomagnetic storm effect on the low and equatorial latitudes *D*-region ionosphere remains sparsely studied ([17]; Maurya et al., 2018). For the December 14–16, 2006 geomagnetic storm, the study of purely low latitude path by analyzing NWC and

NPM to Suva, Fiji TRGCP reported a continuous decrease in the daytime [17]. A similar observation is also reported during St. Patricks day's geomagnetic storm of March 17, 2015. The study reports observation of ~2 hours duration decrease near evening terminator of the NWC signal recorded at low latitude station Allahabad (now Pryagraj) India for several days following the storm onset day (Maurya et al., 2018). An example plot of NWC signal variation during the St. Patrick's Day geomagnetic storm is shown in Fig. 16.2 (lower panel). Fig. 16.2 shows NWC signal variation at Allahabad for five international quiet days mean in March 2015 with a blue color line along with standard deviation (2σ), the day before the storm (16 March) with green color, storm day (17 March) with red color, recovery days (28 and 29 March) with dark green and cyan colors respectively. The red line on the storm main phase day (17 March) around ~16:30 LT, the signal amplitude began decreasing below 2σ and remained low for ~2 hours and then approached the 2σ line after ~18:30 LT. The decrease in the signal amplitude near the evening terminator is ~3.60 dB

Though the proper mechanism of storm effect on the low latitude D-region is not well understood, the most prominent mechanism reported in the above-mentioned work is suggesting the propagation of atmospheric gravity waves (GWs)/traveling ionosphere disturbances to the low latitude D region due to the Joule heating at high latitudes.

16.2.3 Solar flares and D-region ionosphere

As the name is self-explanatory by the word "solar" means Sun and "flare" means a sudden burst of flame or light, thus term "solar flares" coins to a sudden and very intense burst of flame on the surface of the Sun. A solar flare occurs when the magnetic energy that has built up in the solar atmosphere is suddenly released. During a solar flare, energy is released across the entire electromagnetic spectrum, that is, from the radio waves to the X-rays and γ-rays, including the visible spectrum. They are classified based on peak flux of X-ray wavelength region in 0.1–0.8 nm as X, M, C, and B-class flares.

Solar flares are known to cause ionospheric perturbations in the daytime mainly through X-rays and EUV radiation released during the flare event (*e.g.*, [21,59]. The EUV (26–34 nm) radiation mostly affects the upper ionosphere, whereas X-rays (0.1–0.8nm) affect the lower ionosphere/D-region of the ionosphere. This increased ionization which is many folds to the background density produces a different type of sudden ionospheric disturbances phenomenon (SIDs). During flare events, the increased X-ray flux from the Sun in the wavelength range of 0.2–0.8 nm ionizes neutral constituents along with the major species O_2 and N_2 [21,22] at the D-region altitude. This in turn modifies the electron density in the D-region ionosphere and hence changes the propagation conditions of very low frequency (VLF, 3–30 kHz) waves in the EIWG formed between the D-region ionosphere and the Earth's surface and causes well known SID phenomenon [23].

The VLF measurement of D-region variations due to solar flares is one of the most studied phenomena (e.g., [23,24,60]). X-ray flux from solar flares produces well-pronounced effects in the daytime VLF signals. An example of an X-class solar flare that occurred on May 14, 2013 is shown in Fig. 16.3A. Fig. 16.3A (upper panel) shows the X-ray flux variation which shows peak flux ~3.23×10^{-4} w/m^2 indicating an X-class solar flare peaked around 01:12UT. Fig. 16.3A (Middle panel), shows the NWC amplitude at Prayagraj showing a peak around the same time ~01:13UT, and Fig. 16.3A (lower panel) peak in the NWC phase. Though both increase and decrease in the VLF signals are reported most increase in signal amplitude is found over the Indian low latitude region. The possible explanation of such change in the VLF signal is associated with the extra ionization due to solar flare that sharpens the D-region or upper boundary of the EIWG and causes it to descend up to several km depending upon the intensity of solar flare flux [24]. The VLF signal propagates through the region of enhanced electron concentration and finds a sharper boundary which gives a mirror type of reflection [24]

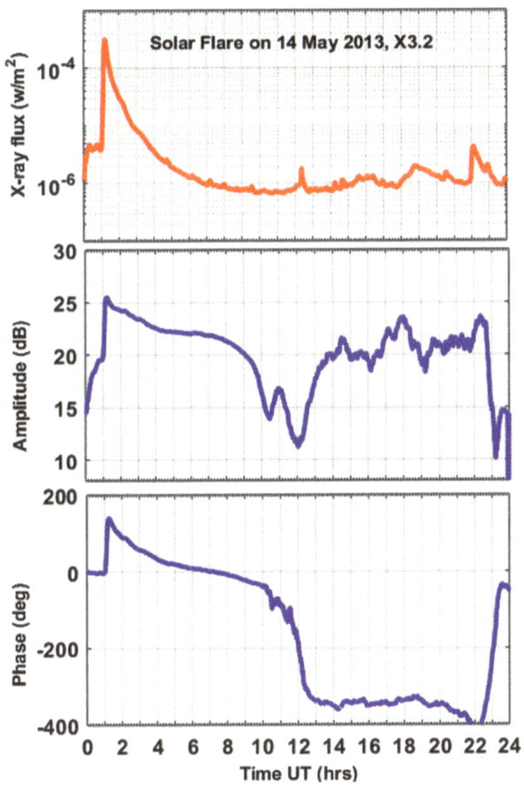

Fig. 16.3a Upper panel showing X-ray flux (in w/m^2) on May 14, 2013, an X3.2 class solar flare can be seen. **Middle panel** NWC signal amplitude at Allahabad showing peak corresponding to solar flare peak time. **Lower panel** NWC signal phase on the same day.

to increase the amplitude of VLF signals recorded at the receiving station. In a few cases, during flare events, the D-region ionization increases but it do not sharpen rather remains defused, and thus VLF signals do not get mirror type reflection rather need to penetrate deeper to get reflection and resulting in a decrease in signal amplitude [24].

The Wait ionospheric parameters: D-region reflection height (H') in km and D-region electron density gradient or sharpness factor (β) in km^{-1} [25] have been extensively used to estimate D-region electron density change due to solar flares. The change in the VLF signal amplitude and phase is being simulated using the VLF subionospheric propagation model known as the long wave propagation capability (LWPC) code (developed by Naval Ocean Systems Center, San Diego, USA) [61]. The LWPC provides h' and β based on the change in VLF signals, their parameters are further used as input to Wait's equation to get the corresponding electron density change in the D-region. The quantification of the D-region reflection height and electron density changes during flare events have been reported by several researchers (e.g., [23]). Fig. 16.3B shows the electron density variation with altitude for a few X-class flare events and is compared with a normal day. The electron density of the D-region during flare events increases to many folds, as can be seen from Fig. 16.3B. The increased ionization due to flares lowers the H' roughly in proportion to the logarithm of the X-ray flares intensity [26]. The solar flare effect on the D-region as sensed by the VLF signals also depends on the seasons and local time [8]. As can be seen from Fig. 16.3B, the electron density due to the X3.2 solar flare is lower than the other lower X class flares. These flares occurred during different seasons and times of the day, thus showing variations. Further, the latitudinal and longitudinal dependence of the solar flare effect

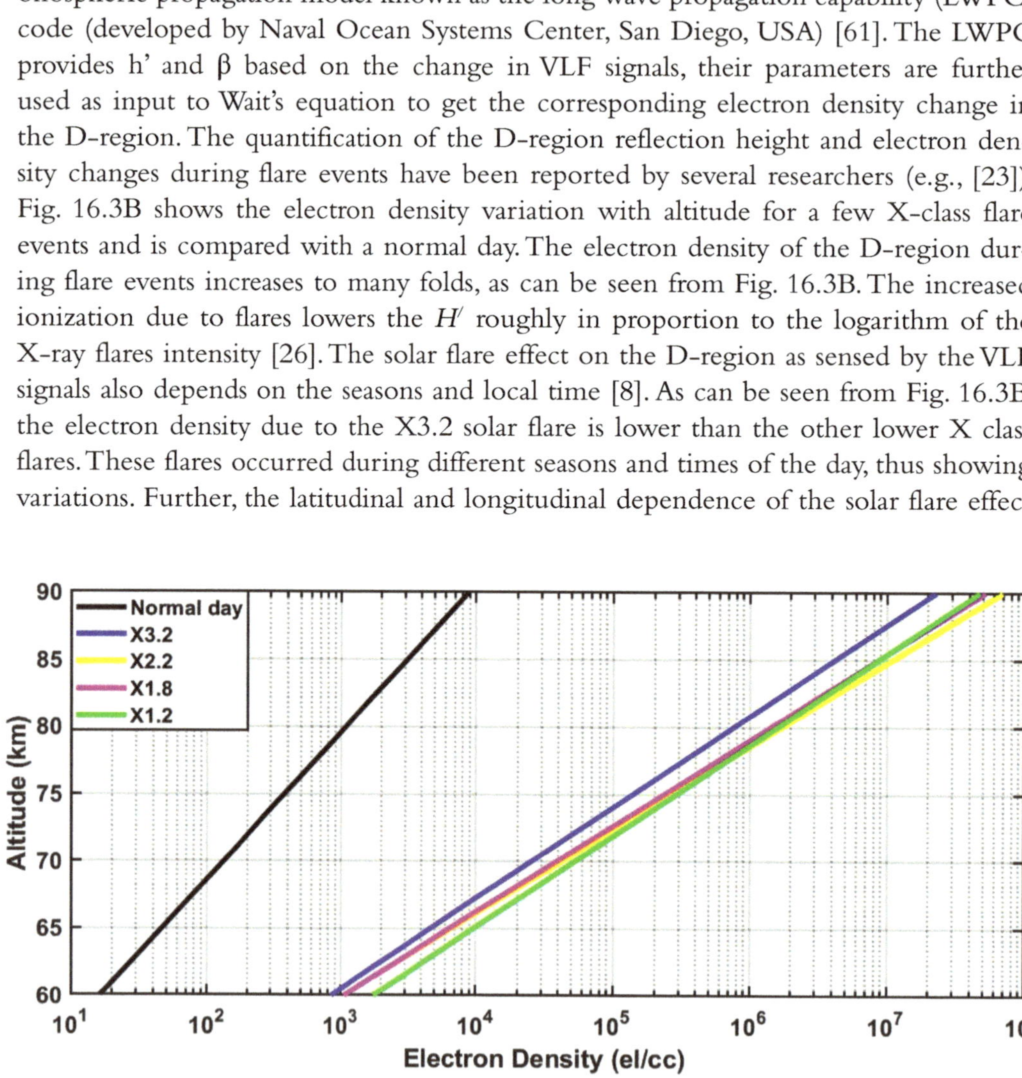

Fig. 16.3b Showing electron density variation comparison for a normal day and X class flare days with altitude.

on the D-region is yet to be studied. As most of the effects on the D-region are associated with electromagnetic waves (X-ray flux), thus any such dependence is very unlikely, but the author proposes a thorough study before any conclusion.

16.2.4 Solar eclipse and D-region ionosphere

A solar eclipse is one of the most popular celestial phenomena since ancient times, which had inspired several scientists. The solar eclipse happens when the Moon comes in between Sun and Earth and blocks solar radiation from falling on the Earth. The solar eclipse could be total when the Moon fully blocks the Sun, it happens when the apparent lunar size is larger than the Sun. A partial eclipse is when the Moon partially blocks the Sun, an annular eclipse is when the Moon almost blocks the Sun but not completely. The solar eclipses always remain of great scientific interest as they provide a unique opportunity to study photochemical, optical, and radio-physical processes in the Earth's environment under the sudden reduced solar radiation conditions almost close to night conditions. This severely affects radio communication supported by the ionosphere and produce error in satellite navigation [27,28].

During a solar eclipse, the decrease in solar flux due to the Moon's shadow causes a sudden change in the D-region physical and chemical processes. During the totality due to the blocking of Lyman-α 1215Å (major D-region ionizing radiation) by the Moon's umbral shadow, the electron density decreases drastically toward the nighttime values [28] which is supported by the observation of nighttime phenomena such as "tweek radio atmospherics" [27]. Apart from a wide range of implications of the solar eclipse on near-Earth atmosphere and ionosphere, the two most important consequences of solar eclipse are ionospheric variability and generation of GWs [29].

The VLF navigational transmitter signals have been extensively used to monitor D-region variation during solar eclipses. The sudden drop in D-region ionization affects VLF propagation conditions and subsequent changes are recorded in VLF signal amplitude and phase. The solar eclipse effect on the D-region is quantified by using LWPC modeling and simulation of VLF signal changes [30,31]. Kumar et al. [30] estimated change in VLF signal (amplitude and phase) during three solar eclipses (July 22, 2009; November 13–14, 2012; and May 9–10, 2013). Their results show both decreases (for November 2012 and May 2013 eclipse) and increase (for July 2009) in VLF signal amplitude. All three eclipses have shown an increase in average D-region reflection height for low-equatorial latitude paths. An interesting observation for the July 22, 2009 total solar eclipse is presented in Fig. 16.4A and B [31,32] using JJI- Allahabad (now Prayagraj), JJI-Varanasi & JJI-Nainital and NWC-Allahabad, NWC-Varanasi & NWC-Nainital transmitter-receiver great circle path. As one can see from Fig. 16.4A, the JJI-TRGCP is along the solar eclipse totality line, whereas NWC-TRGCPs intersect the totality line.

The results for JJI amplitude are presented in Fig. 16.4B upper panels for three receiving stations, both ALD and VNS show an increase in amplitude during the eclipse of

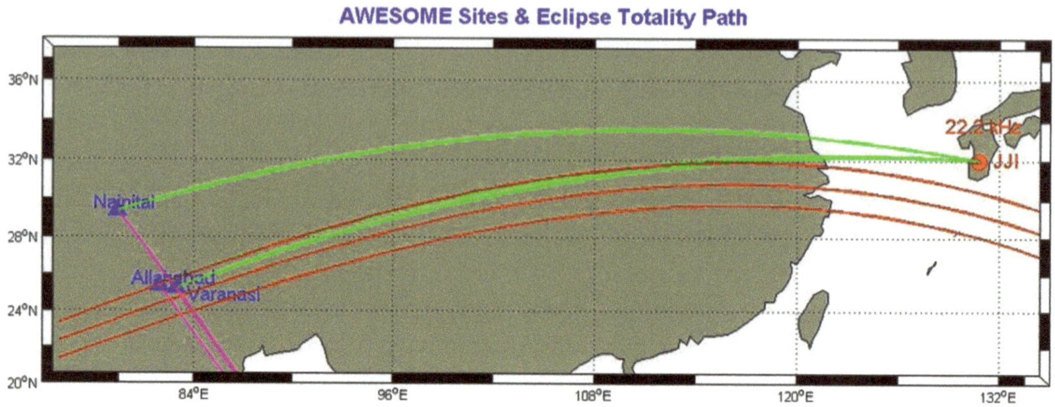

Fig. 16.4a Showing the July 22, 2009 total solar eclipse path (three parallel red lines), the Allahabad, Varanasi, and Nainital with JJI path (green lines), Allahabad, Varanasi, and Nainital with NWC path (magenta lines).

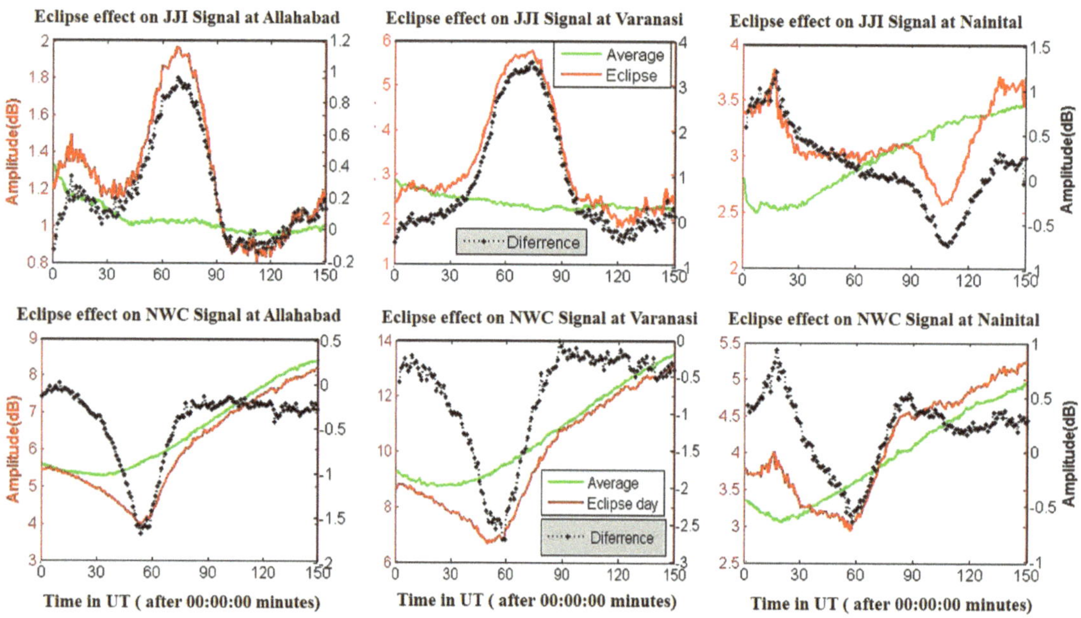

Fig. 16.4b Showing the July 22, 2009 total solar eclipse effect on the JJI signal at three stations (Allahabad, Varanasi, and Nainital), while lower panels show NWC signal amplitude change for three stations. The green line is the control day/normal day VLF signal, the red line shows the eclipse day VLF signal while the black line show difference between eclipse day and control day.

~2–3 dB. The Nainital stations show a peculiar feature, initially increase and then followed by the decrease. The result for NWC amplitude is presented in Fig. 16.4B lower panel for three stations. Here both ALD and VNS show a decrease in amplitude which is minimum exactly close to solar eclipse maxima at the receiving stations. Nainital station again shows peculiar features but the minimum is close to eclipsing maxima. This is the first time both increase and decrease in VLF signal amplitude are reported at the low latitude stations. It was suggested that the observations of both increase or decrease of VLF signal amplitude depend on the interference conditions between sky and ground waves converted at the discontinuity created by the eclipse. If the interference conditions are constructive, there is an increase in VLF signal amplitude, and when there is destructive interference the amplitude decrease [13]. The interference conditions depend mainly on the TRGCP path length, path direction, solar eclipse, and ionospheric conditions along the path [32].

Another important consequence of the solar eclipse is the observation of atmospheric gravity waves (AGWs). The supersonic motion of cooling spot of solar eclipses in the stratosphere (ozone layer) disturbs the heat balance of the atmosphere and has been considered to be the major source of AGWs [29,33] with the period ranging from minutes to a few hours. The observations of AGWs during an eclipse in the low latitude D-region are reported during past solar eclipses. The reported periodicity of AGWs mostly lies in the ~20–90 minutes [30,32]. It was observed that stations in the totality region show relatively shorter period (~20–40 min) AGWs, whereas stations under the partial eclipse region show longer period (~40–80 min) waves. It was concluded that the observed AGWs might be generated by the sudden cutoff of the photo-ionization creating nighttime-like conditions in the D-region of the ionosphere and solar eclipse-induced GWs coming to the D-region from below and above [32]. Though the exact mechanism is not yet clear and needs further investigations by taking several cases of solar eclipses.

16.2.5 Earthquakes and D-region ionosphere

EQs are one of the most devastating forms of natural disasters on Earth. The EQ is shaking the surface of the Earth when seismic waves pass through it. With the recent advancement in geophysical observation techniques and analysis methods, researchers can provide long-term (~100–50 years) EQ predictions but EQ mitigation depends on the short-term (within a few days to hours) EQ prediction [34]. The long-term predictions are conventionally based on the crustal movement measurement which ought to be not of much use for the short-term EQ predictions. Since the last few decades, there has been significant progress in this direction and there has been enough evidence for the existence of electromagnetic phenomena before an EQ (e.g., [34]).

Probably, the first most significant work in the VLF radio sounding for seismic-ionospheric effects was done by Russian colleagues [35], for the famous Spitak EQ by using the VLF propagation technique over a long distance from Reunion to Omsk. They observed a significant propagation anomaly a few days before the EQ. Subsequently, there have been two methods developed mostly by Russian and Japanese researchers, to detect preseismic anomalies in the D-region using VLF remote sensing technique, famously known as "terminator time method or TT method" and "nighttime fluctuations method or NF method" [36].

In the TT method, the focus is at the time of morning and evening terminator (the terminators are the characteristics minima seen in the diurnal variation of VLF signal around the sunrise and sunset of local time, generated due to interference between sky and ground waves) and time shifts are analyzed near the TT before and after the EQs. Hayakawa et al. [37] for the famous Kobe EQ of January 17, 1995 (Mw = 7.2 and depth ~20 km) devised this technique and have reported that the TT method is more suitable for short TRGCP (~2000 km path length). Further, they have also observed that evening terminators shifted toward the night side (late hours) and morning terminator shifts toward the morning side (early hours) a few days before the EQ and became normal a few days after the EQ. They observed that the evening terminator is more indicative of seismic effect and reported a maximum shift of ~45 min in the evening terminator for the Kobe EQ. The predominant shift in terminators is attributed to an increase in D-region electron density which lowers the boundary of the ionosphere by ~1.5–2 km [16]. Later and most recent work shows that the TT method is also applicable to long TRGCPs (~6000 km) [10]. Maurya et al. [10] studied the effect of Nepal 2015 EQs on the VLF signal using NWC-Allahabad TRGCP and after carefully scrutinizing possible contributions from other potential sources, they concluded with the presence of significant shifts of ~45 and ~26 min respectively, in evening terminator time in the NWC VLF amplitude signal, one day before both the major EQs of M7.8 and M7.3 Nepal EQ. Fig. 16.5A shows the detailed location map of April 25, and May 12, 2015 EQs by red stars, the location of the VLF receiving station at Allahabad (now Prayagraj) in India, the location of the NWC VLF transmitter in Australia, the TRGCP of NWC-ALD path as black line and region of EQ preparation zone based on the EQ magnitude of April 25, 2015 (M7.8) by red circle. It is important to note that EQ can affect a given TRGCP only if it is passing under an encircled region which depends on the magnitude of the EQ [10]. Fig. 16.5B shows the detailed analysis of the TT method for the two major aftershocks during the 2015 Nepal EQ. Fig. 16.5B upper panel shows the diurnal amplitude variation of NWC signal at Allahabad during April 10–30, 2015, the EQ day is April 25, 2015 (M7.8) indicated with a red line. The significant shift in the evening terminator can be seen on April 24, 2015, one day before the EQ. Similarly, the May 12, 2015 (M7.3) aftershock results are shown in Fig. 16.5B lower panel, with a significant TT shift again one day before the EQ (on 11 May 2015). Interestingly, there was no significant shift reported in the morning terminator as reported in previous cases.

In the NF analysis method, VLF amplitude time series data for the pure nighttime is considered, that is, the data during the local nighttime portion ~19:00–03:00 LT (to

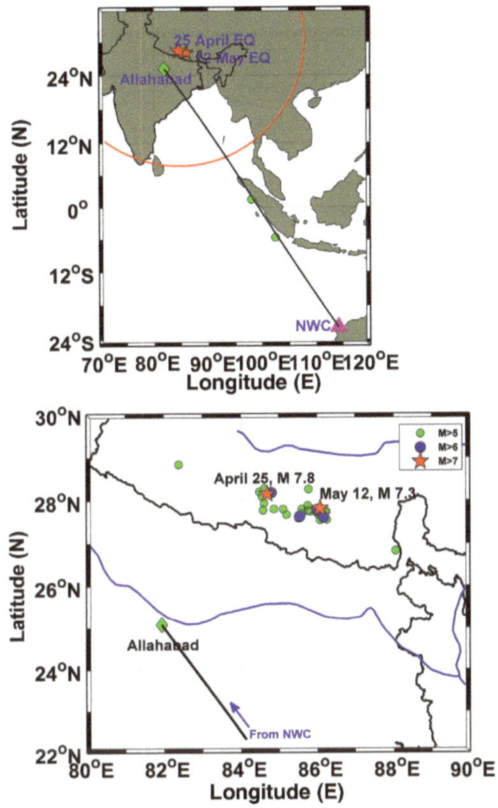

Fig. 16.5a Depicts detailed map of two Nepal EQs (April 25, 2015 and May 12, 2015), location of VLF transmitter and receiver, TRGCP path along with EQ preparation zone as a red circle.

avoid the terminator effect) at the receiving station during which the entire TRGCP are in night condition.

At present, our understanding of how do the ionospheric disturbance evolves due to seismicity is not well understood. The role of the lithosphere–atmosphere–ionosphere (LAI) coupling mechanism, needs to be understood properly, but with the work of the last few decades, it is possible to point out the most successful mechanism at present. There have been three proposed mechanisms for the LAI coupling process (e.g., [38]; and reference therein). These are: (1) chemical channel: controlled by the geochemical quantities, for example, surface temperature, pressure, and radon emanation produces a change in the vertical electric field of the atmosphere which modify lower ionospheric properties, (2) acoustic and GW channel: during EQ preparation period, change in various parameters excites the atmospheric oscillations traveling up to the ionosphere, and (3) electromagnetic channel: generation of radio emission from lithosphere which modifies ionospheric properties. Later several observations supported the second

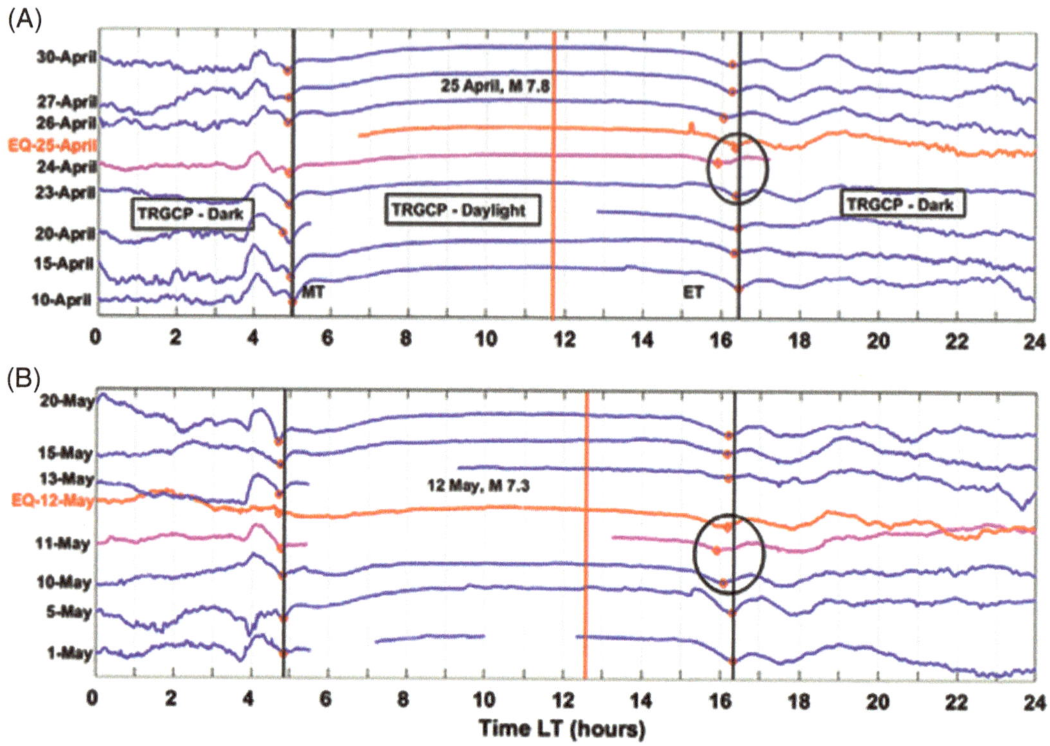

Fig. 16.5b (Upper panel) Sequential plot of NWC (19.8 kHz) signal amplitude variation at Allahabad station for 24 hours local time for April months. The black vertical line shows the morning and evening terminator respectively. (Lower panel) Similar plot but for May 12, 2015 aftershock (adopted from [10]).

mechanism: the acoustic and GW channel. Most of the reports on the EQ precursors signatures using sub-ionospheric VLF/LF signals show direct or indirect evidence of the presence of atmospheric GWs with periods of a few minutes to a few hours [36,38,39]. There are several factors such as ionospheric conditions, soil properties, etc., that can affect the LAI coupling process. Thus, more case studies and pieces of evidence are needed to reach any conclusion.

16.3 Remote sensing using broadband VLF waves

Continuous monitoring and analysis of the VLF waves (such as radio atmospheric, tweeks, whistlers, VLF emissions) yield information about the height of the Earth ionosphere waveguide, equatorial electron density profiles, electron/ion temperature [62], convective electric fields in the magnetosphere [40], duct width, duct lifetime, irregularities in the electron density, LEP [41].

16.3.1 Tweeks and D-region ionospheric parameters

The VLF signals radiated by lightning discharges are called atmospherics or "sferics" in short. These VLF signals travel through the EIWG by the multiple reflections, which are formed by the Earth and the lower ionosphere. Tweeks are formed mainly during night hours, as lightning-generated VLF waves propagate to long-distance in the EIWG with low attenuation [42]. The D region of the ionosphere has been studied using the tweeks analysis. A typical example of higher harmonic tweeks recorded at low latitude station Varanasi on April 16, 2011 is shown in Fig. 16.6.

The cut-off frequency (f_c) of the tweek is measured and used to compute the ionosphere reflection height (h) using the equation $h = c/(2 f_c)$ [43], where "c" is the velocity of light in free space. The ionospheric electron density (n_e) at the reflection height (h) is determined by using the equation $n_e(\text{cm}^{-3}) = 1.365 \times 10^{-2} f_c$. After analysis of various tweeks recorded at low latitudes, the electron density was computed in the range of 20–28 cm^{-3} at the ionospheric reflection height of 80–95 km [2,3]. Tweeks have also been used to estimate their propagation distance along the propagation path which varies between 1000 and 5000 km [42].

16.3.2 Whistlers and ionospheric/magnetospheric parameters

VLF waves under certain conditions penetrate the ionosphere and propagate along geomagnetic field lines from one hemisphere to the other without appreciable attenuation. This mode of propagation is called the whistler-mode and the waves are termed whistlers [44]. Whistlers are right-hand polarized waves having wave frequency smaller than the electron gyrofrequency and electron plasma frequency [44]. The whistler waves propagate through the ionosphere, plasmasphere, and deep into the magnetosphere.

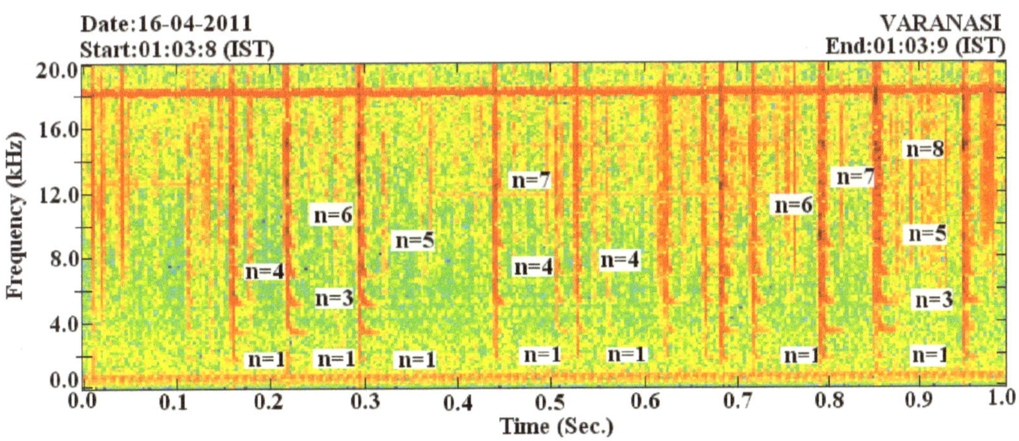

Fig. 16.6 ***Typical examples of Tweeks of a different order of mode recorded at Varanasi.***

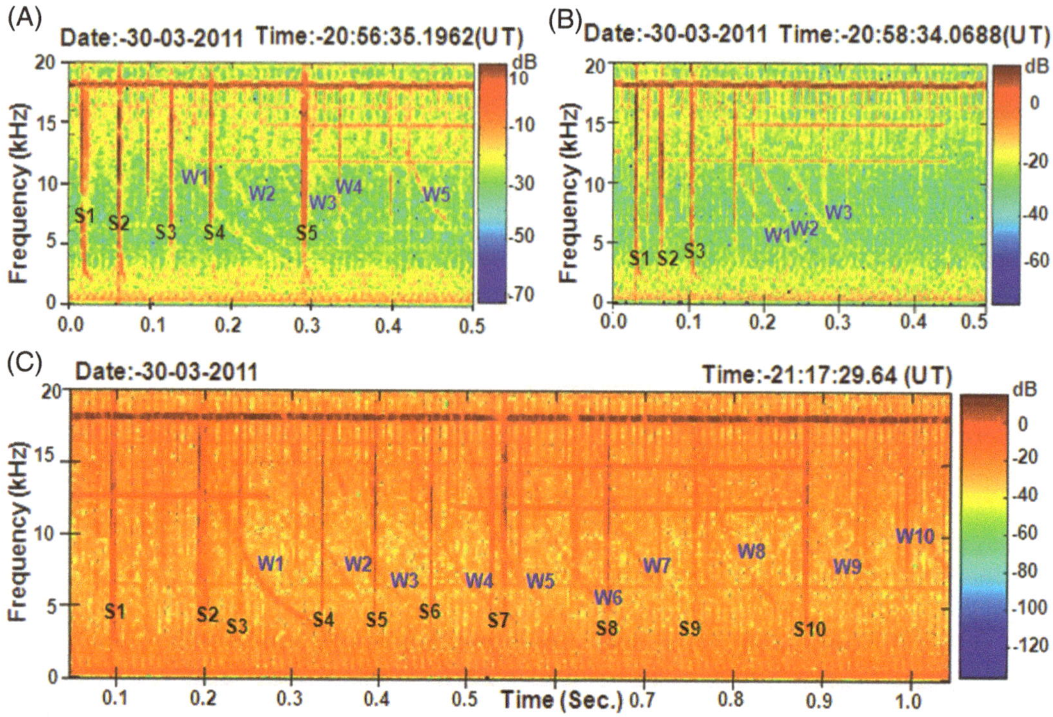

Fig. 16.7 ***Different types of whistlers were observed at Varanasi station on March 30, 2011.***

During their propagation through the ionized medium, it is dispersed, that is, different frequencies travel with different speeds and a typical whistler spectrum is observed. The whistlers spectrogram (i.e., frequency versus time diagram) is controlled by the distribution of electron density and the magnetic field along the path of propagation. Whistlers are generally a high- and mid-latitude phenomenon and are rarely observed at low-latitude ground stations when favorable conditions for whistler occurrence exist [5,6]. Typical examples of various types of whistlers recorded at low latitude station Varanasi on March 30, 2011 are shown in Fig. 16.7. The whistlers are marked by W1, W2, W3, etc., and the causative lightning sferics are marked by S1, S2, S3, etc. Most of the recorded whistlers are accompanied by associated causative lightning sferics.

Various workers [6,45] reported that whistlers occurrence at low latitude Indian stations is low and sporadic. It is also reported that the whistler's occurrence rate is maximum from January to April when the lightning activity is highest in the magnetic conjugate hemisphere.

The dispersion D of the whistlers is given by $D = t f^{1/2}$, where t is the time delay of the signal at frequency f. The dispersion is dependent upon the electron density and the McIlwain parameter L-value of the duct. Various studies at low latitude Indian ground stations have reported the dispersion of the whistlers in the range from 8 $s^{1/2}$ to70 $s^{1/2}$

([5,6] and references therein). The dispersion analysis of recorded whistlers yields information about the ionospheric/plasmaspheric/magnetospheric parameters such as electron density and total electron content of a flux tube [5,6,63], and electron temperature (Sazhin et al., 1992). Large-scale electric fields present in the magnetosphere are also estimated during continuous whistlers recording if a change in nose frequency with time is observed [46]. Singh et al. [6] reported that the ionospheric columnar electron contents of the observed whistlers at Varanasi vary between 13.21 TECU and 56.57 TECU. The average duct lifetime is found to be 30 minutes to 2 hours, while duct lifetimes as long as 1–2 days have been also reported. Whistler traces have a diffuse character that provides information regarding duct width (~100 km). The ionospheric/plasmaspheric/magnetospheric parameters derived from whistler measurements compare well with direct rocket and satellite measurements (Singh et al., 1998). The whistler's diagnostic technique is comparatively low cost and technically easy which also provides long-term and continuous variations of the various parameters.

16.3.3 VLF emissions

VLF emissions are whistler-mode waves that have complex structured as well as unstructured dynamic spectra. Unstructured emissions have a band spectrum like hiss emissions. Singh et al. [47] has reported that hiss emissions may be recorded at repeated time intervals in the form of pulsing hiss. Structured emissions show discrete characteristics on the dynamic spectrum such as chorus, periodic emissions, and quasiperiodic emissions [44]. Simultaneous measurements of VLF waves and charged particles have shown that the energetic electrons in the magnetosphere are the main source of energy for VLF emissions [48]. Depending on the different spectral shapes, VLF emissions are classified into the following characteristic types such as:

(a) Hiss emission.
(b) Chorus emissions.
(c) Periodic and quasiperiodic emissions.
(d) Triggered emissions.

Fig. 16.8 shows the typical examples of different types of VLF emission observed at low latitude Indian stations.

(a) Hiss emissions: The spectral form of hiss emission is a nondispersive having a limited band. The dynamic spectrum shows that hiss emissions can be classified as continuous hiss and banded hiss. The generation and propagation of hiss are controlled by magnetospheric conditions. It is supposed that whistler-mode waves propagating back and forth along geomagnetic field lines sometimes undergo interference and diffraction processes which may produce a hiss emission.

(b) Chorus emissions: Chorus emissions are VLF emissions that occur in the frequency range from a few hundred Hz to a few kHz and have complex structures. They were generated due to plasma instabilities in the Earth's magnetosphere [44]. Using cluster

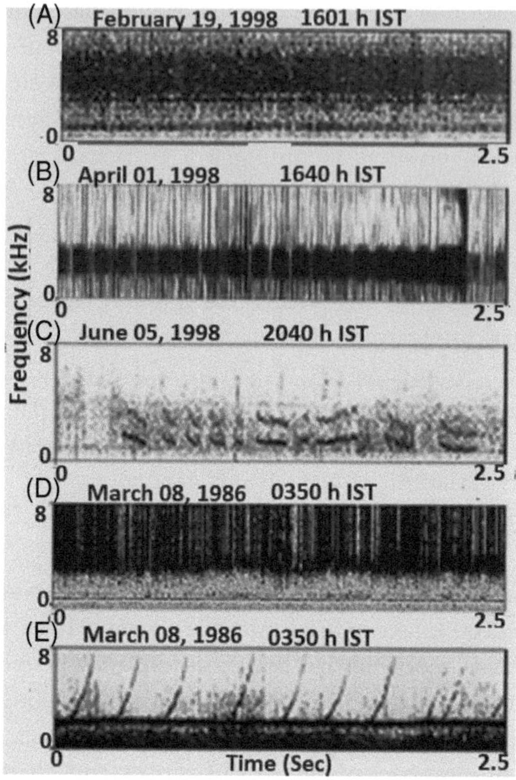

Fig. 16.8 Typical examples of various types of VLF emissions recorded at low latitude Indian Stations (A) hiss, (B) pulsing hiss, (C) hissler, (D) chorus, and (E) riser emissions.

satellite measurements, Pickett et al. [49] have reported different types of risers, fallers, and hooks in the 1.5–3.5 kHz frequency range. Chorus is also noticed at low latitude Indian ground stations [45]. A rare observation of hisslers was reported by Singh et al. [45] at low latitude Indian ground station. Chorus emissions are important because it is believed that these waves accelerate electrons outside the plasmasphere. The concept that electrons are accelerated by chorus emissions during wave-particle interactions is supported by flat top pitch angle distributions [50], and energy dependence in particle spectrum [51].

(c) Periodic and quasiperiodic emissions: VLF emissions that occur on a regular or semiregular basis both on the ground and on-board satellites at regular intervals of a few seconds are known as periodic emissions. They are divided into two categories: dispersive and nondispersive. Between the bursts, the period varies regularly with frequency in the dispersive type, but there is little or no discernible systematic variation in period with frequency in the nondispersive type [44]. VLF emissions with periods of

20–50 seconds are referred to as quasiperiodic emissions (QP) [44]. Smith et al. (1998) studied data from the South Pole and Halley and discovered a strong link between periodic emissions (PE) and the presence of QP at the same time.

(a) Triggered emissions: Triggered emissions are VLF emissions having complex types of dynamic spectral shapes which follow a source either a whistler [52], discrete emissions [53], VLF transmitter signal [53], the upper-frequency boundary of hiss [54], or power line radiation. The triggering cause may even be unnoticeable sometimes. The triggering phenomenon has been explained by several nonlinear processes [55].

16.4 Summary

In this chapter, we have presented the narrowband and broadband VLF waves and their probing potentials for remote sensing the upper atmosphere. We have discussed various geophysical and solar events that influence the D-region of the ionosphere from above and below. We have explained how the VLF navigational transmitter signals form a cost-effective tool for continuous monitoring of the D-region affected due to these events. A detailed discussion followed by serval examples shows how these events create variability in the low latitude D-region of the ionosphere. The long-range communication with precession and accuracy of modern-day navigation depends on the state of the D-region which is influenced by both terrestrial and solar sources. The celestial phenomenon of the solar eclipse also produces a significant change in the D-region. Further, it is also found that pre-EQ signals can affect the D-region and VLF waves can be effectively used to detect preseismic anomalies. Thus, a dense network of VLF transmitters and receivers will be helpful to understand and monitor D-region conditions continuously.

Lightning-generated broadband VLF waves like tweeks, whistlers, and VLF emissions were observed over low-latitude and their propagation and generation mechanisms have been discussed. Analysis of tweeks recorded at low latitudes provided the electron density of lower ionosphere in the range of 20–28 cm^{-3} at the ionospheric reflection height of 80–95 km. The propagation distance along the propagation path varies between 1000 and 5000 km. The analysis of the whistlers computes various propagation and ionospheric/plasmaspheric parameters. Various studies at low latitude Indian ground stations have reported the dispersion of the whistlers in the range from $8s^{1/2}$ to $70s^{1/2}$. The dispersion analysis shows that the observed whistlers have propagated in ducted mode and after exiting from the duct they penetrate the ionosphere and receive at Indian low latitude stations. The ionospheric/plasmaspheric/magnetospheric parameters derived from whistler measurements compare well with direct rocket and satellite measurements. Thus, VLF waves are a comparatively low-cost and technically easy remote sensing tool to study upper atmospheric parameters which also provide long-term and continuous variations of the various parameters.

16.5 Recommendations

At present several empirical models have been developed to explain ionospheric variability, but the D-region ionosphere especially at the low latitude remains the least studied and no model data can predict its variability accurately. This is mostly because of the absence of D-region data which can be input to various models and simulations for comparisons. The VLF navigational transmitters are very limited, and the receivers are sparsely located and thus provide very sparse coverage of D-region parameters. Due to this, the effect of geophysical and solar events could not be correctly quantified. Thus, it is suggested and recommended to build a dense network ofVLF transmitter-VLF receivers across the low latitude region. Further, a focused and dedicated model/simulation must be developed for the low latitude D-region ionosphere and upper atmosphere.

Acknowledgments

This work is supported by the Science and Education Research Board financial support under Ramanujan Fellowship (File No SB/S2/RJN-052/2016) and Faculty Recharge Program (FRP) of the University Grant Commission (UGC) (ID FRP62343), New Delhi. The work is partially supported by SERB, New Delhi, under the CRG project (File No: CRG/2021/001322) and (File No: CRG/2019/000573). One of the authors S.B. Singh is thankful to the DST, New Delhi, for DST-PURSE (5050) grant for providing Research Associateship. The authors would like to extend gratitude to the Director of the Indian Institute of Geomagnetism for providing necessary support and encouragement.

References

[1] R. Barr, D.Li. Jones, C.J. Rodger, ELF and VLF radio waves, J. Atmos. Sol.-Terr. Phys. 62 (2000) 1689–1718.

[2] A.K. Maurya, R. Singh, B. Veenadhari, S. Kumar, M.B. Cohen, R. Selvakumaran, P. Pant, A.K. Singh, D. Siingh, U.S. Inan, Morphological features of tweeks and nighttime D region ionosphere at tweek reflection height from the observations in the low-latitude Indian sector, J. Geophys. Res. 117 (2012) A05301.

[3] A.K. Maurya, B. Veenadhari, R. Singh, S. Kumar, M.B. Cohen, R. Selvakumaran, S. Gokani, P. Pant, A.K. Singh, U.S. Inan, Nighttime D region electron density measurements from ELF-VLF tweek radio atmospherics recorded at low latitudes, J. Geophys. Res. 117 (2012) A11308.

[4] S.A. Gokani, R. Singh, M.B. Cohen, S. Kumar, K. Venkatesham, A.K. Maurya, R. Selvakumaran, J. Lichtenberger, Very low latitude (L = 1.08) whistlers and correlation with lightning activity, J. Geophys. Res. Space Physics 120 (2015) 6694–6706.

[5] R. Singh, M.B. Cohen, A.K. Maurya, B. Veenadhari, S. Kumar, P. Pant, R.K. Said, U.S. Inan, Very low latitude (L=1.08) whistlers, Geophys. Res. Lett. 39 (2012) L23102.

[6] A.K. Singh, S.B. Singh, R. Singh, S. Gokani, A.K. Singh, D. Siingh, J. Lichtenberger, Whistlers detect-edandanalyzedbyAutomaticWhistlerDetector (AWD) at low latitude Indian stations, J. Atmospheric Solar-Terrestrial Phys. 121 (2014) 221–228.

[7] B. Singh, R. Singh, R. Singh, Whistler triplets, bands, and fine structure observed in low latitude ground station, Geophys. Res. Lett. 24 (20) (1997) 2507–2510.

[8] R. Selvakumaran, A.K. Maurya, S.A. Gokani, B. Veenadhari, S. Kumar, K. Venkatesham, D.V. Phanikumar, A.K. Singh, R. Singh, Solar flares induced D-region ionospheric and geomagnetic perturbations in the Indian sector, J. Atmos. Sol. Terr. Phys. 123 (2015) 102–112.

[9] A.K. Maurya, R. Singh, B. Veenadhari, S. Kumar, A.K. Singh, Sub-ionospheric VLF perturbations associated with the 12 May 2008 M7.9 Sichuan earthquake, Nat. Hazards Earth Syst. Sci. 13 (2013) 2331–2336.
[10] A.K. Maurya, K. Venkatesham, P. Tiwari, K. Vijaykumar, R. Singh, A.K. Singh, D.S. Ramesh, The 25 April 2015Nepal Earthquake: investigation of precursor in VLF sub-ionospheric signal, J. Geophys. Res. Space Phys. 121 (2016) 10403–10416.
[11] H.Y. Fukunishi, Y. Takahashi, M. Kubota, K. Sakanoi, U.S. Inan, W.A. Lyons, Elves: lightning-induced transient luminous events in the lower ionosphere, Geophys. Res. Lett. 23 (1996) 2157–2160.
[12] M.B. Cohen, U.S. Inan, R.K. Said, T. Gjesteland, Geolocation of terrestrial gamma-ray flash source lightning, Geophys. Res. Lett. 37 (L02) (2010) 801, doi:10.1029/2009GL041753 rm.
[13] M.A. Clilverd, C.J. Rodger, R.T. Neil, J. Lichtenberger, P. Steinbach, P. Cannon, M.J. Angling, Total solar eclipse effects on VLF signals: observation and modeling, Radio Sci. 36 (2001) 773–788.
[14] J.R. Wait, The mode theory of VLF ionospheric propagation for finite ground conductivities, Proc. IRE 45 (1957) 760–768.
[15] K.J.W. Lynn, Propagation effects of very low-frequency radio waves: Proceedings of the 1st International Conference on Science with Very Low-Frequency Radio Waves—Theory and Observations, VLF waveguide propagation: The basics, AIP Conf. Proc. 1286 (1) (2010) 3–41.
[16] M. Yoshida, T. Yamauchi, T. Horie, M. Hayakawa, On the generation mechanism of terminator times in sub-ionospheric VLF/LF propagation and its possible application to seismogenic effects, Nat. Hazards Earth Syst. Sci. 8 (2008) 129–134.
[17] S. Kumar, A. Kumar, F. Menk, A.K. Maurya, B. Veenadhari, R. Singh, Response of the low latitude D-region ionosphere to extreme space weather event of 14–16 December 2006, J. Geophys. Res. Space Physics 120 (2015) 788–799.
[18] W.D. Gonzalez, J.A. Joselyn, Y. Kamide, H.W. Kroehl, G. Rostoker, B.T. Tsurutani, V.M. Vasyliunas, What is a magnetic storm? J. Geophys. Res. 99 (A4) (1994) 5771–5792.
[19] N.G. Kleimenova, O.V. Kozyreva, A.A. Rozhnoy, M.S. Soloveva, Variations in the VLF signal parameters on the Australia-Kamchatka radio path during magnetic storms, Geomagnetism Aeronomy 44 (2004) 354–361.
[20] J.W. King, J.L. Fooks, Long-lasting storm effects in the ionospheric D-region, J. Atmospheric Solar—Terrestrial Phys. 30 (4) (1968) 639–643.
[21] A.P. Mitra, Ionospheric Effects of Solar Flares, Springer, New York, NY, 1974.
[22] J.K. Hargreaves, The Solar-Terrestrial Environment, Cambridge University Press, New York, NY, 1992, https://doi.org/10.1017/CBO9780511628924.
[23] N.R. Thomson, C.J. Rodger, M.A. Clilverd, Large solar flares and their ionospheric D region enhancements, J. Geophysical Res. 110 (A9) (2005) 6306.
[24] D. Grubor, D. Sulic, V. Zigman, Influence of solar X-ray flares on the Earth-ionosphere waveguide, Serbian Astronomical J. 171 (2005) 29–35.
[25] J.R. Wait, K.P. Spies, Characteristics of the Earth-ionosphere waveguide for VLF radio waves, Tech. Note 300, Natl. Bur. of Stand., Boulder, CO, 1964.
[26] W.M. McRae, N.R. Thomson, Solar flare induced ionospheric D-region enhancements from VLF phase and amplitude observations, J. Atmos. Sol. Terr. Phys. 66 (2004) 77–87.
[27] R. Singh, B. Veenadhari, A. Maurya, M. Cohen, S. Kumar, R. Selvakumaran, P. Pant, A. Singh, U. Inan, D-region ionosphere response to the total solar eclipse of 22 July 2009 deduced from ELF-VLF tweek observations in the Indian sector, J. Geophys. Res. 116 (2011) A10301.
[28] L.G. Smith, Rocket observations of solar UV radiation during the eclipse of 7 March 1970, J. Atmos. Terr. Phys. 34 (4) (1972) 601–611.
[29] G. Chimonas, Internal gravity-wave motion induced in the Earth's atmosphere by a solar eclipse, J. Geophysical Res. 18 (1970) 5545–5551.
[30] S. Kumar, A. Kumar, A.K. Maurya, R. Singh, Changes in the D region associated with three recent solar eclipses in the South Pacific region, J. Geophys. Res. Space Physics 121 (2016) 5930–5943.
[31] K. Venkatesham, R. Singh, A.K. Maurya, A. Dube, S. Kumar, D.V Phanikumar, The 22 July 2009 total solar eclipse: modeling D region ionosphere using narrowbandVLF observations, J. Geophysical Res.: Space Phys. 124 (2019) 616–627.

[32] A.K. Maurya, D.V. Phanikumar, R. Singh, S. Kumar, B.Veenadhari, Y.-S. Kwak, et al., Low-mid latitude D region ionospheric perturbations associated with 22 July 2009 total solar eclipse: wave-like signatures inferred from VLF observations, J. Geophysical Res.: Space Phys. 119 (2014) 8512–8523.
[33] G.C. Chimonas, O. Hines, Atmospheric gravity waves induced by a solar eclipse, J. Geophysical Res. 75 (1970) 857–875.
[34] M. Hayakawa, Y. Kasahara, T. Nakamura, F. Muto, T. Horie, S. Maekawa, Y. Hobara, A.A. Rozhnoi, M. Solovieva, O.A. Molchanov, A statistical study on the correlation between lower ionospheric perturbations as seen by sub ionospheric VLF/LF propagation and earthquakes, J. Geophys. Res. 115 (2010) A09305.
[35] M.B. Gokhberg, I.L. Gufeld, A.A. Rozhnoi, V.F. Marenko, V.S. Yampolsky, E.A. Ponomarev, Study of seismic influence on the ionosphere by super long wave probing of the Earth-ionosphere waveguide, Phys. Earth Planet. Inter. 57 (1989) 64–67.
[36] M. Hayakawa, VLF radio sounding of ionospheric perturbations associated with earthquakes, Sensors 7 (2007) 1141–1158.
[37] M. Hayakawa, O.A. Molchanov, T. Ondoh, E. Kawai, On the precursory signature of Kobe earthquake in sub-ionospheric VLF propagation, J. Comm. Res. Lab. 43 (1996) 169–180.
[38] A.V. Shvets, On some aspects of monitoring of the lower ionosphere using tweek atmospherics, Paper 572 presented at 17th International Wroclaw Symposium on EMC, Wroclaw, Wroclaw Univ. of Technol., 2004 29 June to 1 July.
[39] O.A. Molchanov, M. Hayakawa, K. Miyaki, VLF/LF sounding of the lower ionosphere to study the role of atmospheric oscillations in the lithosphere-ionosphere coupling, Adv. Polar Upper Atmos. Res. 15 (2001) 146–158.
[40] D.L. Carpenter, K. Stone, J.C. Siren, T.L. Crystal, Magnetospheric electric field deduced from drifting whistler paths, J. Geophys. Res. 77 (1972) 2819–2836.
[41] U.S. Inan, S.A. Cummer, R.A. Marshall, A survey of ELF and VLF research on lightning-ionosphere interactions and causative discharges, J. Geophysical Res. 115 (2010) A00E36.
[42] A.K. Singh, R.P. Singh, Propagation features of higher harmonic tweeks at low latitude, Earth Moon Planet 73 (1996) (1996) 277–290.
[43] M. Yamashita, Propagation of tweek atmospherics, J. Atmos. Terr. Phys. 40 (1978) (1978) 155–156.
[44] R.A. Helliwell, Whistlers and Related Ionospheric Phenomena, Stanford University Press, Stanford, 1965.
[45] D.K. Singh, R.P. Singh, A.K. Kamra, The electrical environment of the Earth's atmosphere: a review, Space Sci. Rev. 113 (2004) 375–408.
[46] L.C. Bernard, A new nose extension method for whistlers, J. Atmos. Terres. Phys. 35 (1973) 871–880.
[47] A.K. Singh, R. Singh, K. Singh, R.P. Singh, An explanation of the observation of pulsing hiss at low latitude, Adv. Space Res. 41 (2008) 1695–1698.
[48] R.P. Singh, R.P. Patel, R. Singh, R.N. Singh, Lightning produced nitrogen oxides in the lower atmosphere-an overview, Indian J. Radio Space Phys. 34 (2005) 248–254.
[49] J.S. Pickett, O. Santolik, S.W. Kahler, A. Masson, M.L. Adrian, D.A. Gurnet, T.F. Bell, H. Laakso, M. Parrot, P. Decreau, A. Fazakerley, N. Cornilleau-Wehrlin, A. Balogh, M. Andre, Multi-point cluster observations of VLF risers, fallers and hooks at and near the plasmapause, Multiscale Processes Earth's Magnetosphere 178 (2004) 307–328.
[50] R.B. Horne, S.A. Glauert, R.M. Thorne, Resonant diffusion of radiation belt electrons by whistler-mode chorus, Geophys Res Lett 30 (9) (2003) 1493.
[51] D. Summers, C. Ma, N.P. Meredith, R.B. Horne, R.M. Thorne, D. Heynderickx, R.R. Anderson, Model of the energization of outer zone electrons by whistler-mode chorus during the October 9, 1990 geomagnetic storm, Geophys. Res. Lett. 29 (24) (2002) 2174.
[52] D. Nunn, A.J. Smith, Numerical simulation of whistler-triggered VLF emissions observed in Antarctica, J. Geophys. Res. 101 (1996) 5261–5277.
[53] T.F. Bell, J.P. Luette, U.S. Inan, ISEE-1 observation of VLF line radiation in the Earth's magnetosphere, J. Geophys. Res. 87 (1982) 3530–3536.
[54] A.K. Singh, K. Ronnmark, Generation mechanism for VLF chorus emissions observed at a low latitude ground station, Ann. Geophys. 22 (2004) 2067–2072.

[55] Y. Omura, D. Nunn, H. Matsumoto, M.J. Rycroft, A review of observational, theoretical and numerical studies of VLF triggered emissions, J. Atmos. Terr. Phys. 53 (1991) 351–368.

[56] W.B. Peter, U.S. Inan, A quantitative comparison of lightning-induced electron precipitation and vlf signal perturbations, J. Geophys. Res. 112 (A12212) (2007), doi:10.1029/2006JA012,165.

[57] A.K. Maurya, K. Venkatesham, R. Singh, S. Kumar, P. Tiwari, A.K. Singh, Effects of St. Patrick's Day geomagnetic storm of March 2015 and of June 2015 on low-equatorial D-region ionosphere, Journal of Geophysical Research-Space Physics 123 (2018) 6836–6850.

[58] W.B. Peter, M.W. Chevalier, U.S. Inan, Perturbations of midlatitude sub-ionospheric VLF signals associated with lower ionospheric disturbances during major geomagnetic storms, J. Geophys. Res. 111 (2006) A03301.

[59] B.T. Tsurutani, K. Shibata, SI. Akasofu, O. Mitsuo, A two-step scenario for both solar flares and magnetospheric substorms: Short duration energy storage, Earth Planet Sp 61 (2009) 555–559.

[60] P. Pant, Relation between VLF phase deviations and solar X-ray fluxes during solar flares, Astrophys. Space Sci. 209 (1993) 297–306.

[61] D.G. Morfitt, C.H. Shellman, MODESRCH: An improved computer program for obtaining ELF/VLF/LF propagation data, Tech. Rep. NOSC/FR 141, Nay. Ocean Syst. Cent. (1976) San Diego, Calif.

[62] S.S. Sazhin, A.J. Smith, E.M. Sazhina, Can magnetospheric electron temperature be inferred from whistler dispersion measurements? Ann. Geophys. 8 (1990) 273.

[63] S.S. Sazhin, H.F. Balmforth, R.J. Moffett, Y. Rippeth, Modified models of electron distribution in the magnetosphere at L = 2.3, Planetary and Space Science 40 (5) (1992) 671–679.

CHAPTER 17

Remote Sensing of Ionospheric Irregularities

Kalpana Patel
SRM Institute of Science and Technology, Delhi-NCR Campus, Modinagar, Ghaziabad, Uttar Pradesh, India

17.1 Introduction

Ionosphere is the ionized layer of Earth's upper atmosphere whose altitudinal extend is from 60 to 1000 km. The electron density in the ionosphere is not evenly distributed and it is observed that there is irregular distribution of electron density in the form of electron bubble and polar patches which is known has ionospheric irregularities [1,2]. During an irregularity electron density fluctuation occurs with respect to space and time which is known as "nonfrozen irregularities" and that which does not change with time is called as "frozen irregularities" [2]. Equatorial plasma instability which is caused by the effect of gravity, eastward electric field, and vertically downward neutral wind in the presence of vertically upward density gradient is a popular phenomenon [2–4]. The age and development of these inconsistencies are contemplated utilizing multisatellite signals and recording them at multi-stations [5,6]. There is a wide range of irregularities scaling from centimeters to tens of kilometers [1]. The irregularities are separated into various scale sizes [7]:

(i) Terrestrial scale (>1000 km).
(ii) Average scale (10–1000 km).
(iii) Intermediary scale (1–10 km).
(iv) Middle scale (10–100 m).
(v) Small scale (<10 m).

There are two regions in the whole globe where the irregularities are dominant:

(i) Equatorial irregularity

In the equatorial region, the solar radiations are dominant which produces enhance ionization in the daytime and just after sunset the different ions and electrons start to recombine due to which there is electron density depletion. Thus, in the equatorial region of the ionosphere plasma bubbles are formed just after the sunset known as equatorial plasma bubbles (EPBs) [4].

(ii) Polar patches

In the polar regions, 70°–80° north and south poles, the radiation belt plasma particles interact with the ionospheric constituents and give them energy to produce

Atmospheric Remote Sensing
DOI: https://doi.org/10.1016/B978-0-323-99262-6.00003-1

aurora and they are precipitated to the ionosphere which produces irregular distribution of electron density in small patches which is known as polar patches.

Depending upon the regions where they are generated, these irregularities are divided into two types: (1) F-region irregularities and (2) E-region irregularities.

17.1.1 F-region irregularities

The plasma irregularities generated in F-layer of ionosphere mostly just after sunset is known as spread F irregularity. These irregularities are generated due to plasma instability in the equatorial region mainly Rayleigh–Taylor instability (RTI) and are called as equatorial spread-F (ESF). A typical example of spread F irregularities is shown in Fig. 17.1 which shows the range–time intensity variation. The S/N ratio bar shows the proportion of desired signal level with respect to background noise level. The figure also

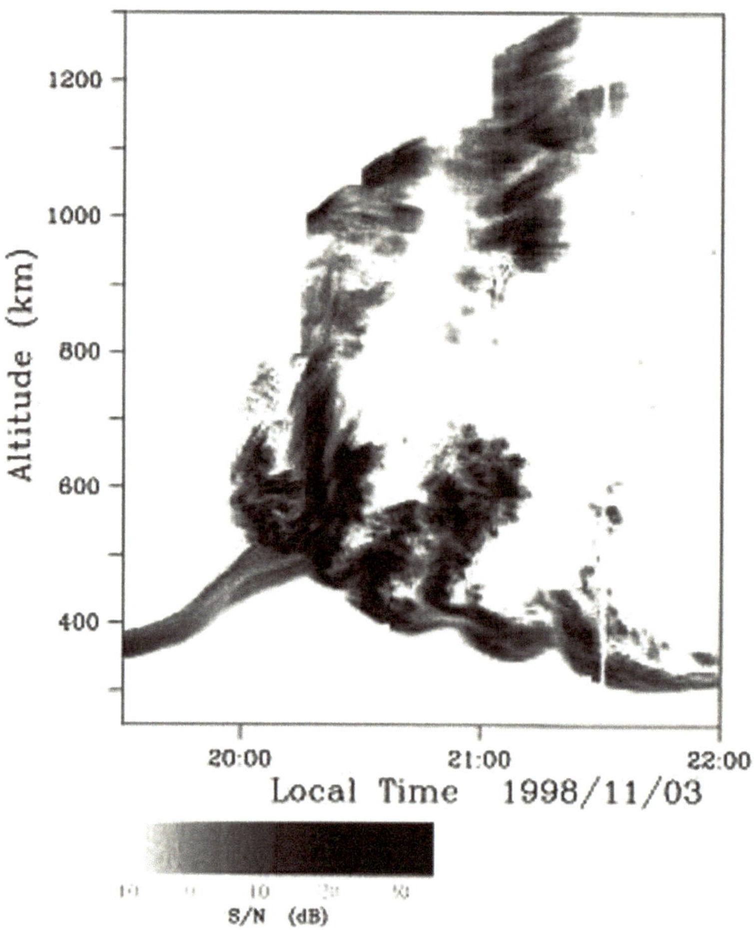

Fig. 17.1 ***A typical example of F-region irregularities showing the range–time intensity variation.***

shows that phenomena mainly take place during night hours and enhance temporally and spatially both.

ESF irregularities occurrence is due to the internal gravity waves which arrange the low and high layer plasma density regions with the same horizontal wavelength of gravity waves [1]. Due to such disturbances, the generalized R–T instability can take over and causes irregularities to occur and enhance [8]. ESF is categorized as "range spread" which is the spreading along the horizontal path of the phenomena developing an ambiguous range and "frequency spread" which is the spreading in the section of the trace that traverse upwards and creates ambiguity in calculating critical frequency. Basu and Kelley [09] used the ground–based and satellite data to give a sensible model of equatorial irregularities that can explain the classic features of equatorial scintillations. It was inferred that the irregularities basically reveal power law spectra advising a liquid turmoil-like behavior. Their results showed that discontinuity was due to the splitting-up of the sharp and slanting range extents. On the basis of simulations of various case studies, Ossakow et al. [10] showed the results that collisional Rayleigh–Taylor instability makes linear up-rise of the bottom side of the F-layer which creates the plasma depletions occurring due to the polarization of $\mathbf{E} \times \mathbf{B}$ motion. Yeh and Liu [11] evaluated the hypothesis of ionospheric scintillations depending on theoretical and experimental aspects with the help of measurements made through satellite, radar, and experiments. They inferred that huge instability in amplitude were noticeable because of the scattering and diffraction in close proximity of steep slopes obtained at GHz. Makela [12] studied about the low latitude discontinuity procedure and exposed that R–T instability (RTI) is because of the growth in the depleted emission area ranging from 50 km to number of 100 kms in the eastern and western directions. For the north–south orientation, the discontinuity ranges around 2000 kms when one moves from the equatorial line toward pole.

17.1.2 E-region irregularities

The irregularities that take place during daytime in the E-region of the ionosphere are known as E-region irregularities. The E- regions have isolated structures of ionization which has different types of shapes and sizes. They are called so because they are taking place quasirandomly on a day-to-day basis [13]. They do not have diurnal or seasonal variations. E-region irregularities are also called as Sporadic-E (E_s or SpE) irregularities. Sporadic-E is a transient, thick ionized slab that reflects radio waves in the range of 25 MHz to 225 MHz. Electron density profile changes due to the thick layer of electrons in the clouds of E_s [1]. Sporadic-E can be dense or thin which may appear as a blanket or transparent layer to the radio waves [14].

Makela [12] reported that EPBs are the outcome due to the enhancement of equatorial region irregularities under the state of RTI. EPBs are formed when there is no ionization due to Sun, which forms a low-density layer that rises up through the convection process through highly ionized layer. Plasma bubbles are plasma reductions in the ionosphere and are the significant source of electron density irregularities in the innermost F-region [15–17]. Kil et al. [18] explained the formation of a plasma depletion shell in the equatorial ionosphere.

Nowadays, it is important to study about the ionosphere and its irregularities because it affects the radio wave propagation to far-off places existing on Earth. This can be used to enhance communication and surveillance system for both civilian and military purpose. It is useful in improving the communication system, in monitoring and controlling the pseudo-range errors in GPS dual frequency system. Further, the amplitude fluctuations due to VHF waves traversing through the ionosphere is helpful in determining the ionospheric irregularities [19]. Through this chapter, one will come to know about the various techniques and instruments involved in the analysis of ionospheric irregularities through ground-based technique as very high frequency (VHF) and in-situ techniques including various instruments like global positioning system (GPS), Defense Meteorological Satellite Program (DMSP), Stretched Rohini Satellite Series, Communications/Navigation Outage Forecasting System (C/NOFS) and the morphology accompanying ionospheric irregularities as discussed in the chapter further.

17.2 Techniques to study ionospheric irregularities

17.2.1 Ground-based techniques

17.2.1.1 Ionospheric scintillations

Whenever the radio signal propagates through ionosphere, the irregular nature of it brings changes in the amplitude, phase, and angle of arrival of the signals at the receiver which is collectively known as scintillation [6,20]. The scintillation method is an important technique to study the characteristics of ionospheric irregularities. Scintillation phenomena can be studied with the help of three possible theories given [21]:

(1) Ray optics
(2) Scattering theory
(3) Diffraction theory.

It is further divided into the following types depending upon the phase screen scattering model:

(i) Shallow screen model.
(ii) Strong Screen model.

17.2.1.1.1 Very high frequency (VHF) scintillations

VHF wave is represented as a portion of the electromagnetic range consolidating radiation with a frequency in the range of 30–300 MHz. As we realize that when electromagnetic waves pass through a medium their phase, adequacy, and showing up point show changes and by looking at these boundaries, we can come to think about the medium qualities. Keeping this view, waves from the transmitter of VHF artificial satellite is utilized to study the ionospheric irregularities.

The VHF scintillation recording system consists of an 11-element Yagi-Uda antenna, a VHF receiver AKASH and a z-fold path paper strip recorder to record the radio signal transmitted from geostationary satellite [22]. The amplitude scintillations of the 250 MHz signal transmitted through geostationary orbiter FLEETSAT/UFO-02 parked at 72°E latitude were consistently observed at low latitude station Varanasi [22]. The recoded data were analyzed by the scintillation technique depicted by Basu and Basu [23] to study the different characteristics of ionospheric irregularities. Typical examples of various types of VHF amplitude scintillations recoded at low latitude station Varanasi is shown in Fig. 17.2.

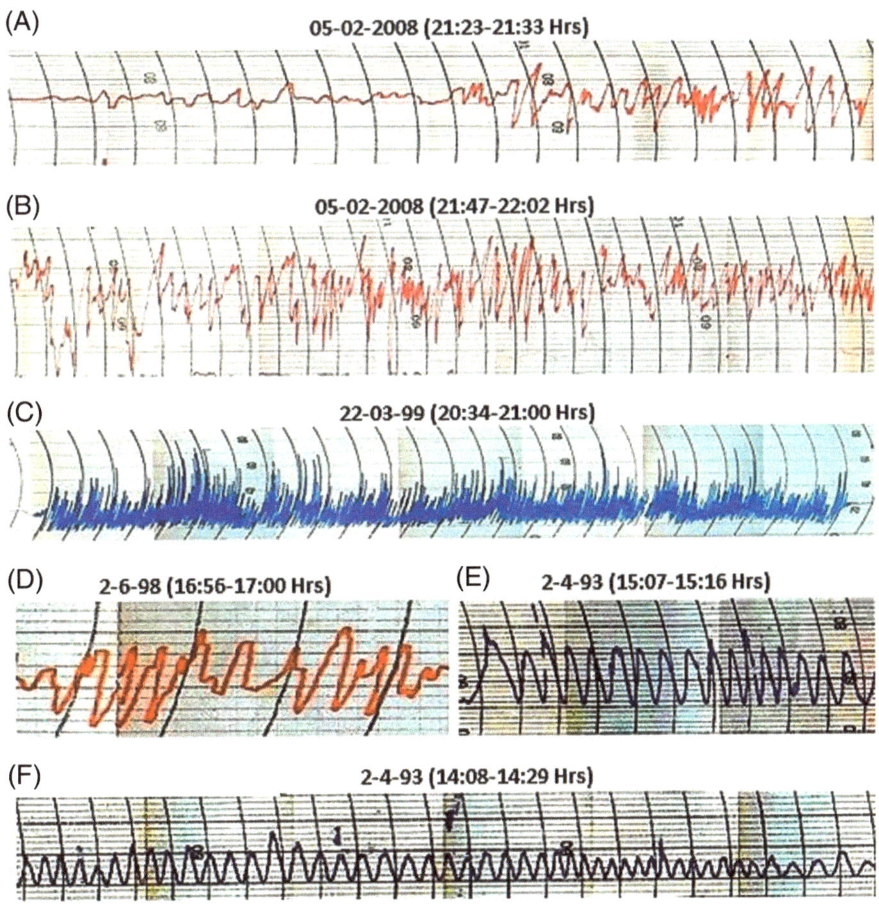

Fig. 17.2 ***Typical examples of various types of amplitude scintillation patches recorded at low latitude station Varanasi.***

From Fig. 17.2, it is clearly seen the structure of ionospheric irregularities observed over low latitude. The following different types of scintillation patches were recorded at low latitude:

a) Intense or saturation type scintillations with high fading rates. It occurred now and often before 12:00 am with maximum amplitude value as 14–16 dB (Fig. 17.2C).
b) Intense scintillation with low fading rate. They are obtained after 12:00 am hours with maximum amplitude as 13 dB.
c) Moderate scintillations with high fading rates. They are in general obtained before 12:00 am with maximum amplitude as 5–8 dB (Fig. 17.2B).
d) Moderate scintillations with low fading rates which occurred after 12:00 am hours with amplitude near to 5 dB (Fig. 17.2A and D).
e) Long duration scintillations of time span larger than 2 hours and majorly obtained after 12:00 am. They have low fading rates. (Fig. 17.2E and F).
f) Slowly building up scintillations with rare occurrence that took place at Varanasi station. It took 20 minutes to reach up to 13 dB of amplitude.

Apart from the above describes scintillation patches quasiperiodic scintillations have been also observed over low latitude station Varanasi which shows the structured structure of irregularities. Fig. 17.3 shows the typical examples of seven forms of quasiperiodic scintillation received at Varanasi; type 1: primary fade out followed by a ringing pattern and not ringing beforehand (Fig. 17.2A); type 2: primary fade out accompanied by short ringing before and long extended ringing afterward (Fig. 17.2B); type 3: primary fade out accompanied by extended ringing before and short ringing after (Fig. 17.2C); type 4: burst-like scintillations (Fig. 17.2D); type 5: valley-like scintillations (V-type) (Fig. 17.2E); type 6: spike-type scintillations with periodicity (Fig. 17.2G); and type 7: long duration ringing pattern (Fig. 17.2F). From these variants of quasiperiodic scintillation patches formed and recorded for station Varanasi it is observed that type 1 was of 2.7%, 4.8 % of type 2, type 3 was 4.3%, type 4 was of 33%. Valley type was of 25%, 7% was of spike type and 10% was of long extended form and 12% were identified because their variant was not defined. These patterns of quasiperiodic scintillations were obtained around 75% in night and 25% in day hours [1].

To analyze the recorded scintillation patches, the scintillation index S_4 has been computed which features out the level of signal fluctuation and gives the spectral feature of the ionospheric irregularity.

$$S_4^2 = \langle I^2 \rangle - \langle I \rangle^2 / \langle I \rangle^2 \tag{17.1}$$

Here, the signal strength, brackets <> show the commonness of the term. The S_4 has been assessed for each 60 seconds.

To compute the scale length of the irregularities, the autocorrelation function has been computed which informs us regarding the relationship of the amplitude inside a similar signal. This can be defined as:

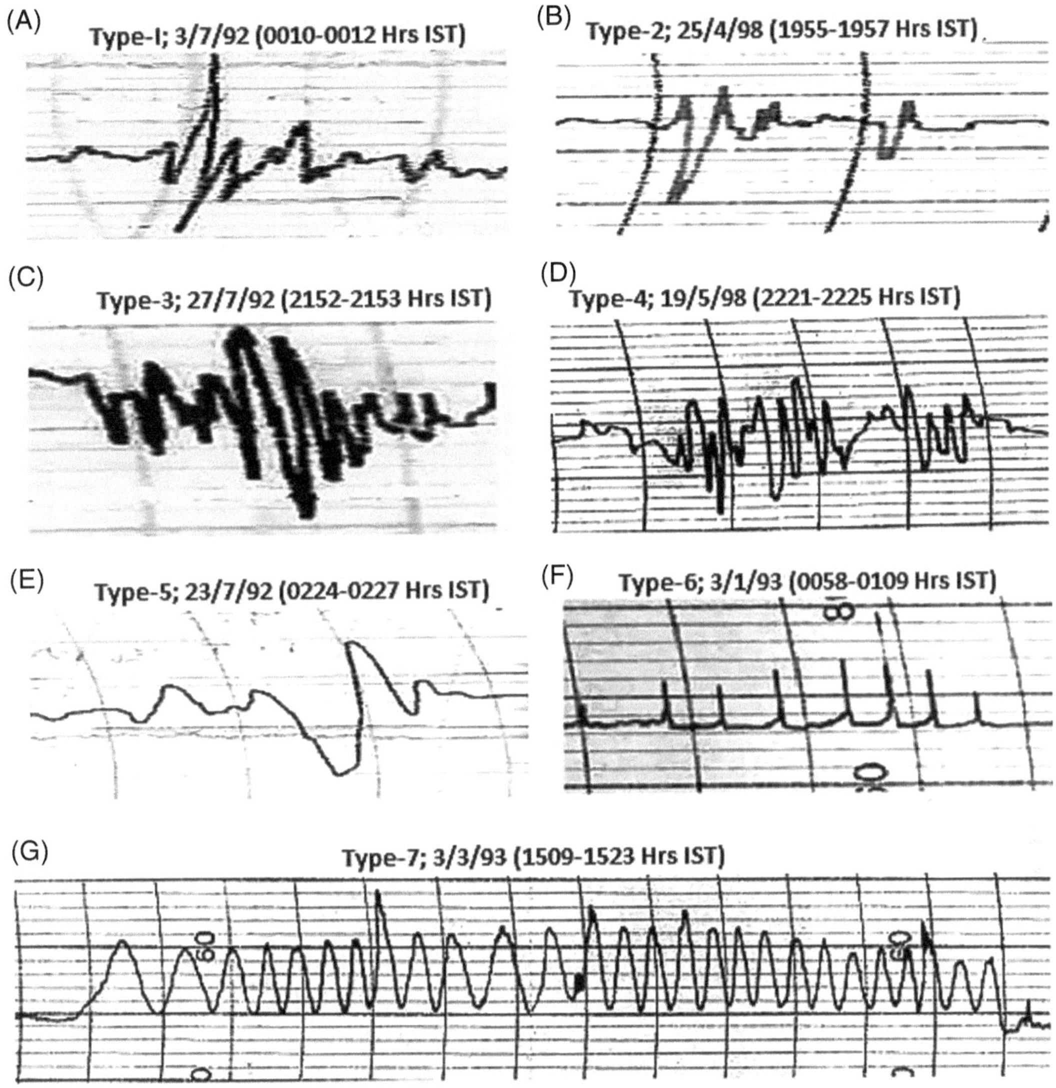

Fig. 17.3 ***Typical examples of the different forms of quasiperiodic scintillation patches obtained in Varanasi.***

$$\Gamma_k = \frac{\sum_{t=k+1}^{T} \left\{ \left(X_t - \bar{X}\right)\left(X_{t-k} - \bar{X}\right) \right\}}{\sum_{t=1}^{T} \left(X_t - \bar{X}\right)^2} \tag{17.2}$$

Here, X_t is the amplitude of the exact duration and $\bar{X}$, the normalness from the given information. With the assistance of an autocorrelation function plot, the size of the irregularity and the magnitude of the ionospheric irregularities can be computed. The size of the irregularity is equivalent to the length at which the autoconnection function fall downs to 0.5 [24], when the abnormality is moving with a known float speed. The period when autocorrelation function tumbles to 0.5 is called half de-correlation time (τ). The power spectrum of an irregularity is obtained from beginning to end of analysis of the advanced scintillation information. The horizontal speed of the scintillation producing ionospheric irregularities depending upon the phase-changing screen model is given as [25]:

$$V = (\lambda z)^{1/2} f_{min} \tag{17.3}$$

where z is the height of irregularity phase screen, λ is the signal frequency, and f_{min} is the primary Fresnel minima in the given spectra [19,20].

17.2.1.1.2 GPS scintillation

The GPS is the new advance technique of global navigation satellite systems, which has improved the study of ionospheric exploration [26]. The GPS is a mainly used for positioning and navigation. It radiates radio signals of frequency L1= 1.57542 GHz and L2 = 1.2276 GHz (~20 cm wavelength) which is used for ionospheric studies. As GPS signal passes through the atmosphere, there is delay in its signal. This delay is due to the presence of ions end electrons in the ionosphere and water vapor in troposphere [27]. The analysis of the GPS signal delay provides information about the total electron contents (TECs) of the ionosphere as well as tropospheric water vapor contents [28]. The GPS is also used for L-band scintillation study. Rathore et al. [29] studied the diurnal and seasonal variations of ionospheric irregularities from 2009 to 2014 after analyzing the amplitude scintillation index S_4 calculated from the dual frequency receiver installed at the low-latitude station Varanasi (Lat. 25.31° N, Long. 82.97° E). Aol et al. [30] compared the ionospheric irregularities and scintillations obtained from in-situ data and ground-based L- band scintillations from the year 2014–2018.

17.2.2 Satellite-based techniques

17.2.2.1 Defense meteorological satellite program (DMSP)

DMSP spacecraft are tri-axis unbiased shuttle that flutters circularly, solar-simultaneous glacial rings (tendency 98.7°) at a height of ~840 km [31,32]. Every spacecraft conveys a set-up containing detectors called the special sensor ions, electrons, and scintillation (SSIES) for quantifying the thickness, warmth, and glide movements within the area of particles and electrons [32]. SSIES contains a Globular Langmuir probe build upon a 0.8 m blast for estimating thickness and warmth of surrounding electrons, alongside three detached detectors accumulated upon leading disc looking for ram route. The detectors are (1) a particle lock-in for estimating complete particle thickness, (2) a particle glide scale for quantifying level (V_H) and vertical (V_V) traverse-pathway section of plasma glide, (3) a delaying/retarding potential analyzer (RPA) for estimating particle temperature and in-track segments of plasma float $V_{\|}$ [33].

The trajectory period of DMSP satellites is around ~104 min permit 14 rounds for every date in a normal way. In between the trajectory-increasing hubs, DMSP spacecraft relapse ~25° in longitude. The DMSP provides information into 24 longitude bins, each 15° in the span [34]. The in-situ plasma thickness estimations from the SSIES instruments on DMSP space apparatus has been analyzed to study the occurrence characteristics of EPBs events [34].

17.2.2.2 Stretched Rohini satellite series (SROSS-C2)

SROSS-C2 was the second satellite effectively circled by launch vehicle as Augmented Satellite Launch Vehicle started on 04/05/1994 at SHAR, Sri Harikota, India. It was utilized for the correspondence with the band as S_0 band and VHF band. It has two payloads as gamma-ray bursts and RPA. It was put in the circle of 430 × 600 km at a leaning point of 45^0. To examine the unique highlights of the ionospheric irregularities (plasma bubble) over an area, the electron density is recorded by RPA probe of India, near to the ground coursing spacecraft SROSS-C2 is utilized [35]. The SROSS-C2 instrument comprises a primary and repetitive scintillation working in the energy range of 20–3000 keV. The width of primary crystals was 76 mm plus that of excess was 37 mm. Each was of thickness 12.5 mm. The RPA payload comprises of two sensors, namely, electron and particle sensors and related hardware [35]. Next to this, a Langmuir test is incorporated and utilized as a test for assessing the variety of shuttle potential as the satellite twists. The electron and particle sensors both have planar calculation and contain multiframe Faraday cups with a collector cathode.

Satellite in-situ estimations of plasma thickness changes give the data close to the construction and morphology of anomalies which are liable for scintillations of broadcasting waves across the ionosphere linkages [36]. The chosen datasets of ion thickness differing with time (second) are used for computation of scintillation index (S_4), disturbance strength (C_s), phase/period change ($\delta\phi$) [36]. Following the methodology given by Kill and Heelis [37], the general abnormality amplitudes ($\Delta N/N$) have been scaled. An example of 6 min (1371-1357 UT) dataset showing (1) the intense irregular pattern of the ion thickness, (2) relative irregularity amplitude ($\Delta N/N$), is less than equal to 5%, (3) power spectrum with 2.84 as gradient, and (4) the autocorrelation function obtained on October 29, 1994 with SROSS-C2 path has been shown in Fig. 17.4. The power spectrum tells one to get the variation of fluctuation in electron density required to get the power law index describing irregularities.

17.2.2.3 Communication navigation outage forecasting system (C/NOFS)

The aim of the Communication/Navigation Outage Forecasting System (C/NOFS) is of determining and figuring the tropical ionosphere, including spread *F* and ionospheric scintillations [38,39]. C/NOFS contains six instruments mentioned as follows.

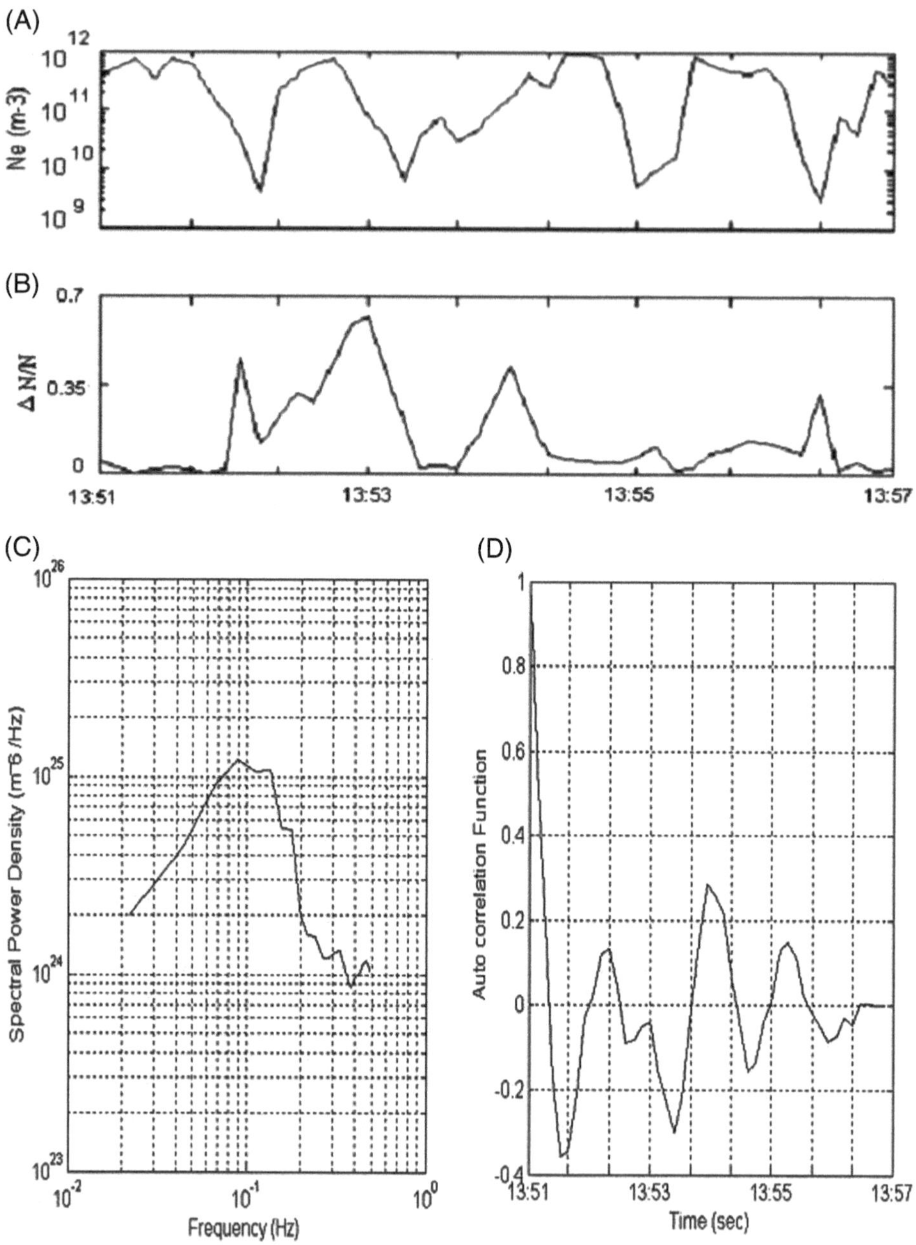

Fig. 17.4 ***On October 29, 1994, a typical 6 minutes ion density data segment obtained from the SROSS-C2 satellite.*** (A) The intense irregular pattern of the ion thickness as Y-axis with time in UT on Y-axis, (B) relative irregularity amplitude on Y-axis and time in UT on Y-axis, (C) power spectrum (m^{-6}/Hz) on Y-axis and frequency (Hz) as X-axis, (D) the autocorrelation function on Y-axis with time (seconds, UT) on X-axis.

17.2.2.3.1 The planar Langmuir probe (PLP)

PLP screens plasma thickness plus electron warmth present in the ionosphere. It gives the proportion of surrounding plasma just like sky-scraping instance-aim thickness abnormality estimations.

17.2.2.3.2 The ion velocity meter (IVM)

IVM involves two detectors, a particle glide scale plus RPA. IVM straightforwardly estimates particle glide vector, particle warmth plus significant particle structure, in the company of the spatial aim of around 4 km all along with spacecraft tracking. Particle glide meter gives estimations for nearby upright plus even particle glide segments through the spatial aim of 500 m.

17.2.2.3.3 The neutral wind meter (NWM)

NWM straightforwardly estimates impartial breeze vector through the spatial aim of around 8 km next to spacecraft tracking. Estimations must be made over a confined height array among perigees and around 450 Km elevation because the impartial environment should be adequately thick.

17.2.2.3.4 The vector electric field instrument (VEFI)

VEFI comprises tri-symmetrical 20 m tip-to-tip twofold test receiving wires.VEFI estimates DC plus AC field of force. A flux valve gauss meter, ocular beaming finder plus stabilized-angle Langmuir tester completes the VEFI apparatus bundle.

17.2.2.3.5 The coherent electromagnetic radio tomography (CERTO)

CERTO is a three-group (150 MHz, 400 MHz, and 1067 MHz) radio guide. CERTO is used within multitudinous-recurrence scintillation that concentrates through a leveled receiver in the direction of determining plasma forms among the area of C/NOFS plus Earth. This is used in imaging the reproduction of electron thickness outlines.

17.2.2.3.6 The C/NOFS occultation receiver for ionospheric sensing and specification (CORISS)

CORISS is the GPS double recurrence recipient. CORISS quantifies TEC in the line of scene amid C/NOFS plus GPS satellites. Attachment outlines of TEC can be rearranged in creating upright outlines for electron thickness.

Using the above instruments on board C/NOFS satellite observations of various characteristics of plasma density and electric field irregularities have been reported [40].

17.3 Characteristics of ionospheric irregularities over Indian region

17.3.1 Occurrence characteristics

The percentage of events of irregularities is highest in the low temperature (winter season) season, moderate in the equinoctial months, and lowest in the summer months [8].

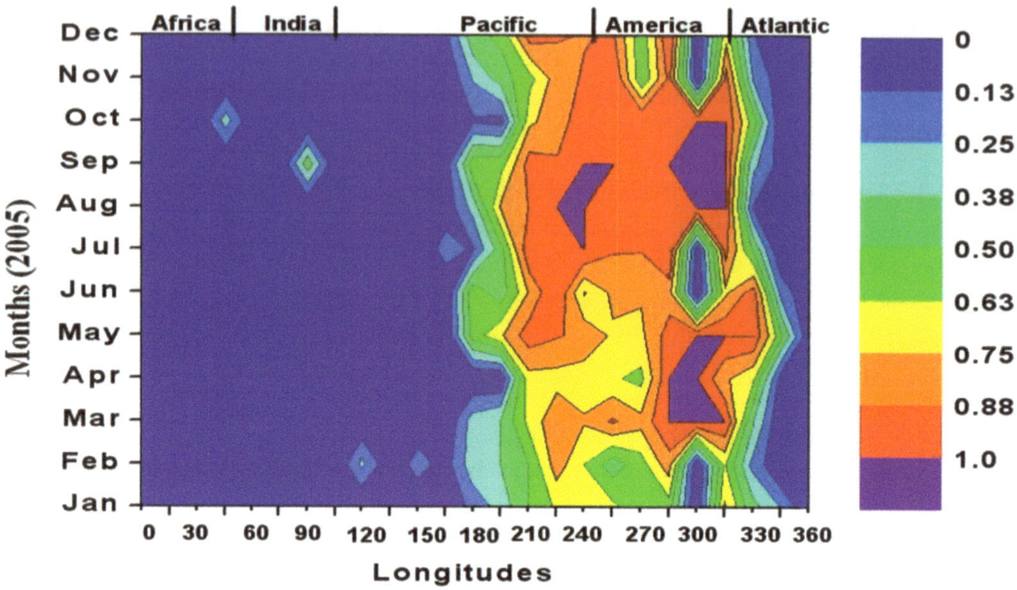

Fig. 17.5 ***Equatorial plasma bubble rates on a contour plot for the solar minimum year of 2005.*** The outline pattern, X-axis has longitudinal value and Y-axis shows months of the year 2005.

Yearly, at low latitude the maximum occurrence is around 1545 hours (Indian Standard Time) (IST) due to the E_s sheet present in the late afternoon hours [41]. At Varanasi, the percentage occurrence of daytime scintillation is phenomena less than that of the nighttime occurrence [1]. Regardless to solar activity, the maximum occurrence of irregularities was reported in equinox months and thereafter in winter with fewer occurrences in summer [42]. At Varanasi, the scintillation patch duration was around < 20 minutes [1]. DasGupta and Kersley [43] reported around 45% of scintillation patches with the time ranging to be 15 minutes or less. During winter and equinox months the scintillation were of intense nature before midnight [44,45] and that in summer was of weak and slow nature and with short interval of time [46,47]. Related to quasiperiodic scintillation, showed maximum occurrence in the nighttime at the station Varanasi due the major recordings done in the nighttime only [22]. Seasonally, the summer months have shown the maximum occurrence than in winter for Varanasi station [48]. From the DMSP data analysis, Fig. 17.5 shows the seasonal changes along with longitudinal bins for occurrence of EPBs. Longitude ranges from 0^0 to 360^0 which was divided in 24 bins with 15^0 bins. For the easiness, the globe was divided in various sectors as African, American, Indian, Pacific, and Atlantic Oceans colors in the color bar show the EPB rates. Largest EPB rates in the American–Atlantic region are seen during March–April and August–October, indicating both seasonal and longitudinal impacts. The Indian-Pacific region had the lowest rates of EPB. Less confined hot spots are there in the Indian–Pacific area [49]. The distribution of GPS scintillation index S_4 occurrence characteristics during 2009–2014

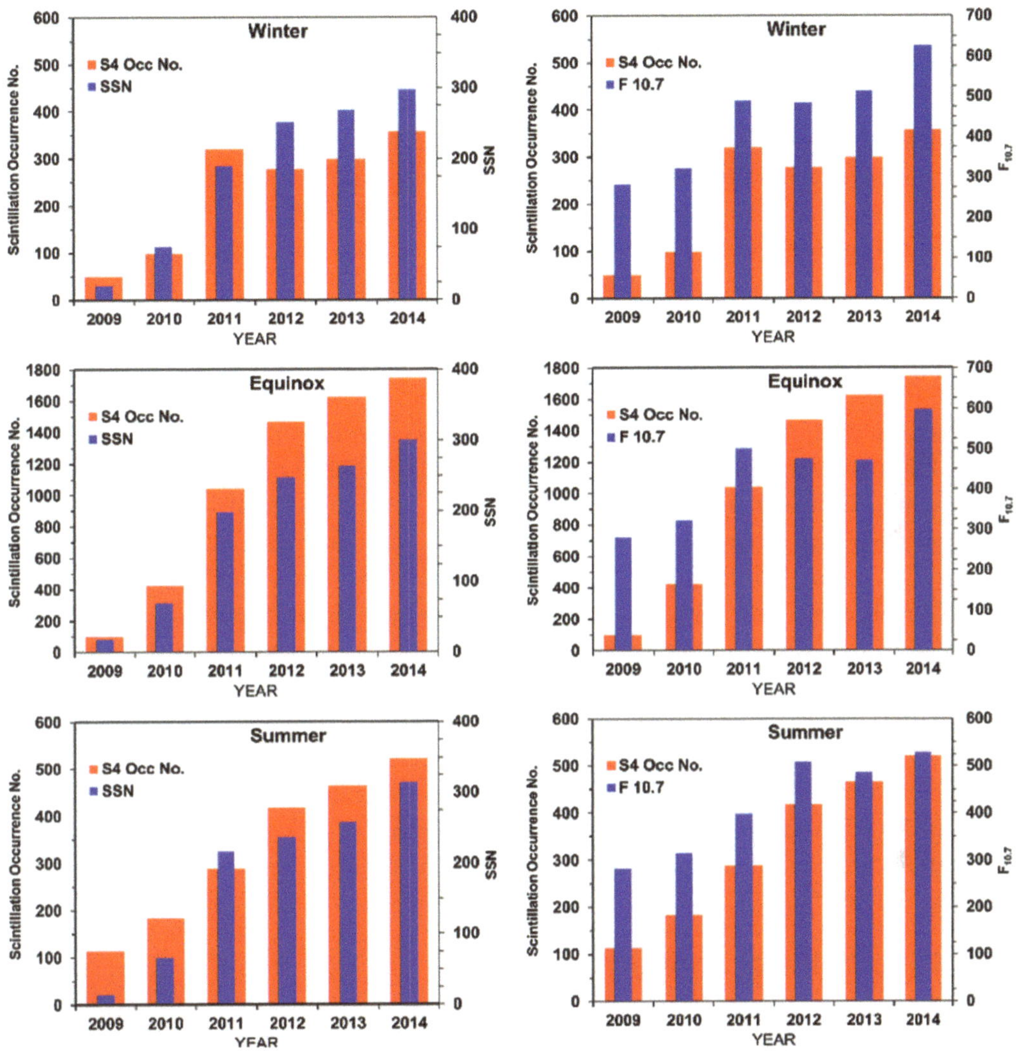

Fig. 17.6 ***The left panel shows the seasonal scintillation occurrence number (S_4) represented as red bars, smoothed sunspot number (SSN) with blue bars.*** The right panel of the graph shows variation of seasonal scintillation occurrence number (S_4) represented as red bars, solar flux (F10.7) with blue bars during the ascending period of solar cycle 24 from 2009 to 2014 [31].

over low latitude station Varanasi is shown in Fig. 17.6. From the C/NOFS data analysis, Ref. [50] informed that in equinox months, the EPBs were obtained for about 12 hours or longer and latitude wise covers Asian, African, and Atlantic sectors at around 1900 hours IST. On the other hand almost throughout nighttime 12,000 km longitudinally was covered. The ground-based amplitude scintillations were improved for huge counting's of Swarm obtained ionospheric irregularities. Obtained amplitude scintillation was low for Swarm B than for A and C and less in agreement with the examined range [32].

17.3.2 Scale size

The temporal scale length of the irregularities obtained from VHF at low latitude is around 100–1100 m for 75 m/s speed and around 400 to 3000 m for 200 m/s speed. This shows that it is of intermediate scale length [1,48,51]. For quasiperiodic scintillation, the computed and obtained values for different parameters are in good agreement. The scale size for E- and F-region irregularities is of intermediate length with values as 200 and 650 m for drift speed of 100 m/s [52]. The scale size of E_s irregularities over low latitude is reported at around 300 m [53]. Somayajulu et al. [54] reported the scale size ranging between 800 and 924 m for E-region irregularities.

17.3.3 Spectral analysis

The estimated scintillation index S_4 for Varanasi, ranges between 0.2 and 0.9, which shows that irregularities are from weak to sensible level [52]. The spectral index for the low latiyude Indian station Varanasi was reported between -2 and -9 with the mean value as -4 [48,52]. From the in-situ measurements, the obtained spectral indices ranges between -1.7 and -5 at frequencies more than 60 Hz and less than 60 Hz, respectively [55]. The observed and computed spectral parameters for quasiperiodic scintillation observed at low latitude is tabulated in Table 17.1 [52]. It is clearly seen that the computed spectral index varies between -1.3 and -8.7 and Scintillation index (S_4) between 0.4 to 0.8, scale size between 200 and 650 m and half decorrelation time between 2.0 and 6.4 seconds which is found to be in good agreement with the obersed parameters [1,52]. From the SROSS-C2 data analysis, power spectrum with spectral slope calculated as 2.84. Turbulence strength (C_s) increase with the increase in electron density fluctuations and decreases with the increase of *Lo* (outer scale of irregularity) and υ (spectral index) [56].

17.3.4 Effect of solar activity

The effect of solar activity changes the occurrence characteristics of ionospheric irregularities. Fig. 17.6 shows the seasonal variation of scintillation index (S_4) with smoothed sunspot number (SSN) and solar flux (F10.7 cm) during the ascending phase of solar

Table 17.1 The observed and computed characteristics parameters of quasiperiodic scintillations over low latitude station Varanasi.

S. no.	Quasiperiodic scintillation quality parameters	Obtained	Computed
1	Half decorrelation time (τ) (in sec)	1.0–13.1	2.0–6.4
2	Scale length (in meter)	100–1300	200–650
3	Spectral Index (p)	-1 to -8	-1.3 to -8.7
4	Scintillation Index (S_4)	0.2–0.8	0.4–0.8

cycle 24 from 2009 to 2014. It is observed that in equinoctial months S_4 increases with the increase in SSN and in summer months it increase with the increase in solar flux F10.7 [31]. The figure shows the similar variation for the SSN and solar flux (F10.7). From the figure, it is observed that occurrence of irregularity is high for high solar activity years of 2013 and 2014 compared to low solar activity period (2009). The lifetime of the plasma bubbles at high solar activity was, in general, shorter than 3.3 hours. Plasma bubbles occur in the midnight-dawn sector at low solar activity, but strong plasma bubbles exist primarily in the evening sector at high solar activity [50]. The lifetime of plasma bubbles is long (7 hours or longer) at low solar activity but becomes much shorter (3 hours) at high solar activity. Broad plasma depletions occur near dawn at low solar activity, but the wide low-density regions with multiple plasma bubbles occur in the evening sector at high solar activity [50]. At Varanasi, the variation of irregularity showed increase with increase in sunspot number (SSN) except during the summer months. For the quasiperiodic scintillation, it increases with the sunspot minimum [31,45,57].

17.4 Conclusions

In this chapter, we have summarized the remote sensing estimations of the ionospheric irregularities and their characteristics parameters mainly over low latitudes. Based on the studies done, the conclusions made are as follows:

1. The irregularity percentage occurrence showed maximum occurrence during equinoctial months, moderate in winter months and lowest in the summer months. On the basis of solar activity, major occurrence was in equinox months and later on in winter with lesser events in summer.
2. From the DMSP data analysis, larger EPBs rates were in the American–Atlantic section during March-April and August-October, indicating both seasonal and longitudinal impacts. The Indian-Pacific region has little rates of EPB.
3. From the C/NOFS data analysis, the EPBs were obtained for about 12 hours or longer in equinox months and latitude-wise covered Asian, African, and Atlantic sectors at around 1900 hours IST.
4. The temporal scale length of irregularities obtained from VHF scintillation analysis at low latitude is around 100 m to 1100 m for 75 m/s speed and around 400 m to 3000 m for 200 m/s speed showing the intermediate scale length. E-region irregularities for daytime scintillation vary from 20 m to 600 m.
5. Scintillation index (S_4) ranges between 0.2 and 0.9 for low latitude stations which informs that observed irregularities are of weak. Spectral index low latitude station Varanasi was reported between -2 and-9 with the mean value as -4. The in-situ measurements showed spectral indices ranging from -1.7 to -5.
6. Studies from quasiperiodic scintillation were obtained giving spectral index as -1.3 to -8.7 as scintillation index (S_4) index as 0.4 to 0.8.

7. Plasma bubbles take place in the midnight-dawn sector during low solar activity period, whereas strong plasma bubbles exist mainly in the evening hours during the high solar activity period. The lifespan of plasma bubbles is long at around 7 hours or longer during low solar activity time. On the other hand, it is much shorter for about 3 hours during high solar activity time.
8. The changes of irregularity increases with an increase in sunspot number (SSN) except during summer months.

References

[1] K. Patel, Study of Ionospheric Irregularities at low latitude, Atmospheric Research Laboratory, Department of Physics, Banaras Hindu University, Varanasi, 2009.

[2] G. Haerendal, Theory of equatorial Spread-F, Report, Max-Planck-Institute, Space Plasma Physics of Near-Earth Environment, MPI for Extraterrestrial Physics, Max Planck Society, ou_159897, Fur. Phys. Und. Astrophys. Garching, West Germany (1974).

[3] M.C. Kelley, R.A. Heelis, The Earth's Ionsophere, Plasma Physics and Electrodynamics, International Geophy. Series, 43, Academic Press Inc., California, 1989.

[4] B. Basu, On the linear theory of the equatorial plasma instability: comparison of different descriptions, J. Geophys. Res. 107 (A8) (2002), doi:10.1029/2001JA000317.

[5] R.S. Dabas, D.R. Lakshmi, B.M. Reddy, Effect of geomagnetic disturbances on the VHF nighttime scintillation activity at equatorial and low latitudes, Rad. Sci. 24 (4) (1989) 563–573, doi:https://doi.org/10.1029/RS024i004p00563.

[6] J. Aarons, The role of the ring current in the generation or inhibition of equatorial F layer irregularities during magnetic storms, Radio Sci. 26 (4) (1991) 1131–1149.

[7] M.C. Kelley, Recent results and outstanding problems of equatorial spread-F, J. Atmos. Solar. Terr. Phys. 47 (745) (1985).

[8] K. Patel, A.K. Singh, R.P. Patel, R.P. Singh, Characteristics of low latitude ionospheric E-region irregularities linked with daytime VHF scintillations measured from Varanasi, J. Earth. Sys. Sci. 118 (6) (2009) 721–732, doi:https://doi.org/10.1007/s12040-009-0058-x.

[9] Basu, Su., and Kelley, M. C., Review of equatorial scintillation phenomena in light of recent developments in the theory and measurements of equatorial irregularities, J. Atmos, Terres. Phys., 39, 1229, 1977.

[10] S.L. Ossakow, S.T. Zalesok, B.E. McDonald, P.K. Chaturvaedi, Non-equatorial Spread-F: dependence on the altitude of the F peak and bottom side background electron density gradient scale length, J. Geophys. Res. 84 (17) (1979).

[11] K.C. Yeh, C.H. Liu, Radio wave scintillations in the ionosphere, Proc. IEEE. 70 (1982) 324–360.

[12] J.J. Makela, A review of imaging low-latitude ionospheric irregularity processes, J. Atmospheric Solar-Terrestrial Phys. 68 (13) (2006) 1441–1458 ISSN 1364-6826.

[13] J.F. Goodman, Space Weather and Telecommunications, Radio Propagation services, Inc., Alexandria, Springer, 2005 printed in U.S.A.

[14] Tascione, T.F., Introduction to Space Environment, 2nd edition, 2nd Ed. 1994, Reprint Ed. 2010, 172 pp. Krieger publishing company, ISBN: 978-0-89464-071-1,1994.

[15] M.A. Abdu, Equatorial ionosphere thermosphere system: electrodynamics and irregularities, Adv. Space Res. 35 (2005) 771–787, doi:10.1016/j.asr.2005.03.150.

[16] M.C. Kelley, The Earth's Ionosphere: Plasma Physics and Electrodynamics, International Geophysics Series 96, Academic press, San Diego, CA, 2009.

[17] H. Takahashi, C.M. Wrasse, C.A.O.B. Figueiredo, D. Barros, M. Ali Abdu, Y. Otsuka, K. Shiokawa, Equatorial plasma bubble seeding by MSTIDs in the ionosphere, Prog. Earth Planet Sci. 5 (32) (2018).

[18] H. Kil, R.A. Heelis, L.J. Paxton, S.J. Oh, Formation of a plasma depletion shell in the equatorial ionosphere, J. Geophys, Res. 114 (2009) A11302 doi.org/10.1029/2009JA014369.

[19] A.K. Singh, R.P. Singh, Observations and modeling of nocturnal ionospheric irregularities in low latitude region, J. Geomag. Geoelec. 49 (1997) 1115–1129.

[20] C.L. Rino, The Theory of Scintillation with Applications in Remote Sensing, IEEE Press, Wiley, Inc., Hoboken, New Jersey, USA, 2011.
[21] B.H. Briggs, Brief Review of Scintillation Studies. Radio Science, Rad. Sci. 1 (10) (1966) 1163–1167, doi:10.1002/rds19661101163.
[22] R.P. Singh, R.P. Patel, A.K. Singh, Effect of solar and magnetic activity on VHF scintillations near the equatorial anomaly crest, Ann. Geophys. 22 (2004) 2849–2860.
[23] S. Basu, Su. Basu, Scintillation technique for probing ionospheric irregularities, in: C.H. Liu (Ed.), World Ionospheric/Thermospheric Studies (WITS) Handbook, 2, SCOSTEP University of III Urbana-Champaign, Boston, 1989, pp. 128–130.
[24] S.R. Khastgir, R.N. Singh, The size of moving irregularities in the F-region and spread angle of the radio waves scattered from them, J. Atmos. Terr. Phys. 18 (1960) 123–126.
[25] S.J. Franke, C.H. Liu, Observations and modeling of multi-frequency VHF and GHz scintillations in the equatorial region, J. Geophysical Res. *88* (A9) (1983) 7075–7085.
[26] S Kumar, S. Priyadarshi, G. Seemala, A.K. Singh, GPS-TEC variations during low solar activity period (2007-2009) at Indian low latitude stations, Astrophys. Space Sci. 339 (2012) 165–178, doi:10.1007/s10509-011-0973-6.
[27] T.P. Yunck, C.H. Liu, R. Ware, A history of GPS sounding, terrestrial, Atmospheric and Oceanic Sciences 11 (1) (2000) 1–20, doi:10.3319/TAO.2000.11.1.1(COSMIC).
[28] V.S. Rathore, S. Kumar, A.K. Singh, A statistical comparison of IRI TEC prediction with GPS TEC measurement over Varanasi, India, J. Atmospheric Solar Terrestrial Phys. 124 (2015) 1–9.
[29] V.S. Rathore, S. Kumar, A.K. Singh, Occurrence characteristics of ionospheric irregularities over Indian low latitude region Varanasi during ascending phase of solar cycle 24, Ann. Geophys. 61 (1) (2018) PA113.
[30] S. Aol, S. Buchert, E. Jurua, Ionospheric irregularities and scintillations: a direct comparison of in situ density observations with ground-based L-band receivers, Earth Planets Space 72 (164) (2020), https://doi.org/10.1186/s40623-020-01294-z.
[31] W.J. Burke, A.G. Rubin, N.C. Maynard, L.C. Gentile, P.J. Sultan, F.J. Rich., O de La Beaujardiere, C.Y. Huang, G.R. Wilson, Ionospheric disturbances observed by DMSP at middle to low latitudes during the magnetic storm of June 4–6, 1991, J. Geophys. Res. 105 (2000) 18391–18405.
[32] B.S Erjiang Fu, An Investigation of GNSS Radio Occultation Atmospheric Sounding Technique for Australian Meteorology (Thesis), School of Mathematical and Geospatial Sciences College of Science, Engineering and Health, RMIT University, Melbourne, Australia, 2011.
[33] F.J. Rich, M. Hairston, Large-scale convection patterns observed by DMSP, J. Geophys. Res. 99 (1994) 3827.
[34] C.Y. Huang, W.J. Burke, J.S. Machuzak, L.C. Gentile, P.J. Sultan, Equatorial plasma bubbles observed by DMSP satellites during a full solar cycle: Toward a global climatology, J. Geophys. Res. 107 (A12) (2002) 1434, doi:10.1029/2002JA009452.
[35] S.C. Garg, U.N. Das, Aeronomy experiment on SROSS C2, J. Spacecr. Technol. 5 (11) (1995).
[36] A.W. Wernik, L. Alfonsi, M. Materassi, Scintillation modeling using in situ data, Radio Sci. 42 (2007), doi:10.1029/2006RS003512.
[37] H. Kil, R.A. Heelis, Global distribution of density irregularities in the equatorial ionosphere, J. Geophys, Res. 103 (1998) 407–417.
[38] O.de La. Beaujardière, C/NOFS: a mission to forecast scintillations, J. Atmospheric Solar-Terrestrial Phys. 66 (17) (2004) 1573–1591. VolumeIssuePagesISSN 1364-6826, https://doi.org/10.1016/j.jastp.2004.07.030.
[39] O.de La. Beaujardière, D. Cooke, J. Retterer, C/NOFS: A Satellite Mission to Forecast Equatorial Ionospheric Scintillation, American Inst. Aeronautics Astronautics 0269 (2005).
[40] W.J. Burke, O. de La Beaujardière, L.C. Gentile, D.E. Hunton, R.F. Pfaff, P.A. Roddy, G.R. Wilson, C/NOFS observations of plasma density and electric field irregularities at post-midnight local times, Geophysical Res. Lett. 36 (2009), doi:10.1029/2009gl038879.
[41] H. Chandra, R.G. Rastogi, Blanketing sporadic E layer near the magnetic equator, Geophys. Res. 80 (1975) 149–153.
[42] P.V.S. Rama Rao, P.T. Jayachandran, P. Sriram, B.V. Ramana Rao, D.S.V.V.D. Prasad, K.K. Bose, Characteristics ofVHF Radio wave Scintillations Over a Solar Cycle (1983−1993) at a Low-Latitude Station: Waltair (17.7°N, 83.3°E), Ann. Geophys. 15 (1997) 729–733.

[43] A. Das Gupta, L. Kersley, Summer daytime scintillations and sporadic-E, J. Atmos. Terr. Phys. 38 (1976) 615–618.
[44] B.M. Pathan, R.G. Rastogi, D.R.K. Rao, On the Width and Complexities of the Equatorial Nighttime Radio Wave Scintillation Belt in the Indian Region, J. Geomag. Geolectr. 44 (1992) 129–142.
[45] S. Kumar, A.K. Gwal, P.V.S. Rama Rao, P.T. Jayachandran, S.V.V.D. Prasad, R.P. Singh, U.P. Singh, A. Dasgupta, K. Basu, R. Sethuraman, B.M. Pathan, D.R.K. Rao, S. Banola, P.S. KesavaRao, A. Naidu, T.R. Tyagi, P.N. Vijaykumar, H. Chandra, G.D. Vyas, B. Singh, P. Chauhan, K.N. Iyer, K.N. Pathak, C.S. Shalgaonkar, B.M. Vyas, R.G. Rastogi, Co-ordinated observations of VHF scintillations in india during February-March 1993, Ind. J. Radio Space Phys. 29 (22) (2000).
[46] A. Das Gupta, A. Maitra, Sa. Basu, Occurrence of Nighttime VHF Scintillations near the Equatorial Crest in the Indian Sector, Rad. Sci. 16 (1981) 1455–1458.
[47] S.K. Chakraborty, A. Das Gupta, S. Ray, S. Banerjee, Long term observations of VHF scintillation and total electron content near the crest of the equatorial anomaly in the Indian, Longitude Zone, Radio Sci. 34 (1999) 241–255.
[48] A.K. Singh, R.P. Patel, R.P. Singh, Statistical features of overhead ionospheric irregularities and its generation mechanism at low latitude, J. Atmos. and Solar Terr. Phys. 68 (2006) 1116–1124.
[49] K. Patel, A.K. Singh, Characteristic of plasma bubbles observed by DMSP in the topside ionosphere during the year 2005, J. Earth System Sci. 119 (2010) 211–220, doi:10.1007/s12040-010-0012-y.
[50] C.S. Huang, O.de La. Beaujardiere, P.A. Roddy, D.E. Hunton, J.O. Ballenthin, M.R. Hairston, Generation and characteristics of equatorial plasma bubbles detected by the C/NOFS satellite near the sunset terminator, J. Geophys. Res. 117 (2012) A11313.
[51] K. Patel, A.K. Singh, R. Patel, R.P. Singh, Ionospheric scintillations by sporadic-E irregularities over low latitude, Bull. Astronomical Soc. India 35 (2007) 625–630.
[52] K. Patel, A.K. Singh, R.P. Singh, Characteristics of quasi-periodic scintillations observed at low latitude, Radio Sci 44 (2009) RS6007, doi:10.1029/2008RS003975.
[53] R.G. Rastogi, K.N. Iyer, Ionospheric scintillations induced by cloud of intense sporadic-E layer, Curr. Sci. 45 (1976) 685–686.
[54] Y.V. Somayajulu, S.C. Garg, P.N.Vijay Kumar, Scintillation studies using ATS-6 radio beacons at Delhi, Indian J. Radio Space Phys. 6 (1977) 250–256.
[55] J.M. Jahn, J. LaBelle, Rocket measurements of high-altitude Spread-F irregularities at the magnetic dip equator, J. Geophys. Res. 103 (A10) (1998) 23427–23441.
[56] K. Patel, A.K. Singh, P. Subrahmanyam, A.K. Singh, Modeling of ionospheric scintillation at low-latitude, Adv. Space. Res. 47 (2010) 515–524, doi:10.1016/j.asr.2010.09.017.
[57] K.N. Pathak, R.D. Jivrajani, H.P. Joshi, K.N. Iyer, Characteristics of VHF scintillations in the equatorial anomaly crest region in India, Ann. Geophys. 13 (1995) 730–739.

CHAPTER 18

Characteristics of tropical cyclones through remote sensing-based observational platforms

Jagabandhu Panda, Debashis Paul, Subodh Kumar
Department of Earth and Atmospheric Sciences, National Institute of Technology Rourkela, Rourkela, Odisha, India

18.1 Introduction

Tropical cyclones (or TCs) are large-scale low-pressure disturbances that form over warm tropical waters and are characterized by high cyclonic winds, structured convection, and heavy rainfall. The high winds and heavy rain produced by TCs cause major devastation to property and human life during its land falling. Those warm core systems are driven by the latent heat energy produced in the center areas of convective clouds. These warm core systems characterized by massive vortices of swirling winds, usually draw the energy from the ocean. Due to the Coriolis Effect [1–4], TCs rotate clockwise in the southern hemisphere and counterclockwise in the northern hemisphere. In the year 1848, Henry Piddington, a British Indian sea captain coined the name of such phenomenon as "cyclone" while traveling through a ship. Notably, these cyclonic systems play a key role in global air circulation by transporting heat and energy from the tropics toward temperate latitudes [5–7].

There are seven global basins, namely, North Atlantic Ocean, eastern and western parts of North Pacific Ocean, south-western pacific, south-western and south-eastern Indian Ocean and North Indian Ocean (NIO), where TCs are produced. Accordingly, TCs are categorized as hurricanes, typhoons, tropical storms, cyclonic storms, and tropical depressions depending on their location and severity. The average TC frequency, season, and intensity varies from one basin to the other. NIO basin mainly experiences a bimodal TC season with maximum frequency during the post-monsoon months (i.e., October–December) followed by the premonsoon months (i.e., April–May). Usually, postmonsoon TCs are stronger than premonsoon ones. Although the NIO TCs are relatively smaller in size and less intense compared to that of other basins of the world, this region is quite important in view of the population density with poor socioeconomic conditions. Although smaller in size, the NIO basin is unique in terms of shape, bathymetry and large river systems. Generally, the NIO basin experiences about 5–6 TCs annually, with Bay of Bengal (BOB) being more active as compared to the Arabian Sea. The average life period of NIO TCs is 3–5 days in contrast to those observed in other ocean basins.

Atmospheric Remote Sensing
DOI: https://doi.org/10.1016/B978-0-323-99262-6.00001-8

Researchers worldwide are mainly concerned about the understanding of TC progression, track prediction, strength, and size [8–15]. However, the mechanics of TC genesis, intensification, and landfalling still need to be understood more. Furthermore, it is unclear how TC features, such as strength, location, frequency, and size, change with changing climate especially over NIO. Few studies in these directions have improved our understanding in past two decades [12–21]. However, improving the understanding of TCs via modeling can aid India and other nations across the world in better forecasting them [22–25].

The current understanding indicates that the TCs are associated with relatively low pressure in the troposphere, where the largest pressure agitations occur near the surface at low altitudes [5, 26, 27]. The environment near the center of a TC is warmer than the surroundings at all altitudes. The TC environmental setting enables producing surface winds higher than 17 m/s, with winds exceeding 90 m/s in severe situations. The critical factors such as maximum sustained winds (i.e., V_{max} or MSW), minimum sea level pressure, and potential wind devastation close to the TC center are useful in forecasting the track [9, 21, 28–32].

Besides numerical modeling, remote sensing (RS) products are quite useful in advancing the understanding of physics and dynamics associated with TCs over ocean. Although the satellite data products are more suitable for this purpose, other RS measurements play a supportive role. The satellite products mainly help in detecting the occurrence of a TC, its size, determining the center, estimating the intensity, knowing the life cycle stages, and finally the track. They help in real-time forecasting of TCs, besides the research works including the impact of wind observations. Also, the satellite products could be used for determining the precipitation, sea surface temperature (SST), and moist variables associated with TCs. In the recent past, it is realized that the Doppler weather radar (DWR) systems, aircraft measurements and light detection and ranging (LIDAR) observations are useful in understanding, and prediction of TCs too.

This chapter emphasizes upon various aspects to illustrate how RS observations have aided in TC research and forecasting. Accordingly, the chapter is divided into six sections. The following section describes the satellite RS application. The third section elucidates about the radar observations and associated techniques aiding to the TC research. The fourth section illustrates the TC characteristics through LIDAR observations. The fifth section highlights about the aircraft and other RS observations and tools used for the TC research and forecasting. The last section summarizes about the review and future prospects.

18.2 Satellite remote sensing application

Due to the advancement in RS, fewer TCs nowadays go unobserved, and with improvement in observation and forecasting, we can understand the associated processes.

Fig. 18.1 ***The first image obtained from the GOES-1 satellite, October 25, 1975 1645 GMT.*** The figure is adapted from NOAA (National Ocean and Atmospheric Administration, USA) photo library (http://www.photolib.noaa.gov/).

However, in the case of in situ data, it is not so as the TCs spend most of their life over the ocean. Moreover, the satellite observations help in the awareness programs and forecasting of TCs by various agencies such as India Metrological Department (IMD), Joint Typhoon Warning Centre (JTWC), National Hurricane Centre (NHC), etc. [33]. The first satellite used for TC monitoring and tracking was the television infrared observation satellite or TIROS (Sadler, 1961) launched in 1960. Similarly, the US agencies used geostationary operational environmental satellite or GOES (launched in 1975) for weather forecasting, severe storm tracking, and meteorological research (Fig. 18.1). Likewise, various countries such as China, Europe, India, and Japan started their own meteorological satellite programs for the improvement on weather forecasting and monitoring. The Indian National Satellite (INSAT) launched in 1982 as a multipurpose geostationary satellite, is used both for meteorological purpose and communication payloads.

18.2.1 Atmospheric and ocean characteristics observed through satellites

Observing and monitoring TCs is a very important application of meteorological satellites. The TC related studies using satellite products include both observational (e.g. [34,12,15,16,30,35–37] and modeling (e.g., [5,29,186,22,38–41]). These studies explored numerous aspects related to TC characteristics (namely, size, structure, dynamics) using observational data. Also, the observational datasets are incorporated in numerical modeling studies of TCs to analyze the associated atmospheric and ocean characteristics (e.g., [16,26,32,42,43]). Some of these features are described in the following subsections.

18.2.1.1 Meteorological characteristics observed through satellites

Generally, it is seen that most of the weather agencies depend more or else on satellite measurements for observation and forecasting of TCs. The cyclone monitoring and observation is based on various factors, which include the characteristics of sensors, orbital parameters, orbital altitude and inclination, swath and the spatial resolution of observed data. In the case of meteorological satellites, the visible (0.4–0.7 mm), infrared (1–100 mm), and microwave (0.3–30 cm) region of electromagnetic spectrum is used for obtaining various observations related to terrestrial and atmospheric parameters. Various studies [44–47] have shown the importance of satellite imagery in TC research and forecasting. For instance, Kalsi (1999 and 2002) has shown extensive use of satellite imagery for TCs developing over the NIO.

The first satellite observations were obtained utilizing visible/infrared wavelengths (VIS/IR), and these observations are still considered crucial for operational meteorological applications today. Almost all geostationary satellites employ visible/infrared sensors to obtain frequent observations of the earth–atmosphere system. Operational meteorological polar satellites, such as National Oceanic and Atmospheric Administration (NOAA), make worldwide observations in visible and infrared channels, with greater spatial and spectral precision but lower temporal resolution than geostationary satellites. These observations are most commonly employed to outline clouds in the atmosphere, and also used to detect cyclogenesis along with the position, movement, and areal extent of cyclones. Using the VIS imagery Dvorak [9] developed a technique which is a method used worldwide for estimating TC strength based on observations of specific characteristics associated with TC cloud fields. Later on, IR imagery was also incorporated to it for accuracy. The SST, which might suggest where severe convection may occur or a moving convective system may strengthen or weaken, is another valuable parameter obtained by VIS/IR observations. SST determination usually necessitates the use of more than one infrared channel (e.g., a split window channel: 10.5–11.5 and 11.5–12.5 m). However, the presence of clouds has a significant impact on the accuracy and availability of VIS/IR-based SST, which in turn makes it difficult to collect SST data in the area of TCs. Thus, microwave methods are employed to measure SST in the presence of clouds.

Several researchers across the globe have developed novel concepts for detecting cyclogenesis using satellite data during the past years. One of these approaches [48] makes use of a succession of satellite infrared measurements (available every 30 minutes from geostationary satellites). This method is based on the idea that in a tropical low-pressure system, mid-level mesoscale vortices interact. Because mesoscale vortices are usually accompanied by deep convection, the aforementioned phenomenon may be seen in IR satellite images. Correspondingly, an index called the "merger index" is defined as follows [48]:

$$\mu = r^*(n)/(R_{CG}/L)$$

where $r^*(n)$ is the critical radius for merger, R_{CG} is the average distance of mesoscale vortices from their common centroid, and L is the mean radius of the vortices system.

In comparison to VIS/IR imagery, microwave (MW) imagery offers substantial advantage as it can determine atmospheric parameters both in clear as well as in cloudy conditions. These important features of MW sensors make them very useful observational tool for the study of TCs. They are either passive microwave (PMW) type or active microwave (AMW) type sensors. Multi-frequency scanning microwave radiometer (MSMR, on board Indian OceanSAT-1), TRMM (tropical rainfall measuring mission) microwave imager (TMI), special sensor microwave/imager (SSM/I) are some of the examples of PMW sensors. They are mainly used for detection of SST as they operate in very low range of frequency, that is, 6–150 GHz, besides the estimation of surface wind speed and rainfall rates, and found to be accurate [49]. Thus the availability of SST even during cloudy conditions is very much helpful in TC studies, as it aides knowing the SST variation along the track of a TC to relate with the course of its evolution. Vertically integrated water vapor also called as precipitable water (PW) is another important parameter that can be retrieved using PMW sensors. PW with combined patterns of surface wind proves to be a good indicator of moisture flux transport which is very much important for cyclogenesis and intensification.

Analyzing surface-level wind fields obtained by Microwave scatterometers is another intriguing technique to capture cyclogenesis. Low level cyclonic vorticity, which emerges around 3–4 days before a low-level storm develops into a cyclone, is one of the first signs of cyclogenesis. Microwave frequencies are able to penetrate clouds even in the presence of light precipitation, making microwave measurements of surface winds useful for cyclogenesis research. The method is based on the idea that developing systems have certain similarities in low-level wind circulation that may be used to identify them from non-developing systems. The microwave energy transmitted in the case of an AMW sensor is transferred back to the receiving antennae and is processed. Scatterometer and altimeters are the most common examples of satellite AMW sensors. It is well known that storm

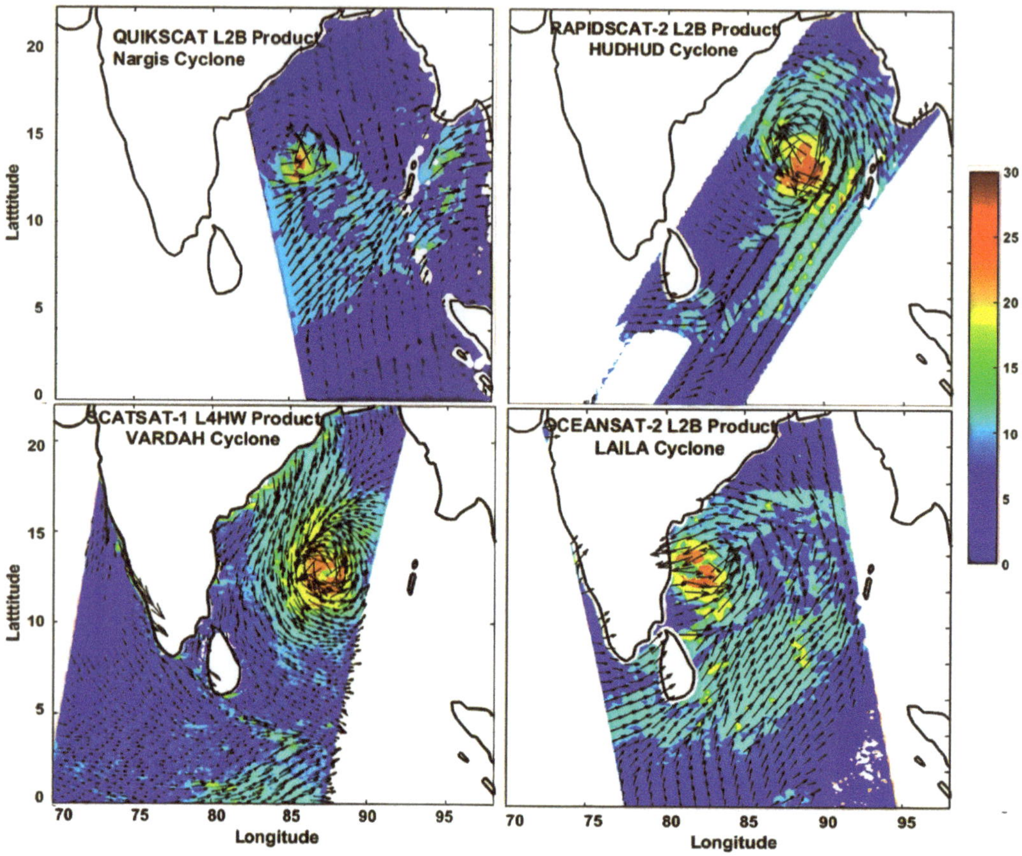

Fig. 18.2 Spatial representation of ocean surface wind field (speed and vectors) derived from scatterometers of various satellites, namely, (A) QUICKSCAT, (B) RAPIDSCAT, (C) SCATSAT, and (D) OCEANSAT.

surges caused by TCs are very much dangerous. In this regard, wind vector is a crucial parameter that can be used for the prediction of such TC-induced storm surges for which continuous observations of surface wind field over the oceans become utmost necessity. The scatterometer is the only device which can facilitate with ocean surface wind vectors for TC (e.g., Fig. 18.2) studies. It operates within the frequency range of 5–14 GHz. Some of the examples of scatterometers are ASCAT (Advanced Scatterometer) on board European satellite METOP (Meteorological Operational Satellite), the ones in QUIKSCAT (Quick Scatterometer) and RAPIDSCAT (Rapid Scatterometer), and Indian Space Research Organization's OSCAT (OceanSat Scatterometer) on board OCEANSAT-2 (Second Indian RS Satellite for Oceanographic Applications) and OCEANSAT-3 (Third Indian RS Satellite for Oceanographic Applications) and the one in SCATSAT (Scanning Scatterometer Satellite). The ocean surface wind products have been used by various researchers all over the world to study observed characteristics of

TCs [11, 15, 36, 50–52]. This resemblance of wind patterns has been quantified by calculating the "matching index" between a particular wind pattern and wind signatures of emerging systems based on previous observations of scatterometers such as QuikScat derived wind vectors.

Besides scatterometer, "radar altimeter" is another significant AMW sensor. This type of AMW sensor is specially developed for the measurement of parameters related to oceans such as significant wave height, surface wind speed, sea surface elevation (signified by sea level anomaly in several instances), ocean currents, surface geoids and tides. The said parameters can become very much useful for the estimation of intense winds and sea surface undulations caused by TC movements. TOPEX (Topography Experiment), JASON (Joint Altimetry Satellite Oceanography Network) series, ENVISAT (Environmental Satellite) and ERS-2 (European RS satellite-2) are some of the altimeter missions that provide altimeter data. The sea level height measurements provided by the TOPEX satellite is found to be accurate by 5–6 cm.

Further, "sounders" are also employed (e.g., high-resolution infrared sounder (HIRS) on board of NOAA satellites) for determining temperature sounding and water vapor concentration. Sounders are instruments which are used for obtaining vertical profiles of various atmospheric parameters such as temperature, humidity, etc. These instruments when put into satellites facilitate with three-dimensional distribution of these atmospheric parameters. However, outside the cyclone, the retrievals are possible, and are very useful in determining the environmental influences which may affect the cyclone track and intensity.

Besides the above-mentioned aspects, TC characteristics include the T. or CI number, outgoing long-wave radiation (OLR), vorticity measurements from wind field, precipitation estimates, etc. [11, 15, 23, 52–54]. OLR can be termed as a proxy for inherent convection. Usually, the OLR values are minimum in the center of a TC (e.g. Fig. 18.3B), which could be less than 220 W/m^2 at the region of deep convection associated with TC centers [55]. For instance, a study regarding the cyclone PHET by Panda et al. [23] indicated that the central region was associated with OLR values $< 240\,W/m^2$. Similarly, the low-level relative vorticity is usually maximum at the center, and precipitation estimates are generally in higher side around the center of a TC (e.g., Fig. 18.3A). The T or CI numbers derived from satellites typically range up to 7 and signify the intensity of a TC.

Since TCs develop over open ocean waters, description of satellites and sensors employed to measure various parameters mentioned here, also included several ocean features. However, the following subsection describes the ocean characteristics specifically observed through satellites as RS platforms to further emphasize on this aspect.

18.2.1.2 Ocean characteristics observed through satellites

The climate-oriented sustained ocean observing system (e.g., [56]) and observational process studies have greatly aided our knowledge of upper-ocean processes that lead to

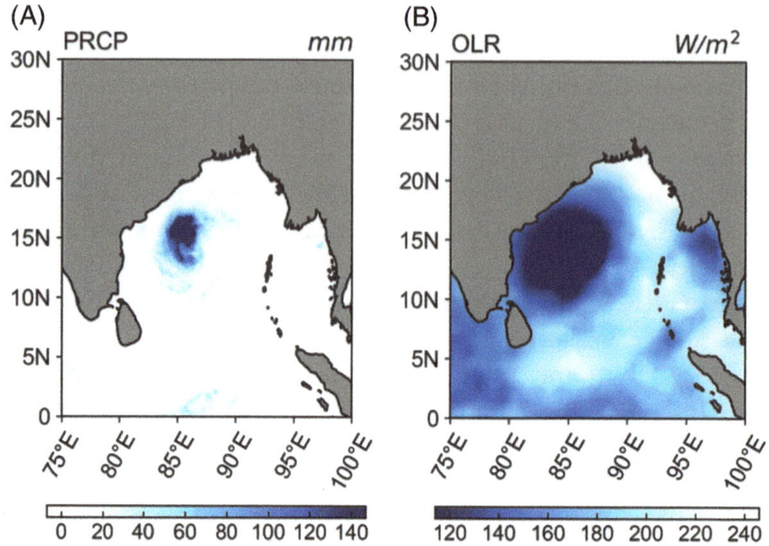

Fig. 18.3 (A) Integrated multisatellite retrievals for GPM (IMERG) precipitation data and (B) high resolution infrared radiation sounder (HIRS) OLR data for May 18, 2020 during SUCS Amphan.

extreme weather events such as TCs. Ocean observations are essential in TC research and forecasting since they allow researchers to investigate the complexities of air–sea interaction mechanisms that might contribute to the genesis and intensification of TCs [57–61]. During OceanObs' 09 [62], the specific suggestions were put forward concerning to the critical areas to focus in support of TC studies and predictions, which have mostly been answered by recent improvements in research and ocean observing activities. On a larger scale, recent advancements in ocean observations in support of TC research have been complemented by an overall increase in worldwide coverage based on in situ and satellite ocean observations (e.g., satellite coverage enabled substantial advances in satellite altimetry [63] and satellite-derived salinity measurements [64], which is available with the launch of the aquarius and the soil-moisture active passive missions, etc.). In addition, community-wide projects like the global ocean data assimilation experiment [65], the Mercator-ocean prediction system [66], and other related efforts have resulted in substantial improvements in oceanic forecasting capabilities.

The potential for TC formation and intensification has been estimated using a variety of factors obtained from in situ and remotely sensed data that characterize energy available in the upper ocean. The tropical cyclone heat potential (TCHP) is one such factor which is dimensionless, and is defined as the excess heat content in the water column from the sea surface to the depth of the 26°C isotherms [59]. The NOAA NHC and the Joint Typhoon Warning Center (JTWC) routinely use SST and TCHP fields for subjective TC intensity forecasts as well as quantitatively in the Statistical Hurricane Intensity Prediction Scheme (SHIPS; [67,68]) and rapid intensification aids [69,70].

With a few assumptions, Ali et al. [71] estimated ocean mean temperature (OMT) using satellite-derived TCHP and the depth of the 26°C isotherms (D26). OMT and ocean heat content (OHC) are similar variables, which signify the total heat content from the surface to a particular depth. Higher TCHP of OHC values indicate more potential of having cyclogenesis over a particular area of Ocean (e.g., Fig. 18.4 shows TCHP and OHC values in the genesis region of Amphan). The OMT estimate helped in predicting the Indian monsoon rainfall better than SST [71,72]. The use of OMT in numerical models instead of SST has the potential to improve TC forecasting too. The subsurface ocean influences TCs thermodynamically by controlling TC-induced cold SST wakes. When the cold wake is modest (< 0.5°C), TCHP outperforms other predictors such as SST and vertically averaged temperature in predicting TC intensification, whereas when the cold wake is strong (> 0.5°C), ocean dynamic temperature (Tdy) performs much better [73]. Tdy is defined as the ocean temperature averaged from the surface to the poststorm mixed layer depth, which depends on the upper-ocean stratification as well as the TC intensity and translation speed [73]. When coupled with subsurface in situ data, satellite-based sea surface salinity measurements can give additional information on salinity stratification as well. When combined with Tdy, this additional parameter might lead to even better statistical TC prediction schemes, as well as more relevant validation of operational ocean analysis and predictions.

18.2.2 Structural and radial characteristics of tropical cyclones

Long-term or climatological studies on cyclonic disturbances have been very interesting to undertake for inherent scientific reasons and also for disaster mitigation purposes, by assuming a symmetrical storm routinely monitored throughout the TC's life cycle [74–76]. Routine TC intensity analysis from imagery began soon after the first satellite weather data were available. Until the significant work of Dvorak in 1972, determining the TC's center location was the only reliably generated operational use of satellite data-based analysis. Dvorak created an empirical model based on cloud structure in the visible channel to diagnose TC intensity. The amplitude of the brightness temperature, and the curvature of the banding or distortion of the cloud pattern were the factors influencing the storm's intensity. Today, the Dvorak approach is still used extensively in operational analysis, although some attempts are still being made to improve the accuracy of the approach, such as updating the wind pressure relationship that is used to determine the intensity [77]. In the process, it is realized that accurate estimation of wind field is quite important to determine the associated vorticity, radial and structural parameters, and the intensity.

There are numerous additional ways for diagnosing TC structure and intensity from infrared and visible channels besides the Dvorak technique. For instance, Kossin et al. [78] put an effort to illustrate the usability of infrared band observations for determining the operational storm magnitude variables (i.e., radius of maximum winds and

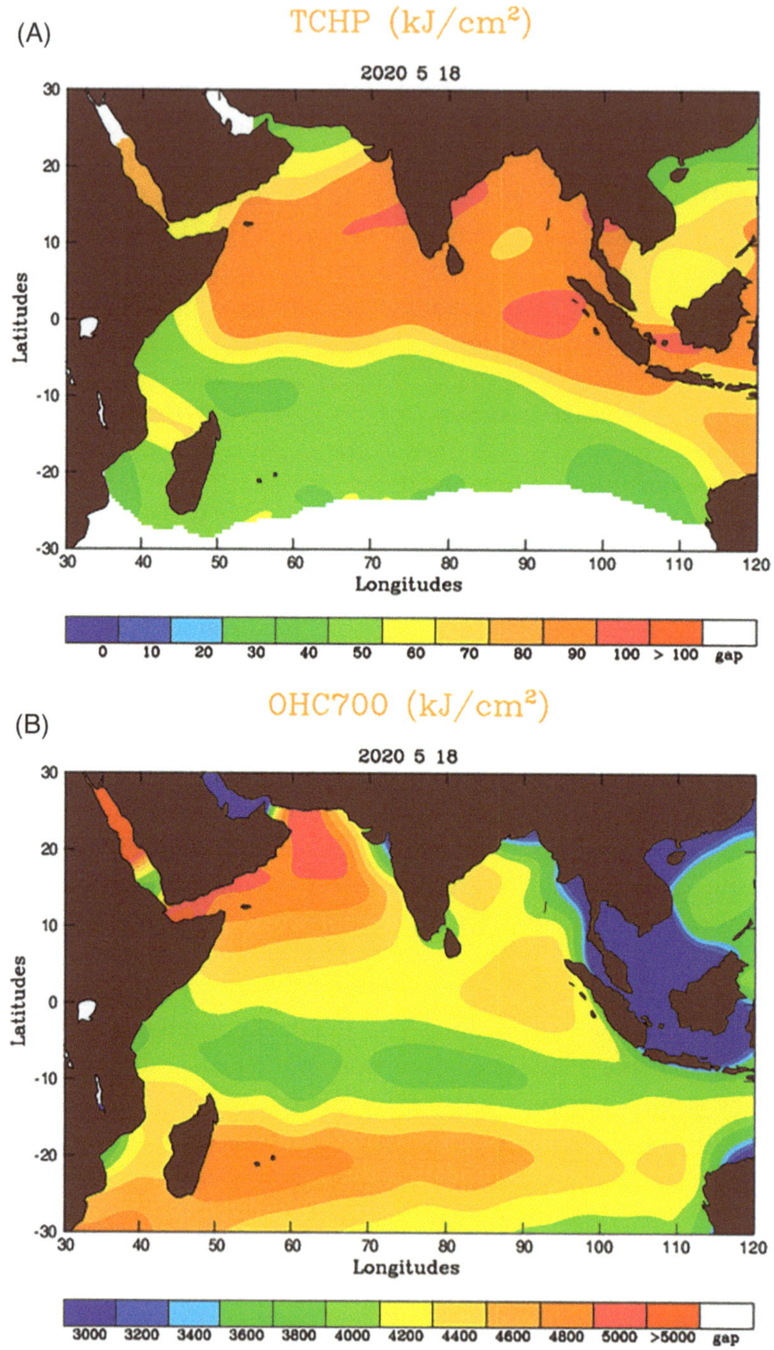

Fig. 18.4 (A) Tropical cyclone heat potential (TCHP) and (B) ocean heat content (OHC) up to 700 m depth from the surface over NIO on May 18, 2020 during the super cyclone Amphan. The figures are taken from NRSC Bhuvan data archive (https://bhuvan-app3.nrsc.gov.in/data/download/index.php).

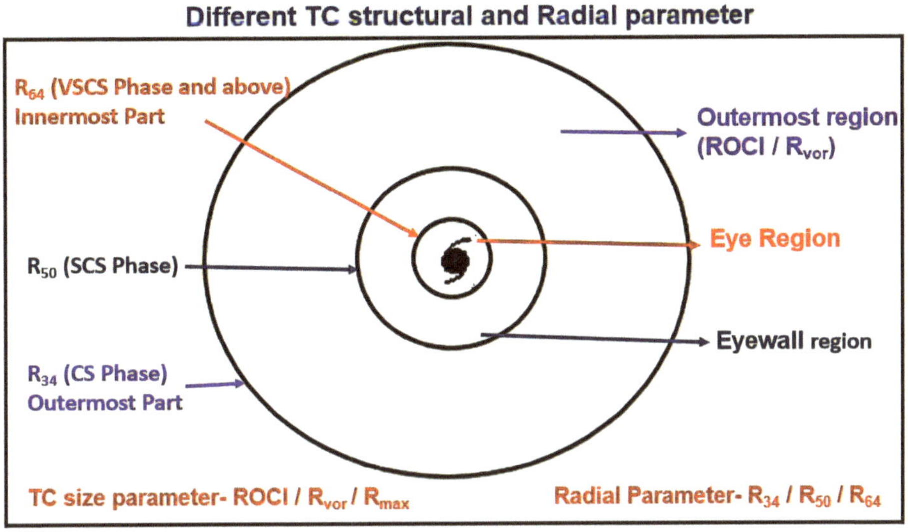

Fig. 18.5 ***Schematic representation of TC size and radial parameters.***

radii of the 34, 50, and 64 kt winds). Similarly, Knaff et al. [79] highlighted identical use of infrared imaging for the determination of TC sizes. Other RS frequencies have proved effective for monitoring of TCs in addition to the conventional infrared and visible channels. The NHC operational endeavors have benefited greatly from the deployment of the AMW radars like scatterometers in the satellite platforms to assess storm magnitude and intensity by surface wind analysis [33]. Scatterometers send out pulses that bounce off the ocean surface, and the fluctuation in backscatter caused by wind roughening allows for the extraction of surface wind vector estimations. Using a wind structure model, QuikSCAT scatterometer data were utilized to construct climatology of storm sizes at the 23 kt radius as well as an outer radius [80]. Several studies have been carried out to show the usability of scatterometer wind estimates for TC structure analysis [11,15,36,52,80]. A schematic representation of various radial and size parameters associated with TCs is shown in Fig. 18.5. An empirical relationship is also developed by some of the researchers by adopting the linear regression method and using satellite data products. For instance, Takagi and Wu [81] formulated an empirical relation to compute the size of TCs over Western North Pacific through the structural parameters, namely R50 and R_{max}, by following Knaff and Zehr [26] and Willoughby and Rahn [53]. A recent effort in this direction was done by Kumar et al. [15] to setup the empirical formulation $R_{max} = 0.59 \times$ R50, which is valid for NIO basin. They adopted the vorticity-based approach for all cyclonic phases and used scatterometer observations from four satellites during the years 1999–2017 for this purpose. Besides the horizontal structural analysis, vertical structure is also quite important. Notably, several observational studies have documented the vertical profiles and eyewall replacement process of TCs [82–84], related Doppler radar analysis of a concentric eyewall [187,188] and vertical structure of the primary and outer eyewall ([85–87]; Houze et al., 2007) over different world basins.

While progress has been made in detecting concentric eyewall cycles after they have begun [88], forecasting the beginning of concentric eyewall cycles and the extent to which they affect intensity are still topics of interest for exploring. The distribution of different essential structural parameters of TCs, such as R_{max} (radius of maximum winds) and ROCI (radius of close isobars), as well as the relevant methodologies, can be investigated using multiple satellite data [5,15,16,30,80,89]. Some of the studies performed a detailed study on the variations of structural parameters, namely, wind radii of 34, 50, and 64 kts [16,42,50,78,90,91]. Mostly R_{34} is also used for TC size representation whereas R_{50} and R_{64} represent the inner core (found near eyewall region). Also, these correspond to the different stages of TCs in terms of 34-50-64 knots winds representing the CS-SCS-VSCS phases. An illustration concerning to these radial parameters is given in Fig. 18.5.

In general, while defining TC size, ROCI is paid a lot of attention. The ROCI is a barrier between the ambient sea level pressure (SLP) and the reduction in pressure associated with a TC, with the whole of it included inside the ROCI. It was initially developed by Brand [92] to investigate the magnitude of TCs in the Western North Pacific or WNP [90,93,94]. It's a quadrant-based measurement that calculates the average ROCI by averaging the radius in the four cardinal directions [30]. This parameter is useful for estimating the actual size of the TC, more resilient than other size metrics, and is seen to be a realistic illustration of the TCs' structure [95]. It is also an excellent precipitation analysis parameter since a TC's rain area is frequently found within the ROCI [96,97]. It is also utilized as a parameter in the studies of TC intensification and model improvement [79,98–100]. Recent studies have discovered a convenient way for determining TC size utilizing a vorticity-based approach [15,52,101,102]; however, this technique is not documented in JTWC and other best-track data. It is termed as "R_{vor}." It is worth noting that the R_{vor} 's range closely resembles the ROCI values for a given cyclone, which have been validated against ROCI values of JTWC and have been used to investigate their seasonal variations [15,52] too.

18.2.3 Numerical modeling approach using satellite observations

In TC research, operational monitoring, and forecasting, observations from satellites play a vital role. Due to the lack of accurate data over wide open oceans, forecasting TC development has always been a difficult task. These days, available satellite observations are assimilated by numerical weather prediction (NWP) models (global and mesoscale models) across the world. The ability to forecast cyclogenesis has substantial operational significance because it gives operational forecasters and emergency managers the time to plan and prepare for potential hazards.

One of the most significant uses of satellite data is to accurately describe the position, strength, and magnitude of TCs in NWP models' initial condition. Mesoscale models use coarse resolution assessments supplied by global models to produce high-resolution initial conditions. Vortex initialization is such a method for representing the dynamic and thermodynamic structure of a cyclonic vortex as accurately as possible in a high-resolution initial condition for mesoscale simulations. However, satellite observations give relatively precise cyclone position and strength estimations that can be used to initiate vortex formation. A number of studies have found that properly representing the cyclonic vortex in models improves the prediction of cyclone track and intensity substantially [40,102–112,113,114].

Although the TC intensity forecast has improved very less compared to the TC track forecast, the TC intensity forecast has clearly improved after 2001, when additional satellite measurements have been routinely incorporated in NWP simulations [115–119]. One of the most significant breakthroughs is the incorporation of satellite data directly into cloud-resolving models. For instance, Zou et al. (2013), included Advanced Technology Microwave Sounder (ATMS) data into the HWRF system to study the influence on TC strength and track forecasts. There are more satellite observations that are available than are incorporated into the models. More research and validations are needed to ensure proper use of all these satellite data which will have a beneficial influence on forecasting skills. Besides, satellite observations are used as an input to the numerical models for improving the predictability of TCs and associated meteorological features [39–41]. Thus, it is essential to improve the accuracy of high-resolution satellite data, which is useful for the study of structural characteristics, different phases of the TC life cycle, and associated meteorological features [15,39–41]. Therefore, these satellite datasets need to be validated with other observations, so that the associated limitations can be determined.

18.3 Radar observations and associated techniques

Radar has made a substantial contribution to TC operational surveillance, as well as the research and advancement in understanding of the associated characteristic features. Radar technology was developed to identify airplanes during World War II. When there were considerable weather changes in the region, it was quickly evident from the radar observations. The National Weather Service developed the first radar network in the United States in 1957 to monitor the coastline and provide early warning of an incoming TC. In this section, relevant TC characteristics visible through radar are described.

18.3.1 Observational aspects

US Navy made the first documented radar sighting of a TC from an airplane in 1945. The spiral structure of the rain bands could only be seen for the first time from the radar [189]. The spatial pattern of precipitation was studied using aerial and ground radar in many TCs in the Atlantic and Pacific, and a structural model was developed by Rockney [120]. Raghavan [183] gave the typical horizontal structure of a well-developed TC obtained by compositing radar echoes from a number of cyclones in the BOB. All these authors categorized the radar observations of a TC into five signature observations, namely, precyclone squall lines, outer convective activity, spiral rain bands and rain shield, eyewall (sometimes a double wall), and streamers.

Some of these signatures may be seen developing gradually and changing from time to time as the TC develops from an initial disturbance—generally a diffuse low-pressure region in the ocean. Despite this, because the TC is hundreds of kilometers across, a single stationary radar is unlikely to observe all of the characteristics at the same time. In this case, a temporal composite of observations with the same radar or a composite of images from various radars with overlapping coverage would provide a more detailed depiction of the whole TC. Because of range, height, and signal strength restrictions, land-based radars will still be unable to view sections of a TC that are far away. Intervening heavy precipitation might obscure the sights beyond. When the TC is far away and the weather around the radar is clear, radio-wave propagation circumstances may be favorable. Because of the precipitation and the rise in humidity at all altitudes as the TC approaches, propagation may become subnormal. As a result, it is possible that the TC is noticed well when it's farther away than when it's closer (Fig. 18.6).

Radars may occasionally offer a distorted view of the TCs. A "10 cm radar (S-band)" would be excellent for detecting TCs from land locations due to heavy precipitation spreading over long distances. Notably, there is a necessity for quantitative readings that are not corrupted by attenuation. Therefore, some nations utilize "5 cm radars (L-band)" due to the detection benefits. And the WSR-88D, a 10 cm Doppler system, is utilized in the United States for this purpose. The 10-cm or 5-cm signify the wavelength of the beam associated with the corresponding radar. Along the coast of India, a network of "10 cm radars" known as "Cyclone Detection Radars" has been built by IMD. Employing DWRs can give a high-resolution, spatiotemporal view of land falling TCs [121]. For instance, HUDHUD, which occurred in BOB during October 2014 could be visualized through the Visakhapatnam DWR (Fig. 18.6).

The radars installed in a satellite platform can also be useful for the observations of TC characteristics. For instance, the TRMM precipitation radar and TMI in association with lightning flash density, can be used to examine variations in frequency thresholds between different parts of a TC [122]. Synthetic Aperture Radar (SAR) can be utilized to observe extremely fine mesoscale features of a TC to subjectively categorize ocular attributes such as spatial extent, shape, and wavenumber. Unlike the visible and infrared channels, SAR TC retrievals are weakly linked to physical processes, and radar retrievals from space are rare, but PMW may directly define TC structure by piercing through the nonraining cloud tops [123].

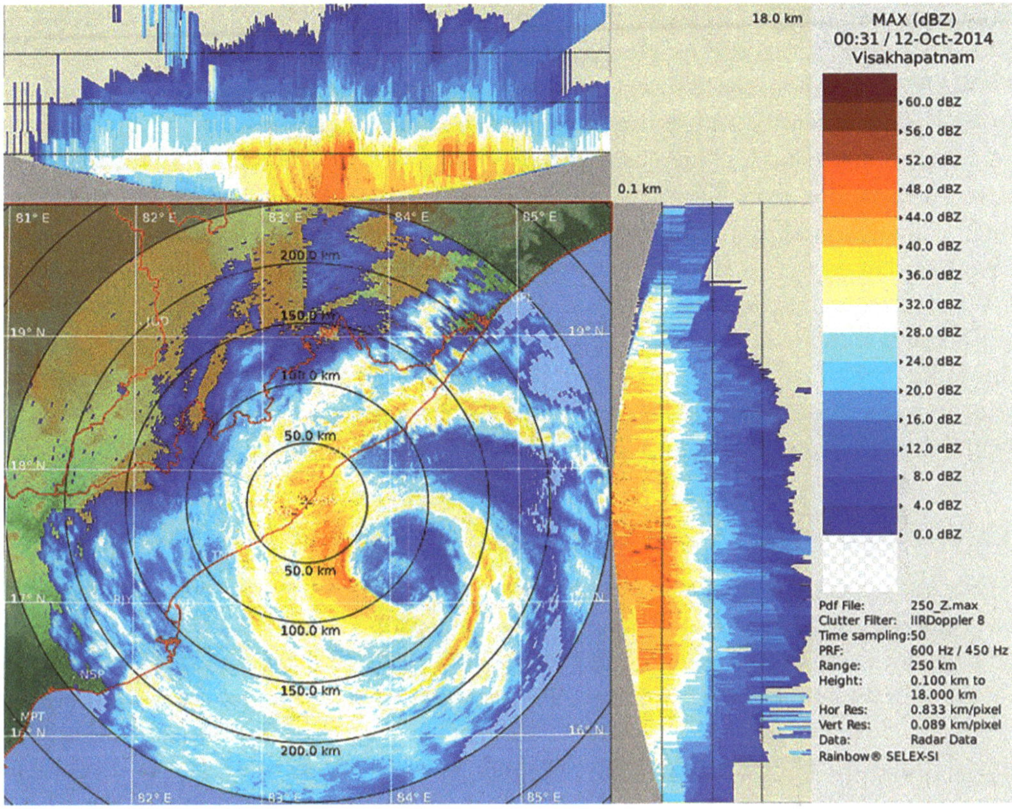

Fig. 18.6 ***The landfalling of HUHUD TC seen through the Visakhapatnam DWR.*** Image courtesy: India Meteorological Department (IMD).

18.3.2 Modeling-based analysis using radar observations

Significant improvement has been made in the short-range forecasts of TCs over the previous few decades, particularly in track forecasting, forecasting of TC formation and intensity changes [124]. At the meso- and micro-scales, DWR-based high-resolution reflectivity and radial velocity fields are critical for weather research and forecasting. The reflectivity and velocity fields from DWR can reveal substantial information of the TC eyewall, inner-core winds, and hydrometeor structures when the TC is within the DWR's range [121,125]. Even if the TC is outside of the DWR range, it can still give potentially important data on the TC environment. But, even with the assimilation of radar observations, our ability to initiate a TC with dynamically consistent shape and intensity remains restricted, despite advancements in employing advanced data assimilation methods with or without initial vortex bogussing (e.g., [111,126,127]).

Previous research works demonstrated that high-resolution radar inner-core data have an overall positive effect on analysis and prediction, particularly over the United

States and for landfalling TCs in the Atlantic basin (e.g., [185,128]). Xiao et al. [129], and Zhao and Jin [130] employed three-dimensional variational data assimilation (3DVAR) systems in the context of several regional/mesoscale models to enhance storm course, intensity, and structure prediction. When radial velocity and reflectivity are coupled, track and intensity assumptions are improved compared to the independent data. Furthermore, ensemble Kalman filter could be used to assimilate DWR measurements (both ground and aerial based) to enhance TC [131,132] and severe convection forecasts. Routray et al. [133,134] and Prasad et al. [135] demonstrated that including DWR radial velocity and reflectivity in a numerical modeling framework through data assimilation techniques can help improving the predictability of convective events over the Indian region. Govindankutty et al. [136] found that DWR radial wind and global telecommunication system (GTS) data had a beneficial influence on the severity and spatial distribution of landfalling TCs.

18.4 Tropical cyclone characteristics through Lidar observations

While the quality and quantity of meteorological observations have increased, there are still significant regions in a TC where the data are sparse [137–139]. The boundary layer and precipitation-free zones, such as moats between the eyewall and outer rain bands or the eyewall and secondary eyewall, are examples of such regions (Fig. 18.7; [182]). Understanding the physical processes and dynamics in these locations, and eventually linking them to TC genesis and development in order to enhance TC predictions, is a top priority for researchers [140–142]. During a concurrent TC overpass, the space-based polar orbiting radars termed as scatterometers in ASCAT and SCATSAT-1 satellites take a picture of the surface wind speeds occasionally. Surface wind speed estimates are also provided by NASA's newly launched Cyclone Global Navigation Satellite System; however, validation of these data is still under progress [143]. Hurricane reconnaissance and research aircraft, when available, use several observing platforms to give wind data. Despite the fact that GPS dropsondes are commonly used during reconnaissance flights, they only monitor discrete profiles with limited spatiotemporal coverage. The most comprehensive coverage of the TC wind field is provided by the tail Doppler wind radar (TDR); nevertheless, TDR data are inaccurate below 500 m and are unable to give readings in precipitation-free regions [144]. In order to close these observational gaps, an additional sensor known as an Airborne Doppler Wind Lidar (ADWL) can be used to get more frequent and precise wind profiles in the near-storm and inner-core TC regions.

Depending on the kind of ADWL, it detects the Doppler shift of the backscattered return pulse along the laser beam induced by the motion of aerosols or molecules. Coherent detection and direct detection are the two main types of ADWL systems [145]. The measurement of Mie backscatter from aerosol or cloud particles is used

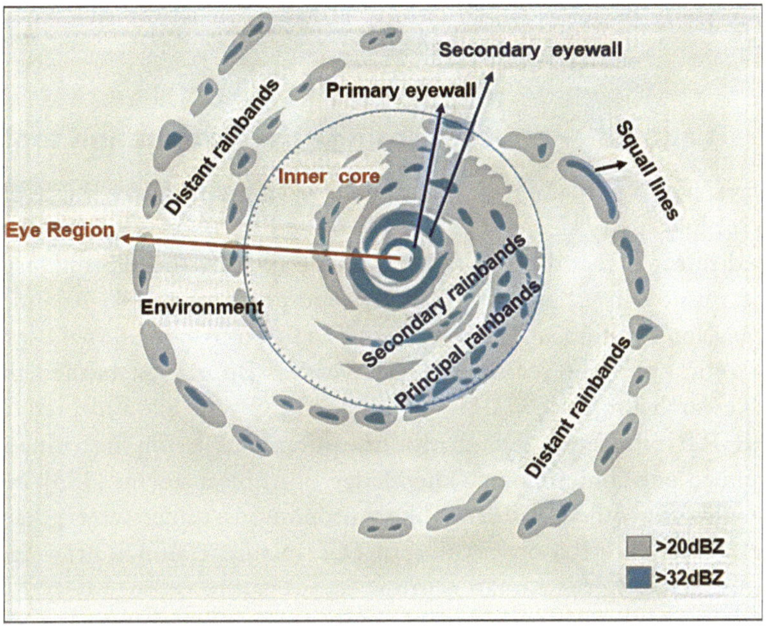

Fig. 18.7 ***Schematic illustration of spiral rainbands, eyewall, and the eye in a tropical cyclone.*** The figure is reproduced using COREL DRAW and adopted from Houze [184].

in coherent detection whereas direct detection uses Mie and Rayleigh backscatter from particles and smaller air molecules to measure the winds. Given the presence of aerosols, clouds, or molecules, this device offers the benefit of more precise wind measurement height assignments within an atmospheric column. The first ADWLs to fly in the TC environment was in 2008 as part of the Pacific Asian Regional Campaign of the Observing System Research and Predictability Experiment (THORPEX) (T-PARC). The WP-3D Orion of the US Naval Research Laboratory was fitted with a coherent ADWL and orbited Typhoon Nuri [146,147]. During NASA's Genesis and Rapid Intensification Processes (GRIP) field program, the first Atlantic hurricane coherent ADWL was flown into Hurricane Earl aboard a DC-8 aircraft in 2010. According to the study by Bucci et al. [148], by collecting data in locations generally unseen by conventional instruments, the ADWL can give both complementary and distinct observations in the TC environment. These devices, when used in combination with the tail Doppler wind radar (TDR), can capture wind data in both precipitation and clear air.

In India, National Atmospheric Research Laboratory, a unit of Indian Space Research Organization leads the research works relating to Lidar technology. It hosts the Indian Lidar Network program. It operates Rayleigh/Mie Lidar, Boundary-Layer Lidar, and Sodium Lidar. The Indian Lidar network has been used mainly in the study of aerosols and clouds. However, it is not well known for any TC-related research.

18.5 Aircraft and other remote sensing observations and tools

Since the first flight piloted by United States Army Air Force Lt. Col. Joseph P. Duckworth on July 17, 1943, aircrafts have been used for observing TCs [149]. And the first dedicated research flight into a TC was undertaken by the US military [150,151], and a second mission investigated the upper troposphere of a 1947 Atlantic TC [152]. Aircrafts have a long history in TC research, in addition to operational uses. Following the 1954 Atlantic hurricane season, the US Weather Bureau established the National Hurricane Research Project (NHRP) to enhance TC research and predictions. Soon after, the NHRP pioneered quantitative measurements using instrumented planes, which influenced our contemporary knowledge of tropical storms [153]. Not only the aircrafts, but also the other platforms like Autonomous underwater gliders, Drifters, profilers, etc. are used for RS observation of TCs. In this section, a brief description of such platforms including the aircraft is given.

18.5.1 Aircraft platforms

The NOAA bought two WP-3D (P-3) aircraft in the mid-1970s, which are modified US Navy maritime surveillance aircraft that are commercial Lockheed Electra variants [154]. Observations of TC structure and dynamics, involvement in TC modification experiments, and monitoring of TC formation were all proposed as possible applications of the aircraft, all with the objective of enhancing the TC prediction and warning process to reduce damage and loss of life. Since then, the NOAA Aircraft Operations Center has kept the P-3s as one of the world's best meteorological research planes. The P-3s can also acquire upper ocean observations in addition to collecting atmospheric data [155–159]. With all of the sophisticated instruments mounted, two P-3 aircrafts are the most modern and technologically advanced aircrafts in operation today for meteorological and ocean observations.

In 1982, TC numerical modeling was still in its infancy, and statistical track assistance was the most precise track guidance available to forecasters. NOAA began testing the concept that more vertical wind and thermodynamic profiles in the TC environment (the "synoptic flow") would enhance analyses and numerical model predictions that give human forecasters operational direction. Through 1996, data collected on P-3 research missions helped to considerably minimize inaccuracies in TC track estimates from global and regional dynamical models [160]. The success of this research effort led to the creation of a new generation of dropwindsonde [161] and the procurement of the Gulfstream-IV-SP (G-IV) as an operational platform for performing synoptic surveillance flights. Dropwindsondes are generally used to measure wind profile, vertical pressure, humidity, and temperature as they descend on a parachute while being deployed from aircrafts. These operational flights, along with ongoing research into the optimum targeting and sampling techniques, as well as data assimilation, have helped the National Centers for Environmental Prediction considerably improve operational global model track forecasts [162].

The Dropwindsonde Observations for Typhoon Surveillance near the Taiwan Region began in 2003, encouraged by the partial success of the NOAA G-IV [163,164]. The track model estimates have been significantly improved because to data from dropwindsondes launched by the Advanced Stealth Technology Reconnaissance Aircraft [164].

The 53rd Weather Reconnaissance Squadron of the United States Air Force Reserve operates a fleet of ten Lockheed-Martin C-130J aircraft that are primarily used for operational hurricane fixing (finding the center location and storm intensity) in the Atlantic, Eastern, and Central Pacific basins. These aircrafts, like the NOAA P-3s, collect and transmit flight-level wind, pressure, temperature, and moisture data, as well as stepped frequency microwave radiometer surface wind speed measurements and dropwindsonde data [165].

In India, aircraft platforms are used mainly for aerosol and cloud related measurements. Indian Institute of Tropical Meteorology uses such a platform for conducting Cloud Aerosol Interaction and Precipitation Enhancement Experiment. It was carried out in three phases during 2009–2015 for the study of aerosol-cloud-precipitation interaction over India [166]; https://www.tropmet.res.in/~caipeex/). However, there is no well-known evidence of conducting such experiments using aircrafts for the TC-related research.

18.5.2 Other remote sensing platforms

Autonomous underwater gliders [167] offer temperature and salinity readings in the upper 100 m of the ocean and are quickly becoming important components of the ocean observing system [168,169]. Gliders can be configured to monitor current profiles, bio-optical characteristics, dissolved oxygen, and turbulent microstructure in addition to the conventional measures of temperature, salinity, and depth-average currents (e.g., [170]). They cover key observational gaps in the ocean observing system due to their adaptability and flexibility [168], notably in terms of TC intensity forecasts. Gliders are especially useful for studying ocean dynamics on continental shelves before and during TC impact, where other ocean data are sparse (e.g., [171,172]). The bulk of underwater glider data gathered in support of Atlantic Hurricane research and forecasts is delivered in real-time to the GTS and the US Integrated Ocean Observing System underwater glider data assembly center (U.S. Glider Data Assembly Center: https://gliders.ioos.us/data/), where it is made immediately available to operational forecast centers.

Drifters of various types provide data on ocean current velocity, SST, and SLP, which is used to enhance weather forecasting of extreme events such as TCs. Continuous global drifter observations are used to constrain satellite SST errors and biases (e.g., [173]), and assimilation of in situ SLP observations has improved global weather forecasting [174–176]. Horányi et al. [175] demonstrated that, in the event of cyclogenesis, SLP measurements from drifters resulted in significant reductions in prediction errors, on the other hand, the impact of SLP drifter observation on TC track forecasting, has yet to be investigated.

Besides, air deployed profiling instruments and profiling floats are also used to study the ocean characteristics and conditions before and after the occurrence of TCs. In the Atlantic and Pacific basins, airborne profiling equipment are frequently deployed in targeted sampling mode ahead of TCs. In the Atlantic, the NOAA Hurricane Field Program uses Airborne eXpendable Bathythermographs (AXBTs), Airborne eXpendable Conductivity Temperature and Depth, and Airborne eXpendable Current Profilers to survey ocean conditions in front of and beneath TCs [177–179]. Profiling floats (e.g., [180,181]) have the benefit of giving a long-term, large-scale record of the majority of the world's oceans. Observations of temperature and salinity from Argo floats are frequently included into operational ocean models (e.g., [65]) that are used to initialize the ocean component of hurricane forecast models.

In India, Indian National Centre for Ocean Information Services Hyderabad operates the Argo floats, various types of Buoys (namely, moored, drifting, and rama), gliders and profilers (namely, XBT), which provide the upper ocean in-situ observations for the NIO basin (https://incois.gov.in/portal/datainfo/insituhome.jsp). These observations may be used for the TC-related research. However, sophisticated drifters, profilers and underwater gliders are not known to be in use explicitly for TC research although used in general.

18.6 Concluding remarks

The TC research and operational forecasting have surely got significant benefits from satellite observations, which has helped us in better understanding of TC physics and dynamics. Numerous satellite observations which are obtained utilizing VIS/IR and MW (e.g., those in SCATSAT, OCEANSAT, etc.) sensors are considered quiet crucial for operational meteorological applications today. Other RS platforms like radar, lidar, and aircraft are also used for TC-related research in many countries. Ocean observations on the other hand are essential for this purpose as they allow researchers to investigate the complexities of air–sea interaction mechanisms that might contribute to the genesis and intensification of TCs. Furthermore, data obtained from underwater gliders, profilers, drifters, and buoy (such as ARGO) are also used in TC research. Besides operational forecasting, these RS observations help in studying the meteorological, radial, structural, and ocean characteristics associated with TCs.

It is pertinent that the satellite RS observations have played a significant role in the TC research and forecasting, though other platforms like ocean-based equipment, radar,

LIDAR, and aircraft mostly act as complementary sources of observations. Most of these complementary sources are used extensively for TC research in advanced countries such as the USA. Although India has progressed quite extensively in acquiring and usage of such datasets for TC research, it will take some time for India to reach the level achieved by other advanced countries. Nevertheless, India is doing quite well in terms of using the satellite RS observations for the TC research and operational forecasting during last two decades. And, in the last decade, it has put substantial resources for improving the ocean-based observational sources and radar network. In a similar time-frame, that is, post-2010, it has improved its computational resources too. Such improvements have helped in data acquisition and processing from satellite, ocean-based, and radar platforms.

References

[1] W.M. Gray, Global view of the origin of tropical disturbances and storms, Monthly Weather Review 96 (10) (1968) 669–700, doi:10.1175/1520-0493(1968)096%3C0669:GVOTOO%3E2.0.CO;2.
[2] W.H. Schubert, J.J. Hack, Inertial stability and tropical cyclone development, Journal of the Atmospheric Sciences 39 (8) (1982) 1687–1697, doi:10.1175/1520-0469(1982)039%3C1687:ISATCD%3E2.0.CO;2.
[3] M. DeMaria, J.D. Pickle, A simplified system of equations for simulation of tropical cyclones, Journal of the atmospheric sciences 45 (10) (1988) 1542–1554, doi:10.1175/1520-0469(1988)045%3C1542:ASSOEF%3E2.0.CO;2.
[4] D.S. Nolan, For tropical cyclogenesis, Australian Meteorological Magazine 56 (4) (2007).
[5] G.J. Holland, An analytic model of the wind and pressure profiles in hurricanes, Monthly weather review *108* (8) (1980) 1212–1218, doi:10.1175/1520-0493(1980)108%3C1212:AAMOTW%3E2.0.CO;2.
[6] K.E. Trenberth, The extreme weather events of 1997 and 1998, Consequences 5 (1) (1999) 3–15.
[7] K.E. Trenberth, A. Dai, R.M. Rasmussen, D.B. Parsons, The changing character of precipitation, Bulletin of the American Meteorological Society 84 (9) (2003) 1205–1218, doi:10.1175/BAMS-84-9-1205.
[8] V.F. Dvorak, Tropical clouds and cloud systems observed in satellite imagery: Tropical cyclones, 2, Workbook, 1995.
[9] V.F. Dvorak, Tropical cyclone intensity analysis using satellite data. In: NOAA Technical Report NESDIS 11, NTIS, Springfield, VA 22161, 1984, pp. 45.
[10] S. Quiring, A. Schumacher, C. Labosier, L. Zhu, Variations in mean annual tropical cyclone size in the Atlantic, Journal of Geophysical Research: Atmospheres 116 (D9) (2011), doi:10.1029/2010JD015011.
[11] N. Jaiswal, D.T.T. Ha, C.M. Kishtawal, Estimation of size of tropical cyclones in the North Indian Ocean using Oceansat-2 scatterometer high-resolution wind products, Theor. Appl. Climatol. *136* (1) (2019) 45–53, https://link.springer.com/article/10.1007/s00704-018-2464-y.
[12] K. Singh, J. Panda, S.S. Rath, Variability in landfalling trends of cyclonic disturbances over North Indian ocean region during current and pre-warming climate, Theor App Climatol 137 (1) (2019) 417–439, doi:10.1007/s00704-018-2605-3.
[13] K. Singh, J. Panda, M. Sahoo, M. Mohapatra, Variability in tropical cyclone climatology over North Indian ocean during the period 1891 to 2015, Asia-Pacific J Atmos Sci 55 (2) (2019) 269–287, doi:10.1007/s13143-018-0069-0.
[14] K. Singh, J. Panda, M. Mohapatra, Robustness of best track data and associated cyclone activity over the North Indian ocean region during and prior to satellite era, J Earth Syst Sci 129 (1) (2020) 1–20, doi:10.1007/s12040-020-1344-x.
[15] S. Kumar, J. Panda, K. Singh, B.K. Guha, S. Kant, Structural characteristics of North Indian Ocean tropical cyclones during 1999–2017: a scatterometer observation-based analysis, Theor. Appl. Climatol. *143* (1) (2021) 227–240, doi:10.1007/s00704-020-03431-w.

[16] C.S. Lee, K.K. Cheung, W.T. Fang, R.L. Elsberry, Initial maintenance of tropical cyclone size in the western North Pacific, Monthly Weather Review *138* (8) (2010) 3207–3223, doi:10.1175/2010MWR3023.1.
[17] G. Villarini, J.A. Smith, M.L. Baeck, T. Marchok, G.A. Vecchi, Characterization of rainfall distribution and flooding associated with US landfalling tropical cyclones: Analyses of Hurricanes Frances, Ivan, and Jeanne (2004), Journal of Geophysical Research: Atmospheres 116 (D23) (2011), doi:10.1029/2011JD016175.
[18] H. Murakami, Y. Wang, H. Yoshimura, R. Mizuta, M. Sugi, E. Shindo, …, A. Kitoh, Future changes in tropical cyclone activity projected by the new high-resolution MRI-AGCM, Journal of Climate 25 (9) (2012) 3237–3260, doi:10.1175/JCLI-D-11-00415.1.
[19] M.S. Park, M. Kim, M.I. Lee, J. Im, S. Park, Detection of tropical cyclone genesis via quantitative satellite ocean surface wind pattern and intensity analyses using decision trees, Remote sensing of environment 183 (2016) 205–214, doi:10.1016/j.rse.2016.06.006.
[20] F. Samrin, I. Irwana, N. Hasanah, Analysis of the Meteorological Condition of Tropical Cyclone Cempaka and Its Effect on Heavy Rainfall in Java Island, IOP Conference Series: Earth and Environmental Science, 303, IOP Publishing, 2019, July 012065.
[21] N. Jaiswal, C.M. Kishtawal, P.K. Pal, Cyclone intensity estimation using similarity of satellite IR images based on histogram matching approach, Atmos. Res. *118* (2012) 215–221, http://dx.doi.org/10.1016/j.atmosres.2012.07.006.
[22] U.C. Mohanty, K.K. Osuri, S. Pattanayak, P. Sinha, An observational perspective on tropical cyclone activity over Indian seas in a warming environment, Natural hazards 63 (3) (2012) 1319–1335, doi:10.1007/s11069-011-9810-z.
[23] J. Panda, H. Singh, P.K. Wang, R.K. Giri, A. Routray, A qualitative study of some meteorological features during tropical cyclone PHET using satellite observations and WRF modeling system, J. Indian Soc. Remote Sens. *43* (1) (2015) 45–56, http://dx.doi.org/10.1007/s12524-014-0386-4.
[24] K.K. Osuri, R. Nadimpalli, U.C. Mohanty, D. Niyogi, Prediction of rapid intensification of tropical cyclone Phailin over the Bay of Bengal using the HWRF modelling system, Quarterly Journal of the Royal Meteorological Society 143 (703) (2017) 678–690, doi:10.1002/qj.2956.
[25] R. Nadimpalli, K.K. Osuri, U.C. Mohanty, A.K. Das, A. Kumar, S. Sil, D. Niyogi, Forecasting tropical cyclones in the Bay of Bengal using quasi-operational WRF and HWRF modeling systems: an assessment study, Meteorology and Atmospheric Physics 132 (1) (2020) 1–17, doi:10.1007/s00703-019-00669-6.
[26] J.A. Knaff, R.M. Zehr, Re-examination of tropical cyclone wind–pressure relationships, Weather and Forecasting *22* (1) (2007) 71–88, doi:10.1175/WAF965.1.
[27] J.A. Knaff, B.A. Harper, Tropical cyclone surface wind structure and wind-pressure relationships. Seventh Int, Workshop on Tropical Cyclones, La Reunion, France, WMO, KN, 2010.
[28] V.F. Dvorak, Tropical cyclone intensity analysis and forecasting from satellite imagery, Mon. Weath. Rev. 103 (1975) 420–430, doi:10.1175/1520-0493(1975)103< 0420:TCIAAF>2.0.CO;2.
[29] R.A. Anthes, Tropical cyclones: structure, computer simulation models, and operational prediction, Contemp. Phys. *22* (6) (1981) 643–680.
[30] S.B. Cocks, W.M. Gray, Variability of the outer wind profiles of western North Pacific typhoons: classifications and techniques for analysis and forecasting, Monthly weather review *130* (8) (2002) 1989–2005, doi:10.1175/1520-0493(2002)130%3C1989:VOTOWP%3E2.0.CO;2.
[31] K. Bessho, M. DeMaria, J.A. Knaff, Tropical cyclone wind retrievals from the Advanced Microwave Sounding Unit: Application to surface wind analysis, Journal of applied meteorology and climatology 45 (3) (2006) 399–415, doi:10.1175/JAM2352.1.
[32] M. Mohapatra, M. Sharma, Characteristics of surface wind structure of tropical cyclones over the north Indian Ocean, Journal of Earth System Science 124 (7) (2015) 1573–1598, https://www.ias.ac.in/article/fulltext/jess/124/07/1573-1598.
[33] E.N. Rappaport, J.L. Franklin, L.A. Avila, S.R. Baig, J.L. Beven, E.S. Blake, A.N. Tribble, Advances and challenges at the National Hurricane Center, Weather and Forecasting *24* (2) (2009) 395–419, doi:10.1175/2008WAF2222128.1.
[34] W.R. Holland, P.B. Rhines, An example of eddy-induced ocean circulation, J. Phys. Oceanogr. *10* (7) (1980) 1010–1031, doi:10.1175/1520-0485(1980)010%3C1010:AEOEIO%3E2.0.CO;2.

[35] K. Kikuchi, B. Wang, H. Fudeyasu, Genesis of tropical cyclone Nargis revealed by multiple satellite observations, Geophysical Research Letters 36 (6) (2009), doi:10.1029/2009GL037296.
[36] K.T. Chan, J.C. Chan, Global climatology of tropical cyclone size as inferred from QuikSCAT data, Int. J. Climatol. *35* (15) (2015) 4843–4848, doi:10.1002/joc.4307.
[37] Singh, K., Panda, J., Osuri, K. K., & Vissa, N. K. (2016). Progress in tropical cyclone predictability and present status in the North Indian Ocean region. in recent developments in tropical cyclone dynamics, prediction, and detection, 193-215. doi: 10.5772/64333
[38] U.C. Mohanty, M. Mandal, S. Raman, Simulation of Orissa super cyclone (1999) using PSU/NCAR mesoscale model, Natural Hazards 31 (2) (2004) 373–390, doi:10.1023/B:NHAZ.0000023358.385 36.5d.
[39] J. Panda, R.K. Giri, A comprehensive study of surface and upper-air characteristics over two stations on the west coast of India during the occurrence of a cyclonic storm, Natural Hazards *64* (2) (2012) 1055–1078, http://dx.doi.org/10.1007/s11069-012-0282-6.
[40] J. Panda, R.K. Giri, K.H. Patel, A.K. Sharma, R.K. Sharma, Impact of satellite derived winds and cumulus physics during the occurrence of the tropical cyclone Phyan, Indian Journal of Science and Technology *4* (8) (2011) 859–875, http://dx.doi.org/10.17485/ijst/2011/v4i8/30885.
[41] J. Panda, H. Singh, P.K. Wang, R.K. Giri, A. Routray, A qualitative study of some meteorological features during tropical cyclone PHET using satellite observations and WRF modeling system, Journal of the Indian Society of Remote Sensing 43 (1) (2015) 45–56, doi:10.1007/s12524-014-0386-4.
[42] J.L. Demuth, M. DeMaria, J.A. Knaff, Improvement of advanced microwave sounding unit tropical cyclone intensity and size estimation algorithms, Journal of applied meteorology and climatology *45* (11) (2006) 1573–1581, doi:10.1175/JAM2429.1.
[43] M. Balaji, A. Chakraborty, M. Mandal, Changes in tropical cyclone activity in north Indian Ocean during satellite era (1981–2014), International Journal of Climatology 38 (6) (2018) 2819–2837, doi:10.1002/joc.5463.
[44] B. Johnson, S. Thomas, J.S. Rani, A novel framework for objective detection and tracking of TC center from noisy satellite imagery, Adv. Space Res. *62* (1) (2018) 44–54, http://dx.doi.org/10.1016/j.asr.2018.04.017.
[45] R. Senthilkumar, K. Porkumaran, A new approach to low density cloud removal (LDCR) of TC in geostationary meteorological satellite imagery, 2014 International Conference on Green Computing Communication and Electrical Engineering (ICGCCEE), IEEE, 2014, pp. 1–3.
[46] H. Hu, F. Weng, Y. Han, Y. Duan, Remote sensing of tropical cyclone thermal structure from satellite microwave sounding instruments: Impacts of background profiles on retrievals, Journal of Meteorological Research 33 (1) (2019) 89–103, doi:10.1007/s13351-019-8094-1.
[47] C. Wang, Q. Xu, X. Li, Y. Cheng, CNN-Based Tropical Cyclone Track Forecasting from Satellite Infrared Images, IGARSS 2020-2020 IEEE International Geoscience and Remote Sensing Symposium, IEEE, 2020, pp. 5811–5814 10.1109/IGARSS39084.2020.9324408.
[48] T.N. Venkatesh, J. Mathew, Prediction of tropical cyclone genesis using a vortex merger index, Geophys. Res. Lett. *31* (4) (2004), http://dx.doi.org/10.1029/2003GL019005.
[49] D.B. Chelton, F.J. Wentz, C.L. Gentemann, R.A. de Szoeke, M.G. Schlax, Satellite microwave SST observations of transequatorial tropical instability waves, Geophysical Research Letters 27 (9) (2000) 1239–1242, doi:10.1029/1999GL011047.
[50] K.T. Chan, J.C. Chan, Size and strength of tropical cyclones as inferred from QuikSCAT data, Monthly weather review *140* (3) (2012) 811–824, doi:10.1175/MWR-D-10-05062.1.
[51] N. Jaiswal, C.M. Kishtawal, Structural analysis of tropical cyclone using INSAT-3D observations, in: Eastwood Im, Raj Kumar, Song Yang, Remote Sensing of the Atmosphere, Clouds, and Precipitation VI, 9876, International Society for Optics and Photonics, 2016, https://doi.org/10.1117/12.2223508.
[52] N. Jaiswal, C.M. Kishtawal, Prediction of tropical cyclogenesis using scatterometer data, IEEE Trans Geosci Remote Sens 49 (2018) 12, doi:10.1109/TGRS.2011.2153862.
[53] H.E. Willoughby, M.E. Rahn, Parametric representation of the primary hurricane vortex. Part I: Observations and evaluation of the Holland (1980) model, Monthly Weather Review 132 (12) (2004) 3033–3048, doi:10.1175/MWR2831.1.
[54] S.K. Deb, C.M. Kishtawal, P.K. Pal, Impact of Kalpana-1-derived water vapor winds on Indian Ocean tropical cyclone forecasts, Monthly Weather Review 138 (3) (2010) 987–1003, doi:10.1175/2009MWR3041.1.

[55] S.K. Mandke, A.K. Sahai, Twin tropical cyclones in the Indian Ocean: the role of equatorial waves, Natural Hazards 84 (3) (2016) 2211–2224, doi:10.1007/s11069-016-2546-z.
[56] D.M. Legler, H.J. Freeland, R. Lumpkin, G. Ball, M.J. McPhaden, S. North, M.A. Merrifield, The current status of the real-time in situ global ocean observing system for operational oceanography, Journal of Operational Oceanography 8 (sup2) (2015) s189–s200, doi:10.1080/1755876X.2015.1049883.
[57] J. Dong, R. Domingues, G. Goni, G. Halliwell, H.S. Kim, S.K. Lee, L. Pomales, Impact of assimilating underwater glider data on Hurricane Gonzalo (2014) forecasts, Weather and Forecasting 32 (3) (2017) 1143–1159, doi:10.1175/WAF-D-16-0182.1.
[58] A. Kuwano-Yoshida, S. Minobe, Storm-track response to SST fronts in the north western Pacific region in an AGCM, Journal of Climate 30 (3) (2017) 1081–1102, doi:10.1175/JCLI-D-16-0331.1.
[59] D.F. Leipper, D. Volgenau, Hurricane heat potential of the Gulf of Mexico, (1972). doi: 10.1175/1520-0485(1972)002%3C0218:HHPOTG%3E2.0.CO;2
[60] M. Mainelli, M. DeMaria, L.K. Shay, G. Goni, Application of oceanic heat content estimation to operational forecasting of recent Atlantic category 5 hurricanes, Weather and Forecasting 23 (1) (2008) 3–16, doi:10.1175/2007WAF2006111.1.
[61] F. Sanders, J.R. Gyakum, Synoptic-dynamic climatology of the "bomb, Monthly Weather Review 108 (10) (1980) 1589–1606, doi:10.1175/1520-0493(1980)108%3C1589:SDCOT%3E2.0.CO;2.
[62] G. Goni, M. DeMaria, J. Knaff, C. Sampson, J. Price, A. Mehra, K. Kang, The ocean observing system for tropical cyclone intensification forecasts and studies, Proceedings of the" OceanObs, 9, 2010.
[63] S. Verrier, P.Y. Le Traon, E. Remy, J.M. Lellouche, Assessing the impact of SAR altimetry for global ocean analysis and forecasting, Journal of Operational Oceanography 11 (2) (2018) 82–86, doi:10.1080/1755876X.2018.1505028.
[64] T. Meissner, F.J. Wentz, D.M. Le Vine, The salinity retrieval algorithms for the NASA Aquarius version 5 and SMAP version 3 releases, Remote Sensing 10 (7) (2018) 1121, doi:10.3390/rs10071121.
[65] E.P. Chassignet, H.E. Hurlburt, E.J. Metzger, O.M. Smedstad, J.A. Cummings, G.R. Halliwell, …, J. Wilkin, US GODAE: global ocean prediction with the HYbrid Coordinate Ocean Model (HYCOM), Oceanography 22 (2) (2009) 64–75.
[66] M. Drévillon, R. Bourdallé-Badie, C. Derval, J.M. Lellouche, E. Rémy, B. Tranchant, L. Parent, The GODAE/Mercator-Ocean global ocean forecasting system: results, applications and prospects, Journal of Operational Oceanography 1 (1) (2008) 51–57 10.1080/1755876x.2008.11020095.
[67] M. DeMaria, J. Kaplan, A statistical hurricane intensity prediction scheme (SHIPS) for the Atlantic basin, Weather and Forecasting 9 (2) (1994) 209–220, http://dx.doi.org/10.1175/1520-0434(1994)009%3C0209:ASHIPS%3E2.0.CO;2.
[68] M. DeMaria, M. Mainelli, L.K. Shay, J.A. Knaff, J. Kaplan, Further improvements to the statistical hurricane intensity prediction scheme (SHIPS), Weather and Forecasting 20 (4) (2005) 531–543, doi:10.1175/WAF862.1.
[69] J. Kaplan, M. DeMaria, J.A. Knaff, A revised tropical cyclone rapid intensification index for the Atlantic and eastern North Pacific basins, Weather and forecasting 25 (1) (2010) 220–241 10.1175/2009WAF2222280.1.
[70] J.A. Knaff, C.R. Sampson, K.D. Musgrave, An operational rapid intensification prediction aid for the western North Pacific, Weather and Forecasting 33 (3) (2018) 799–811 10.1175/WAF-D-18-0012.1.
[71] M.M. Ali, P.V. Nagamani, N. Sharma, R.T. Venu Gopal, M. Rajeevan, G.J. Goni, M.A. Bourassa, Relationship between ocean mean temperatures and Indian summer monsoon rainfall, Atmos. Sci. Lett. 16 (3) (2015) 408–413, doi:10.1002/asl2.576.
[72] T. Venugopal, M.M. Ali, M.A. Bourassa, Y. Zheng, G.J. Goni, G.R. Foltz, M. Rajeevan, Statistical evidence for the role of southwestern Indian Ocean heat content in the Indian summer monsoon rainfall, Sci. Rep. 8 (1) (2018) 1–10, doi:10.1038/s41598-018-30552-0.
[73] K. Balaguru, G.R. Foltz, L.R. Leung, S.M. Hagos, D.R. Judi, On the use of ocean dynamic temperature for hurricane intensity forecasting, Weather and Forecasting 33 (2) (2018) 411–418, doi:10.1175/WAF-D-17-0143.1.
[74] J. Pennington, M. DeMaria, K. Williams, Development of a 10-year Atlantic basin tropical cyclone wind structure climatology, (2000).

[75] P.S. Chu, Large-scale circulation features associated with decadal variations of tropical cyclone activity over the central North Pacific, Journal of Climate 15 (18) (2002) 2678–2689, doi:10.1175/1520-0442(2002)015%3C2678:LSCFAW%3E2.0.CO;2.
[76] C.R. Sampson, J.A. Knaff, A consensus forecast for tropical cyclone gale wind radii, Weather and Forecasting 30 (5) (2015) 1397–1403, doi:10.1175/WAF-D-15-0009.1.
[77] K.R. Knapp, M.C. Kruk, D.H. Levinson, H.J. Diamond, C.J. Neumann, The international best track archive for climate stewardship (IBTrACS) unifying tropical cyclone data, Bull. Am. Meteorol. Soc. 91 (3) (2010) 363–376, doi:10.1175/2009BAMS2755.1.
[78] J.P. Kossin, K.R. Knapp, D.J. Vimont, R.J. Murnane, B.A. Harper, A globally consistent reanalysis of hurricane variability and trends, Geophys. Res. Lett. 34 (4) (2007).
[79] J.A. Knaff, S.P. Longmore, D.A. Molenar, An objective satellite-based tropical cyclone size climatology, Journal of Climate 27 (1) (2014) 455–476, doi:10.1175/JCLI-D-13-00096.1.
[80] D.R. Chavas, K.A. Emanuel, A QuikSCAT climatology of tropical cyclone size, Geophys. Res. Lett. 37 (18) (2010), doi:10.1029/2010GL044558.
[81] H. Takagi, W. Wu, Maximum wind radius estimated by the 50 kt radius: improvement of storm surge forecasting over the western North Pacific, Natural Hazards and Earth System Sciences 16 (3) (2016) 705–717, doi:10.5194/nhess-16-705-2016.
[82] M.L. Black, H.E. Willoughby, The concentric eyewall cycle of Hurricane Gilbert, Monthly weather review 120 (6) (1992) 947–957, doi:10.1175/1520-0493(1992)120%3C0947:TCECOH%3E2.0.CO;2.
[83] J.L. Franklin, S.J. Lord, S.E. Feuer, F.D. Marks Jr., The kinematic structure of Hurricane Gloria (1985) determined from nested analyses of dropwindsonde and Doppler radar data, Mon. Wea. Rev. 121 (1993) 2433–2451, doi:10.1175/1520-0493(1993)121%3C2433:TKSOHG%3E2.0.CO;2.
[84] P. Dodge, R.W. Burpee, F.D. Marks Jr, The kinematic structure of a hurricane with sea level pressure less than 900 mb, Monthly Weather Review 127 (6) (1999) 987–1004, doi:10.1175/1520-0493(1999)127%3C0987:TKSOAH%3E2.0.CO;2.
[85] L.J. Shapiro, H.E. Willoughby, The response of balanced hurricanes to local sources of heat and momentum, Journal of the Atmospheric Sciences 39 (2) (1982) 378–394, doi:10.1175/1520-0469(1982)039%3C0378:TROBHT%3E2.0.CO;2.
[86] H.E. Willoughby, J.A. Clos, M.G. Shoreibah, Concentric eye walls, secondary wind maxima, and the evolution of the hurricane vortex, Journal of Atmospheric Sciences 39 (2) (1982) 395–411, doi:10.1175/1520-0469(1982)039%3C0395:CEWSWM%3E2.0.CO;2.
[87] H.E. Willoughby, Temporal changes of the primary circulation in tropical cyclones, Journal of Atmospheric Sciences 47 (2) (1990) 242–264, doi:10.1175/1520-0469(1990)047%3C0242:TCOTPC%3E2.0.CO;2.
[88] J.P. Kossin, M. Sitkowski, An objective model for identifying secondary eyewall formation in hurricanes, Monthly Weather Review 137 (3) (2009) 876–892, http://dx.doi.org/10.1029/2006GL028836.
[89] L. Dean, K.A. Emanuel, D.R. Chavas, On the size distribution of Atlantic tropical cyclones, Geophys. Res. Lett. 36 (14) (2009), doi:10.1029/2009.
[90] R.T. Merrill, A comparison of large and small tropical cyclones, Monthly Weather Review 112 (7) (1984) 1408–1418, doi:10.1175/1520-0493(1984)112%3C1408:ACOLAS%3E2.0.CO;2.
[91] K.J. Mueller, M. DeMaria, J. Knaff, J.P. Kossin, T.H. Vonder Haar, Objective estimation of tropical cyclone wind structure from infrared satellite data, Weather and forecasting 21 (6) (2006) 990–1005, doi:10.1175/WAF955.1.
[92] S. Brand, Very large and very small typhoons of the western North Pacific Ocean, J. Meteor. Soc. Japan 50 (1972) 332–341 10.2151/jmsj1965.50.4_332.
[93] G.J. Holland, R.T. Merrill, On the dynamics of tropical cyclone structural changes, Q. J. R. Meteorolog. Soc. 110 (465) (1984) 723–745, doi:10.1002/qj.49711046510.
[94] S.K. Kimball, M.S. Mulekar, A 15-year climatology of North Atlantic tropical cyclones. Part I: Size parameters, Journal of Climate 17 (18) (2004) 3555–3575, doi:10.1175/1520-0442(2004)017%3C3555:AYCONA%3E2.0.CO;2.
[95] T.B. McKenzie III, A climatology of tropical cyclone size in the western North Pacific using an alternative metric. In: Doctoral dissertation, The Florida State University, Ann Arbor, MI, 2017, 1346, https://www.proquest.com/openview/564a4d639dca3399db24dcd2342f643b/1?pq-origsite=gscholar&cbl=18750.

[96] Y. Lin, M. Zhao, M. Zhang, Tropical cyclone rainfall area controlled by relative sea surface temperature, Nat. Commun. 6 (1) (2015) 1–7, http://dx.doi.org/10.1038/ncomms7591.
[97] C.J. Matyas, A geospatial analysis of convective rainfall regions within tropical cyclones after landfall, International Journal of Applied Geospatial Research (IJAGR) 1 (2) (2010) 71–91, http://dx.doi.org/10.4018/jagr.2010020905.
[98] C.A. Carrasco, C.W. Landsea, Y.L. Lin, The influence of tropical cyclone size on its intensification, Weather and Forecasting 29 (3) (2014) 582–590, doi:10.1175/WAF-D-13-00092.1.
[99] F. Mouton, O. Nordbeck, P. Peduzzi, Cyclone Database Manager, A tool for converting point data from cyclone observations into tracks and wind speed profiles in a GIS, UNEP/GRID-Europe, Geneva, 2002 Technical Report, 67.
[100] H.C. Weber, On the pressure-wind relationship in tropical cyclones, Preprints, 27th Conf. on Hurricanes and Tropical Meteorology, Monterey, CA, Amer. Meteor. Soc., Vol. 6, 2006 P14A.
[101] K.S. Liu, J.C. Chan, Size of tropical cyclones as inferred from ERS-1 and ERS-2 data, Monthly Weather Review 127 (12) (1999) 2992–3001, doi:10.1175/1520-0493(1999)127%3C2992:SOTCAI%3E2.0.CO;2.
[102] Liu, Q., Marchok, T., Pan, H. L., Bender, M., & Lord, S. (2000). Improvements in hurricane initialization and forecasting at NCEP with global and regional (GFDL) models. US Department of Commerce, National Oceanic and Atmospheric Administration, National Weather Service, Office of Meteorology, Science Division.
[103] T. Fujita, Pressure distribution within typhoon, Geophys. Mag. 23 (1952) 437–451.
[104] Y. Kurihara, M.A. Bender, R.J. Ross, An initialization scheme of hurricane models by vortex specification, Monthly weather review 121 (7) (1993) 2030–2045, doi:10.1175/1520-0493(1993)121%3C2030:AISOHM%3E2.0.CO;2.
[105] Y. Kurihara, M.A. Bender, R.E. Tuleya, R.J. Ross, Improvements in the GFDL hurricane prediction system, Monthly Weather Review 123 (9) (1995) 2791–2801, doi:10.1175/1520-0493(1995)123%3C2791:IITGHP%3E2.0.CO;2.
[106] Q. Xiao, X. Zou, B. Wang, Initialization and simulation of a landfalling hurricane using a variational bogus data assimilation scheme, Monthly weather review 128 (7) (2000) 2252–2269, doi:10.1175/1520-0493(2000)128%3C2252:IASOAL%3E2.0.CO;2.
[107] Q. Xiao, Y.H. Kuo, Y. Zhang, D.M. Barker, D.J. Won, A tropical cyclone bogus data assimilation scheme in the MM5 3D-Var system and numerical experiments with Typhoon Rusa (2002) near landfall, Journal of the Meteorological Society of Japan. Ser. II 84 (4) (2006) 671–689, doi:10.2151/jmsj.84.671.
[108] M.A. Bender, I. Ginis, R. Tuleya, B. Thomas, T. Marchok, The operational GFDL coupled hurricane–ocean prediction system and a summary of its performance, Monthly Weather Review 135 (12) (2007) 3965–3989, doi:10.1175/2007MWR2032.1.
[109] L.F. Hsiao, C.S. Liou, T.C. Yeh, Y.R. Guo, D.S. Chen, K.N. Huang, J.H. Chen, A vortex relocation scheme for tropical cyclone initialization in Advanced Research WRF, Monthly Weather Review 138 (8) (2010) 3298–3315, doi:10.1175/2010MWR3275.1.
[110] Liou, C. S. (2004). Improving forecast of rainfall and strong wind associated with typhoons approaching Taiwan (p. 23). *Technical Report MOTC-CWB-93-3M-01.*
[111] Z.X. Pu, S.A. Braun, Evaluation of bogus vortex techniques with four-dimensional variational data assimilation, Monthly Weather Review 129 (8) (2001) 2023–2039, http://dx.doi.org/10.1175/1520-0493(2001)129%3C2023:EOBVTW%3E2.0.CO;2.
[112] C.C. Wu, K.H. Chou, Y. Wang, Y.H. Kuo, Tropical cyclone initialization and prediction based on four-dimensional variational data assimilation, J. Atmospheric Sci. 63 (9) (2006) 2383–2395, doi:10.1175/JAS3743.1.
[113] A. Routray, V. Singh, A. Gupta, D. Dutta, J.P. George, Impact of vortex initialization in prediction of tropical cyclones over Bay of Bengal with NCUM Model, Marine Geodesy 42 (2) (2019) 201–226, doi:10.1080/01490419.2018.1552219.
[114] L. Wang, S. Ma, Effect of initial vortex intensity correction on tropical cyclone intensity prediction: A study based on GRAPES_TYM, Journal of Meteorological Research 34 (2) (2020) 387–399, doi:10.1007/s13351-020-9093-y.

[115] J.P. Cangialosi, E. Blake, M. DeMaria, A. Penny, A. Latto, E. Rappaport, V. Tallapragada, Recent progress in tropical cyclone intensity forecasting at the National Hurricane Center, Weather and Forecasting 35 (5) (2020) 1913–1922, doi:10.1175/WAF-D-20-0059.1.
[116] J.S. Goerss, Prediction of consensus tropical cyclone track forecast error, Monthly weather review 135 (5) (2007) 1985–1993, doi:10.1175/MWR3390.1.
[117] J. Kaplan, M. DeMaria, Large-scale characteristics of rapidly intensifying tropical cyclones in the North Atlantic basin, Weather and forecasting 18 (6) (2003) 1093–1108, doi:10.1175/1520-0434(2003)018< 1093:LCORIT>2.0.CO;2.
[118] J. Kaplan, C.M. Rozoff, M. DeMaria, C.R. Sampson, J.P. Kossin, C.S. Velden, …, J.E. Solbrig, Evaluating environmental impacts on tropical cyclone rapid intensification predictability utilizing statistical models, Weather and Forecasting 30 (5) (2015) 1374–1396, doi:10.1175/WAF-D-15-0032.1.
[119] M.D. Powell, S.D. Aberson, Accuracy of United States tropical cyclone landfall forecasts in the Atlantic basin (1976–2000), Bull. Am. Meteorol. Soc. 82 (12) (2001) 2749–2768, doi:10.1175/15200477(2001)082.
[120] V.D. Rockney, Hurricane detection by radar and other means, Proc. of the tropical cyclone symposium, Brisbane, 1956 179–197.
[121] F.D. Marks, State of the science: radar view of tropical cyclones. In: Radar and Atmospheric Science: A Collection of Essays in Honor of David Atlas, American Meteorological Society, Boston, MA, 2003, pp. 33–74, http://dx.doi.org/10.1007/978-1-878220-36-3_3.
[122] H. Jiang, E.M. Ramirez, Necessary conditions for tropical cyclone rapid intensification as derived from 11 years of TRMM data, Journal of Climate 26 (17) (2013) 6459–6470, https://www.jstor.org/stable/26192766.
[123] J.D. Hawkins, T.F. Lee, J. Turk, C. Sampson, J. Kent, K. Richardson, Real-time Internet distribution of satellite products for tropical cyclone reconnaissance, Bull. Am. Meteorol. Soc. 82 (4) (2001) 567–578, http://dx.doi.org/10.1175/1520-0477(2001)082%3C0567:RIDOSP%3E2.3.CO;2.
[124] R.L. Elsberry, T.D. Lambert, M.A. Boothe, Accuracy of Atlantic and eastern North Pacific tropical cyclone intensity forecast guidance, Weather and Forecasting 22 (4) (2007) 747–762, doi:10.1175/WAF1015.1.
[125] F.D. Marks, L.K. ShayPDT-5., Landfalling tropical cyclones: Forecast problems and associated research opportunities, Bull. Am. Meteorol. Soc. 79 (2) (1998) 305–323, http://dx.doi.org/10.1175/1520-0477(1998)079%3C0305:LTCFPA%3E2.0.CO;2.
[126] Z. Xiao, S.G. Xiao, Y.Q. Hao, D.H. Zhang, Morphological features of ionospheric response to typhoon, Journal of Geophysical Research: Space Physics (A4) (2007) 112, doi:10.1029/2006JA011671.
[127] X. Zou, Q. Xiao, Studies on the initialization and simulation of a mature hurricane using a variational bogus data assimilation scheme, J. Atmospheric Sci. 57 (6) (2000) 836–860, doi:10.1175/1520-0469(2000)057%3C0836:SOTIAS%3E2.0.CO;2.
[128] K. Zhao, M. Xue, Assimilation of coastal Doppler radar data with the ARPS 3DVAR and cloud analysis for the prediction of Hurricane Ike (2008), Geophys. Res. Lett. 36 (12) (2009), doi:10.1029/2009GL038658.
[129] Q Xiao, YH Kuo, J Sun, WC Lee, E Lim, YR Guo, DM Barker, Assimilation of Doppler Radar observations with a regional 3 DVAR system: impact of Doppler velocities on forecast of heavy rainfall case, J Appl Meteor 44 (2005) 768–788, doi:10.1175/JAM2248.1.
[130] Q Zhao, Y Jin, High-resolution radar data assimilation for hurricane Isabel (2003) at landfall, Bull Am Meteorol Soc 89 (2008) 1355–1372, doi:10.1175/2008BAMS2562.1.
[131] A. Aksoy, S.D. Aberson, T. Vukicevic, K.J. Sellwood, S. Lorsolo, X. Zhang, Assimilation of high-resolution tropical cyclone observations with an ensemble Kalman filter using NOAA/AOML/HRD's HEDAS: Evaluation of the 2008–11 vortex-scale analyses, Monthly weather review 141 (6) (2013) 1842–1865, doi:10.1175/MWR-D-12-00194.1.
[132] A. Aksoy, S. Lorsolo, T. Vukicevic, K.J. Sellwood, S.D. Aberson, F. Zhang, The HWRF Hurricane Ensemble Data Assimilation System (HEDAS) for high-resolution data: the impact of airborne Doppler radar observations in an OSSE, Monthly weather review 140 (6) (2012) 1843–1862, doi:10.1175/MWR-D-11-00212.1.

[133] A. Routray, U.C. Mohanty, K.K. Osuri, S.K. Prasad, Improvement of monsoon depressions forecast with assimilation of Indian DWR data using WRF-3DVAR analysis system, Pure Appl. Geophys. 170 (12) (2013) 2329–2350, http://dx.doi.org/10.1007/s00024-013-0648-z.
[134] A. Routray, U.C. Mohanty, S.R.H. Rizvi, D. Niyogi, K.K. Osuri, D. Pradhan, Impact of Doppler weather radar data on numerical forecast of Indian monsoon depressions, Q. J. R. Meteorolog. Soc. 136 (652) (2010) 1836–1850, doi:10.1002/qj.678.
[135] S.K. Prasad, U.C. Mohanty, A. Routray, K.K. Osuri, S.S.V.S. Ramakrishna, D. Niyogi, Impact of Doppler weather radar data on thunderstorm simulation during STORM pilot phase—2009, Natural hazards 74 (3) (2014) 1403–1427, doi:10.1007/s11069-014-1250-0.
[136] M. Govindankutty, A. Chandrasekar, D. Pradhan, Impact of 3DVAR assimilation of Doppler Weather Radar wind data and IMD observation for the prediction of a tropical cyclone, Int. J. Remote Sens. 31 (24) (2010) 6327–6345, doi:10.1080/01431160903413689.
[137] J.J. Cione, E.A. Kalina, E.W. Uhlhorn, A.M. Farber, B. Damiano, Coyote unmanned aircraft system observations in Hurricane Edouard (2014), Earth and Space Science 3 (9) (2016) 370–380, doi:10.1002/2016EA000187.
[138] R.F. Rogers, P.D. Reasor, J.A. Zhang, Multiscale structure and evolution of Hurricane Earl (2010) during rapid intensification, Monthly Weather Review 143 (2) (2015) 536–562, doi:10.1175/MWR-D-14-00175.1.
[139] J.A. Zhang, R.F. Rogers, D.S. Nolan, Marks Jr, D F., On the characteristic height scales of the hurricane boundary layer, Monthly Weather Review 139 (8) (2011) 2523–2535.
[140] R.K. Smith, M.T. Montgomery, N.Van Sang, Tropical cyclone spin-up revisited, Quarterly Journal of the Royal Meteorological Society: A journal of the atmospheric sciences, applied meteorology and physical oceanography 135 (642) (2009) 1321–1335, doi:10.1002/qj.428.
[141] J.A. Zhang, R.F. Rogers, V. Tallapragada, Impact of parameterized boundary layer structure on tropical cyclone rapid intensification forecasts in HWRF, Monthly Weather Review 145 (4) (2017) 1413–1426, doi:10.1175/MWR-D-16-0129.1.
[142] P. Zhu, Z. Zhu, S. Gopalakrishnan, R. Black, F.D. Marks, V. Tallapragada, …, C. Gao, Impact of subgrid-scale processes on eyewall replacement cycle of tropical cyclones in HWRF system, Geophys. Res. Lett. 42 (22) (2015) 10–027, doi:10.1002/2015GL066436.
[143] C.S. Ruf, R. Atlas, P.S. Chang, M.P. Clarizia, J.L. Garrison, S. Gleason, V.U. Zavorotny, New ocean winds satellite mission to probe hurricanes and tropical convection, Bull. Am. Meteorol. Soc. 97 (3) (2016) 385–395, doi:10.1175/BAMS-D-14-00218.1.
[144] J.F. Gamache, Evaluation of a fully three-dimensional variational Doppler analysis technique, *Conference on Radar Meteorology, 28 th, Austin, TX* (p. 1997, 1997.
[145] W.E. Baker, R. Atlas, C. Cardinali, A. Clement, G.D. Emmitt, B.M. Gentry, J.G. Yoe, Lidar-measured wind profiles: The missing link in the global observing system, Bull. Am. Meteorol. Soc. 95 (4) (2014) 543–564, doi:10.1175/BAMS-D-12-00164.1.
[146] G.D. Emmitt, K. Godwin, S. Greco, Airborne DopplerWind Lidar data impacts on tropical cyclone track and intensity forecasting: the data processing, interpretation and assimilation, Proceedings of the 15th Conference on Integrated Observing and Assimilation Systems for the Atmosphere, Oceans, and Land Surface, Seattle, WA, USA, 22–27 January 2011, Boston, MA, USA, American Meteorological Society, 2011.
[147] Z. Pu, L. Zhang, G.D. Emmitt, Impact of airborne Doppler wind lidar profiles on numerical simulations of a tropical cyclone, Geophys. Res. Lett. (5) (2010) 37, http://dx.doi.org/10.1029/2009GL041765.
[148] L.R. Bucci, C. O'Handley, G.D. Emmitt, J.A. Zhang, K. Ryan, R Atlas, Validation of an airborne doppler wind lidar in tropical cyclones, Sensors 18 (12) (2018) 4288, doi:10.3390/s18124288.
[149] H.C. Sumner, North Atlantic hurricanes and tropical disturbances of 1943, Monthly Weather Review 71 (11) (1943) 179–183.
[150] H. Wexler, The structure of the September, 1944, hurricane when off Cape Henry, Virginia, Bull. Am. Meteorol. Soc. 26 (5) (1945) 156–159, http://www.jstor.org/stable/26257175.
[151] AC Wood, A flight into the September, 1944, hurricane off Cape Henry, Virginia, Bull. Am. Meteorol. Soc. 26 (5) (1945) 153–156, doi:10.1175/1520-0477-26.5.153.
[152] R.H. Simpson, Hurricanes, Sci. Am. 190 (6) (1954) 32–37.

[153] N.M. Dorst, The National Hurricane Research Project: 50 years of research, rough rides, and name changes, Bull. Am. Meteorol. Soc. 88 (10) (2007) 1566–1588, http://dx.doi.org/10.1175/BAMS-88-10-1566.
[154] S.D. Aberson, M.L. Black, R.A. Black, R.W. Burpee, J.J. Cione, C.W. Landsea, F.D. Marks Jr, Thirty years of tropical cyclone research with the NOAA P-3 aircraft, Bull. Am. Meteorol. Soc. 87 (8) (2006) 1039–1056, http://dx.doi.org/10.1175/BAMS-87-8-1039.
[155] J.D. Boyd, Improved depth and temperature conversion equations for Sippican AXBTs, J. Atmos. Oceanic Technol. 4 (3) (1987) 545–551, doi:10.1175/1520-0426(1987)004%3C0545:IDATCE%3E 2.0.CO;2.
[156] J.D. Boyd, R.S. Linzell, Evaluation of the Sparton tight-tolerance AXBT, J. Atmos. Oceanic Technol. 10 (6) (1993) 892–899, doi:10.1175/1520-0426(1993)010%3C0892:EOTSTT%3E2.0.CO;2.
[157] T.B. Sanford, Motionally induced electric and magnetic fields in the sea, J. Geophys. Res. 76 (15) (1971) 3476–3492, doi:10.1029/JC076i015p03476.
[158] E.J. Walsh, C.W. Wright, D. Vandemark, W.B. Krabill, A.W. Garcia, S.H. Houston, Marks Jr, D F., Hurricane directional wave spectrum spatial variation at landfall, J. Phys. Oceanogr. 32 (6) (2002) 1667–1684, doi:10.1175/1520-0485(2002)032%3C1667:HDWSSV%3E2.0.CO;2.
[159] C.W. Wright, E.J. Walsh, D. Vandemark, W.B. Krabill, A.W. Garcia, S.H. Houston, F.D. Marks, Hurricane directional wave spectrum spatial variation in the open ocean, J. Phys. Oceanogr. 31 (8) (2001) 2472–2488, doi:10.1175/1520-0485(2001)031%3C2472:HDWSSV%3E2.0.CO;2.
[160] R.W. Burpee, J.L. Franklin, S.J. Lord, R.E. Tuleya, S.D. Aberson, The impact of Omega dropwindsondes on operational hurricane track forecast models, Bull. Am. Meteorol. Soc. 77 (5) (1996) 925–934, doi:10.1175/1520-0477(1996)077%3C0925:TIOODO%3E2.0.CO;2.
[161] T.F. Hock, J.L. Franklin, The NCAR GPS dropwindsonde, Bull. Am. Meteorol. Soc. 80 (3) (1999) 407–420, doi:10.1175/1520-0477(1999)080%3C0407:TNGD%3E2.0.CO;2.
[162] S.D. Aberson, Targeted observations to improve operational tropical cyclone track forecast guidance, Monthly weather review 131 (8) (2003) 1613–1628, doi:10.1175//2550.1.
[163] C.C. Wu, P.H. Lin, S. Aberson, T.C. Yeh, W.P. Huang, K.H. Chou, C.H. Liu, Dropwindsonde observations for typhoon surveillance near the Taiwan region (DOTSTAR), Bull. Am. Meteorol. Soc. 86 (6) (2005) 787–790, http://dx.doi.org/10.1175/BAMS-86-6-787.
[164] Wu, C. C., Chou, K. H., Lin, P. H., Aberson, S. D., Peng, M. S., & Nakazawa, T. (2007). The impact of dropwindsonde data on typhoon track forecasts in DOTSTAR. Weather and Forecasting, 22(6), 1157-1176. doi: 10.1175/2007WAF2006062.1
[165] S. D. Aberson, J. Cione, C. C. Wu, M. M. Bell, J. Halverson, C. Fogarty, M. Weissmann, Aircraft observations of tropical cyclones, Global Perspectives on Tropical Cyclones: From Science to Mitigation, World Scientific, 2010, pp. 227–240, https://doi.org/10.1142/9789814293488_0008.
[166] J.R. Kulkarni, R.S. Maheskumar, S.B. Morwal, B.P. Kumari, M. Konwar, C.G. Deshpande, ..., B.N. Goswami, The cloud aerosol interaction and precipitation enhancement experiment (CAIPEEX): overview and preliminary results, Curr. Sci. (2012) 413–425, https://www.jstor.org/stable/24083888.
[167] D.L. Rudnick, Ocean research enabled by underwater gliders, Annual review of marine science 8 (2016) 519–541, doi:10.1146/annurev-marine-122414-033913.
[168] T. Liblik, J. Karstensen, P. Testor, P. Alenius, D. Hayes, S. Ruiz, E. Mauri, Potential for an underwater glider component as part of the Global Ocean Observing System, Methods in Oceanography 17 (2016) 50–82, doi:10.1016/j.mio.2016.05.001.
[169] P. Testor, B. de Young, D.L. Rudnick, S. Glenn, D. Hayes, C.M. Lee, D. Wilson, OceanGliders: a component of the integrated GOOS, Frontiers in Marine Science 6 (2019) 422, doi:10.3389/fmars.2019.00422.
[170] R.E. Todd, D.L. Rudnick, J.T. Sherman, W.B. Owens, L. George, Absolute velocity estimates from autonomous underwater gliders equipped with Doppler current profilers, J. Atmos. Oceanic Technol. 34 (2) (2017) 309–333, doi:10.1175/JTECH-D-16-0156.1.
[171] S.M. Glenn, T.N. Miles, G.N. Seroka, Y. Xu, R.K. Forney, F. Yu, J. Kohut, Stratified coastal ocean interactions with tropical cyclones, Nat. Commun. 7 (1) (2016) 1–10, http://dx.doi.org/10.1038/ncomms10887.
[172] T. Miles, G. Seroka, S. Glenn, Coastal ocean circulation during Hurricane Sandy, J. Geophys. Res.: Oceans 122 (9) (2017) 7095–7114, doi:10.1002/2017JC013031.

[173] H.M. Zhang, R.W. Reynolds, R. Lumpkin, R. Molinari, K. Arzayus, M. Johnson, T.M. Smith, An integrated global observing system for sea surface temperature using satellites and in situ data: Research to operations, Bull. Am. Meteorol. Soc. 90 (1) (2009) 31–38, doi:10.1175/2008BAMS2577.1.
[174] L. Centurioni, A. Horányi, C. Cardinali, E. Charpentier, R. Lumpkin, A global ocean observing system for measuring sea level atmospheric pressure: Effects and impacts on numerical weather prediction, Bull. Am. Meteorol. Soc. 98 (2) (2017) 231–238, doi:10.1175/BAMS-D-15-00080.1.
[175] A. Horányi, C. Cardinali, L. Centurioni, The global numerical weather prediction impact of mean-sea-level pressure observations from drifting buoys, Q. J. R. Meteorolog. Soc. 143 (703) (2017) 974–985, doi:10.1002/qj.2981.
[176] B. Ingleby, L. Isaksen, Drifting buoy pressures: impact on NWP, Atmos. Sci. Lett. 19 (6) (2018) e822, doi:10.1002/asl.822.
[177] P.C. Meyers, L.K. Shay, J.K. Brewster, B. Jaimes, Observed ocean thermal response to Hurricanes Gustav and Ike, J. Geophys. Res. Oceans 121 (2015) 162–179, doi:10.1002/2015JC010912.
[178] B. Jaimes, L.K. Shay, J.K. Brewster, Observed air-sea interactions in tropical cyclone Isaac over Loop Current mesoscale eddy features, Dyn. Atmos. Oceans 76 (2016) 306–324, doi:10.1016/j.dynatmoce.2016.03.001.
[179] J.A. Zhang, R. Atlas, G.D. Emmitt, L. Bucci, K. Ryan, Airborne Doppler wind lidar observations of the tropical cyclone boundary layer, Remote Sensing 10 (6) (2018) 825, doi:10.1175/MWR-D-10-05017.1.
[180] S.C. Riser, H.J. Freeland, D. Roemmich, S. Wijffels, A. Troisi, M. Belbéoch, S.R. Jayne, Fifteen years of ocean observations with the global Argo array, Nature Climate Change 6 (2) (2016) 145–153, doi:10.1038/nclimate2872.
[181] D. Roemmich, M.H. Alford, H. Claustre, K. Johnson, B. King, J. Moum, I. Yasuda, On the future of Argo: a global, full-depth, multi-disciplinary array, Frontiers in Marine Science 6 (2019) 439, doi:10.3389/fmars.2019.00439.
[182] A.C. Didlake, G.M. Heymsfield, P.D. Reasor, S.R. Guimond, Concentric eyewall asymmetries in Hurricane Gonzalo (2014) observed by airborne radar, Monthly Weather Review 145 (3) (2017) 729–749, doi:10.1175/MWR-D-16-0175.1.
[183] S. Raghavan, Structure of tropical cyclones in the Bay of Bengal, Mausam 41 (2) (1990) 184–187, doi:10.54302/mausam.v41i2.2619.
[184] R.A. Houze Jr, Clouds in tropical cyclones, Monthly Weather Review 138 (2) (2010) 293–344.
[185] Q. Xiao, Y.H. Kuo, J. Sun, W.C. Lee, D.M. Barker, E. Lim, An approach of radar reflectivity data assimilation and its assessment with the inland QPF of Typhoon Rusa (2002) at landfall, Journal of Applied Meteorology and Climatology 46 (1) (2007) 14–22, doi:10.1175/JAM2439.1.
[186] D.V. Bhaskar Rao, K. Ashok, Simulation of tropical cyclone circulation over Bay of Bengal using the Arakawa-Schubert cumulus parametrization. Part I—Description of the model, initial data and results of the control experiment, pure and applied geophysics 156 (3) (1999) 525–542, doi:10.1007/s000240050311.
[187] A.C. Didlake Jr, R.A. Houze Jr, Kinematics of the secondary eyewall observed in Hurricane Rita (2005), Journal of the atmospheric sciences 68 (8) (2011) 1620–1636, doi:10.1175/2011JAS3715.1.
[188] R.A. Houze Jr, S.S. Chen, B.F. Smull, W.C. Lee, M.M Bell, Hurricane intensity and eyewall replacement, Science 315 (5816) (2007) 1235–1239, doi:10.1126/science.1135650.
[189] R.H. Simpson, Hurricanes, Scientific American 190 (6) (1954) 32–37.

CHAPTER 19

Significance of remote sensing in tropical cyclone prediction and disaster management: Indian perspective

Vivek Singh[a], Atul Kumar Srivastava[a], Abhay Kumar Singh[b], Rajeeb Samanta[c]
[a]Indian Institute of Tropical Meteorology, Ministry of Earth Sciences, Government of India, New Delhi, India
[b]Atmospheric Research Laboratory, Department of Physics, Institute of Science, Banaras Hindu University, Varanasi, Uttar Pradesh, India
[c]Department of Geography, Prabhat Kumar College, Purba, Medinipur, West-Bengal, India

19.1 Introduction

Cyclones are huge air circulations that rotate anticlockwise around a low-pressure region over the warm Oceanic surface in the northern hemisphere. In contrast, the rotation of cyclone winds is "clockwise" in the southern hemisphere oceans. Unfortunately, every year almost all the coastal countries across the globe are severely affected by these huge cyclonic episodes. Once cyclones reach the shores of peninsular nations, they wreak havoc on property and human-lives. They are called by different names across different parts of the world basins [1]. These storms are known as "Typhoons" when they originate in the "China Sea" and the "Pacific Ocean," "Hurricanes" in the Atlantic Ocean, the West Indian Islands, and the Caribbean Sea, "tropical cyclones (TCs)" in the Bay of Bengal (BoB), and the Arabian Sea, Willy-Willies in North-Western Australia, "Cordonzo" in Mexico, and "Tornadoes" in the southern USA and Guinea lands of West Africa.

A colonial officer "Henry Piddington," deputed as a curator in Kolkata during British rule in India, was the first person who coined the word "cyclone" in the year 1848 [2]. The term "cyclone" is adopted from the Greek word "*Kuklos*" meaning "circle" and "*kykloma*" meaning "wheel or coil of a snake" [3], Chakrabarti [4]. Every year, approximately 85 tropical storms form over the world's warm tropical oceans (https://public.wmo.int/en/our-mandate/focus-areas/natural-hazards-and-disaster-risk-reduction/tropical-cyclone). The average life cycle of TCs is about 6 days. Over the North Indian Ocean (NIO), region, four to five cyclones occur on an average each year especially in the BoB and the Arabian Sea. Every year, TCs develop over the oceanic region during premonsoon (April–May) and the postmonsoon period (October–December). However, because of larger values of difference in magnitudes of lower and upper tropospheric winds(known as "wind shear") during the Indian summer monsoon season (June–September), TCs are less likely to develop. TCs are generally associated with very dense convective clouds and strong surface winds extending up to 10–15 km in height

Atmospheric Remote Sensing
DOI: https://doi.org/10.1016/B978-0-323-99262-6.00008-0

and 150–800 km in diameter. However, the central part of a TC is generally "cloud-free" and calm. This central part is known as the "eye" of a cyclone. In a fully matured TC, the eye ranges from 30–50 km in diameter. The typical structure of "eye" and "eye-wall" in a well-organized TC is depicted in Fig. 19.1A. The satellite imagery showing total rainfall and observed path of a very severe cyclonic storm (VSCS) "Vardah" on May 2, 2008, is shown in Fig. 19.1B.

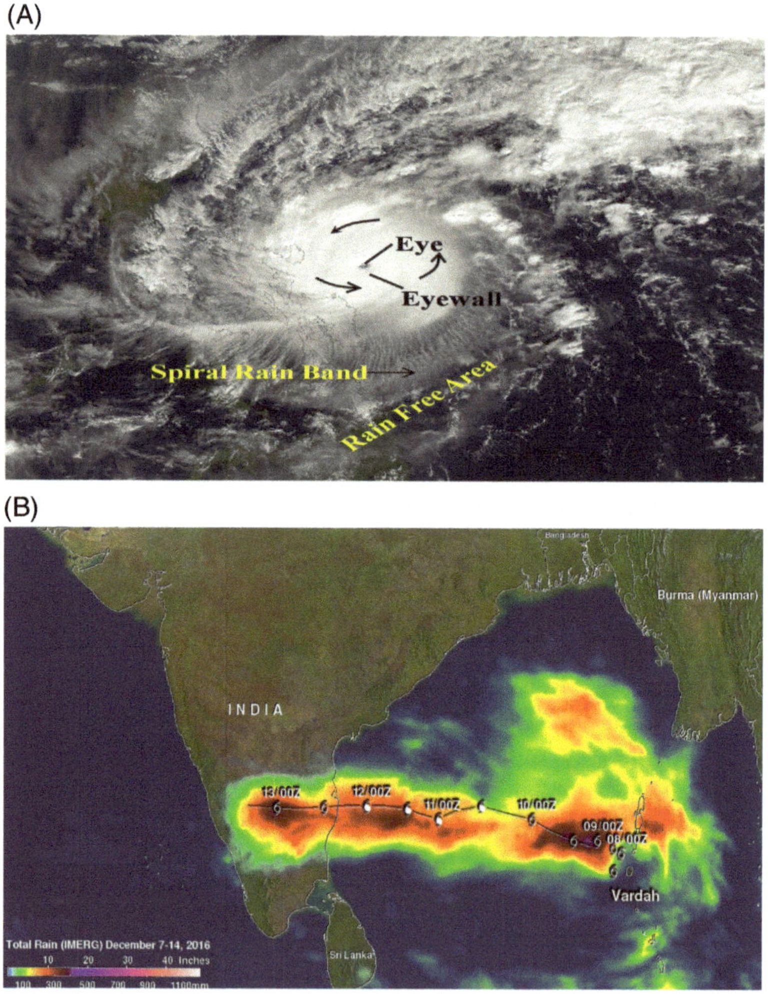

Fig. 19.1 (A) The typical structure of "eye" and "eye-wall" in a well-organized super typhoon "Nock-ten" (2016) formed over Northwestern Pacific Ocean, on Dec. 25 at 0210 UTC, using NASA's Terra satellite. (B) The figure shows the total accumulated rainfall along the track of VSCS "Vardah" during December 7–14, 2016 (Figure courtesy: NASA).

Thus, a cyclone can be described as a low-pressure system that draws air into its eye area. The development of large-scale weather patterns, like cyclones, is due to the Coriolis effect (Lighthill [5]). During long-distance travel around the globe, objects that are not firmly anchored to the ground will experience the Coriolis effect in the form of deflection from their routed path. The Earth rotates faster at the Equator than at the poles, and this is taken into consideration by the "Coriolis effect." Its impact is most felt with long distances and higher speeds, depending on the speed and direction of both the Earth and the object that is traveling. TCs are highly connected with the thermal regime of sea surface temperature (SST), humidity, and other Coriolis parameters. Other than winds blowing at high speeds, TCs can also be associated with storm surges, falls in pressure, wind stress, and tides affecting the landfall areas in the coastal region.

Atmospheric conditions cannot be controlled and the only way to reduce their disastrous effects is by predicting such extreme events well in advance for better preparation to deal with them. Since cyclones are extremely destructive, it is deemed as a necessary factor to predict them well in advance, so that their effects on human lives and property can be minimized. Accurate and timed prediction of the track and intensity of TCs can provide policymakers with enough time to plan evacuation or mitigation strategies. However, these natural meteorological phenomena originate over the oceanic regions, where observational data coverage is sparse in nature; which makes forecasting them quite challenging. The only way of predicting these phenomena is by having a constant monitoring system, and a milestone in this regard was achieved with the launch of the Television and InfraRed Observation Satellite (TIROS-1) by The National Aeronautics and Space Administration (NASA) of the United states of America (USA) on April 1, 1960 [6] and [7]. The present chapter aims to analyze the importance and role of various remote sensing satellite data products in predicting the track and intensity of TCs, majorly in the Indian context.

Further detailed discussion on the present research topic is arranged under different sections. Section 19.2 deals with the application of remote sensing data for TC identification, tracking, forecasting, evacuation, and loss-estimation by various sensors introduced by India and other leading countries. Section 19.3 will shed light on the use of India-prepared TropFlux data (https://incois.gov.in/tropflux/), and its utility for the prediction of the genesis, track, and adverse weather conditions associated with the VSCS "Phailin" (2013) over the BoB. Section 19.4 includes the result and discussion of various parameters used to monitor the VSCS "Phailin," and also documents its formation, diurnal tracking, landfall, and devastations in details. Further, Section 19.5 concludes the study and discusses about the possible future prospects of remote sensing techniques for TC track and intensity predictions.

19.2 Various applications of remote sensing techniques for cyclone track and intensity prediction

Remote sensors, mounted on satellites or aircraft, are used for collecting data by detecting the energy which gets reflected from the earth, function in both active and passive ways. They are used for a majority of purposes, including the assessment of hazards like cyclones, earthquakes, erosions, or floods. Data collected by them can be used to better assess the impact of natural disasters and prepare strategies (US Department of Commerce, National Oceanic and Atmospheric Administration (NOAA) report, 2020). TCs mostly develop over warm ocean regions and are nonfrontal in nature, unlike cyclones of the middle and high latitudes [1]. An increase in the understanding of TCs was seen after the probes by the Air Force & Navy aircraft and also the National Hurricane Research Project of America. According to Palmen and Newton [1], even after forming several hypotheses about the formation of cyclones, a final solution had been difficult because of the inadequate observations at that time. Since the development of cyclones is not spontaneous, finding out about the initial formation of TCs used to be the major difficulty. Also, intense cyclones always rose from smaller or less intense turbulence or disturbances, which might have existed for a long time. Whatever knowledge was possessed about the winds over oceanic bodies before about a hundred years, all depended on the facts collected by traveling ships. The sparsity of these observational pieces of information kept the resolution restricted to more than hundreds of kilometers; and often, vast areas which lay outside of standard routes of ships remained uncharted. The relationship with both SST and surface wind speed (SWS) was formerly thought to be negative for a long time (Frankignoul [8], Mantua et al. [9], Okumura et al. [10]). This is because for a long time, analysis depended on data collected by ships and modeling of climate, all of which was on vague and on rough scale. Other than the tropical regions, this interaction on large scale was interpreted as that the ocean was passively responding to latent and sensible heat fluxes caused by the wind, which was a way of atmospheric forcing of the ocean [11]. The Arrival of satellite-borne microwave radar scatterometers capable of measuring the global surface wind fields with spatial resolutions of up to 25 km and microwave radiometers with spatial resolutions of about 50 km that could measure SST in almost all weather conditions, brought forward the revelation that on the oceanic mesoscale of 10–1000 km, the ocean–atmosphere interaction is distinct. [11]. Later, it was found that SWS was locally higher over warmer waters and lower over cooler waters, which meant that a positive correlation was found between the two factors, which was opposite to what was previously believed [11]. In further course of time, it was found that a positive correlation lies between SST and surface winds on a smaller scale only, which brought the understanding that the interactions between the ocean and the atmosphere is driven by spatial variations of SST. The reason behind the opposite correlation that is seen on larger scale may be the adjustment time of equilibrium of the marine atmospheric layer. It also differs regionally, which is dependent on factors like wind speed, etc. In places where the SST is unstable and ever changing, the modifications of the surface wind field cause converging and diverging surface winds. Also, vertical motions set by pressure variations can invade into the troposphere. This in turn influences global weather patterns. Wind stress variations induced by SST also have a feedback effect on the ocean surface in the form of turbulent heat flux and mixing caused by the wind, and wind stress curl-driven upwelling that generate ocean currents and has the ability to modify the SST itself [11].

Previously, due to the sensor resolution, repeatability, the data processor was limited and needed many corrections in collected data. But in the last 40 years, TC research has reached the present heights and precision with gradual development in satellite sensors [12]. India started monitoring TCs with the launching of the Indian National Satellite (INSAT) in the year 1982 [13,14]. The meteorological sensors can detect imaging in visible, infrared, and microwave regions of the electromagnetic spectrum. The Indian Meteorological Department (IMD) extensively uses satellite imagery-based Dvorak Techniques [15], to analyze the vortex, track, and development stages of the cyclones. Kalsi [16] used Dvorak Techniques in the study of the Super Cyclone that struck Orissa, and successfully detected the location and development of the cyclone using visible and EIR bands. The NOAA of the United States of America was founded in the year 1970 to observe the origin and locations of TCs using the facilities of visible and infrared bands [7]. High-resolution infrared sounder was also used for the measurement of water vapor whose accuracy depended on the thermal properties of the phenomenon [17]. Since visible (0.4 μm–0.7 μm) and IR (2.5–16 μm) bands are highly sensitive to clouds, their efficiency gets restricted and microwaves play a significant role in monitoring the meteorological phenomena.

After the Second World War (WWII), the application of active microwave remote sensing in forecasting systems was emphasized. At present, the Advanced Scatterometer (ASCAT) and OceanSAT-2 Scatterometer are capable of successfully measuring wind speeds and direction with the accuracy of 2 m/s. All over the world, the numerical weather prediction (NWP) model is used to demarcate the position and magnitude of cyclones with precision. But the NWP is unable to do so at the formative stage of the cyclones, due to the low resolution of images and also that it can only measure the dynamic form and structure of cyclones. The Ozone Satellite data was also incorporated in the NWP model to help in accurate measurement of the location, intensity, and mapping the movement track of the cyclones [18].

After the launch of the TIROS-1, (the first weather satellite of the world) in 1960, a new era of space-based observations started. Classifications of the observations from satellites can be based on sensor characteristics. Meteorological satellites use various regions of the electromagnetic spectrum (i.e., visible, wavelength ~0.4–0.7 μm; infrared, wavelength ~1–100 μm; and microwave, having wavelength ~0.3–30 cm) to obtain observations of various atmospheric and terrestrial phenomena [7]. In the earlier times, observations from satellites were done with the use of visible/infrared wavelengths however, they are seen as important even in the present day. Every half an hour, the geostationary satellites located near the equator of the Earth, use these sensors to observe the Earth–Atmosphere system. In addition to worldwide observations in visible and infrared channels, the satellites in the polar region also make global observations in visible and infrared channels. They provide better spatial and spectral resolutions, but their temporal resolutions are poorer than those of the geostationary satellites. The primary purpose of these observations is to discover cyclogenesis areas, as well as the location, motion, and geographical extent of cyclones. A principal technique that is very much in use in determining the intensity of cyclones is making observations of fields of clouds in TCs. Originally, this technique only used visible imagery (Rao et al., 1993, Ottenbacher, 1997, [19]). In the later course of time, infrared (IR) and other sensor data were also incorporated in it (Brad et al., 2002, [20]). It is capable of monitoring how the TC develops on a day-to-day basis, changes in cloud patterns, the environment, etc. The main difference between the visual and infrared techniques lies in the features of clouds used in analyzing and the way of measuring them. For predicting how intense a storm is going to be, the superior and digital infrared techniques depend on data from the eye and the clouds surrounding the eye. Other data collected by these techniques are based on the movement of clouds, cloud top temperature and cloud-level winds among many others. These parameters are determined regularly from geostationary satellites, and help in monitoring TCs. Sensors are instruments that measure vertical profiles of atmospheric parameters such as temperature and humidity. When sensors are placed on satellites, they help provide three-dimensional distribution of these atmospheric parameters. [7].

Microwave frequencies are more advantageous than visible/infrared frequencies, since the clouds can be seen as transparent in them and the oceans act as a cold background. When microwave frequencies are used, it is feasible to determine atmospheric circulations in both clear and overcast weather circumstances. Microwave radiation interacts substantially with precipitation raindrops, either by absorption or emission at lower frequencies and scattering at higher frequencies. In contrast to visible/infrared methods, rainfall rate estimate using microwave radiation is nearly physically direct. Microwave sensors are a highly valuable instrument in the observational research of TCs for these reasons.

There are two types of microwave sensors: passive and active. Passive microwave sensors employ microwave radiation from the ground and atmosphere to calculate factors such as “wind velocity” and sea level. The special sensor microwave/imager (SSM/I), multifrequency scanning microwave radiometer (MSMR, onboard Indian OceanSAT-1), TRMM microwave imager (TMI), and MADRAS onboard Indian low earth satellite Megha-Tropiques are all passive microwave sensors. These sensors operate on a frequency ranging from 6 GHz to 150 GHz. The lower ones like 6 and 10 GHz are used to detect SST, but they also help to measure the SWS and amount of rainfall. If SST is available in cloudy situations, it is rendered extremely helpful in the study of cyclones; the variation of SST along the path of the cyclone may help in the determination of the evolution of the cyclone. When it comes to cyclones though, such information is often not available to retrieve, within 200 km from the eye of the storm [7]. Along with this, as the emission of microwave radiation saturates rapidly at higher SWSs, the surface winds which are retrieved may end up showing a huge difference from what the actual values may be, during some strong winds that are generally associated with TCs. The active microwave sensors transmit pulsed energy to illuminate the area of interest. The same antenna is used to receive and process the signal that returns. Active microwave sensors include the scatterometer and the altimeter, to name a few. A wind scatterometer, with working frequencies ranging from 5 to 14 GHz, is the sole instrument that can determine wind vectors over the ocean surface. Measurement of wind vectors, in addition to other factors, is very critical for predicting storm surges generated by cyclones striking the coasts. Storm surges account for a large portion of the damage produced by TCs. It is critical to take repeated observations of surface wind fields over broad areas of seas The ASCAT, boarded on the European satellite METOP and Indian Space research organization's (ISRO) OceanSAT-2 Scatterometer (OSCAT) is currently operational and is capable of measuring ocean surface winds with speed accuracy of 2 m/s and direction accuracy of 20°. [7]. The radar altimeter, which measures time return echoes from the water surfaces, is another active microwave sensor. It can monitor wave amplitude, SWS, and sea surface elevation, as well as undulations in the sea surface and strong winds induced by TC movements. They can also be used to measure the vertical structure of the oceans, when used in complement with data of the historical hydrographic nature. At present, operational altimeters are JASON, Environmental Satellite (ENVISAT), ERS-2, and TOPEX, which measures sea level height to an accuracy of about 5–6 cm. [7].

The tropical rainfall measuring mission (TRMM) precipitation radar is an active microwave sensor equipment. It is the first space-borne equipment that can form three-dimensional maps of the structure of storms (Kozu et al. [21]). These measurements help to provide knowledge about various parameters of rainfall like distribution, type, and intensity. For improving atmospheric circulation models globally, estimating the data of the heat releasing into the atmosphere at various heights is used. The precipitation radar can give vertical profiles of rain up to an elevation of 20 km from the ground surface. It can also detect light rainfall rates, even less than 1 mm per hour. In case of heavier and more intense rainfall rates, newer ways of processing data have been developed. The data and measurements from the TRMM rain radar has been proved to be of great importance for calibrating and validating other sensors that measure rainfall, and also for analyzing three-dimensional rain structures from within the cyclone. It has the potential of high significance to understand the physical processes that may be responsible for the intensification of TCs. Microwave sounders make use of radiometric measurements in the 60 GHz complex or 118 GHz line of oxygen absorption. The microwave sounding unit (MSU) and the advanced microwave sounding unit (AMSU-A) boarded on the NOAA satellites are used for sounding temperature. Observations at levels of 1803 GHz water vapor rebound are used to obtain humidity profiles, and the AMSU-B is an example of a humidity sounder [7].

The hurricane weather research and forecast (HWRF) model started functioning in the year 2007 which could forecast and predict the location, intensity, and the track of hurricanes. The TC model HWRF requires cloud parameterization to control the radiation, precipitation, and dissipation of clouds [22]. The assimilation of remotely sensed satellite data in the Weather Research and Forecasting (WRF) Model have been found positively impactful in predicting track and intensity of TCs forming over the Indian Seas (Srinivas et al., [23], Singh et al. [24], Singh et al. [25])

The turbulence-induced by air–sea interaction is yet another disturbance in the atmospheric imaging system that needs to be simulated into the TC model. The physical processes of air–sea flux, sea surface roughness, length of roughness, and spray condition of sensible and latent heat flux are parameterized for NWP models [12,22,26]. In this direction, passive sensors like the special sensor microwave (SSM/I), TRMM, Indian OceanSAT-1 which have frequencies between 6 GHz and 10 GHz are used in accurately measuring SST, wind speed, evaporation, and rainfall. Additionally, observational investigations of the fine-resolution, 2D horizontal structure of mesoscale SST effect on low-level winds necessitate the application of satellite imageries for accurate measuring of wind and SST [11].

The next section will shed light on the data used and the methodology adopted in the present study for understanding the causes of intense nature of the VSCS "Phailin" (2013) over BoB. The advanced satellite data analysis of the genesis, track, and associated adverse weather conditions associated with the VSCS "Phailin" (2013) helped in the accurate prediction of the TC in NWP models during its occurrence.

19.3 Data and methodology

In the present study, the sensible heat flux, SST, latent heat flux, and wind speed associated with the TC are analyzed using TropFlux reanalysis data [27] and [28]. The TropFlux product data are distributed by the Indian National Center for Ocean Information Services (INCOIS), Hyderabad, India (https://incois.gov.in/tropflux/). The reanalysis data can be easily downloaded after successful registration and request from the concerned authority. The data have been prepared using the COARE v3 bulk algorithm to estimate heat fluxes. It also uses ISCCP and ERA-I, that is, 2-m air and sea temperature, 10-m winds, downward radiative fluxes, and 2-m air relative humidity. The TropFlux, in which all the bias corrections are achieved after comparing with data from the Global Tropical Moored Buoy Array, provides 1° × 1° spatial resolution data for the entire 30°N–30°S region. A collaborative effort between the Laboratoire d'Océanographie: Expérimentation et Approches Numériques from Institut Pierre Simon Laplace, France and National Institute of Oceanography/CSIR, India, can be credited for the development of the TropFlux product. The TropFlux data provides a simple estimation of surface shortwave radiation from the OLR data, and acts as a near real-time alternative to the ISCCP data, which is more accurate but is not updated regularly.

For the present study, the heat fluxes, SST, and wind speed have been analyzed on an daily interval during the occurrence of TC over the BoB region. The next part of the study will focus on the observational features of the VSCS "Phailin" as measured by various remote sensing instruments. Following that, results and discussions of the study are presented.

19.4 Remote sensing of the VSCS "Phailin"

In the year 2013, Phailin crossed India and it was the most powerful cyclone since the 1999 Odisha Super Cyclone. Its name was given by Thailand. It occurred during October 8, 2013 to October 14, 2013 over BoB. (https://rsmcnewdelhi.imd.gov.in/uploads/report/26/26_38a1d4_phailin.pdf). "Phailin," a VSCS as per meteorological nomenclature, formed in the South China Sea. It began as a low-pressure region over the Tenasserim coast on October 6, then grew into a well-marked low-pressure system over the northern Andaman Sea on October 7. It has condensed into a depression in the same region by October 8, 2013. The scatterometry data showed an indication that there was a cyclonic circulation over the region and the wind speed associated with it was about 25–30 knots on October 8, 2013. The speed of wind was considerably higher in the southern sector as compared with the cross-equatorial flow. Further, on 8 October, analyzing the observations from various satellites, intense to very intense convection was noticed over the Andaman Sea and the nearby areas to the Tenasserim coast. From there, while advancing in the west-northwest direction, it developed and took the rage of a deep depression by 9 October, and then on the same day intensified into a cyclonic storm (CS), and was named "Phailin." Further ahead in the timeline, moving in the northwest direction, it intensified into a severe cyclonic storm (SCS) by October 10, 2013 and then into a VSCS on the same day, over the east-central BoB. Rapidly intensifying, from 45 knots to 115 knots, an increase of around 70 knots, starting from the 10 October (morning), the cyclonic system reached its maximum intensity of 115 knots on 11 October (morning). The VSCS then kept moving in the north-west direction at a speed of almost 15 kmph, and crossed Gopalpur, Odisha, and Andhra Pradesh on 12th October (11:30 pm). The northerly component of the system kept increasing gradually for about 12 hours before it made landfall. The VSCS "Phailin" sustaining a maximum SWS of 115 knots or 200–210 kmph to 220 kmph made landfall near Gopalpur, Odisha, and the north coastal area of Andhra Pradesh around 11:30 pm on 12 October. The maximum and intense wind speed produced heavy rainfall all over Odisha and later as a result it developed and took the rage of floods and strong gale winds. Due to all this extensive production, large-scale damage occurred to the structure and coastal inundation. The maximum rainfall occurred over the northeast area of the system center, when the cyclone made a landfall, and a maximum cumulative rainfall of 38 cm in 24 hours was recorded in Cuttack, Odisha. A maximum storm surge of 2–2.5 m above the astronomical tide was recorded in the low lying areas of Ganjam, Odisha and the in-land inundation went up to almost 1 km from the coastline. In all this, the IMD came as a rescue as with their intelligence and technologies, they were able to predict the genesis, intensity, and where the track of the cyclone led to. Along with this, the time and point were also accurately and timely determined for the landfall. After making a landfall, the system kept moving in the north-northeast direction for some time, followed by moving in the north direction and then moved toward the northeast direction, reaching southwest Bihar, India. By 13 October (8:30 am), the cyclonic system weakened into an SCS and then into a CS over southern Odisha on 13 October (11:30 pm). It kept weakening, forming a deep depression over north Chhattisgarh, nearby areas of Odisha, and Jharkhand on the 13th October (5:30 pm), and then transformed into a depression over southwest Bihar by 14 October (8:30 am), weakening further into a well-marked low-pressure area over southwest Bihar and nearby areas on 14 October (2:30 pm). The IMD was involved in continuous prediction and monitoring of the VSCS "Phailin" since its inception. Accuracy and advance prediction of the genesis, track, intensity helped the parties associated with disaster management duties. Even the point and time of making landfall, weather conditions, like rainfall, gale wind and storm surge were predicted accurately by the authorities.

Satellite observations along with the meteorological buoys and coastal island observations were used to monitor the cyclone when it was in the genesis state. These remotely sensing observation techniques were also used while TC was moving away from the Andaman & Nicobar Islands. The Doppler Weather Radar (DWR) at Visakhapatnam, India, was used to monitor the cyclonic system, on the 12 October (6:30 am to 11:30 pm), when it was in its range, that is, about 310 km east-southeast of the Visakhapatnam coast. Along with these, hourly data from coastal observation observatories, satellites, and Automatic Weather Stations, half hourly images from INSAT/Kalpana, DWR images every 10 minutes were used to keep track of "Phailin." All available scatterometry and microwave imagery products were used in the process. Satellite observations were very vital in tracking the cyclonic system, since the buoys were away from its track. Even after the cyclone made a landfall, the Satellite Application Unit of the Satellite Meteorology Division, in an attempt to better record the cyclonic system, provided knowledge about the eye and the intensity of the storm. To monitor the system, mostly images received every half hour from the Kalpana-1 were used. Apart from that, even images from MeteoSAT-7 and MTSAT, which are international geostationary satellites, were also used. Further, microwave and high-resolution images from the Defense Meteorological Satellite Program (DMSP), NOAA series, TRMM, and Metops, which are polar-orbiting satellites, also helped in the tracking of the cyclonic process.

From the genesis stage of the cyclone on the 8 October to the 10 (morning), the estimated central pressure of the system saw a gradual decrease. Since the system rapidly intensified, a further sharp decrease was seen till 11 October (morning). The lowest central pressure was recorded at 940 hPa on 11 October (8:30 am) till the cyclone made landfall. As compared to the genesis stage, the rise in central pressure was more rapid while the cyclone was decaying. There was a gradual increase in the maximum wind speed from the 8–10 October (morning); and then a rapid increase was seen till 11 October (8:30 am), when it reached 115 knots and this continued till the cyclone made landfall.

19.5 Results and discussion

Figs. 19.2–19.6 are the pictorial depictions of the various parameters estimated, using TropFlux data and ocean altimetry satellites, in the process of monitoring the progress of the VSCS "Phailin." The thick black line in the figures depicts the observed track of the VSCS "Phailin," which started on October 9, 2013 and dissipated by October 13, 2013. The following Fig. 19.2A–E shows the observed Latent Heat Flux (watts/meter^2) variation during the whole lifecycle of the "Phailin" cyclone.

The color bar can be seen to understand the latent heat flux of that area during that period. According to Adler & Rodgers (1977), the more Latent Heat is released, the stronger the storm becomes in intensity. As seen, on October 9, 2013 (Fig. 19.2A), the latent heat flux was in the range of 125–150 W/m^2. As the latent heat flux kept rising, the cyclonic system kept maturing reaching up to 200 W/m^2 on October 12, 2013 (Fig. 19.2D). The cyclone appears to have been powerful based on the significant observed latent heat flow. On October 13, 2013, it dropped to 125–150 W/m^2 (Fig. 19.2E), indicating that the storm was dissipating.

Fig. 19.3A–E shows the rainfall distribution during the life cycle of VSCS "Phailin" using TRMM satellite data. The rainfall is depicted in the unit "cm" in these figures. The color-coded bar shown in the figure can be used to understand the severity of rainfall recorded on each of the days of the cyclone. As seen, on 9 October

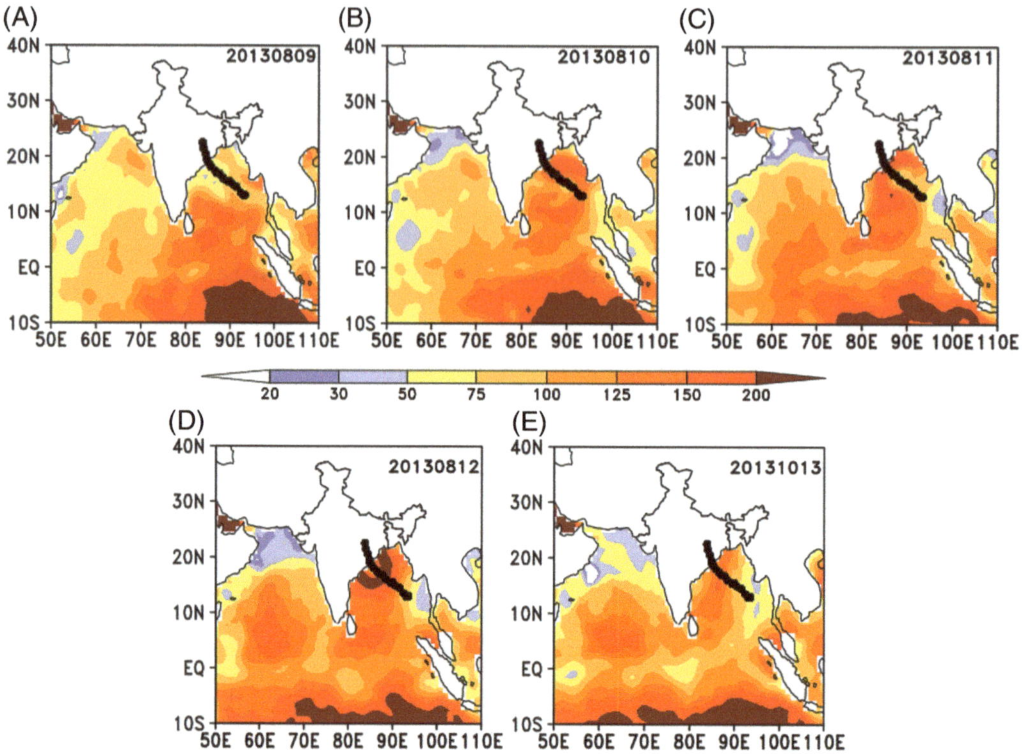

Fig. 19.2 (A–E) Latent heat flux (W/m^2) from the October 9–132013 for VSCS "Phailin." The black curve in the figure shows observed track of the TC during its lifetime. The latent heat flux has been estimated using TropFlux data [27] and [28].

(Fig. 19.3A), the recorded rainfall was about 4–8 cm, and on October 10, 2013, it kept increasing in the area, as the yellow-colored area can be seen growing in Fig. 19.3B.

The increasing rainfall owing to the strengthened cyclone can be observed on 11 October (Fig. 19.3C), reaching upto 8–16 cm in magnitude. The storm began to organize on October 12, 2013 (Fig. 19.3D), with the rainfall area shrinking closer to the beach and ultimately reaching landfall on 13 October (Fig. 19.3E), with rainfall remaining in the 8–16 cm distribution range.

Fig. 19.4A–E shows the sensible heat flux, which is measured in the unit W/m^2, similar to that of latent heat flux. The color bar in Fig. 19.4A through 4E can help you understand how the sensible heat flux changed over the course of five days during the "Phailin" cyclone. The major element in heating the boundary layer intake in TCs is sensible heat fluxes. Also, these fluxes are crucial for the formation and intensification of

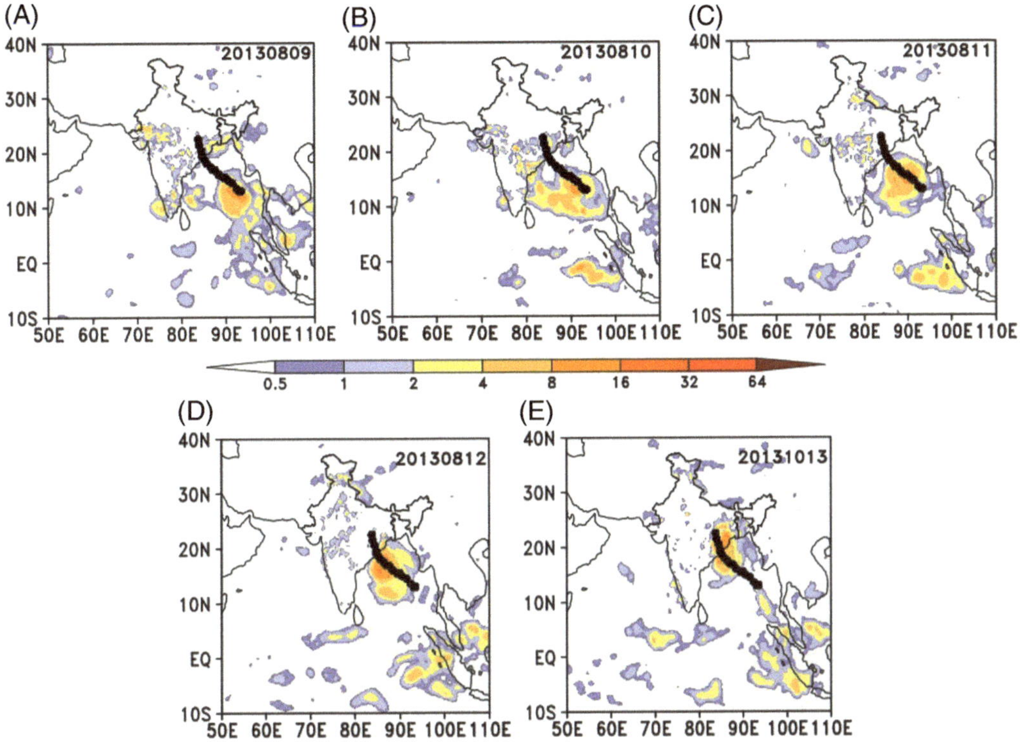

Fig. 19.3 (A–E) Observed daily rainfall from the October 9–13, 2013 for the VSCS "Phailin." The observed rainfall (cm) data has been plotted using tropical rainfall measurement mission (TRMM) data of NASA, USA.

the TCs over the BoB. As can be seen in Fig. 19.4A, the sensible heat flux was around 15–20 W/m^2.

As the cyclonic system matured further, the sensible heat flux kept rising and reached up to 30–40 W/m^2 on 11 October (Fig. 19.4C). By the 12 October (Fig. 19.4D), it can be seen ranging from about 40–50 W/m^2.

Fig. 19.5A–C shows the sea surface height anomaly (SSHA, (m)) obtained from ocean altimetry satellites. The Ocean Topography Experiment (TOPEX) and Jason-1 altimeter provide the sea surface height anomaly [29]. The heat storage in the upper layer of seas is depicted by SSHAs.

The anomaly is calculated with respect to a mean sea surface. To calculate SSHA, we subtract the satellite estimated sea surface height from the mean climatological value of sea surface height. Generally, a positive anomaly is associated with and westward motion

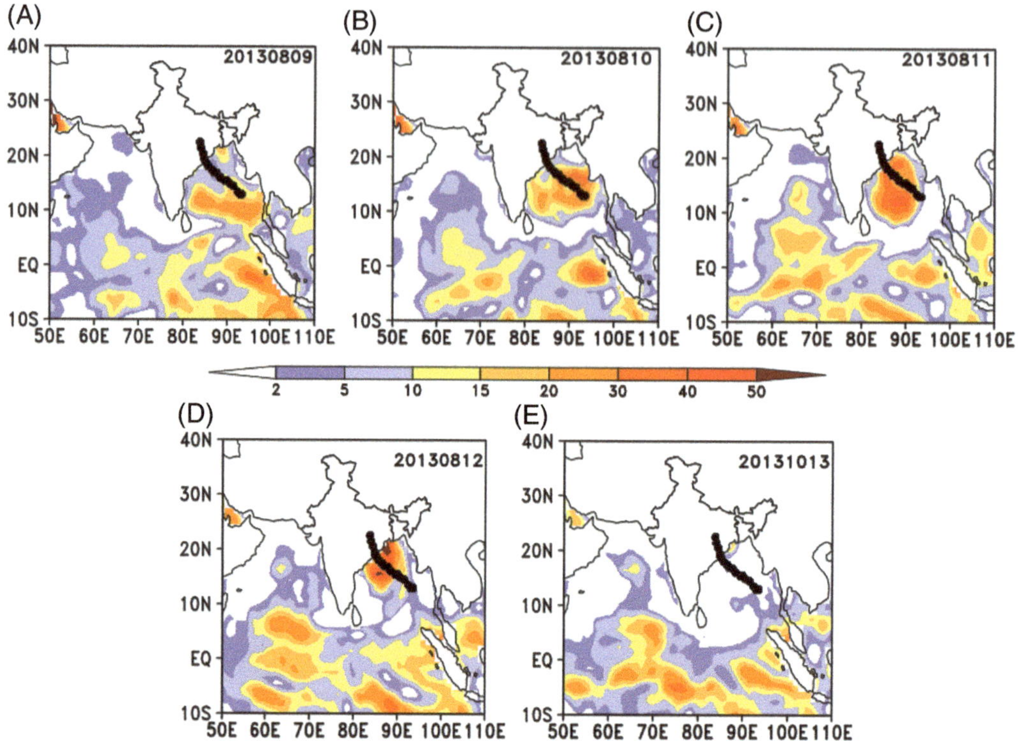

Fig. 19.4 (A–E) Same as Fig. 19.2 (A–E) but for sensible heat flux.

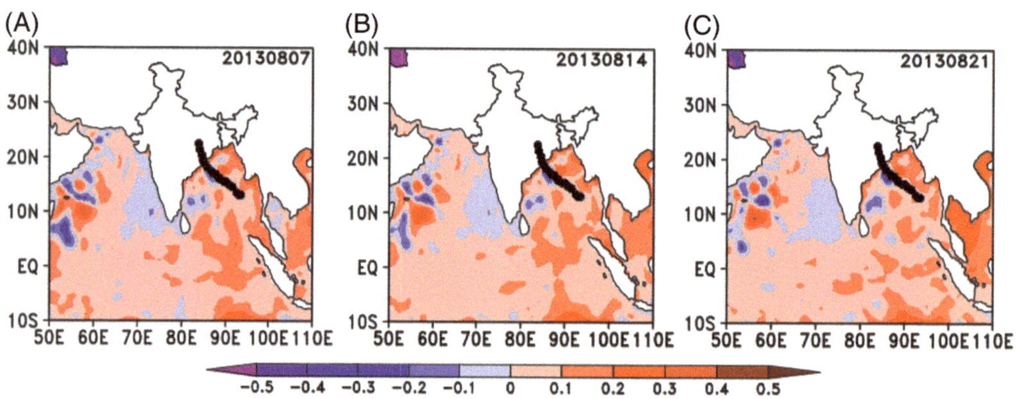

Fig. 19.5 (A–C) Sea surface height anomaly (m) obtained from ocean altimetry satellites for the VSCS Phailin (2013).

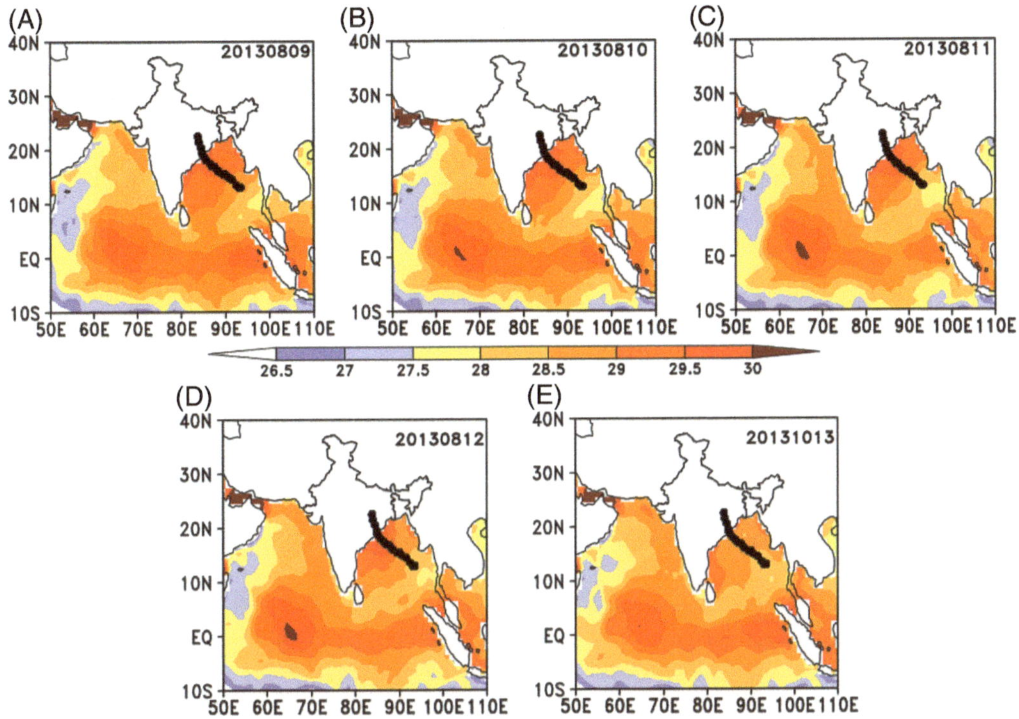

Fig. 19.6 (A–E) Same as Fig. 19.2 (A–E) but for sea surface temperature (in °C).

and intensification of TCs. In all three subfigures above, the positive SSHA can be seen in the rightward direction of the observed TC track (black curve).

Fig. 19.6A–E shows the SST, measured in °C. The strengthening of the TCs requires prevailing of warmer SSTs. From the 9 (Fig. 19.6A) to the 11 (Fig. 19.6C) October, 2013, the SST further raised and was lying between 29–29.5°C along the beach, the region had shrunk by the 12 (Fig. 19.6D), and the temperature dropped to 28.5–29°C by the 13 (Fig. 19.6E). The warmer SST during the occurrence of the TC could have fueled the rapid intensification of the TC over BoB.

The Wind Speed, measured in m/s, is shown in Fig. 19.7A–E. The figure has been plotted using TropFlux data. On 9, the wind speed ranged between 0–2 m/s and 2–4 m/s, as can be observed by the color (Fig. 19.7A), and it kept strengthening after that. The red colored area showing wind speed of 10–15 m/s can be seen growing from the (Fig. 19.7B) 10–12 October (Fig. 19.7D). The area where strong wind speeds could be observed diminishing by the 13 October (Fig. 19.7E), as the cyclonic system dissipated.

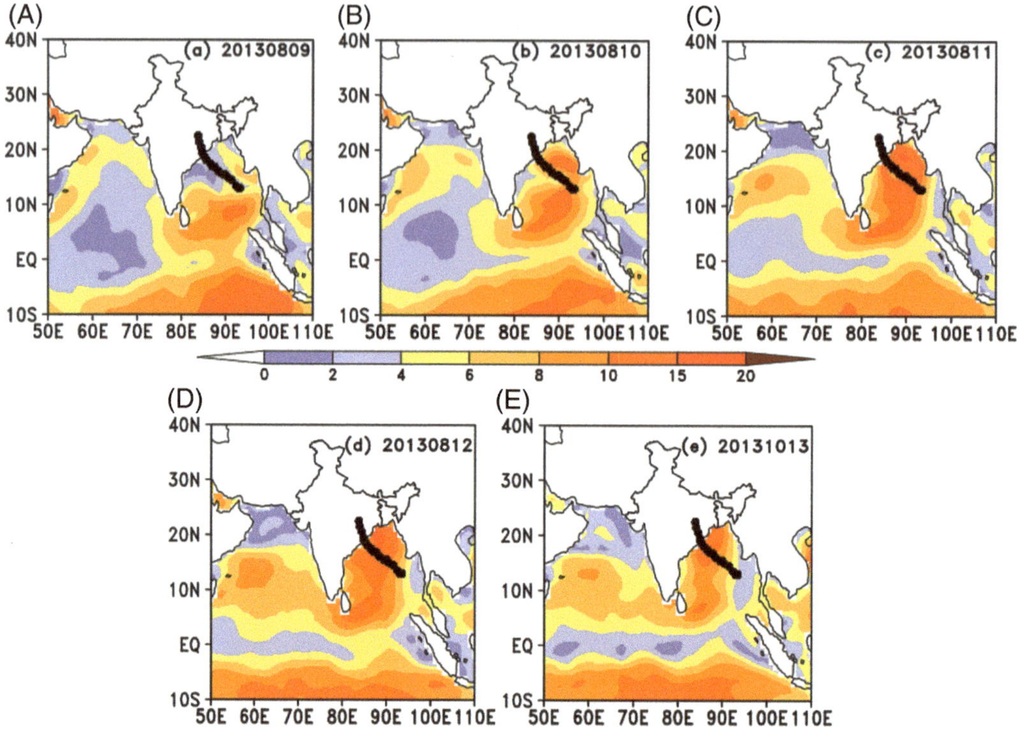

Fig. 19.7 (A–E) Same as Fig. 19.2 (A–E) but for wind speed (m/s).

19.6 Conclusion

The study of forecasting TCs has come a long way in the past four decades. After the launch of the INSAT in 1982, India started monitoring TCs as well. Predicting the course of TCs is important to reduce the loss of lives and property when these adverse weather conditions hit coastal areas. Indian government prepares TropFlux data, after collecting data from various techniques of remote sensing, which helps the parties associated with disaster management prepare for natural disasters like TCs. The advancements in remote sensing have led to the ability to monitor cyclones from their genesis stage, and predict their whole life cycle, their track, intensity, associated weather conditions, and also the time and place of making landfall. In this study, the authors focused on the various equipment and techniques used for observing and collecting data about TCs. The study aimed to analyze how accurate the observations and predictions based on satellite observations were during the occurrence of the TC. The authors reviewed and evaluated the climatic variables that led to the strengthening of the VSCS "Phailin," which slammed the coastal territories of Odisha and Andhra Pradesh in October 2013. The ability to anticipate the position, time, and strength of the TC at the time of landfall was aided by prior information of the TC origin and structure gleaned from remote observations, ensuring that the death toll and damage were kept to a minimum. It was possible to organize evacuations, and about 5.5 lakh people were relocated to safer regions along the coastal regions of Odisha and Andhra Pradesh, making it India's largest successful evacuation in the last 23 years, praised by world leading cyclone monitoring agencies and authorities like World Meteorological Organization and Government of India, etc. The present study successfully found that the TropFlux data products were able to produce relevant observed features during the whole life cycle of October 2013 VSCS "Phailin," which further helped IMD in the accurate prediction of the cyclone, during its occurrence.

Acknowledgment

Without the support of the Indian Government's Ministry of Earth Sciences, this work would not have been possible. The authors wish to express their gratitude to the IMD for sharing the most up-to-date TC's best-track data. We express our gratitude to the Indian National Centre for Ocean Information Services (INCOIS), Hyderabad, for providing the most critical TropFlux data upon request. The authors wish to express their heartfelt appreciation to NASA for providing satellite images of VSCS "Vardah" and super typhoon "Nock-ten."

References

[1] E. Palmen, C.W. Newton, Atmospheric Circulation Systems; Their Structure and Physical Interpretation, Academic Press, London, 1969.
[2] A. Collett, R. McDougall, S Thomas, Tracking the Literature of Tropical Weather: Typhoons, Hurricanes, and Cyclones, Springer, 2017.
[3] N. Pandharinath, The Science of weather and environment, BS Publications, Hyderabad, 2007.
[4] Chakrabarti, Ranjan. "The Calcutta Cyclone of 1737: Cyclone or Earthquake?." In Force of Nature, pp. 123-136. Routledge, 2017.
[5] J. Lighthill, Fluid mechanics of tropical cyclones, Theor. Comput. Fluid Dyn. 10 (1) (1998) 3–21.
[6] A.J. Manna, 25 years of Tiros satellites, American Meteorological Society, Bulletin *66* (1985) 421–423.
[7] C.M. Kishtawal, Use of satellite observations in tropical cyclone studies. In: Advanced Numerical Modeling and Data Assimilation Techniques for Tropical Cyclone Prediction, Springer, Dordrecht, 2016, pp. 35–47.
[8] C. Frankignoul, Sea surface temperature anomalies, planetary waves, and air-sea feedback in the middle latitudes, Rev. Geophys. *23* (4) (1985) 357–390.
[9] N.J. Mantua, S.R. Hare, Y. Zhang, J.M. Wallace, R.C. Francis, A Pacific interdecadal climate oscillation with impacts on salmon production, Bull. Am. Meteorol. Soc. *78* (6) (1997) 1069–1080.
[10] Y. Okumura, S.P. Xie, A. Numaguti, Y. Tanimoto, Tropical Atlantic air-sea interaction and its influence on the NAO, Geophys. Res. Lett. *28* (8) (2001) 1507–1510.
[11] D.B. Chelton, S.P. Xie, Coupled ocean-atmosphere interaction at oceanic mesoscales, Oceanography *23* (4) (2010) 52–69.
[12] S.G. Gopalakrishnan, F. Marks Jr, J.A. Zhang, X. Zhang, J.W. Bao, V. Tallapragada, A study of the impacts of vertical diffusion on the structure and intensity of the tropical cyclones using the high-resolution HWRF system, J. Atmospheric Sci. *70* (2) (2013) 524–541.
[13] R.R. Kelkar, A.V.R.K. Rao, R.C. Bhatia, Recent improvements in cloud motion vector derivation from INSAT, Proc. 2nd Int. Wind Workshop, 1993 65–70.

[14] U.R. Rao, N. Pant, P.P. Kale, K. Narayanan, P. Ramachandran, The Indian national satellite system-INSAT, Space Communication Broadcasting *5* (1987) 339–358.
[15] V.F. Dvorak, Tropical cyclone intensity analysis and forecasting from satellite imagery, Monthly Weather Review *103* (5) (1975) 420–430.
[16] S.R. Kalsi, Orissa super cyclone–a synopsis, Mausam *57* (1) (2006) 1–20.
[17] J Le Marshall, Cloud and water vapour motion vectors in tropical cyclone track forecasting—a review, Meteorol. Atmos. Phys. *65* (3) (1998) 141–151.
[18] X. Zou, Y. Wu, On the relationship between Total Ozone Mapping Spectrometer (TOMS) ozone and hurricanes, Journal of Geophysical Research: Atmospheres *110* (D6) (2005).
[19] I. Kaur, S.K. Deb, C.M. Kishtawal, P.K. Pal, R. Kumar, Low level cloud motion vectors from Kalpana-1 visible images, J. Earth Syst. Sci. *122* (4) (2013) 935–946.
[20] X. Su, T. Li, C. An, G. Wang, Prediction of short-time cloud motion using a deep-learning model, Atmosphere *11* (11) (2020) 1151.
[21] T. Kozu, T. Kawanishi, H. Kuroiwa, M. Kojima, K. Oikawa, H. Kumagai…, K. Nishikawa, Development of precipitation radar onboard the tropical rainfall measuring mission (TRMM) satellite, IEEE Trans. Geosci. Remote Sens. *39* (1) (2001) 102–116.
[22] S.G. Gopalakrishnan, F. Marks Jr, X. Zhang, J.W. Bao, K.S. Yeh, R. Atlas, The experimental HWRF system: A study on the influence of horizontal resolution on the structure and intensity changes in tropical cyclones using an idealized framework, Monthly Weather Review *139* (6) (2011) 1762–1784.
[23] C.V. Srinivas, V. Yesubabu, R. Venkatesan, S.S.V.S. Ramarkrishna, Impact of assimilation of conventional and satellite meteorological observations on the numerical simulation of a Bay of Bengal Tropical Cyclone of November 2008 near Tamilnadu using WRF model, Meteorol. Atmos. Phys. *110* (1) (2010) 19–44.
[24] K.S. Singh, M. Mandal, P.K. Bhaskaran, Impact of radiance data assimilation on the prediction performance of cyclonic storm SIDR using WRF-3DVAR modelling system, Meteorol. Atmos. Phys. *131* (1) (2019) 11–28.
[25] R. Singh, C.M. Kishtawal, P.K. Pal, P.C. Joshi, Assimilation of the multisatellite data into the WRF model for track and intensity simulation of the Indian Ocean tropical cyclones, Meteorol. Atmos. Phys. *111* (3) (2011) 103–119.
[26] J.W. Bao, J.M. Wilczak, J.K. Choi, L.H. Kantha, Numerical simulations of air–sea interaction under high wind conditions using a coupled model: a study of hurricane development, Monthly Weather Review *128* (7) (2000) 2190–2210.
[27] B.P. Kumar, J. Vialard, M. Lengaigne, V.S.N. Murty, M.J. Mcphaden, TropFlux: air-sea fluxes for the global tropical oceans—description and evaluation, Clim. Dyn. *38* (7) (2012) 1521–1543.
[28] B.P. Kumar, J. Vialard, M. Lengaigne, V.S.N. Murty, M.J. Mcphaden, M.F. Cronin…, K.G. Reddy, TropFlux wind stresses over the tropical oceans: evaluation and comparison with other products, Clim. Dyn. *40* (7-8) (2013) 2049–2071.
[29] P.S. Polito, O.T. Sato, Global interannual trends and amplitude modulations of the sea surface height anomaly from the TOPEX/Jason-1 altimeters, Journal of climate *21* (12) (2008) 2824–2834.

CHAPTER 20

Dust storm characteristics over Indo-Gangetic basin through satellite remote sensing

Prashant Kumar Chauhan[a], Akhilesh Kumar[a,b], Vineet Pratap[a], Shivam Kumar Chaubey[a], Abhay Kumar Singh[a,c]
[a]Atmospheric Research Laboratory, Department of Physics, Institute of Science, Banaras Hindu University, Varanasi, Uttar Pradesh, India
[b]Kashi Naresh Government Post Graduate College Gyanpur, Bhadohi, India
[c]DST - Mahamana Centre of Excellence in Climate Change Research, Banaras Hindu University, Varanasi, India

20.1 Introduction

A dust storm is a meteorological phenomenon that happens when the intense wind blows loose sand and debris from a dry surface, which is most common in arid and desert areas [1]. During the premonsoon season, Indo-Gangetic basin (IGB) frequently experiences dust storms originated from mid-arid regions and Thar Desert [2,3]. Dust storms are mainly associated with a large uncertainty because of the different source regions, transportation, and complex atmospheric conditions and also the global climate models suggest the higher value of total global dust emission which is approximately 500–6000 Tg/year [4]. In aerosol loading, Dust is one of the most important constituents of which contributes nearly 30% of global aerosol optical thickness and 75% global aerosol mass load [5]. The dust particles mixed with fine mode anthropogenic aerosols are called brown carbon that increases the concentration of aerosol particles and can affect the biosphere, cryosphere, and eco-system of Earth [6,7]. They are also referred to the fraction of organics that shows light absorption characteristics from the near UV to the visible region. The air quality, atmospheric chemistry, vegetation, public health, and welfare are also affected by these dust particles [8].

Most of the dust storms originate from the desert and semidesert parts of the world and have various impacts on the Earth system and human society [9]. These dust hazards affect the environment, society, and livelihoods [9,10]. More than 100 people's deaths were reported because of very strong lightning and wind associated with an intense dust storm in northwestern India, in May 2018 (https://www.wunderground.com/cat6/fierce-thunderstorm-related-winds-kill-more-100/India). The main reasons for the deaths were many walls and roofs of the houses were collapsing and trees were falling onto the buildings due to the strong dust hazards. There were water shortages and power cuts because thousands of the electricity poles were blown [11]. Later in the same

Atmospheric Remote Sensing
DOI: https://doi.org/10.1016/B978-0-323-99262-6.00007-9

month, thousands of people were killed and millions of people were affected by a series of dust storms that covered a much wider area of central and southwest Asia, which affected the regions of northwestern states of India, northwestern parts of Pakistan, eastern regions of Iran, and some parts of Uzbekistan and Turkmenistan [12].

For the first time, an overview of dust storms was presented by Middleton [13]. Dust particles can perturb the Earth's radiation budget, which usually changes the temperature of the Earth's surface and lead to influence the exchange processes between the Earth's surface, atmosphere, and atmospheric dynamics. Because of their broad spatial distributions and large optical depths, it is very important to measure the accurate estimation of the radiative impact of dust storms [14]. Dust storms are transported over large distance from one place to another place and vertically up to 6–8 km which usually depends on the intensity of wind [15]. The IGB, one of the world's biggest basin, which frequently experiences dust storms during the premonsoon season (April–June) and carries a large number of dust particles, resulting in the enhancement of aerosol optical depth and reducing the solar radiation by nearly 50–100 W/m^2 at the surface [16,17]. Dust is transported from the Arabian Peninsula to the surrounding oceans and subsequently toward the Thar and IGP [2,18–20]. Depending upon meteorological conditions, wind velocity, wind direction, air temperature, boundary layer height, etc., the western parts of India were first affected by these dust storms and then it entered the IGB and was transported toward the eastern parts of India. Remote sensing data has been used to measure severe dust loading across the IGB during the premonsoon season [21].

Dust storm causes serious health risk to human health [22]. The size of the dust particles is the key factor in determining the possible risk to human health. The larger particle of size larger than 10 μm are not breathable, may affect the external organs like eye and skin infection, ocular infection, and conjunctivitis [22]. Smaller particles of size less than 10 μm are inhalable, may cause respiratory disorders like Tracheitis, Asthma, and Allergic rhinitis, etc., because these particles may enter and get trapped in the nose and upper respiratory tract [23]. However, the finer particles can enter the lower respiratory tract and in the bloodstream and may affect the internal organs, causing cardiovascular disorder [24]. There was a global model assessment in 2004, gave an idea of estimation that dust particles cause premature deaths of about 400,000 by cardiopulmonary disease in the 30 and older years age population [25,26].

To study the dust storm characteristics, remote sensing techniques are well accepted. Remote sensing techniques are widely used for the detailed study of dust storm, their detection, transportation, and their path mapping [27]. When we study the dust storm by only use of satellite remote sensing technique, we observe that there is large uncertainty in measuring the accurate information of the dust storm; because there are different types of limitations of satellite remote sensing like cloud contamination, spatial distribution, and data retrievals algorithm, etc. Therefore to reduce these uncertainties and inaccuracy to dust storms, different researchers have studied dust storms by combined

Fig. 20.1 ***View of a dust storm on June 9, 2018 at Uttar Pradesh.*** (https://www.deccanherald.com/national/26-killed-dust-storms-lightning-674149.html)

use of satellite remote sensing, ground-based measurements of properties of dust particles, and modeled data over IGB [2,16,17]. Because dust particles are combined with anthropogenic aerosol particles, which collectively alter the Earth's atmosphere's radiative forcing, therefore it is highly needed to study the dust particles' characteristics over IGB [6,28,29]. For this fact, Prasad and Singh [30] studied long-term (2001–2005) dust storm characteristics over IGB using satellite and ground data. Some other researchers focused on some selected dust storms and studied the different dust characteristics like its path mapping and radiative forcing during these events over the Gangetic–Himalayan region [31], the Western Ghats [32], Delhi [17], Patiala [28,33], Kanpur [16,34]. During the premonsoon season, an enhancement in columnar water vapor is reported which may also increase radiative forcing which leads to net atmospheric warming [16,28]. It is also reported that during the dust events, there was an elevation in dust loading up to 4 km over the IGB [35]. The view of a dust storm that occurred on June 9, 2018 in Uttar Pradesh has been shown in Fig. 20.1 (https://www.deccanherald.com/national/26-killed-dust-storms-lightning-674149.html).

20.2 Data and methodology

20.2.1 AErosol RObotic NETwork (AERONET)

CIMEL Sun/Sky radiometer measurements data are used to find the characteristics of dust particles. In IGP, it is deployed at several locations under the AERONET program of NASA, USA. Kanpur is one of them. This instrument measures direct Sunlight at eight different wavelengths within the range of 0.34–1.02 μm (0.34, 0.38, 0.44, 0.50, 0.67, 0.87, 0.94, and 1.02 μm). At these wavelengths, AOD and columnar water vapor (only at 0.94 μm) are obtained [36]. At four wavelengths (0.33, 0.67, 0.87, and 1.02 μm), it also provides diffuse sky radiance. With the help of this, we can obtain retrieve aerosol columnar volume size distribution (AVSD), phase function, refractive indices of aerosols, single scattering albedo (SSA), and asymmetry parameters (AP) by the inversion method [36,37]. Three different versions (1, 2, and 3) with three different categories, namely, (1) cloud contaminated (level 1.0), (2) cloud screened (level 1.5), and (3) quality assured (level 2.0) exist, respectively, at which data are online available [38]. Under the cloud-free condition, uncertainty in AOD calculation is less than $\pm$ 0.02 for shorter wavelengths and less than $\pm$0.01 for higher wavelengths ($\lambda > 0.44$ μm); while uncertainties up to 10% occur in columnar water vapor [39,40]. For the particles of radius (R) in the range of 0.1–7 μm, the errors are insignificant. As the optical depth decreases, the tendency for errors in the retrieval of refractive index (RI) and SSA tends to increase more than in the volume size distribution [39].

20.2.2 Moderate-resolution imaging spectroradiometer (MODIS)

The moderate resolution imaging spectroradiometer (MODIS) has been installed for monitoring ocean, land, and atmosphere observations. This instrument is attached to the Terra and Aqua satellites, which were activated in December 1999 and May 2002, respectively [41]. MODIS instrument gives daily global data of dust characteristics like Radiance at 36 spectral bands in visible to thermal infrared of wavelength ranging from 0.41 to 14.38 μm [33]. Aerosol characteristics are retrieved within the channels at 0.41 μm to 2.1 μm while in the near-infrared range, information about water vapor is gathered through five channels [42]. Spatial elements are produced by aligning the detectors from the different bands, each containing 81 data channels. The MODIS provides various types of information (having spatial resolution 1 km) such as geodetic latitude and longitude, satellite zenith angle, satellite azimuth height above the earth ellipsoid, range to the satellite, solar azimuth, and solar zenith angle. MODIS scans the entire surface of the earth every 1 to 2 days. Terra and Aqua satellites cross over the Indian region at 10:30 a.m. and 01:30 p.m. IST [43]. MODIS data is accessible at the NASA Giovanni website (https://giovanni.gsfc.nasa.gov/giovanni/) without any cost. It is often used for scanning the earth's atmosphere [44]. Aerosol product by MODIS global level 3 at 1° spatial resolution is derived from level 2 MODIS product having 10 km resolution [42]. Characteristics of aerosol, water vapor, and parameters of ozone can be obtained from MODIS level 3 product in a single HDF file which is important to observe the interaction among aerosol, hydrological cycle, and energy budget [42]. Kumar et al. [16] examined that there was diverse loading of aerosol during dust events over IGB especially from west to east (22.5–32^{0} N & 68–88^{0} E) (Fig. 20.2A). Thick dust plumes are observed by MODIS over the Thar Desert. Initially, these plumes are a shift to the

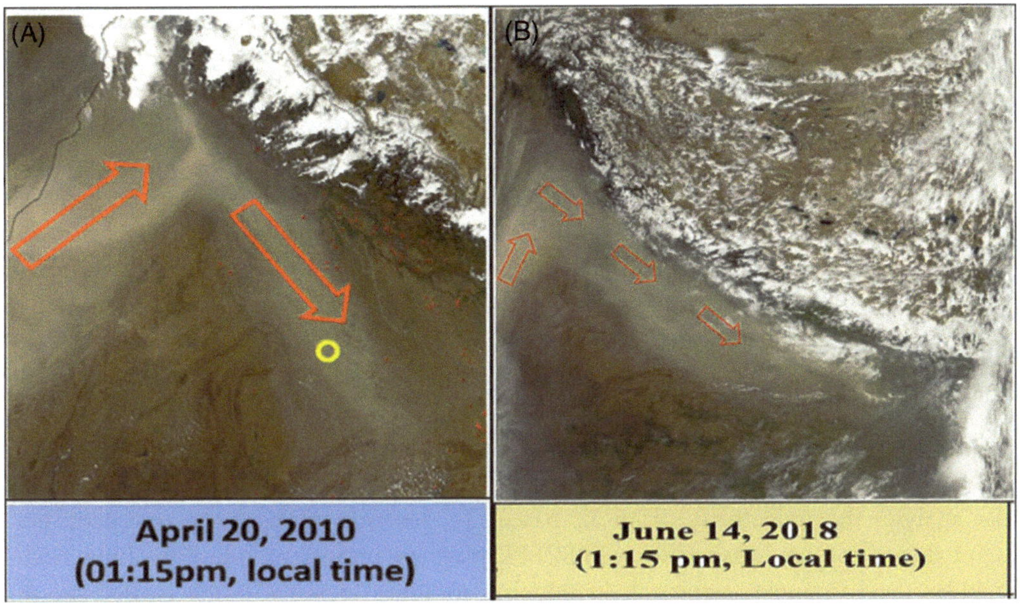

Fig. 20.2 MODIS TERRA true color images showing dust-storm events over IGB (A) on April 20, 2010 [16] and (B) on June 14, 2018 [43]. The dust appears in pale beige color, while clouds are appeared in white color and red arrows show the pathways of the dust plumes and yellow circle represents the location of Kanpur. (Source: http://rapidfire.sci.gsfc.nasa.gov).

northeast and then southeast (Fig. 20.2B). The plenty of dust that was propagating toward the northeast is deposited over the Himalayan foothills through which radiative balance is affected over this region [45].

20.2.3 Cloud-aerosol LIDAR and infrared pathfinder satellite observation (CALIPSO)

The Cloud-Aerosol LIDAR and Infrared Pathfinder Satellite Observations (CALIPSO) is a satellite developed by the National Aeronautics Space Administration (NASA) that measures the vertical profile of dust particles. It monitors atmospheric dust, its transport, and its vertical extent [46]. To study the vertical distribution, optical and physical characteristics of dust particles and clouds at 532 nm and 1064 nm, CALIPSO was launched on April 28, 2006 that discriminates between water and ice cloud and identifies the non-spherical dust particles [47]. Level 1 data of CALIPSO contains LIDAR calibrated profile while level 2 contains cloud layer products. Cloud-Aerosol LIDAR with Orthogonal Polarization (CALIOP) is the main instrument mounted on the CALIPSO. CALIOP measures the ice-water phase with the help of LIDAR backscatter signal depolarization or particle depolarization ratio (PDR) and color ratio (CR) under the sky conditions whereas it cannot observe below thick clouds. As CR measures the size of a particle, it can be calculated by comparing its backscatter coefficient at 1064 nm to that at 532 nm. With the help of PDR, the shape of the particle is measured by dividing the parallel backscatter coefficient by the perpendicular scatter coefficient at 532 nanometers [16]. The presence of large and nonspherical particles is confirmed by the higher values of CR (>0.6) and PDR (>0.2) while smaller values are the indication of small and spherical particles [48,49]. CALIPSO's orbit has a period of 16 days that provides subsatellite paths spaced longitudinally by 172 km at the equator. Spatial coverage by CALIPSO is less compared to MODIS as its observations are restricted to the subsatellite path [43]. CALIOP observations of total attenuated backscatter at 532 nm confirm that there is a higher concentration of dust particles over IGB [43].

20.2.4 Hybrid Single-Particle Lagrangian Integrated Trajectory (HYSPLIT)

The National Oceanic and Atmospheric Administration Air Resources Laboratory's Hybrid Single-Particle Lagrangian Integrated Trajectory Model (HYSPLIT) is a model that analyses the trajectories of an air parcel. To describe atmospheric transport, pollutant deposition, and dispersion, HYSPLIT plays an important role [50]. This model includes the Lagrangian approach in which a moving reference frame is used for the diffusion and advection as the air parcel shift its position and Eulerian methodology in which a fixed three-dimensional grid is used that measures the concentration of pollutant air [51]. HYSPLIT is used for measuring the five days back trajectories of air masses at three different heights of 1000, 1500, and 2000 m above mean sea level in most of the studies [52]. It can be easily calculated at the website http://ready.arl.noaa.gov/HYSPLIT.php. Kumar et al. [16] observed that Arabian Peninsula is the source of air masses at lower altitudes (1000 m), while the Middle East region is the source of air masses at high altitudes (1500 m, 2000 m). Tiwari et al. [43] showed that the Thar Desert is the origin of a higher concentration of dust particles over IGB on June 14, 2018.

20.3 Characteristics of dust particles

20.3.1 Physical and optical characteristics

20.3.1.1 Aerosol optical depth (AOD) and angstrom exponent (AE)

There are two very important parameters AOD and AE to know the characteristics of dust aerosol particles. AOD tells about the attenuation of light due to aerosol particles. Its value is equal to the negative logarithm of the radiation (scattered or absorbed) by the aerosol particles on a vertical path from the surface to the top of the atmosphere. AE gives information about the size of aerosol particles [53]. These parameters can be calculated easily by Eq. (20.1):

$$\tau_a(\lambda) = \beta \cdot \lambda^{-\alpha} \tag{20.1}$$

where τ_a is AOD for the wavelength λ, λ is the wavelength in micrometer, β is the turbidity coefficient, equals to columnar AOD at $\lambda = 1$ μm and α is angstrom exponent

Table 20.1 Aerosol characteristics and radiative forcing in previous studies over Gangetic–Himalayan region during the premonsoon season.

Study location	Dust storm event	AOD	AE	SSA	AVSD	SRF	TOA RF	References
Kanpur	08/05/2001	1.14	0.29	0.91	1.0441			[2]
	09/05/2001	1.28	0.16	0.94	1.086			
	19/05/2002	0.94	–	0.99	1.083			
	20/05/2002	1.81	0.0012	0.97	2.253			
	27/05/2002	1.40	0.04	0.98	1.583			
	09/04/2005	1.251	-0.007	0.963	0.666			
	08/05/2005	1.263	0.137	0.949	0.508			[42]
	03/06/2005	0.853	0.258	0.925	0.397	-87.5	-26	
	10/06/2005	1.446	0.105	0.918	0.993			
	11/06/2005	1.625	0.053	0.961	1.171			
Delhi	08/04/2005	2.99	0.087	–	–	–	–	[65]
Nainital	13/06/2006	0.63	0.04	0.79		-45.0	-30	[31]
Kanpur	20/04/2010	1.38	-0.049	0.919	–	-93.27	-18.16	[16]
	28/05/2010	1.192	0.071	0.904	–	-101.6	-40.95	
	03/06/2010	1.31	0.023	0.969	–	-66.71	-29.58	
Delhi	21/03/2012	0.74	-0.02	–	–	-86.0	–	[66]
Kanpur	17/05/2018	1.11	0.366	0.89	0.694	-160.45	-36.06	[43]
	14/06/2018	2.52	0.067	0.93	1.076	-141.45	-56.51	
Patiala	21/04/2010	1.13	0.31	0.925	–	-75.9	-16.4	[33]
	28/05/2010	1.52	0.14	0.924	–	-102.6	-22.6	
	20/03/2012	1.55	0.35	0.88	–	-96.71	-19.66	[28]

(AE). The more the value of AOD, the higher is the concentration of aerosol particles within a column. The AE is a good indicator of the aerosol particle size distribution and fraction $r < 1$ μm tells about the accumulation mode while $r > 1$ μm tells about the coarse-mode particles quantitatively [54]. To study the different characteristics of dust particles during the major dust storm events, several works have been reported [2,42]. After analyzing the dust events of May and June 2018 over IGB, it was found that the monthly average of AOD in May and June was 0.89 ± 0.19 and 1.05 ± 0.43, respectively, whereas that of AE was 0.71 ± 0.21 and 0.60 ± 0.35, respectively. But during the dust event on May 17, 2018 AOD value became 1.1 which was enhanced by nearly 25% of the monthly mean value and the value of AE got decreased by nearly 50% which clearly showed the enhancement, of course-mode particles. While during the dust event on June 14, 2018, the AOD value increased by 2.4 times the monthly mean value and the value of AE decreased up to nearly zero value (Table 20.1) which tells about the abundance of dust particles into the atmosphere. During the dust storm occurred in 2010 also shows high value of AOD with lower value of AE along with their sources which is shown in Fig. 20.3A and B [16]. Similar studies like a sudden decrease in AE and increase in AOD during dust storm days were reported [2,16].

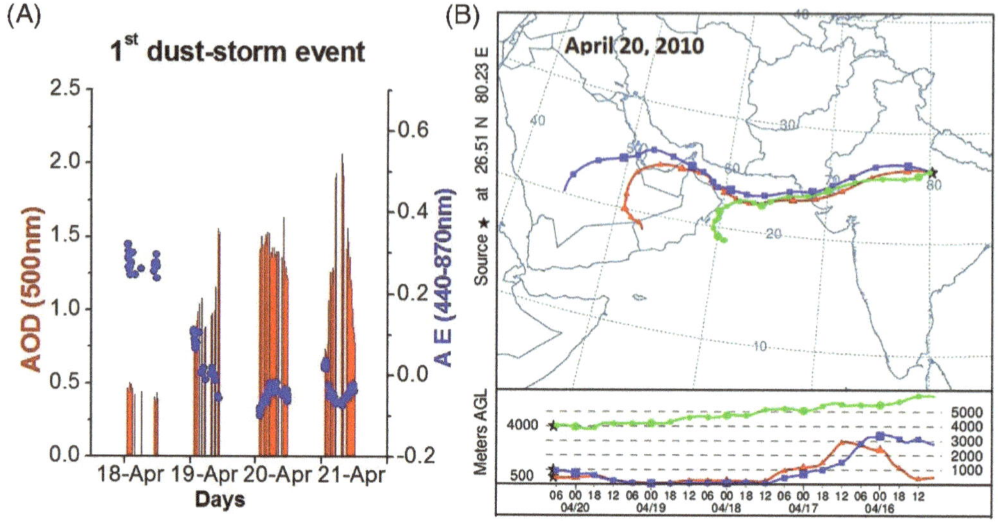

Fig. 20.3 (A) Daily (all point) variation of AOD (500 nm) and Ångström exponent (440–870 nm) at Kanpur AERONET site during the dust-storm events [16], (B) 5-days air mass back trajectory using HYSPLIT model over Kanpur station at three different altitudes.

20.3.1.2 Aerosol volume size distribution (AVSD)

The main symptom that slightly differentiates the optical properties of the dust aerosol from the anthropogenic and biomass burning aerosol is the abundance of coarse particles during the dust storm. Dust aerosols have a stronger absorption of solar and infrared wavelengths compared to aerosols that are influenced (directly or indirectly) by humans, e.g., sulfate.

The aerosol size distribution depends on emission sources, composition, transport patterns, and removal mechanisms. AVSD is a very important parameter that clearly shows a significant difference between dusty and premonsoon days. In Fig. 20.4, we can easily observe the variation of AVSD (dV/dlnr) with respect to the particles' radius

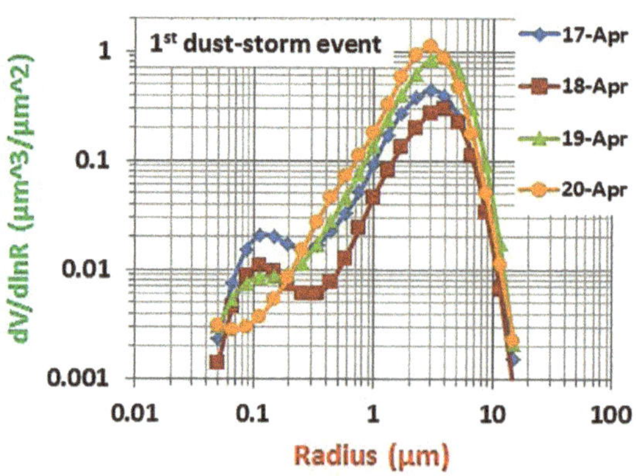

Fig. 20.4 Aerosol volume size distribution for 22 size bins between 0.05 and 15 µm at Kanpur AERONET station during the dust-storm events in premonsoon season [16].

from the fine to coarse mode particles (0.05 to 15 μm). The sun/sky radiometer data according to the approach of Dubovik and King [37] was used for the dust events in Kanpur on April 20, 2010, and are shown in Fig. 20.4 [16]. Significant increase in the concentration of coarse mode particle (as peak of AVSD shifted toward the coarse mode particles) in relation to the anthropogenic particles of fine mode during the dusty days was seen which can also be verified from the very small or even negative value of AE during the dusty days [3,31]. Previous studies had also reported a nearly similar volume size distribution for aerosol particles at the Kanpur station [2], in Karachi [55], and also in Delhi [17] during large dust storms in the premonsoon season.

20.3.1.3 Single scattering albedo (SSA)

One of the foremost vital properties of the dust aerosol properties is SSA which can give the data of approximate decrease in solar as well as terrestrial radiation in contaminated environments. It gives the point of interest data about the absorbing as well as scattering nature of the dust aerosol particles which is one of the key parameters for the calculation of the atmospheric radiative forcing (ARF). SSA has direct relation with ARF in the sense of reflectivity. High surface albedo corresponds to positive radiative forcing while low surface albedo corresponds to negative radiative forcing [56]. The effective method used to calculate SSA is by the scattering optical thickness gotten from the phase function of the normalized dust aerosol particles utilizing diffuse radiance which is measured at diverse angles and has a value ranging from zero to one, where one shows the value for completely scattering types of dust aerosol particles (e.g., sulfate) and zero for completely absorbing types of dust aerosol particles and exceedingly depends on size and composition of the dust aerosol particles [57]. Previously there were several studies done related to the SSA during the dust storm events and it was found that SSA shows an increase with respect to the wavelength and had a very large value (SSA > 0.9) during all the dusty days (two dust events, May 17, 2018; Fig. 20.5A and June 14, 2018; Fig. 20.5B) and had

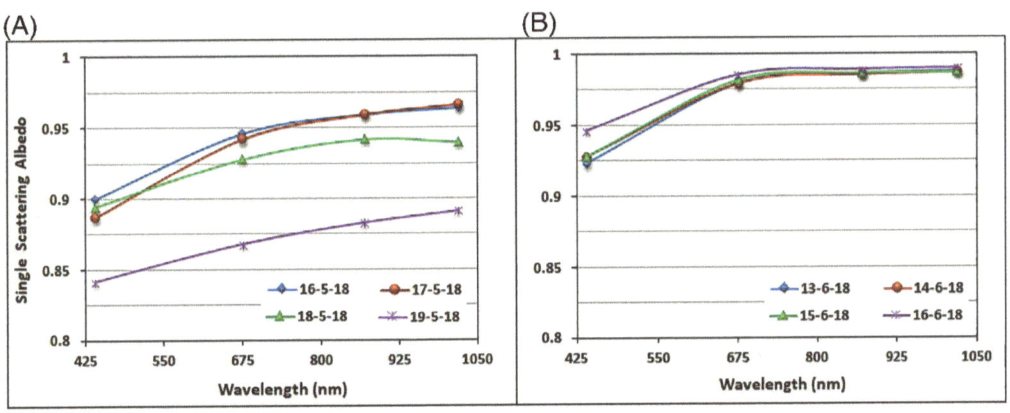

Fig. 20.5 Spectral variation of single scattering albedo (SSA) during the dust storm events (A) on May 18, 2018 and (B) on June 14, 2018 [43].

a very sharp increase at the time of the peak of dust events [2,42,43]. Larger the value of SSA greater than 0.90, alongside increasing fashion with wavelength, are traits of the dominance of dust aerosols particles inside the atmosphere [58].

At shorter wavelengths (440 nm), SSA reveals plenty of decreased values because of the large absorption of dust aerosol particles inside the UV and near-UV spectrum. A sharp upward increase of SSA value in the range of 440–675 nm wavelength clearly shows scattering nature of aerosol particles at a longer wavelength ($\lambda > 670$ nm) due to an increase in the concentration of coarse mode particles (high AOD, low AE) which is likewise shown from the daily variation of AOD (Fig. 20.3A), displaying higher AOD value (>1) at a higher wavelength which also confirms the dominance of coarse mode particles. Several studies have been reported by researchers which show a similar variation of SSA during the different dust events [16,42,43]. Kumar et al. [16] observed a comparable variety of SSA (≤0.95) throughout three specific dust activities in the pre-monsoon season (April 20, 28 May, and June 3, 2010) over Kanpur while over Karachi, Iftikhar et al. [59] observed much larger value of SSA approximately 0.98 throughout an extreme dust event (July 5, 2014).

20.3.2 Radiative characteristics

20.3.2.1 Aerosol radiative forcing (ARF) during major dust storms

Aerosol radiative forcing (ARF) is an essential atmospheric measure that indicates whether the atmosphere is warming or cooling. It is termed as the net difference in solar fluxes in the presence of aerosols and in the absence of aerosols whether it is at top of the atmosphere (TOA) or surface (BOA), respectively [60,61]. Recently, Kumar et al. [16] and Tiwari et al. [43] reported the change in the net radiation fluxes, during the different dust storm events and tried to calculate the ARF at the TOA and SRF with the help of the AERONET data over Kanpur using well-accepted widely used GAME (Global Atmospheric ModEl) radiative model [43] and SBDART model [16]. GAME is a widely used well-accepted radiative model that integrates the spectrum by using a correlated k-distribution based on line-by-line modeling [62], while the SBDART model is most widely used to calculate the change in the net radiation flux (only in the solar spectrum (0.25–4.0 μm) over Kanpur [63], based on very well profound physical models of radiation transmission and is widely used to calculate ARF. The SBDART model was running at 1-hour intervals over 24 hours, and daily ARF values at TOA and SRF were estimated in the days before and during the three dust storms periods (April 17–24, May 25–30, and May 31–June 4, 2010 [16]. The net difference between TOA and SRF force gives rise to the atmospheric forcing (heating or cooling) due to the absorption of light by the aerosol particles [43]. There are very important aerosol parameters that are necessary for calculating the ARF with the help of the SBDART model that involves the SSA, spectral AOD, and asymmetry factor, which could be obtained from the Kanpur AERONET station to achieve larger accuracy within the ARF calculations. But, also there are some parameters utilized in the model that affects the radiative forcing are some of the astronomical factors (which involve sun elevation, azimuth, solar geometry, etc.) as well as the modeled atmosphere for the vertical distribution of temperature, pressure, surface albedo, and the humidity. Kumar et al. [16] found that the daily values for the total column ozone concentrations were calculated from the data extracted from the ozone monitoring instrument (OMI), onboard NASA's Aura satellite, and an average value was used for every dust event. The surface albedo value was obtained 0.15 from the Aura OMI version 3 reflectivity knowledge through the visualization from Giovanni's online data system. The general uncertainty within the calculable radiative forcing because of uncertainties in the input parameters was anticipated to be in the range of ~ 10%–15% [42]. Fig. 20.6 shows the daily variation of ARF during the major two dust-storm events in premonsoon in April 2010 (Fig. 20.6A) and in June 2018 (Fig. 20.6B). With the help of the above formulation, Kumar et al. [16] estimated the ARF values at SRF ranging from -31.85 to -101.60 W/m^2 (average -65.77 ± 21.72 W/m^2), at TOA -3.94 to -40.95 W/m^2 (average -17.99 ± 10.46 W/m^2), and within the atmosphere in the range of +16.87 to +75.11 W/m^2 (average +47.78 ± 16.24 W/m^2), respectively (Fig. 20.6A). But during April 20, May 28, and June 2, 2010, the ARF values were very high at TOA (-18.16, -40.95, and -29.58 $W\ m^{-2}$), at SRF (-93.27, -101.60, and -66.71 W/m^2), and within the atmosphere (+75.11, +60.65, and +37.13 W/m^2), whereas during the non-dusty days the ARF values were lower. Also, Tiwari et al. [43] estimated the ARF values at SRF (ranging from -101 to -191 W/m^2 (May 17, 2018) and -90 to -151 W/m^2 (June 14, 2018)), at TOA (-12 to -37 W/m^2 (May 17, 2018) and -47 to -64 W/m^2 (June 14, 2018)), and within the atmosphere in the range of (ARF_{ATM} = ~ +124 W/m^2 on May 17, 2018 and ARF_{ATM} = ~ +85 W/m^2 on June 14, 2018), respectively (Fig. 20.6B).

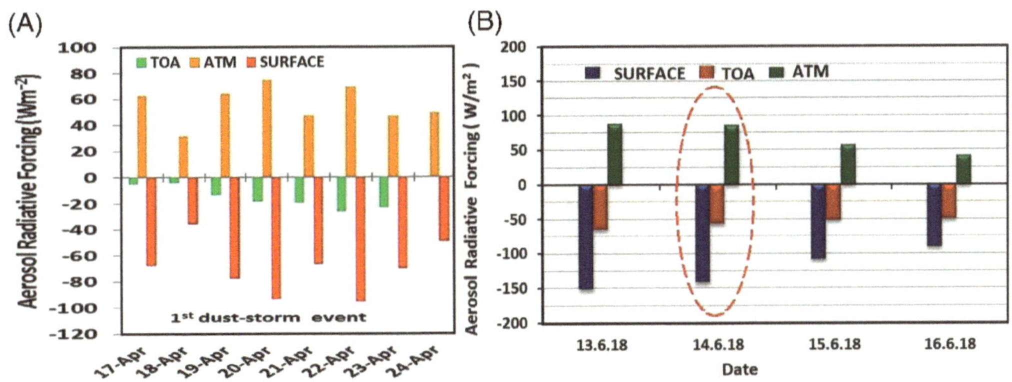

Fig. 20.6 Aerosol radiative forcing values (surface, top of the atmosphere and within the atmosphere in W/m^2) during the major two dust-storm events in premonsoon (A) ARF in April 2010 [16] and (B) ARF in June 2018 [43].

The above results show that because of the sudden increase in dust aerosol, the surface forcing increases (very large), whereas the part heating is very less affected, suggesting that the AOD plays the main role in ARF estimates throughout the dust events. Some of the previous studies are shown in Table 20.1 along with the aerosol characteristics and radiative forcing during different dust event and at different locations.

A significant modification in ARF is determined with the larger value of ARF (both at SRF and TOA) throughout each dusty day suggesting the additional attenuation of solar radiation (coming toward the surface), which can be due to the backscattering because of the dust aerosol particles. However, the positive value of ARF in the atmosphere suggests the net warming result of the atmosphere because of the dust aerosol during each event. Sharma et al. [33] found the ARF value as -100 W/m^2 to -50 W/m^2 at the surface and -25 W/m^2 to -10 W/m^2 at TOA over Patiala during the dust events of April 20–21, 2010 and May 26–28. The calculated ARF was higher than that of Kanpur, indicating an intense dust storm event over the western part of the IGB due to its proximity to dust source areas. Chinnam et al. [64] reported ARF values at SRF (-12 W/m^2) and TOA (+7 W/m^2) during the dusty days (May 12–22, 2004, Kanpur), showing the lower value at the SRF and showing the opposite signs for TOA compared to the above results. Furthermore, Prasad et al. [42] found ARF values in the range of -19 Wm^{-2} and -87 W/m^2 at SRF and +2 W/m^2 and -26 W/m^2 at TOA (during, April–May 2005).

Furthermore, the atmospheric heating rate was obtained by the following equation [16,43]:

$$\frac{\partial T}{\partial t} = \frac{g}{Cp}\frac{\Delta F_{Atmos}}{\Delta P} \tag{20.2}$$

where (in K/day) tells about the atmospheric heating rate, ΔF_{Atmos} describes the atmospheric forcing, g is the gravitational acceleration, Cp is the specific heat capacity of air (at constant pressure), and ΔP shows the pressure difference between top and bottom boundary of each layer of the atmosphere, whereas they had used ΔP as 300 hPa for the layer up to ~3.5 km that correspond to 1000–700 hPa pressure layer. With the help of Eq. (20.1), Kumar et al. [16] estimated the atmospheric heating rates with an average value of 1.33 K/day. The rates for atmospheric heating were reported very high during the dust storm events (April 20, May 28, and June 2, 2010) with the value 2.11 K/day, 1.70 K/day, and 1.04 K/day, respectively, while a very high value of the atmospheric heating rates (1.51–3.88 K/day and 0.9–1.88 K/day in May and June 2018, respectively) were reported by Tiwari et al. [43] showing very serious implications on climate. A similar range of the heating rate was also reported during the dust storm events over Delhi (1.9 K/day) and Jodhpur (2.0 K/day) [34] and over Lahore (1.3–2.2 K/day) and Karachi (0.7 to 1.4 K/day) [59]. Chinnam et al. [64] found anomalies in the rate of warming of the atmosphere in northern India during the premonsoon, mainly due to an increase in the number of dust particles, which contributed to affect the strengthening of the wind system (Indian monsoon). More detailed studies are needed on the effects of increased aerosol warming during the severe dust storm and its effects on regional climate.

20.4 Effect of dust storm

20.4.1 Effect of dust particle on health

Among all the particulate matter, finer one is very important which largely contributes to the adverse effect on human health. It has been noted that among most of the pollutants fine particles are repeatedly reported to be linked with adverse health impacts especially for mortality and morbidity for both globally [22–26,67,68] as well as in India [69–71]. The size of the dust particles is the key factor in determining the possible risk to human health [22]. Comparatively the course particles (size > 10 μm) are not breathable and affect our skin, eye conjunctivitis, and ophthalmic infection [22] whereas the fine particles (size < 10 μm) affect our internal organs causing Asthma, upper respiratory tract, cardiovascular disorder, etc [23,24]. There are very few researchers [68,72–74] have investigated the effect of multiple air pollutants and particulate matter like desert dust particles. Continuously, the concentration of fine particulate matter is increasing in the atmosphere from the natural as well as anthropogenic sources, which has decreased the air quality on regional to global scales [75]. It has been observed that the long-term variation of finer dust particles is associated with adverse health effects like heart disease and lung cancer which increases the risk of premature mortality [25,26]. The desert dust particles are one of the main natural sources of particulate matter which travel in the atmosphere in large amounts [76]. Through whole over the globe, in many regions, the concentration of $PM_{2.5}$ is much larger than the WHO [77] limit (10 $\mu g/m^3$) because of the regular dust events. In the World, the main dust sources (over 60% of the global dust load) are North Africa and the Middle East which causes a higher potential risk to the health of these regions because of their population [22]. South and East Asia have very high population density and these regions are also very often affected by severe dust events, but unfortunately in these regions, very few epidemiological studies have been done and adequate air quality data are lacking in the large amount [78]. Aerosol particles in terms of dust storms have detrimental impact on human health and agricultural activities in Central Asia [79]. It was reported that strong exposures to different particulate matters have major impact on human health [80,81]. However, under the National Air Quality Monitoring Programme, systematic monitoring of air quality indices such as $PM_{2.5}$, PM_{10}, NO_2, SO_2, and O_3 is now available in most Indian cities [82]. Therefore it is much needed to study the effect of atmospheric pollutants especially particulate matter on health.

20.4.2 Effect of a dust particle on climate

Dust particles scatter as well as absorb the solar radiation coming from the Sun. Scattering dust particles cause a cooling effect whereas absorbing dust particles show a warming effect. Dust particles play a vital role in the formation of clouds as they provide cloud condensation nuclei (CCN). A large number of dust particles may increase the number concentration of CCN through which cloud droplet is formed [83]. The cooling and warming effect occurs due to cloud droplets and their characteristics. So dust particles affect the planet's radiation balance and thus changing in climate occur [84]. A dust storm is a natural process but sand dust aerosols produced by man-made sources also affect the climate. Sand dust particles affect the radiation of the local atmosphere as well as the monsoon system, local water cycle, and regional climate environment [85]. Agriculture and the ecosystem are also affected by the dust storm. The size of the dust particle is equivalent to the wavelength of visible light which shows the scattering nature that minimizes the visibility of the atmosphere at the regional level [86]. Apart from its negative health impacts, coarse mode particle is responsible for lowering atmospheric clarity as a key element of smog, diminishing the photosynthesis process through accumulation on the surface of leaves, and trying to influence atmospheric chemistry and climatological processes [87–89]. The resulting decrease in air quality has an influence on social and economic sectors as well as marine and air transportation [90].

20.5 Summary and conclusion

The outcomes introduced in this section show effect of dust on the condition of the climate, environment, and human health. Because of dust storms, diverse atmospheric parameters (precipitation, wind speed, wind direction, humidity, atmospheric temperature, and so forth) are profoundly influenced which debases the air quality. The dust is moved from the source to an excessively far away distance influencing individuals living along the track of dust and the state of the weather conditions. The air quality influences extraordinarily the strength of human health, which is viewed as a drawn-out long-term health hazard. Remote sensing satellites with multisensors cover the globe in a couple of hours, and the satellite information gives data about the various parameters (AOD, AE, AVSD, ARF, and so on) by which condition of the weather can be found and this information can be utilized for observing dust transport from the source to distant locals influenced by dust. This information can be utilized for observing dust storms and offering warnings to individuals about dust and the power of dust storms so that individuals would not be uncovered and experience the ill effects of bad air quality. With the worldwide climate change, land use and land cover are changing which is probably going to build recurrence of dust storms and their force.

Acknowledgment

The work is financially supported by the Institute of Eminence (IoE) (Scheme No: 6031) to BHU, Varanasi.

References

[1] P. Ginoux, J.M. Prospero, T.E. Gill, N.C. Hsu, M. Zhao, Global-scale attribution of anthropogenic and natural dust sources and their emission rates based on MODIS Deep Blue aerosol products, Rev. Geophys. *50* (RG3005) (2012) 1–36.

[2] S. Dey, S.N. Tripathi, R.P. Singh, B.N. Holben, Influence of dust storms on the aerosol optical properties over the Indo-Gangetic basin, Journal of Geophysical Research: Atmospheres *109* (D20211) (2004) 1–13.

[3] R.P. Singh, S. Dey, S.N. Tripathi, V. Tare, B. Holben, Variability of aerosol parameters over Kanpur, northern India, Journal of Geophysical Research: Atmospheres *109* (D23206) (2004) 1–14.
[4] N. Huneeus, M. Schulz, Y. Balkanski, J. Griesfeller, J. Prospero, S. Kinne, S. Bauer, O. Boucher, M. Chin, F. Dentener, T. Diehl, R. Easter, D. Fillmore, S. Ghan, P. Ginoux, A. Grini, L. Horowitz, D. Koch, M.C. Krol, W. Landing, X. Liu, N. Mahowald, R. Miller, J.J. Morcrette, G. Myhre, J. Penner, J. Perlwitz, P. Stier, T. Takemura, C.S. Zender, Global dust model intercomparison in AeroCom phase I, Atmos. Chem. Phys. *11* (15) (2011) 7781–7816.
[5] W. Wang, L. Sheng, H. Jin, Y. Han, Dust aerosol effects on cirrus and altocumulus clouds in Northwest China, Journal of Meteorological Research *29* (5) (2015) 793–805.
[6] S. Tiwari, A.K. Mishra, A.K. Singh, Aerosol climatology over the Bay of Bengal and Arabian Sea inferred from space-borne radiometers and lidar observations, Aerosol and Air Quality Research *16* (11) (2016) 2855–2868.
[7] V. Ramanathan, C. Chung, D. Kim, T. Bettge, L. Buja, J.T. Kiehl, W.M. Washington, Q. Fu, D.R. Sikka, M. Wild, Atmospheric brown clouds: Impacts on South Asian climate and hydrological cycle, Proc. Natl. Acad. Sci. *102* (15) (2005) 5326–5333.
[8] M. Carugno, D. Consonni, G. Randi, D. Catelan, L. Grisotto, P.A. Bertazzi, A. Biggeri, M. Baccini, Air pollution exposure, cause-specific deaths and hospitalizations in a highly polluted Italian region, Environ. Res. *147* (2016) 415–424.
[9] A.S. Goudie, N.J. Middleton, Desert Dust in the Global System, 1st ed., Springer Science & Business Media, Heidelberg, 2006.
[10] Y. Shao, K.H. Wyrwoll, A. Chappell, J. Huang, Z. Lin, G.H. McTainsh, M. Mikami, T.Y. Tanaka, X. Wangh, S. Yoon, Dust cycle: an emerging core theme in Earth system science, Aeolian Res. *2* (4) (2011) 181–204.
[11] T.H. Painter, S.M. Skiles, J.S. Deems, W.T. Brandt, J. Dozier, Variation in rising limb of Colorado River snowmelt runoff hydrograph controlled by dust radiative forcing in snow, Geophys. Res. Lett. *45* (2) (2018) 797–808.
[12] N. Middleton, Dust storm hazards, E3S Web of Conferences, 99, EDP Sciences, 2019, pp. 04001.
[13] N.J. Middleton, A geography of dust storms in South-west Asia, Journal of Climatology *6* (2) (1986) 183–196.
[14] S.K. Mishra, S.N. Tripathi, Modeling optical properties of mineral dust over the Indian Desert, Journal of Geophysical Research: Atmospheres *113* (D23) (2008).
[15] S.C. Tan, J. Li, H. Che, B. Chen, H. Wang, Transport of East Asian dust storms to the marginal seas of China and the southern North Pacific in spring 2010, Atmos. Environ. *148* (2017) 316–328.
[16] S. Kumar, S. Kumar, D.G. Kaskaoutis, R.P. Singh, R.K. Singh, A.K. Mishra, M.K. Srivastava, A.K. Singh, Meteorological, atmospheric and climatic perturbations during major dust storms over Indo-Gangetic Basin, Aeolian Res. *17* (2015) 15–31.
[17] K. Taneja, S. Ahmad, K. Ahmad, S.D. Attri, Impact assessment of a severe dust storm on atmospheric aerosols over an urban site in India, Curr. Sci. *118* (5) (2020) 737.
[18] M.A. Aswini, A. Kumar, S.K. Das, Quantification of long-range transported aeolian dust towards the Indian peninsular region using satellite and ground-based data-A case study during a dust storm over the Arabian Sea, Atmos. Res. *239* (2020) 104910.
[19] K. Suresh, U. Singh, A. Kumar, D. Karri, A. Peketi, V. Ramaswamy, Provenance tracing of long-range transported dust over the Northeastern Arabian Sea during the southwest monsoon, Atmos. Res. *250* (2021) 105377.
[20] A. Kumar, K. Suresh, W. Rahaman, Geochemical characterization of modern aeolian dust over the Northeastern Arabian Sea: implication for dust transport in the Arabian Sea, Sci. Total Environ. *729* (2020) 138576.
[21] P.S. Bhattacharjee, A.K. Prasad, M. Kafatos, R.P. Singh, Influence of a dust storm on carbon monoxide and water vapor over the Indo-Gangetic Plains, Journal of Geophysical Research: Atmospheres *112* (D18203) (2007) 1–14.
[22] D. Giannadaki, A. Pozzer, J. Lelieveld, Modeled global effects of airborne desert dust on air quality and premature mortality, Atmos. Chem. Phys. *14* (2) (2014) 957–968.
[23] C.A. Pope III, M. Ezzati, D.W Dockery, Fine-particulate air pollution and life expectancy in the United States, N. Engl. J. Med. *360* (4) (2009) 376–386.

[24] J. Lepeule, F. Laden, D. Dockery, J. Schwartz, Chronic exposure to fine particles and mortality: an extended follow-up of the Harvard Six Cities study from 1974 to 2009, Environ. Health Perspect. *120* (7) (2012) 965–970.
[25] A.J. Cohen, H.R. Anderson, B. Ostra, K.D. Pandey, M. Krzyzanowski, N. Künzli, K. Gutschmidt, A. Pope, I. Romieu, J.M. Samet, K. Smith, The global burden of disease due to outdoor air pollution, J. Toxicol. Environ. Health Part A *68* (13-14) (2005) 1301–1307.
[26] D. Krewski, M. Jerrett, R.T. Burnett, R. Ma, E. Hughes, Y. Shi, M.C. Turner, C.A. Pope III, G. Thurston, E.E. Calle, M.J Thun, Extended follow-up and spatial analysis of the American Cancer Society study linking particulate air pollution and mortality, 140, Health Effects Institute, Boston, MA, 2009.
[27] S. Namdari, N. Karimi, A. Sorooshian, G. Mohammadi, S. Sehatkashani, Impacts of climate and synoptic fluctuations on dust storm activity over the Middle East, Atmos. Environ. *173* (2018) 265–276.
[28] A. Singh, S. Tiwari, D. Sharma, D. Singh, S. Tiwari, A.K. Srivastava, N. Rastogi, A.K. Singh, Characterization and radiative impact of dust aerosols over northwestern part of India: a case study during a severe dust storm, Meteorol. Atmos. Phys. *128* (6) (2016) 779–792.
[29] S. Tiwari, A.K. Srivastava, A.K. Singh, S. Singh, Identification of aerosol types over Indo-Gangetic Basin: implications to optical properties and associated radiative forcing, Environmental Science and Pollution Research *22* (16) (2015) 12246–12260.
[30] A.K. Prasad, R.P. Singh, Changes in aerosol parameters during major dust storm events (2001–2005) over the Indo-Gangetic Plains using AERONET and MODIS data, Journal of Geophysical Research: Atmospheres *112* (D09208) (2007) 1–18.
[31] A.K. Srivastava, P. Pant, P. Hegde, S. Singh, U.C. Dumka, M. Naja, N. Singh, Y. Bhavanikumar, The influence of a south Asian dust storm on aerosol radiative forcing at a high-altitude station in central Himalayas, Int. J. Remote Sens. *32* (22) (2011) 7827–7845.
[32] G.R. Aher, G.V. Pawar, P. Gupta, P.C.S. Devara, Effect of major dust storm on optical, physical, and radiative properties of aerosols over coastal and urban environments in Western India, Int. J. Remote Sens. *35* (3) (2014) 871–903.
[33] D. Sharma, D. Singh, D.G. Kaskaoutis, Impact of two intense dust storms on aerosol characteristics and radiative forcing over Patiala, northwestern India, Adv. Meteorol., *2012*, Hindawi, 2012, pp. 1–13 956814.
[34] A.K. Srivastava, V.K. Soni, S. Singh, V.P. Kanawade, N. Singh, S. Tiwari, S.D. Attri, An early South Asian dust storm during March 2012 and its impacts on Indian Himalayan foothills: A case study, Sci. Total Environ. *493* (2014) 526–534.
[35] J.M. Prospero, P. Ginoux, O. Torres, S.E. Nicholson, T.E. Gill, Environmental characterization of global sources of atmospheric soil dust identified with the Nimbus 7 Total Ozone Mapping Spectrometer (TOMS) absorbing aerosol product, Rev. Geophys. *40* (1) (2002) 2 -1.
[36] B.N. Holben, T.F. Eck, I. Slutsker, D. Tanré, J.P. Buis, A. Setzer, E. Vermote, J.A. Reagan, Y.J. Kaufman, T. Nakajima, F. Lavenu, I. Jankowiak, A. Smirnov, AERONET—A federated instrument network and data archive for aerosol characterization, Remote Sens. Environ. *66* (1) (1998) 1–16.
[37] O. Dubovik, M.D. King, A flexible inversion algorithm for retrieval of aerosol optical properties from Sun and sky radiance measurements, Journal of Geophysical Research: Atmospheres *105* (D16) (2000) 20673–20696.
[38] A. Smirnov, B.N. Holben, T.F. Eck, O. Dubovik, I. Slutsker, Cloud-screening and quality control algorithms for the AERONET database, Remote Sens. Environ. *73* (3) (2000) 337–349.
[39] O. Dubovik, A. Smirnov, B.N. Holben, M.D. King, Y.J. Kaufman, T.F. Eck, I. Slutsker, Accuracy assessments of aerosol optical properties retrieved from Aerosol Robotic Network (AERONET) Sun and sky radiance measurements, Journal of Geophysical Research: Atmospheres *105* (D8) (2000) 9791–9806.
[40] T.F. Eck, B.N. Holben, J.S. Reid, O. Dubovik, A. Smirnov, N.T. O'Neill, I. Slutsker, S. Kinne, Wavelength dependence of the optical depth of biomass burning, urban, and desert dust aerosols, Journal of Geophysical Research: Atmospheres *104* (D24) (1999) 31333–31349.
[41] K. Alam, T. Trautmann, T. Blaschke, F. Subhan, Changes in aerosol optical properties due to dust storms in the Middle East and Southwest Asia, Remote Sens. Environ. *143* (2014) 216–227.
[42] A.K. Prasad, S. Singh, S.S. Chauhan, M.K. Srivastava, R.P. Singh, R. Singh, Aerosol radiative forcing over the Indo-Gangetic plains during major dust storms, Atmos. Environ. 41 (2007) 6289–6301.
[43] S. Tiwari, A. Kumar, V. Pratap, A.K. Singh, Assessment of two intense dust storm characteristics over Indo–Gangetic basin and their radiative impacts: a case study, Atmos. Res. *228* (2019) 23–40.

[44] R.P. Singh, Dust storms and their influence on atmospheric parameters over the Indo-Gangetic plains. In: Geospatial technologies and climate change, Springer, Cham, 2014, pp. 21–35.
[45] S. Bucci, C. Cagnazzo, F. Cairo, L. Di Liberto, F. Fierli, Aerosol variability and atmospheric transport in the Himalayan region from CALIOP 2007–2010 observations, Atmos. Chem. Phys. *14* (9) (2014) 4369–4381.
[46] D. Liu, Y. Wang, Z. Wang, J. Zhou, The three-dimensional structure of transatlantic African dust transport: A new perspective from CALIPSO LIDAR measurements, Advances in Meteorology *2012* (2012).
[47] D.M. Winker, W.H. Hunt, M.J. McGill, Initial performance assessment of CALIOP, Geophys. Res. Lett. *34* (19) (2007).
[48] A.H. Omar, D.M. Winker, M.A. Vaughan, Y. Hu, C.R. Trepte, R.A. Ferrare, Z. Liu, The CALIPSO automated aerosol classification and lidar ratio selection algorithm, J. Atmos. Oceanic Technol. *26* (10) (2009) 1994–2014.
[49] A.K. Mishra, T. Shibata, Climatological aspects of seasonal variation of aerosol vertical distribution over central Indo-Gangetic belt (IGB) inferred by the space-borne lidar CALIOP, Atmos. Environ. *46* (2012) 365–375.
[50] R.R. Draxler, G.D. Hess, An overview of the HYSPLIT_4 modelling system for trajectories, Australian meteorological magazine *47* (4) (1998) 295–308.
[51] A.F. Stein, R.R. Draxler, G.D. Rolph, B.J. Stunder, M.D. Cohen, F. Ngan, NOAA's HYSPLIT atmospheric transport and dispersion modeling system, Bull. Am. Meteorol. Soc. *96* (12) (2015) 2059–2077.
[52] Draxler, R. R., & Rolph, G. D. (2003). HYSPLIT (HYbrid Single-Particle Lagrangian Integrated Trajectory) model access via NOAA ARL READY (https://www.arl.noaa.gov/ready/hysplit4.html). NOAA Air Resources Laboratory, Silver Spring.
[53] A. Ångström, The parameters of atmospheric turbidity, Tellus *16* (1) (1964) 64–75.
[54] S. Tiwari, D. Kaskaoutis, V.K. Soni, S. Dev Attri, A.K. Singh, Aerosol columnar characteristics and their heterogeneous nature over Varanasi, in the central Ganges valley, Environmental Science and Pollution Research *25* (25) (2018) 24726–24745.
[55] K. Alam, T. Trautmann, T. Blaschke, H. Majid, Aerosol optical and radiative properties during summer and winter seasons over Lahore and Karachi, Atmos. Environ. *50* (2012) 234–245.
[56] T. Takemura, T. Nakajima, O. Dubovik, B.N. Holben, S. Kinne, Single-scattering albedo and radiative forcing of various aerosol species with a global three-dimensional model, Journal of Climate *15* (4) (2002) 333–352.
[57] O. Dubovik, B.N. Holben, Y.J. Kaufman, M. Yamasoe, A. Smirnov, D. Tanré, I. Slutsker, Single-scattering albedo of smoke retrieved from the sky radiance and solar transmittance measured from ground, Journal of Geophysical Research: Atmospheres *103* (D24) (1998) 31903–31923.
[58] O. Dubovik, B.N. Holben, T.F. Eck, A. Smirnov, Y.J. Kaufman, M.D. King, D. Tanre, I. Slutsker, Variability of absorption and optical properties of key aerosol types observed in worldwide locations, J. Atmospheric Sci. *59* (3) (2002) 590–608.
[59] M. Iftikhar, K. Alam, A. Sorooshian, W.A. Syed, S. Bibi, H. Bibi, Contrasting aerosol optical and radiative properties between dust and urban haze episodes in megacities of Pakistan, Atmos. Environ. *173* (2018) 157–172.
[60] K. Alam, T. Trautmann, T. Blaschke, H. Majid, Aerosol optical and radiative properties during summer and winter seasons over Lahore and Karachi, Atmos. Environ. *50* (2012) 234–245.
[61] A. Kumar, V. Pratap, S. Kumar, A.K. Singh, Atmospheric aerosols properties over Indo-Gangetic Plain: a trend analysis using ground–truth AERONET data for the year 2009-2017, Adv. Space Res. *69* (7) (2022) 2659–2670.
[62] N.A. Scott, A direct method of computation of the transmission function of an inhomogeneous gaseous medium—I: Description of the method, J. Quant. Spectrosc. Radiat. Transfer *14* (8) (1974) 691–704.
[63] P. Ricchiazzi, S. Yang, C. Gautier, D. Sowle, SBDART: A research and teaching software tool for plane-parallel radiative transfer in the Earth's atmosphere, Bull. Am. Meteorol. Soc. *79* (10) (1998) 2101–2114.
[64] N. Chinnam, S. Dey, S.N. Tripathi, M. Sharma, Dust events in Kanpur, northern India: chemical evidence for source and implications to radiative forcing, Geophys. Res. Lett. *33* (L08803) (2006) 1–4.

[65] S.K. Srivastava, M.K. Srivastava, A. Saha, S. Tiwari, S. Singh, U.C. Dumka, B.P. Singh, N.P Singh, Aerosol optical properties over Delhi and Manora Peak during a rare dust event in early April 2005, Int. J. Remote Sens. *32* (23) (2011) 7939–7954.
[66] S. Singh, S. Naseema Beegum, Direct radiative effects of an unseasonal dust storm at a western Indo Gangetic Plain station Delhi in ultraviolet, shortwave, and longwave regions, Geophys. Res. Lett. *40* (10) (2013) 2444–2449.
[67] K.R. Smith, M. Jerrett, H.R. Anderson, R.T. Burnett, V. Stone, R. Derwent, W.A. Richard, C. Aaron, B.S.MPH Seth, K. Daniel, P. Arden, J.T.M.D. Micael, G. Thurston, Public health benefits of strategies to reduce greenhouse-gas emissions: health implications of short-lived greenhouse pollutants, Lancet North Am. Ed. *374* (9707) (2009) 2091–2103.
[68] N.A.H. Janssen, G. Hoek, M. Simic-Lawson, P. Fischer, L. Van Bree, B. Harry ten, M. Keuken, R.W. Atkinson, H.R. Anderson, B. Brunekreef, F.R. Cassee, Black carbon as an additional indicator of the adverse health effects of airborne particles compared with PM10 and PM2. 5, Environ. Health Perspect. *119* (12) (2011) 1691–1699.
[69] S. Chowdhury, S. Dey, K.R. Smith, Ambient PM2. 5 exposure and expected premature mortality to 2100 in India under climate change scenarios, Nat. Commun. *9* (1) (2018) 1–10.
[70] K. Balakrishnan, S. Dey, T. Gupta, R.S. Dhaliwal, M. Brauer, A.J. Cohen, J.D. Stanaway, G. Beig, T.K. Joshi, A.N. Aggarwal, Y. Sabde, The impact of air pollution on deaths, disease burden, and life expectancy across the states of India: the Global Burden of Disease Study 2017, The Lancet Planetary Health *3* (1) (2019) e26–e39.
[71] P. Saini, M. Sharma, Cause and Age-specific premature mortality attributable to PM2.5 Exposure: an analysis for Million-Plus Indian cities, Sci. Total Environ. *710* (2020) 135230.
[72] J. Cao, H. Xu, Q. Xu, B. Chen, H. Kan, Fine particulate matter constituents and cardiopulmonary mortality in a heavily polluted Chinese city, Environ. Health Perspect. *120* (3) (2012) 373–378.
[73] X. Wang, R. Chen, X. Meng, F. Geng, C. Wang, H. Kan, Associations between fine particle, coarse particle, black carbon and hospital visits in a Chinese city, Sci. Total Environ. *458* (2013) 1–6.
[74] J. Lelieveld, J.S. Evans, M. Fnais, D. Giannadaki, A. Pozzer, The contribution of outdoor air pollution sources to premature mortality on a global scale, Nature *525* (7569) (2015) 367–371.
[75] H. Akimoto, Global air quality and pollution, Science *302* (5651) (2003) 1716–1719.
[76] N.M. Mahowald, S. Kloster, S. Engelstaedter, J.K. Moore, S. Mukhopadhyay, J.R. McConnell, S. Albani, S.C. Doney, A. Bhattacharya, M.A.J. Curran, M.G. Flanner, F.M. Hoffman, D.M. Lawrence, K. Lindsay, P.A. Mayewski, J. Neff, D. Rothenberg, E. Thomas, P.E. Thornton, C.S. Zender, Observed 20th century desert dust variability: impact on climate and biogeochemistry, Atmos. Chem. Phys. *10* (22) (2010) 10875–10893.
[77] WHO, Air Quality Guidelines, World Health Organization, 2005, http://www.euro.who.int/__data/assets/pdf_file/0005/78638/E90038.pdf Global update.
[78] F. De Longueville, Y.C. Hountondji, S. Henry, P. Ozer, What do we know about effects of desert dust on air quality and human health in West Africa compared to other regions? Sci. Total Environ. *409* (1) (2010) 1–8.
[79] B. Sobhani, V.S. Zengir, Investigation hazard effect of monthly ferrrin temperature on agricultural products in north bar of Iran, The Iraqi Journal of Agricultural Science *50* (1) (2019) 320–330.
[80] K. Balakrishnan, S. Dey, T. Gupta, R.S. Dhaliwal, M. Brauer, A.J. Cohen, J.D. Stanaway, G. Beig, T.K. Joshi, A.N. Aggarwal, Y. Sabde, The impact of air pollution on deaths, disease burden, and life expectancy across the states of India: the Global Burden of Disease Study 2017, The Lancet Planetary Health *3* (1) (2019) e26–e39.
[81] J. Huang, X. Pan, X. Guo, G. Li, Impacts of air pollution wave on years of life lost: a crucial way to communicate the health risks of air pollution to the public, Environ. Int. *113* (2018) 42–49.
[82] N. Singh, A. Mhawish, T. Banerjee, S. Ghosh, R.S. Singh, R.K. Mall, Association of aerosols, trace gases and black carbon with mortality in an urban pollution hotspot over central Indo-Gangetic Plain, Atmos. Environ. *246* (2021) 118088.
[83] S. Twomey, The influence of pollution on the shortwave Albedo of Clouds, Journal of Atmospheric Sciences *34* (7) (1977) 1149–1152.
[84] M.O. Andreae, P.J. Crutzen, Atmospheric aerosols: Biogeochemical sources and role in atmospheric chemistry, Science *276* (5315) (1997) 1052–1058.

[85] K. Zhang, H. Gao, R. Zhang, Y. ZHU, Y. WANG, Sources and movement routes of sand-dust aerosols and their impact probabilities on China seas in 2000-2002, Advances in Earth Science *20* (6) (2005) 627.
[86] O. Uchino, I. Tabata, Mobile lidar for simultaneous measurements of ozone, aerosols, and temperature in the stratosphere, Appl. Opt. *30* (15) (1991) 2005–2012.
[87] D.A. Grantz, J.H.B. Garner, D.W. Johnson, Ecological effects of particulate matter, Environ. Int. *29* (2-3) (2003) 213–239.
[88] M. Lin, J. Tao, C.Y. Chan, J.J. Cao, Z.S. Zhang, L.H. Zhu, R.J. Zhang, Regression analyses between recent air quality and visibility changes in megacities at four haze regions in China, Aerosol and air quality research *12* (6) (2012) 1049–1061.
[89] E. Von Schneidemesser, P.S. Monks, J.D. Allan, L. Bruhwiler, P. Forster, D. Fowler, A. Lauer, W.T. Morgan, P. Paasonen, M. Righi, K. Sindelarova, M.A. Sutton, Chemistry and the linkages between air quality and climate change, Chem. Rev. *115* (10) (2015) 3856–3897.
[90] F. Zhang, Q.R. Yu, J.L. Mao, C. Dan, Y. Wang, Q. He, T. Cheng, C. Chen, D. Liu, Y. Gao, Possible mechanisms of summer cirrus clouds over the Tibetan Plateau, Atmos. Chem. Phys. *20* (20) (2020) 11799–11808.

CHAPTER 21

Remote sensing-based geomorphological mapping of glacial and paraglacial landforms from semiarid and subhumid Himalaya

Subhendu Pradhan[a], Shubhra Sharma[b]
[a]Department of Geography, Institute of Science, Banaras Hindu University, Varanasi, India
[b]Geosciences Division, Physical Research Laboratory, Ahmedabad, India

21.1 Introduction

In the geological past, pronounced oscillations in global climate led to the significant expansion and contraction of ice sheets and glaciers, which in turn modified the landscape [1]. The term paraglacial was first used by Ryder [2] to explain the formation of alluvial fans in south-central British Columbia which were suggested to have formed due to the resedimentation of glacial deposits by fluvial and mass movement processes. Based on the style and age of proglacial sediments, Church and Ryder [1] defined the term paraglacial as a nonglacial process that is directly conditioned by glaciation. They also used the term to categorize landforms that were the product of paraglacial processes, as well as to describe the time-period following deglaciation when landforms were adjusting to the new conditions. According to Ballantyne [3], the term paraglacial subsumes all nonglacial earth surface processes, sediment accumulations, landforms, land systems, and landscapes that are directly conditioned by glaciation and deglaciation. The dominant paraglacial processes are mass movements and fluvial erosion that are particularly important in the resedimentation and transfer of (para)glacial sediments within and beyond the high-mountain landscapes [3,4].

Geomorphologically, glacial and paraglacial areas are dominated by moraines, alluvial fans, valley-fill deposits (outwash gravel terraces), rock glaciers, proglacial lakes, and relict lake (glacio-lacustrine) deposits. The above geomorphic entities would imply that there is no unique geomorphic process associated with the paraglacial zone (environment). This led to the suggestion of calling it as paraglacial time period [3,5]. The mountain glaciers are very sensitive to climatic oscillations and have fluctuated repeatedly, particularly on millennial timescales, throughout the late Quaternary (e.g., [6,7]) leading to the generation of enormous quantities of the paraglacial sediments in form of moraines, alluvial fans, scree and talus fan deposits that have sequestered within the glacier valleys [4,8–10]. Benn and Evans [11] proposed that paraglacial

Atmospheric Remote Sensing
DOI: https://doi.org/10.1016/B978-0-323-99262-6.00011-0

should be defined as "the period of rapid environmental adjustment following glacier retreat" and Ballantyne [3] proposed to recast the definition of paraglacial to "the time scale over which glacially conditioned sediment stores are either exhausted or attain stability."

In recent times, glacier retreat due to atmospheric warming (e.g., [12]) has exposed a large quantity of poorly consolidated sediment prone to slope failure and debris flows. Some of the largest debris flows in Alps in recent years have had sources in newly de-glaciated areas [13]. More so, the processes responsible for the creation of various (para) glacial landforms provide an insight into the past climate variability; therefore, it is important to document the various landforms that evolved after the advancement and recession of glaciers. Importantly, since global warming is projected to impact more severely the higher elevation areas in the Himalayas [14], it becomes extremely important to have adequate documentation of various landforms and their sediment characteristics sequestered in the glacial and paraglacial zone of the Himalaya. The present study, therefore, derives motivation from the above where an attempt is made to map the various paraglacial landforms in the two distinctly different climatic zones viz. the semiarid eastern Zanskar ranges (Sarchu Plain) and the subhumid upper Ganga catchment (Gangotri valley) in order to assess the pattern and causes of the landform variability.

In Himalaya, two major climate systems, namely, the Indian Summer Monsoon (ISM) and mid-latitude westerlies (MLW) are the major sources of precipitation [11]. The relative contribution of these two major precipitation sources in driving the glaciers is still debated (e.g., [6,7,15,16]). For example, the arid eastern Zanskar ranges (Sarchu Plain) are dominated by the MLW with subordinate contribution from the ISM [7]. Whereas the subhumid upper Gangotri (Bhagirathi) valley is dominated by the ISM with subordinate contribution from the MLW [17]. The sequestered sediments are often mobilized by the events of high precipitation driven by these climate systems [18,19]. Hence, the objectives of the present study are to (1) document and map various geomorphic features in the paraglacial zones using remote sensing, (2) estimate ice volume changes associated with each glacial advancement, and (3) estimate the volume of sediments sequestered in the glacial-paraglacial landforms.

21.2 Study area

The two glaciated valleys- Sarchu Plain, eastern Zanskar ranges (32°54′27.40"N, 77°34′56.47"E; 4305 m) and the Gangotri valley, higher Himalaya (30°47′35.86"N, 79° 9′31.81"E, 4802 m) (Fig. 21.1) are selected in this study. These are in different geological, geomorphological, and climate settings and would facilitate comparison to understand the terrain response (of paraglacial zones) to climatically modulated earth surface processes.

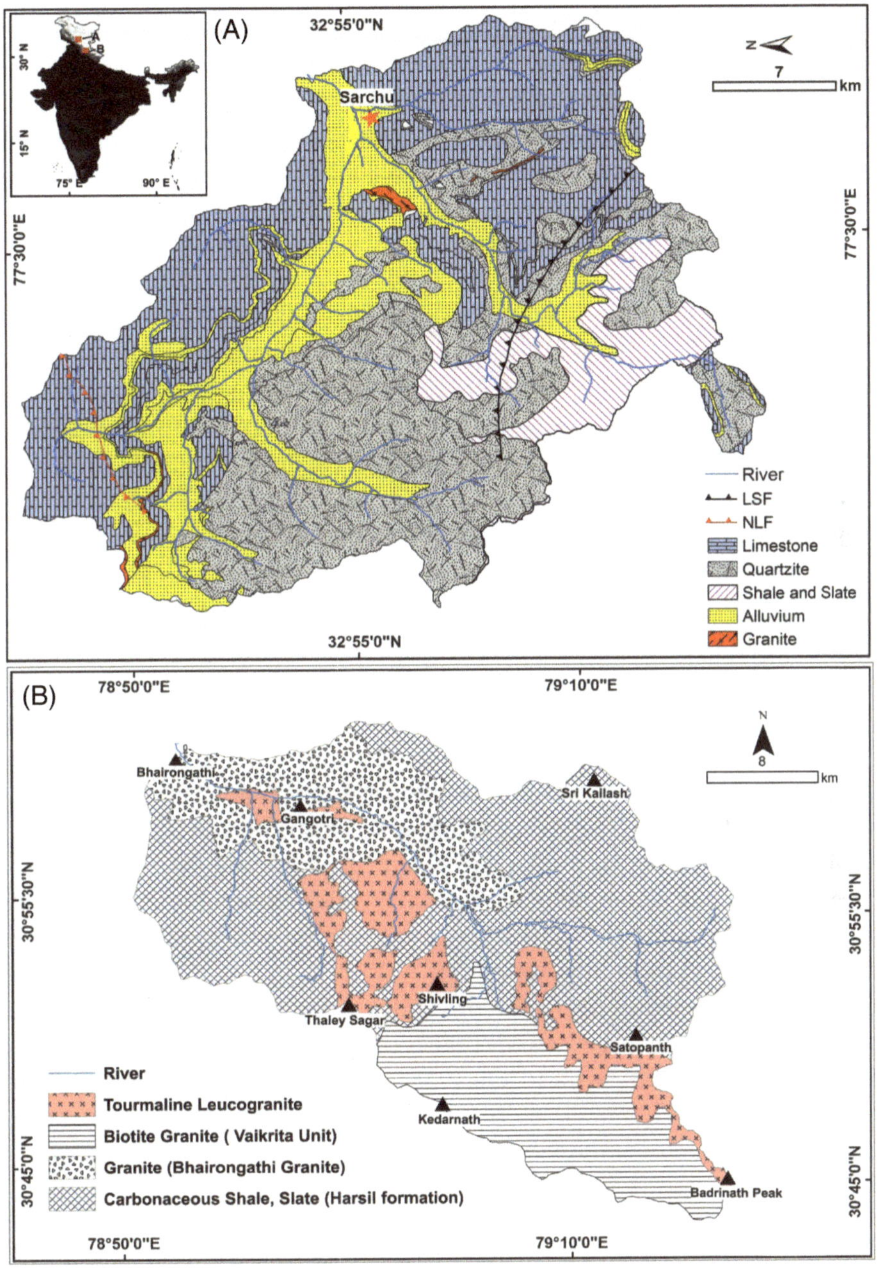

Fig. 21.1 Geological map of (A) Sarchu Plain in the eastern Zanskar with two tectonic structures Lingti-Sarchu Fault (LSF) and North Lahul Fault (NLF). The location of Sarchu Plain is marked with a red star (*modified after*[20]) (B) Gangotri valley with major peaks marked by triangles (*modified after*[54]).

21.2.1 Sarchu Plain

The semiarid Sarchu Plain (32°54′27.40"N, 77°34′56.47"E; 4305 m) lies in the rain shadow zone (north) of the ISM-dominated Higher Himalaya. The Sarchu Plain has an area of ~833 km^2, out of which ~70 km^2 (~8%) is covered with glaciers. The lithology is dominated by the Tethyan sedimentary Sequence (TSS) which consists of Palaeozoic and Mesozoic schist, shale, phyllite, quartzite, dolomite, limestone, and quaternary alluvium. It is bounded by the north-dipping Lingti–Sarchu fault in the south and south-dipping North Lahaul Fault in the north [20,21] (Fig. 21.1A).

The data from Patsio weather station suggests that most of the precipitation (~80%) is received during the winter and spring (December–March) which is contributed by the MLW [7]. The estimation of rainfall variability in the past 30 years (1970–2000) World Clim version 2.1 suggests that the average summer precipitation (June-September) varies between 23 and 48 mm/month, 12 to 23 mm/month in winter (December–February), 11 to 31 mm/month in spring (March–May), and between 6 and 10 mm/month during postmonsoon months (October-November) (Fig. 21.2A–D).

21.2.2 Gangotri valley

The Gangotri valley lies in the Higher Himalaya with a total catchment area up to the snout at Gomukh (4000 m amsl) as 513 km^2, and up to Gangotri (3060 m amsl) as 691 km^2 [22,23]. The total area covered by glaciers is ~277 km^2 (~54%). The Bhagirathi River is sourced from the glacier draining mostly through the Lesser and Central Crystalline rocks [24]. Geologically, the Gangotri glacier system falls in the higher and TSS Himalayan zone [25]. The Gangotri granite is one of the largest bodies of the Higher Himalayan Leucogranite belt located in the Garhwal Himalaya [26–28] and comprises granite (tourmaline rich), sericite schist, mica schist, quartzite, phyllite, sulfide minerals like pyrite, chalcopyrite, and arsenopyrites at the contact of quartz veinlets and massive, banded fine-grained limestone [29].

There is seasonal variability in the precipitation pattern. During summer (June–September) precipitation varies from 118 and 206 mm/month and reduces to 26 and 112 mm/month during winter (December–February). The spring (March–May) precipitation variability lies between 6 and 94 mm/month, whereas during the post-monsoon months (October-November) it varies between 11 and 47 mm/month (Fig. 21.2E–H).

21.3 Methodology

The geomorphic mapping is carried out using the freely available satellite data—the Landsat 8 Operational Land Imager (OLI) (30 m resolution) (Source: USGS Earth Explorer), ALOS Palsar DEM (12.5 m) (Source: Alaska Satellite Facility) and the Google Earth imagery. The Landsat 8 imagery is taken from October 16–20, 2020

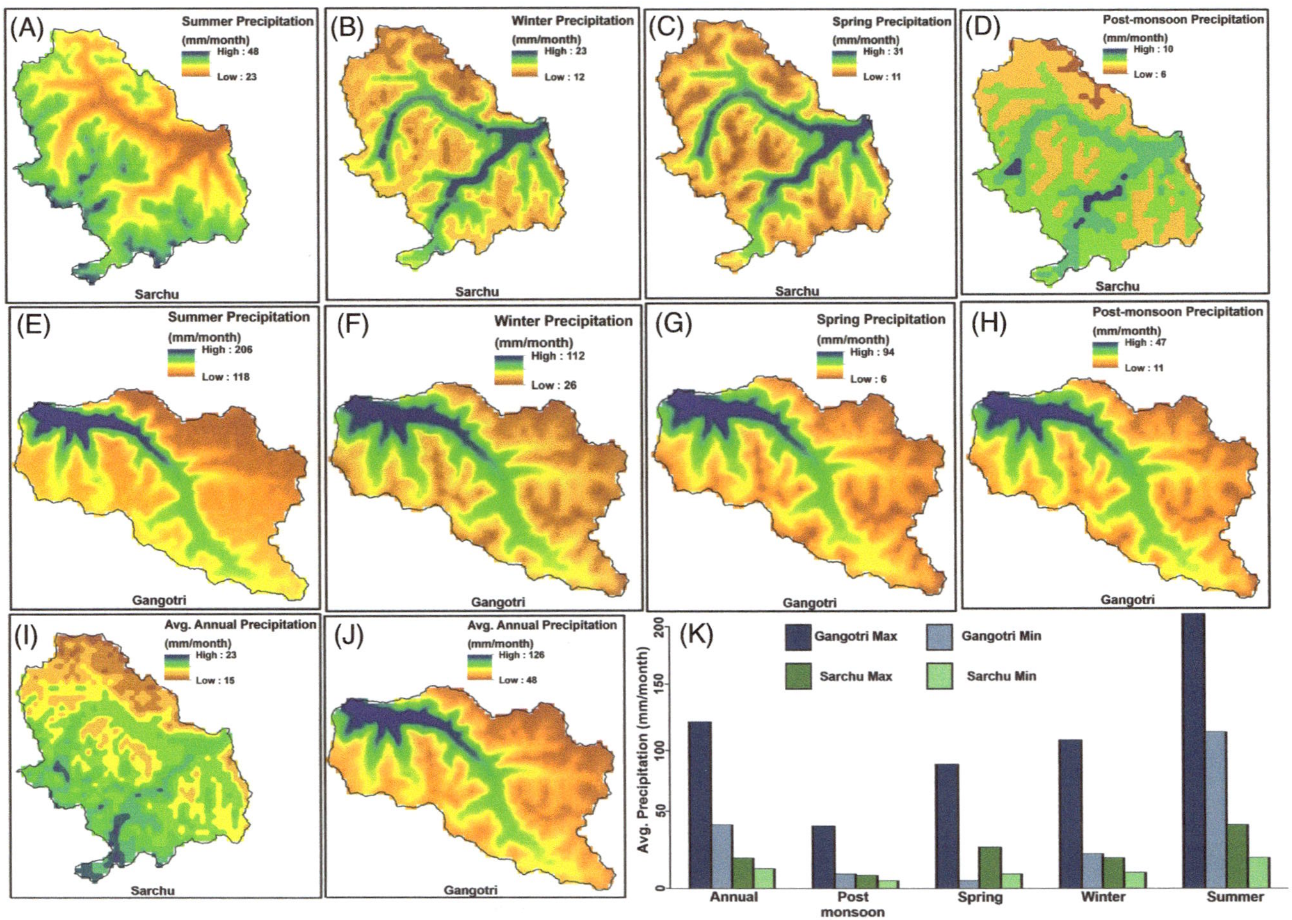

Fig. 21.2 Seasonal average precipitation gradient maps from the year 1970- 2000 of (A–D) Sarchu Plain and (E–H) Gangotri valley for summer (June–September), winter (December–February), spring (March–May), and postmonsoon (October–November) months. (I) and (J) represent annual precipitation in the Sarchu Plain and Gangotri valley respectively. (K) The maximum and minimum precipitation in the Gangotri valley and the Sarchu Plain is represented by blue and green bars (*Source: WorldClim v2* (www.worldclim.org)).

(end of ablation period) (30 m resolution; band combination RGB: 754 (Short-wave infrared) (https://earthexplorer.usgs.gov/) for mapping of the glaciers after Bolch et al., [30]. The images used are with <1.5% cloud cover. The landforms are mapped using google earth imagery which includes lateral and terminal moraines, drumlins, scree cone, alluvial fan, outwash terrace, rock glacier, and moraine-dammed lakes. ArcGIS10.8 was used to prepare the overlay of maps and sediment volume estimation (Fig. 21.3). Errors in ice/sediment volume estimation are at 1 standard deviation and propagated further in final estimations.

For climate data, we used the seasonal average precipitation and temperature maps of the last 30 years (1970–2000) which are generated from WorldClim v2 (www.worldclim.org). WorldClim v2 (30 sec x 30 sec resolution) datasets were modeled with ~86% accuracy with respect to the ground station data for average precipitation and ~99.6% accuracy for average temperature (Fick and Hijmans et al., [31]).

21.3.1 Identification of geomorphic landforms

21.3.1.1 Moraines

Moraines are ridge-like depositional features formed by the accumulation of englacial, sub and supra-glacial debris that are generated by glacial erosion as the glacier advances [32]. These are often classified as frontal/terminal, lateral, and medial moraines depending upon their relative position to the glacial ice. Lateral and frontal moraines were identified using their shape, geometry, and proximal location to the glacier and mapped using Google Earth imagery. These appear as long, narrow, steep, parallel ridges which flank the valley walls or terminate as curvilinear ridges (Figs. 21.4 and 21.5).

21.3.1.2 Scree cones

These are fine rock fragments deposited at the end of steep rock cliffs in the arid subtropical and subpolar climate [33]. The debris is primarily generated due to freeze and thaw weathering of the valley walls [34]. These were identified in Google Earth imagery using their triangular-shaped sediment aprons (broad and round at the bottom and narrow and pointed upward), fine-grained texture, and the location (found at the bottom of the steep rock cliffs) (Figs. 21.4 and 21.5).

21.3.1.3 Glaciers

Glaciers are large, solid masses of ice formed of compressed snow that move slowly under their own weight [35]. The velocity of the ice-flow is greatest at the center and decreases toward the head and terminus [36]. The glaciers appear blue under the band combination 754 of Landsat 8 OLI and possess a smooth texture. The fresh ice appears brighter blue in color and dirty ice (or debris coverd ice) appears in a darker blue tone. The area enclosed by dirty ice shows the melting of the glacier which is marked as an ablation zone using the same band composition.

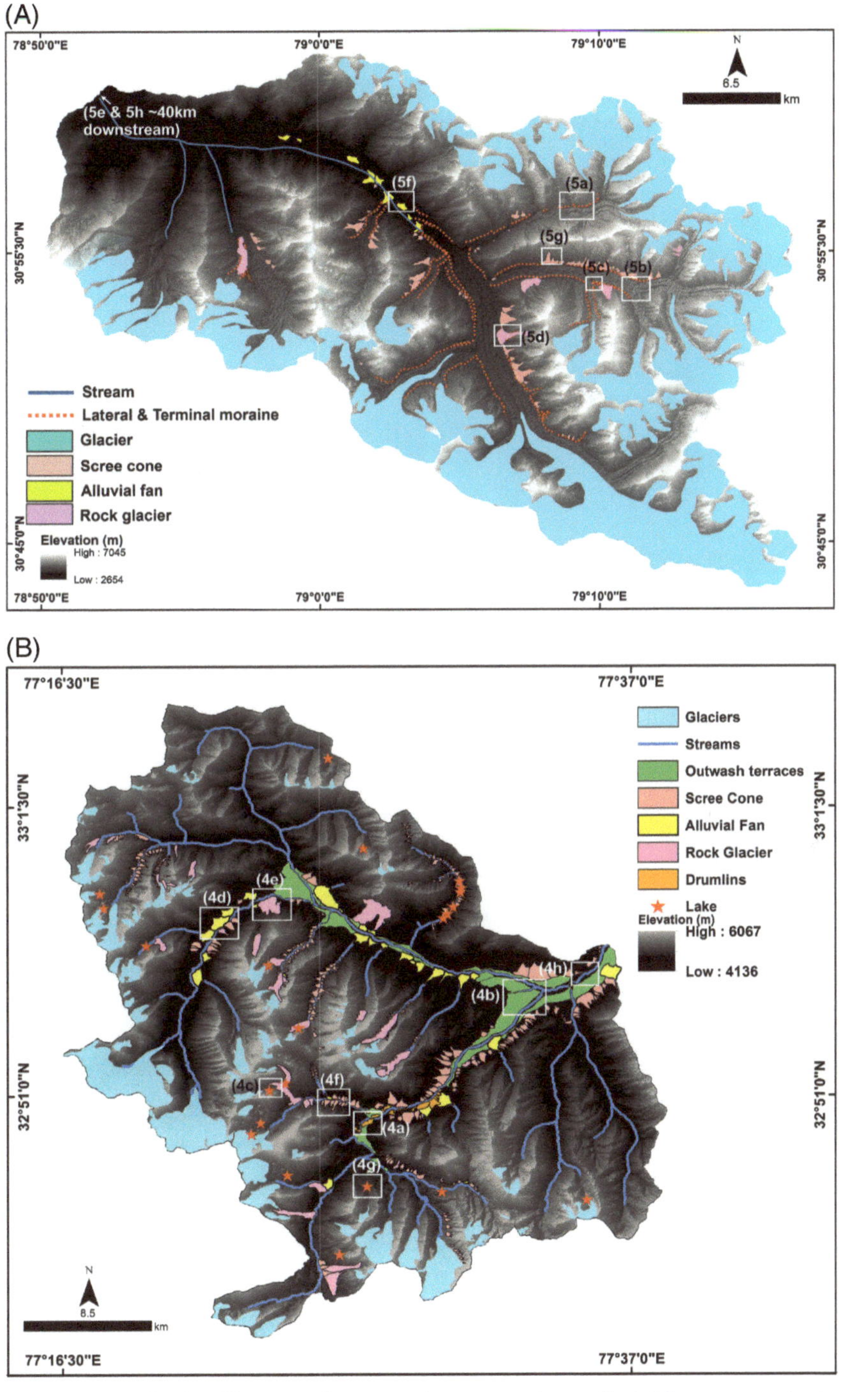

Fig. 21.3 The geomorphic landforms such as moraines, scree cone, alluvial fans mapped using Google Earth Imagery are overlain over Digital Elevation Model (DEM) for (A) Gangotri valley and (B) Sarchu Plain. The locations of close-ups of the landforms given in Figs. 21.4 and 21.5 are also marked.

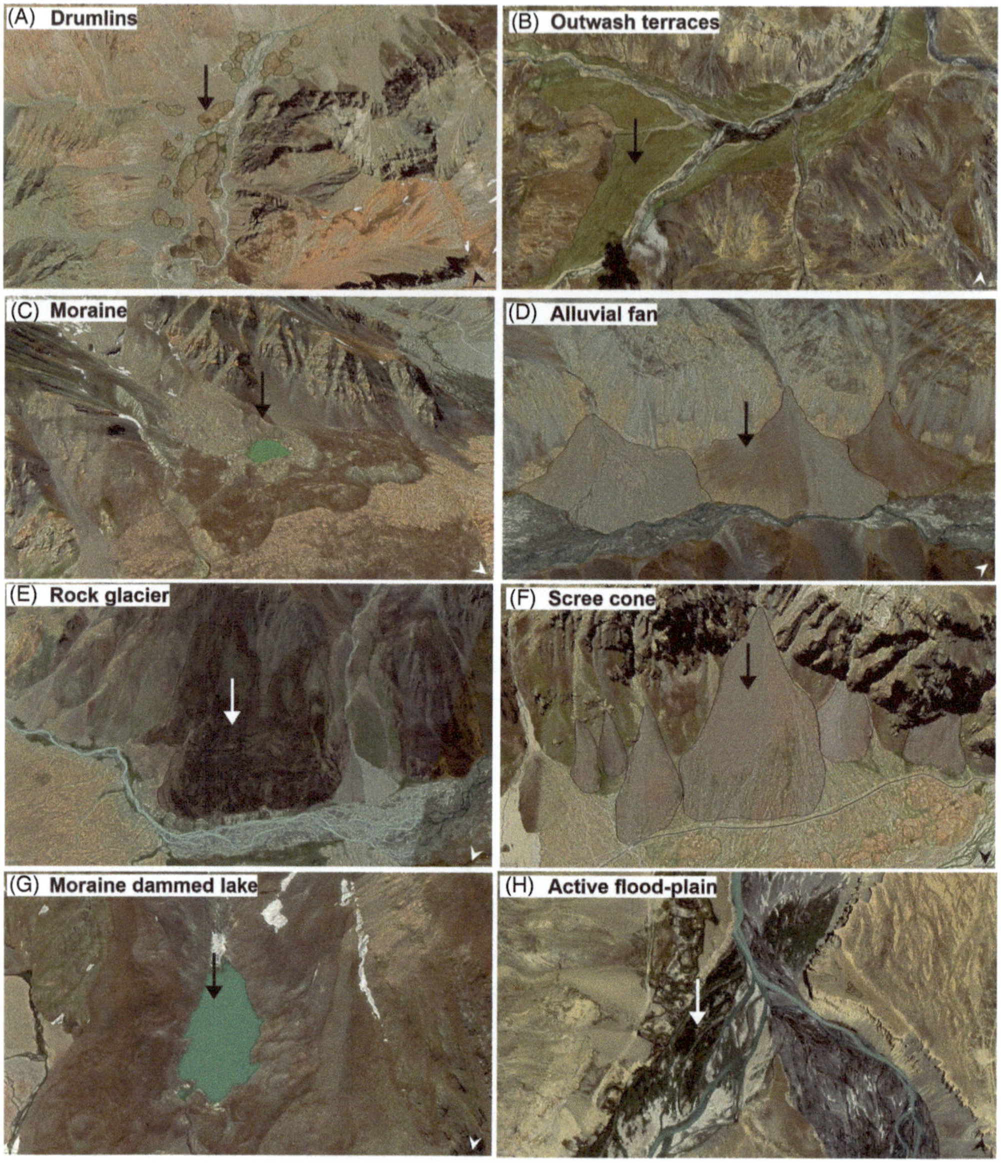

Fig. 21.4 ***Various paraglacial landforms identified in the Sarchu Plain using Google Earth Images.***

21.3.1.4 Estimation of sediment volume

The volume of paraglacial sediment stored in the valleys was estimated by multiplying the surface area of various digitized paraglacial features and the thickness of the sediment deposits. For the estimation of sediment thickness of moraines, the elevation profiles are drawn from the valley floor to the top of moraine's ridge in Google Earth. The thickness of scree cones varies from 10 to 20 m [37], so an average thickness of 15 m

Fig. 21.5 ***Various paraglacial landforms identified in the Gangotri valley using Google Earth Images.***

the taken for estimation of stored sediment volume. Similarly, the average thickness of drumlins is taken as 6.7 m [38] and the average thickness of 105 m (90 to 120m) is taken for alluvial fans [39]. For outwash terrace thickness ranges from 36 to 78 m (average ~48.5 m) that is calculated using Google Earth's elevation profile.

21.3.1.5 Ice volume estimation

The ice volume related to glacial advances during the different time periods is estimated in order to understand the magnitude of glaciation in space and time. To constrain the timing of the advances, published chronology is taken after Saha et al. [40], Sharma et al. [7], and Barnard et al. [4]. Since the glaciers are suggested to respond to the two climate systems (ISM and MLW) on a geological time scale, ice-volume estimation for each documented advance helps to understand the response of the glaciers to the ambient climate conditions (precipitation and temperature). Moraines are used as a proxy for the maximum glacial extent during each advance. The timing and glacial stages are constrained using the published data on moraine stratigraphy and chronology in the Sarchu Plain [7,40] and in the Gangotri valley ([4] and references therein). The ice volume is calculated by using the slope-dependent method suggested by Haeberli and Holzle, [41] and Frey et al. [42], where the thickness of the glacier along central flow line (h_f) is estimated and the mean ice thickness (H_f) is calculated by multiplying (π)/(4) with h_f as,

$$hf = \tau / \left[f \rho g \sin(\alpha) \right]$$

The average basal shear stress along the central flow line (τ) is taken as 150×10^3 N/m^2 [41,42]; shape factor (f) taken as 0.8 for typical valley glacier (Paterson, 2016); density of ice (ρ) is taken as 850 kg/m^3 for Sarchu Plain [42] and 900 kg/m^3 for Gangotri valley [43]. The acceleration due to gravity (g) taken as 9.8 m/s^2, and mean surface slope (α) along the manually digitized flow line is calculated in ArcGIS 10.8.

21.4 Results

21.4.1 Geomorphology

The moraines situated above the valley floor are well preserved along the course of river Yunam Tso that drains into the Sarchu Plain, north of Bara Lacha pass. These are huge, elongated mounds consisting of boulders of irregular shape and size. Between the two flanking moraine ridges, the wide valley is filled with glaciofluvial deposits (Figs. 21.6A and B). It is likely that the wide basin supported a lake till the recent past, as it is still called Yunam Tso (lake). Trails of drumlins followed by hummocky structures are well preserved in the western proximity of the Sarchu Plain (Fig. 21.6B). The wide Sarchu basin is filled with thick fluvial sediments that are incised by the river Yunam, giving rise to a flat-topped elevated surface (Fig. 21.6A). The modern river is a braid-meandering type with the dominance of point bars and distributary channels, implying that the incised sediments were deposited under an improved hydrological regime.

Event stratigraphy in this segment suggests that the river Yunam Tso once supported a valley glacier that scoured a wide U-shaped valley. Following this, sediment-laden meltwater fluxes dominated by fluvial gravel led to the aggradation of the basin, giving rise to the Sarchu Plain. The presence of prominent lateral moraines above the fluvial deposits suggests that the glacial process postdates the formation of the Sarchu Plain (Fig. 21.6).

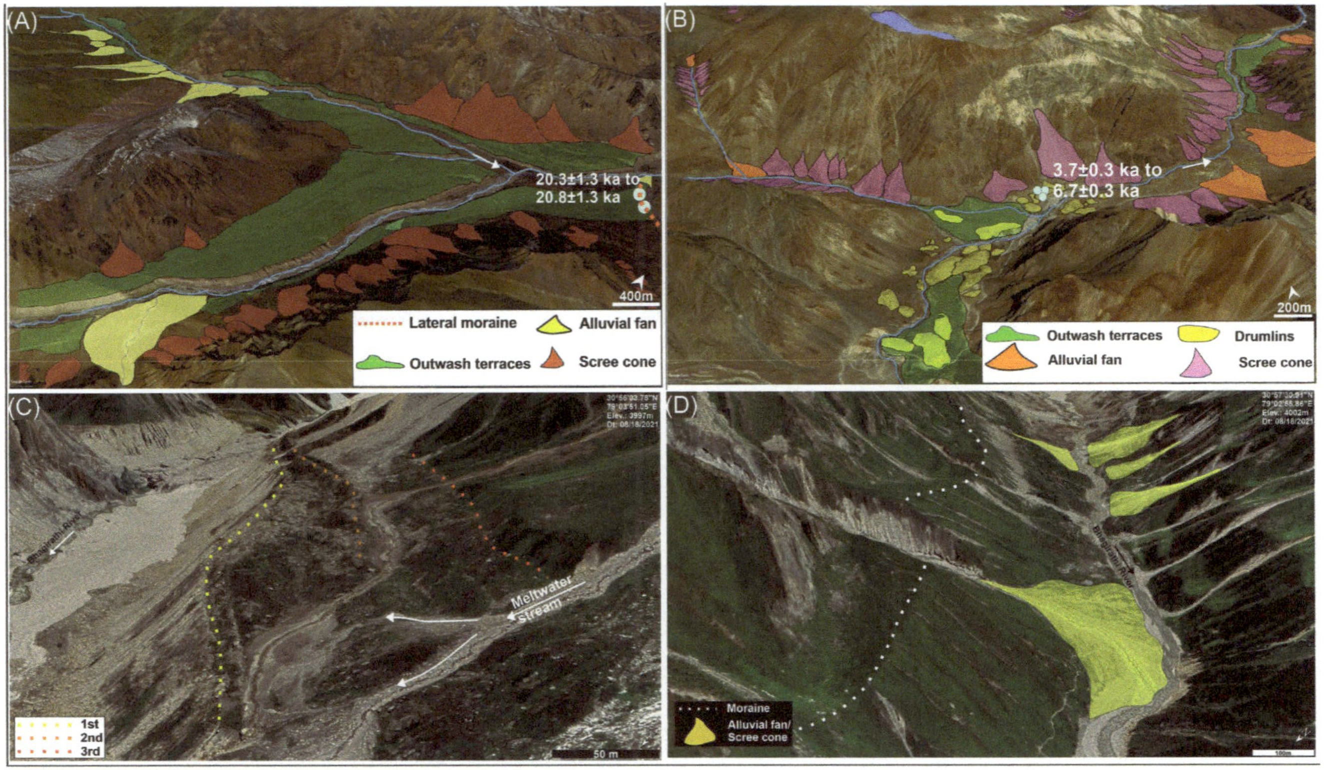

Fig. 21.6 (A) and (B) The (para)glacial landforms in Sarchu Plain are identified and digitized as moraines, outwash terraces, scree cone, alluvial fan, and drumlins draped from the Google Earth imagery. Note that the scree and alluvial fans emanate from the ice-free cirque catchments. The ages (blue circle) for different glacial stages are from Sharma et al. [7]. (C) Three lateral moraines are identified and digitized in the Gangotri valley from the Google Earth imagery. (D) The alluvial fans are emanating from the moraine showing mechanism of sediment remobilization in the Gangotri valley.

However, as the deposits are confined to the northern base of the Bara Lacha pass it indicates small-scale glacial event. Further, trail of drumlins over the fluvial gravel indicates standstill conditions under glacier recession/small-scale readvance. Subsequently, incision by smaller streams led to the development of gravel pinnacles that stand out impressively.

According to Sharma et al., [7], the Sarchu Plain witnessed four major glacial advancements driven by the phases of enhanced MLW followed by phases of deglaciation. Chronology of the oldest moraines [44] indicates that the most extensive and oldest glaciation occurred during cold and humid Marine Isotope Stage (MIS)-4. The second major glaciation event is dated to 20.3 ± 1.3 and 20.8 ± 1.4 by Sharma et al., [7]. However, the exposure age dating of this moraine by Saha et al. [40] yielded an age of ~79–52 ka. The stratigraphically equivalent surface of the moraine dated by Sharma et al., [7] located a few kilometers upstream is dated to 16.7 ± 2.1 and 14.8±1.8 ka (Fig. 21.6A and B).

In the Gangotri valley, the exposure age dating undertaken by Barnard et al. [4] and the optical ages obtained by Sharma and Owen [17] show a complex glacial and paraglacial depositional history. Barnard et al. [4] identified five glacial stages. The oldest (Bhagirathi) glacial stage dated between ~63 and 11 ka is considered the most extensive reaching ~40 km below the present snout at Jhala (below Harshil town) (Fig. 21.7A). Based on chronology, it is suggested that the glacial advance was initiated during the MIS-4/3 and persisted until around the late-glacial interstadial. The second (Kedar) glacial stage is dated to ~7 ka (Fig. 21.7B). During this stage besides the deposition of the moraine, alluvial fans were also deposited in the tributary valleys. The third (Shivling) glacial stage moraine is dated to ~5 ka (Figs. 21.7C and D) and is attributed to the early neoglaciation. The Gangotri stage was dated to ~1 ka (Fig. 21.7E) which led to the deposition of the moraine, alluvial fans, and flood deposits. Climatically, this stage is attributed to the Medieval Warm Period/Little Climatic Optimum. The youngest glacial advance termed as the Bhujbas stage is dated to 200–300 yr BP (Fig. 21.7E) corresponding to the Little Ice age (LIA) which led to the deposition of snout proximal moraines.

21.4.2 Ice volume estimation

21.4.2.1 Sarchu Plain

The maximum ice volume of ~35.5 ± 1.4 km^3 is estimated during the MIS-4 when glaciers covered ~323 km^2 catchment area (Fig. 21.8A). During the MIS-2 (including the Last Glacial Maxima (LGM)), the estimated ice volume was ~24.6 ± 1 km^3 with a total glacierised area of ~223 km^2 (Fig. 21.8B). Similarly, during 8.2 ka area and ice volume occupied by glaciers reduced to ~151 km^2 and ~16.6 ± 0.6 km^3, respectively (Fig. 21.8C). Presently total area occupied by glaciers is ~70 km^2 and the occupied ice volume is ~7.8 ± 0.3 km^2 (Fig. 21.8D; Table 21.1).

21.4.2.2 Gangotri valley

The maximum extent of ice cover was during MIS-4 to early Holocene (~64 to 11 ka) as ~1107 km^2 with an ice volume of ~118 ± 4.6 km^3. During the second advancement

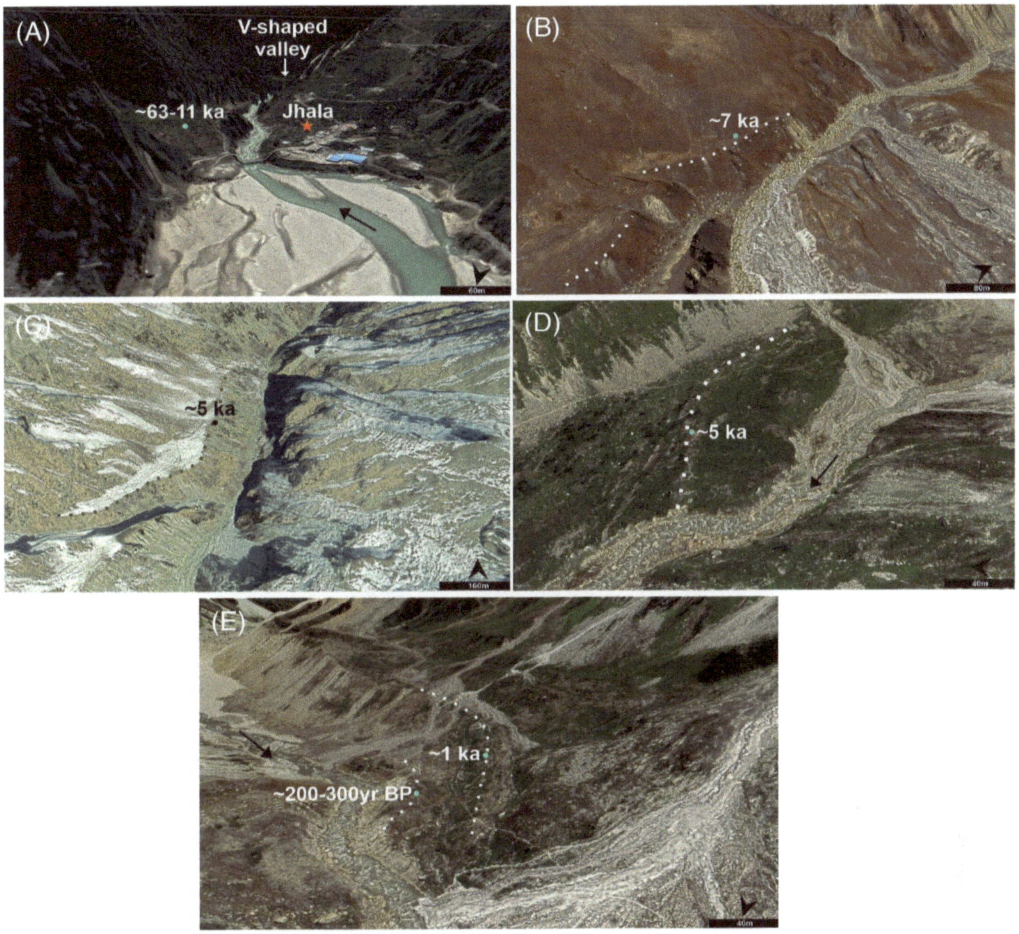

Fig. 21.7 (A–E) Different stages of glacial advances in the Gangotri valley identified and marked using Google Earth (details discussed in the text). The ages after Barnard et al. [4]. Note that the valley above Jhala (A) is a wide U-shape (glacially modified) and in the downstream is a narrow V-shape (fluvially modified).

dated to ~7 ka, the area covered by the glacier was ~363 km^2 and the estimated ice volume was ~38.7 ± 1.5 km^3. Similarly, ~5 ka the glacierized area and ice volume was ~357 km^2 and ~38 ± 1.5 km^3 respectively. Around 1 ka, the total glacierized area was ~285 km^2 and the estimated ice volume was ~30.34 ± 1.2 km^3. Similarly, during ~200–300yr BP the glacierized area and ice volume were reduced to ~284 km^2 and ~30.32 ± 1.2 km^3 respectively. Presently total glacierized area is ~277 km^2 and the associated ice volume is ~29.6 ± 1.1 km^2 (Fig. 21.9; Table 21.2).

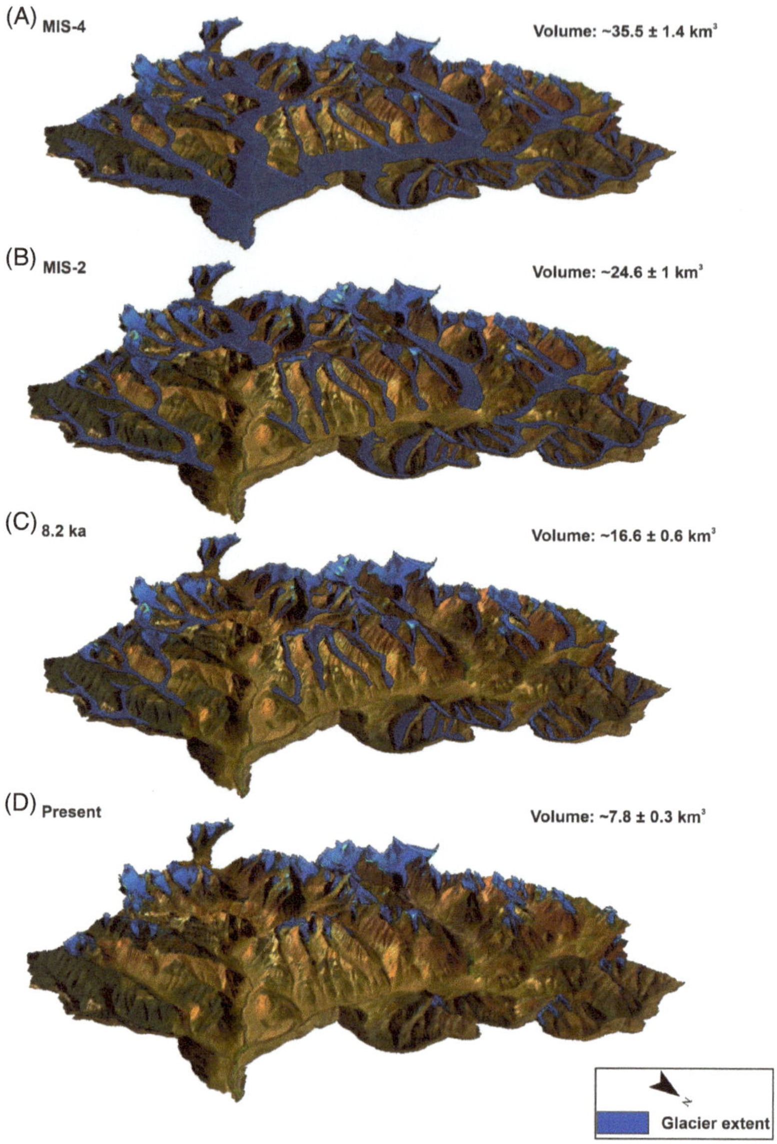

Fig. 21.8 (A–D) The 3-D view of the Sarchu Plain catchment with modelled ice volume for different glacial advances since MIS-4.

Table 21.1 The Sarchu Plain glaciered area and ice volume estimated under different advances.

Age	Area (km^2)	Volume (km^3)	Change in Volume from present (%)
MIS-4	~323	35.5 ± 1.4	~357
MIS-2	~224	24.6 ± 1.0	~217
8.2 ka	~151	16.6 ± 0.6	~114
Present	~71	7.8 ± 0.3	NA

The chronology and hence the stages are constrained after Taylor and Mitchel [44] and Sharma et al. [7].

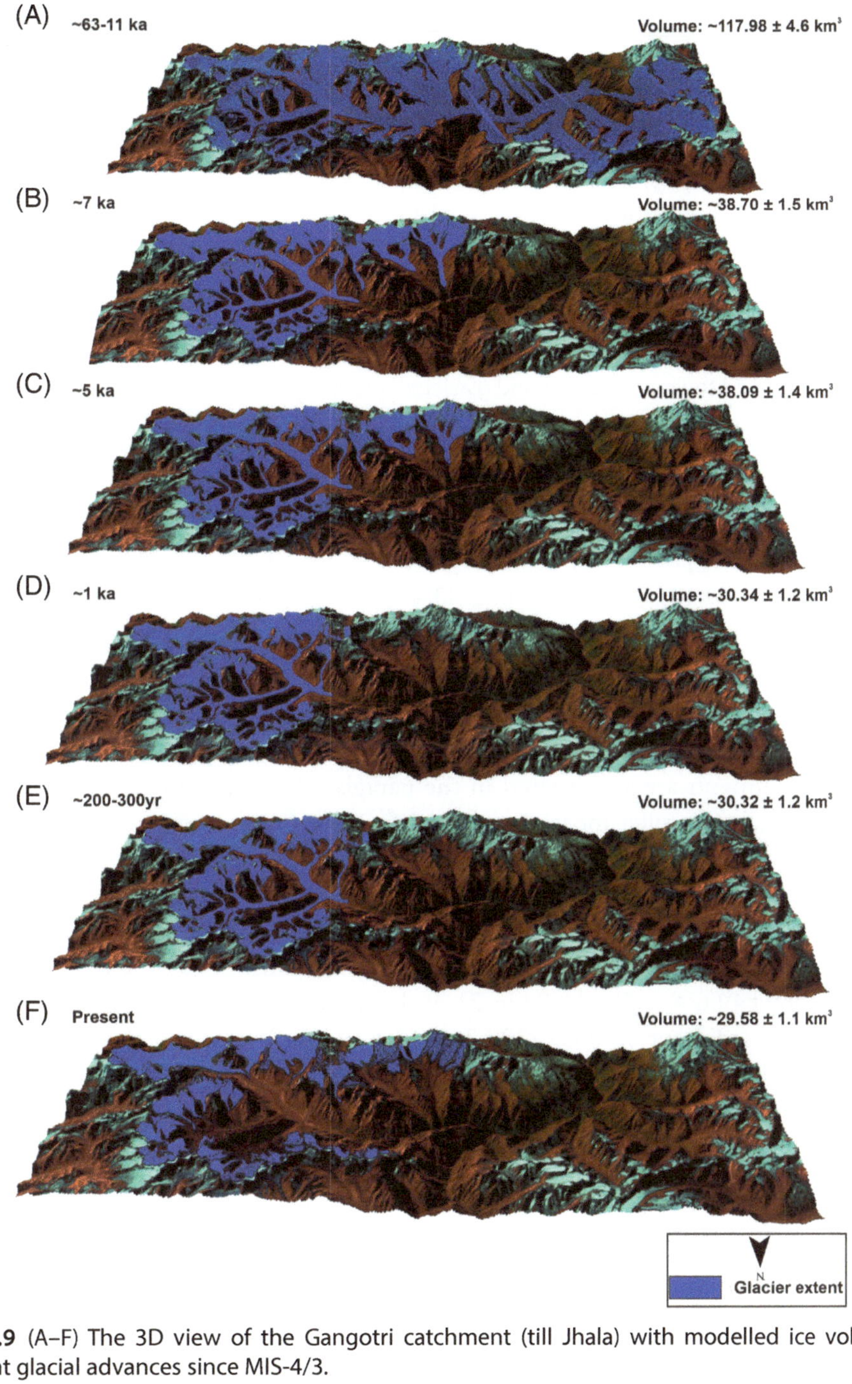

Fig. 21.9 (A–F) The 3D view of the Gangotri catchment (till Jhala) with modelled ice volume for different glacial advances since MIS-4/3.

Table 21.2 The Gangotri valley glaciered area and ice volume estimated under different advances.

Age	Area (km^2)	Volume (km^3)	Change in Volume from present (%)
MIS-4	1106.75	117.98 ± 4.6	~300
~7ka	363.07	38.70 ± 1.5	~31
~5ka	357.40	38.09 ± 1.4	~29
~1ka	284.60	30.34 ± 1.2	~3
~200-300yr BP	284.48	30.32 ± 1.2	~2.5
Present	277.51	29.58 ± 1.1	NA

The chronology and hence the stages are constrained after Barnard et al. [4].

Table 21.3 Estimated sediment volume and glaciered area in the Sarchu Plain and Gangotri valley.

	Sarchu Plain		Gangotri valley	
Landforms	Area (x10^{6}m^2)	Volume (x10^6 m^3)	Area (x10^6 m^2)	Volume (x10^6 m^3)
Moraine	0.05	1.54	5.78	396.2 ± 11.3
Alluvial fan	3.93	413.3 ± 24.1	1.64	172.7 ± 9.6
Scree cone	2.21	303.1 ± 1.5	5.98	89.7 ± 1.6
Outwash terraces	19.42	942 ± 43.4	-	-
Drumlins	1.31	8.7 ± 0.4	-	-
Total	**26.92**	**1669 ± 49.6**	**13.4**	**658.6 ± 14.9**

21.4.2.3 Sediment volume estimation

Largely the sediments are sequestered in the paraglacial valleys of the Sarchu Plain and the Gangotri valley as relict moraines, alluvial fans, talus cones, and outwash gravel terraces. In the Sarchu Plain, the moraines currently sequester least of the paraglacial sediments (1.5 × 10^6 m^3) which are manifested by the absence of well-defined moraines corresponding to MIS-4 unlike in the Gangotri valley. The outwash gravel terraces contribute 942±43.4 × 10^6 m^3. The total sediment volume sequestered in the Sarchu Plain is ~1669±49.6 × 10^6 m^3 (Table 21.3). In the Gangotri valley, moraine contributes maximum sediment volume (396.2±11.3 × 10^6 m^3) followed by the alluvial fan (172.7±9.6 × 10^6 m^3). The total volume of paraglacial sediment stored in the Gangotri valley is 658.6±14.9 × 10^6 m^3 (Table 21.3).

21.5 Discussion and conclusions

The geomorphic features identified using the Google Earth satellite remote sensing data support the earlier studies that the sediments sequestered in paraglacial valleys can be mapped reasonably well. The technique allows us to provide a broad assessment of the temporal changes in ice volume both in the historical and the late Quaternary period. Thus, the technique is emerging as a fast reliable tool for landform mapping and quantification, particularly in the rugged and inaccessible higher Himalayan terrain. In this study, we compared remote sensing-based mapping of the glacial

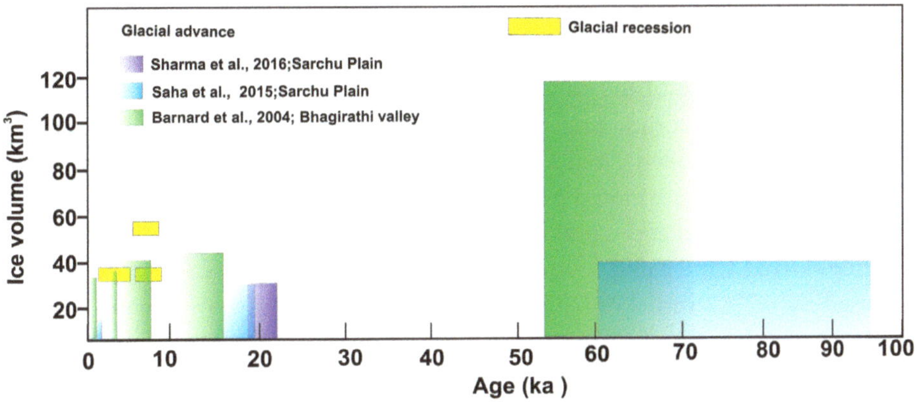

Fig. 21.10 ***A comparison of ice volume during major glacial stages in Gangotri valley and Sarchu Plain.*** Glacial stages are marked after Barnard et al., [4]; Saha et al., (2015) and Sharma et al. [7].

(moraine) and paraglacial landforms (alluvial fans, talus cones, and debris flow) with that of the field-based study from the Gangotri valley [4] and Sarchu Plain [7]. The study indicates that contrary to the conventional wisdom, the subhumid Gangotri valley in monsoon-dominated central Himalaya has preserved a more detailed record of glacier advances compared to the westerly dominated semiarid Sarchu Plain in the NW Himalaya. Similarly, the ice volume estimate based on the valley morphology (Sarchu Plain) and extent of moraines (Gangotri valley) indicate that the latter had larger ice volume during the MIS-4 (Fig. 21.10). The larger ice volume in Gangotri valley is speculatively ascribed to the combined contribution from the enhanced MLW and subdued ISM during the MIS-4.

Contrary to the ice volume, semiarid Sarchu Plain sequestered magnitude higher paraglacial sediments both in the tributary and trunk valleys compared to the subhumid Gangotri valley (Table 21.3). For example, in the Sarchu Plain $\sim 1669 \pm 49.6 \times 10^6$ m^3 of paraglacial sediment are sequestered. These are preserved in the form of outwash terraces, alluvial fans, talus cones, and the moraines. The relict moraine sediment contribution compared to the outwash terraces, alluvial fans, and talus cones (a typical feature of dryland topography) is significantly lower in the Sarchu Plain compared to the Gangotri valley. In the upper Gangotri valley, around $\sim 659 \pm 14.9 \times 106$ m^3 of paraglacial sediments are sequestered which is significantly lower than the Sarchu Plain. We tend to suggest that the low sediment sequestration in the subhumid upper Gangotri valley could be due to (1) high rainfall, (2) terrain instability due to frequent earthquakes, and/or (3) paucity of accommodation space. These factors together led to the high rate of sediment bypassing (low sediment residence time). In comparison, in the semiarid NW Sarchu Plain the higher volume of paraglacial sediment can be attributed to (1) low rainfall (2) relatively higher seismic stability, and (3) gentle topography due to low erosion rate. The resultant of these factors is the subdued geomorphic processes which in turn provide higher sediment residence time.

Assessing the response of the paraglacial zones in terms of sediment mobilization and sequestration further becomes important with the rapidly melting Himalayan glaciers [45,46], where retreat models predict that nearly half of glaciers could disappear by 2100 [47]. The rapid melting is attributed to increasing winter temperature [48], decreasing duration of snow cover season and cloud cover [49] leading to low albedo and strong daytime solar heating during winter [50]. For example, in the western Himalaya mean temperature has increased by ~2°C from 1984/85 to 2007/08 [50]. Similarly, over central and eastern Himalaya, decreased monsoon strength is observed over past 80 years from ice core data [51] which will most likely impact the glacier mass balance. Further, anthropogenic forcing such as black carbon contribution from forest fire during summer and long-distance transport of aerosols during winter [52] are posing serious threat to the health of the glaciers.

The paraglacial zones are sediment surplus and transported limited [53] which with unprecedented global warming (and hence glacial melting) are witnessing a rise in extreme hydro metrological events causing serious damage to the infrastructure and precious human life [53]. The disasters, where mobilization of paraglacial sediment is of major consequence are documented both in the semiarid NW Himalaya [9] and the subhumid central Himalaya [19]. Considering the recent disasters in the NW and Central Himalaya ([18,53] and references therein) where mobilization of paraglacial sediments occurred mostly as debris flow floods, the present study demonstrates that large-scale mapping to quantify the sediments sequestered in multiple Himalayan paraglacial valleys can be effectively used to assess the terrain response to geological processes and therefore the vulnerability arising thereof.

Acknowledgements

The authors are grateful to Science and Engineering Research Board-Department of Science & Technology (SERB-DST) for financial support under Core Research Grant-(CRG/2019/000076).

References

[1] M. Church, J.M. Ryder, Paraglacial sedimentation: a consideration of fluvial processes conditioned by glaciation, Geol. Soc. Am. Bull. 83 (10) (1972) 3059–3072.
[2] J.M. Ryder, Some aspects of the morphometry of paraglacial alluvial fans in south-central British Columbia, Can. J. Earth Sci. 8 (10) (1971) 1252–1264.
[3] C.K. Ballantyne, A general model of paraglacial landscape response, The Holocene 12 (3) (2002) 371–376.
[4] P.L. Barnard, L.A. Owen, R.C. Finkel, Style and timing of glacial and paraglacial sedimentation in a monsoon-influenced high Himalayan environment, the Gangotri Valley, Garhwal Himalaya, Sediment. Geol. 165 (3-4) (2004) 199–221.
[5] D.J. Evans, B.R. Rea, D.I. Benn, Subglacial deformation and bedrock plucking in areas of hard bedrock, Glacial Geol. Geomorphol. (1998) RP04.
[6] S. Sharma, A.D. Shukla, Factors governing the pattern of glacier advances since the Last Glacial Maxima in the transitional climate zone of the Southern Zanskar Ranges, NW Himalaya, Quat. Sci. Rev. 201 (2018) 223–240.

[7] S. Sharma, P. Chand, P. Bisht, A.D. Shukla, S.K. Bartarya, Y.P. Sundriyal, N. Juyal, Factors responsible for driving the glaciation in the Sarchu Plain, eastern Zanskar Himalaya, during the late Quaternary, J. Quat. Sci. 31 (5) (2016) 495–511.
[8] S.N. Ali, R.H. Biswas, A.D. Shukla, N. Juyal, Chronology and climatic implications of Late Quaternary glaciations in the Goriganga valley, central Himalaya, India, Quat. Sci. Rev. 73 (2013) 59–76.
[9] N. Juyal, Cloud-burst triggered debris flow around Leh, Curr. Sci. 99 (9) (2010) 1166–1167.
[10] N. Rana, S. Sharma, S.N. Ali, S. Singh, A.D. Shukla, Investigating the sensitivity of glaciers to climate variability since the MIS-2 in the upper Ganga catchment (Saraswati valley), Central Himalaya, Geomorphology 346 (2019) 106854.
[11] D.I. Benn, L.A. Owen, The role of the Indian summer monsoon and the mid-latitude westerlies in Himalayan glaciation: review and speculative discussion, J. Geol. Soc. 155 (2) (1998) 353–363.
[12] T. Bolch, A. Kulkarni, A. Kääb, C. Huggel, F. Paul, J.G. Cogley…, M. Stoffel, The state and fate of Himalayan glaciers, Science 336 (6079) (2012) 310–314.
[13] M. Chiarle, S. Iannotti, G. Mortara, P. Deline, Recent debris flow occurrences associated with glaciers in the Alps, Global Planet. Change 56 (1-2) (2007) 123–136.
[14] T.P. Sabin, R. Krishnan, R. Vellore, P. Priya, H.P. Borgaonkar, B.B. Singh, A. Sagar, Climate change over the Himalayas. In: Assessment of Climate Change Over the Indian Region, Springer, Singapore, 2020, pp. 207–222.
[15] J.M. Dortch, L.A. Owen, M.W. Caffee, Quaternary glaciation in the Nubra and Shyok valley confluence, northernmost Ladakh, India, Quat. Res. 74 (1) (2010) 132–144.
[16] A. Ganju, Y.C. Nagar, L.N. Sharma, S. Sharma, N. Juyal, Luminescence chronology and climatic implication of the late quaternary glaciation in the Nubra valley, Karakoram Himalaya, India, Palaeogeogr. Palaeoclimatol. Palaeoecol. 502 (2018) 52–62.
[17] M.C. Sharma, L.A. Owen, Quaternary glacial history of NW Garhwal, central Himalayas, Quat. Sci. Rev. 15 (4) (1996) 335–365.
[18] N. Rana, S. Sharma, Y. Sundriyal, S. Kaushik, S. Pradhan, G. Tiwari, N &Juyal, A preliminary assessment of the 7th February 2021 flashflood in lower Dhauli Ganga valley, Central Himalaya, India, J. Earth Syst. Sci. 130 (2) (2021) 1–10.
[19] Y.P. Sundriyal, A.D. Shukla, N. Rana, R. Jayangondaperumal, P. Srivastava, L.S. Chamyal, N. Juyal, Terrain response to the extreme rainfall event of June 2013: Evidence from the Alaknanda and Mandakini River Valleys, Garhwal Himalaya, India, Episodes 38 (3) (2015) 179–188.
[20] S.V. Srikantia, O.N. Bhargava, Tectonic evolution of the Himachal Himalaya, Misc. Publ. Geol. Surv. Ind 34 (1976) 217–236.
[21] S.V. Srikantia, O.N. Bhargava, The Jutogh Group of metasediments of the Himachal Himalaya: its lithostratigraphy, J. Geol. Soc. India (Online Archive from Vol 1 to Vol 78) 32 (4) (1988) 279–294.
[22] Arora, Manohar (2008) Seasonal characterization of ablation, storage and drainage of melt runoff and simulation of streamflow for the Gangotri Glacier, Project Report-1/2008, National Institute of Hydrology, Roorkee.
[23] U.K. Haritashya, P. Singh, N. Kumar, R.P Gupta, Suspended sediment from the Gangotri Glacier: quantification, variability and associations with discharge and air temperature, J. Hydrol. 321 (2006) 116–130.
[24] N.C. Agarwal, G. Kumar, Geology of the Gangotri and Yamuna valleys, Uttarkashi District, Kumaun Himalaya, Himalayan Geol. 3 (1973) 1–23.
[25] K. Kumar, M.S. Miral, S. Joshi, N. Pant, V. Joshi, L.M. Joshi, Solute dynamics of meltwater of Gangotri glacier, Garhwal Himalaya, India, Environ. Geol. 58 (2009) 1151–1159.
[26] A. Gansser, Geology of the Himalayas, Wiley Interscience, London, (1964), p. 289.
[27] A. Heim, A. Gansser, Central Himalaya. Geological observations of the Swiss expedition 1936, M6m. Soc. Helv. Sci. Nat. 73 (1) (1939) 245.
[28] P Le Fort, Himalayas: the collided range. Present knowledge of the continental arc, Am. J. Sci. 275 (1) (1975) 1–44.
[29] B.K. Bhatt, Preliminary study of the Bhagirathi basin between Uttarkashi and Gomukh, In Proc. Nat. Symp. On Himalayan Geology, Calcutta, Miscl. Publ. No.15, Geol. Soc. Ind., (1963), pp. 1–8.
[30] T. Bolch, B. Menounos, R. Wheate, Landsat-based inventory of glaciers in western Canada, 1985–2005, Remote Sens. Environ. 114 (1) (2010) 127–137.
[31] S.E. Fick, R.J. Hijmans, WorldClim 2: new 1-km spatial resolution climate surfaces for global land areas, Int. J. Climatol. 37 (12) (2017) 4302–4315.

[32] B. Gomez, R.J. Small, Genesis of englacial debris within the lower Glacier de Tsidjiore Nouve, Valais, Switzerland, as revealed by scanning electron microscopy, Geografiska Annaler: Series A, Phys. Geogr. 65 (1-2) (1983) 45–51.

[33] A. Rapp, R.W. Fairbridge, Talus fan or cone; scree and cliff debris, in: R.W. Fairbridge (Ed.), Encydopedia of Geomorphology, Dowden, Hutchinson & Ross, Pennsylvania, 1968, pp. 1106–1109.

[34] A. Jahn, Periglacial talus slopes. Geomorphological studies on Spitsbergen and in Northern Scandinavia, Polar Geogr. 8 (3) (1984) 177–193.

[35] R.J. Huggett, Fundamentals of Geomorphology, Routledge, (2016), p. 216.

[36] W.S.B. Paterson, Physics of Glaciers, 3rd ed., Elsevier, Oxford, UK, (1994), p. 3.

[37] P. Bisht, S.N. Ali, A.D. Shukla, S. Negi, Y.P. Sundriyal, M.G. Yadava, N. Juyal, Chronology of late Quaternary glaciation and landform evolution in the upper Dhauligangavalley,(Trans Himalaya), Uttarakhand, India, Quat. Sci. Rev. 129 (2015) 147–162.

[38] M. Spagnolo, C.D. Clark, A.L. Hughes, Drumlin relief, Geomorphology 153 (2012) 179–191.

[39] M.P. Sah, R.A.K. Srivastava, Morphology and facies of the alluvial-fan sedimentation in the Kangra Valley, Himachal Himalaya, Sediment. Geol. 76 (1-2) (1992) 23–42.

[40] S. Saha, M.C. Sharma, M.K. Murari, L.A. Owen, M.W. Caffee, Geomorphology, sedimentology and minimum exposure ages of streamlined subglacial landforms in the NW Himalaya, India, Boreas 45 (2) (2016) 284–303.

[41] W. Haeberli, M. Hölzle, Application of inventory data for estimating characteristics of and regional climate-change effects on mountain glaciers: a pilot study with the European Alps, Ann. Glaciol. 21 (1995) 206–212.

[42] H. Frey, H. Machguth, M. Huss, C. Huggel, S. Bajracharya, T. Bolch..., M. Stoffel, Estimating the volume of glaciers in the Himalayan–Karakoram region using different methods, The Cryosphere 8 (6) (2014) 2313–2333.

[43] D. Farinotti, M. Huss, A. Bauder, M. Funk, M. Truffer, A method to estimate the ice volume and ice-thickness distribution of alpine glaciers, J. Glaciol. 55 (191) (2009) 422–430.

[44] P.J. Taylor, W.A. Mitchell, The Quaternary glacial history of the Zanskar Range, north-west Indian Himalaya, Quat. Int. 65 (2000) 81–99.

[45] Y. Ageta, Study project on the recent rapid shrinkage of summer-accumulation type glaciers in the Himalayas, 1997-1999, Bull. Glaciol. Res. 18 (2001) 45–49.

[46] K. Hewitt, The Karakoram anomaly? Glacier expansion and the 'elevation effect,' Karakoram Himalaya, Mt. Res. Dev. 25 (4) (2005) 332–340.

[47] J. Oerlemans, Quantifying global warming from the retreat of glaciers, Science 264 (5156) (1994) 243–245.

[48] X. Liu, B. Chen, Climatic warming in the Tibetan Plateau during recent decades, Int. J. Climatol.: A Journal of the Royal Meteorological Society 20 (14) (2000) 1729–1742.

[49] K. Rikiishi, H. Nakasato, Height dependence of the tendency for reduction in seasonal snow cover in the Himalaya and the Tibetan Plateau region, 1966–2001, Ann. Glaciol. 43 (2006) 369–377.

[50] M.S. Shekhar, H. Chand, S. Kumar, K. Srinivasan, A. Ganju, Climate-change studies in the western Himalaya, Ann. Glaciol. 51 (54) (2010) 105–112.

[51] K. Duan, T. Yao, L.G. Thompson, Low-frequency of southern Asian monsoon variability using a 295-year record from the Dasuopu ice core in the central Himalayas, Geophys. Res. Lett. 31 (L16209) (2004), doi:10.1029/2004GL020015.

[52] S Negi Pyar, P Pandey Chhavi, N. Singh, Black carbon aerosols in the ambient air of Gangotri Glacier valley of north-western Himalaya in India, Atmos. Environ. 214 (2019) 116879.

[53] S. Sharma, S.P. Sati, Y.P. Sundriyal, H. Dobhal, The 23rd April'21 Snow Avalanche, Girthi Ganga post the 7th February'21 Rishi Ganga Flash Flood: Are these Events Linked to Climate Warming in the Western Himalaya? J. Geol. Soc. India 97 (9) (2021) 975–979.

[54] K.S. Valdiya, Outline of the structure of the Kumaun Himalaya, Tectonic geology of the Himalaya 1 (1978) 14.

CHAPTER 22

Machine learning in remote sensing data—a classification case study

Megha Sharma[a], Anil Kumar[b], M. Supriya[a], Vivek Singh[c], S. Kishore[d]
[a]Department of Computer Science and Engineering, Amrita School of Computing, Amrita Vishwa Vidyapeetham, Bengaluru, India
[b]Photogrammetry and Remote Sensing Department Indian Institute of Remote Sensing ISRO, Govt. of India, New Delhi, India
[c]Indian Institute of Tropical Meteorology, Ministry of Earth Sciences, Government of India, New Delhi, India
[d]Calaver Technologies Pvt. Ltd., Bengaluru, India

22.1 Introduction

There are a variety of practical uses for remote sensing data processing, including monitoring cities for crime, detecting wildfires, mapping crops, and calculating things like temperature and ocean salinity [1,2]. Earth observation on both a local and global scale is becoming more and more popular at this time. The monitoring of urban growth, the detection of oil spills, the estimation of temperature or ocean salinity, and the recognition of surface objects are all made possible through the use of remote sensing. Materials in a scene can reflect, absorb, and emit electromagnetic radiation depending on their molecular composition and structure. Remote sensing can be used in a variety of methods to view a scene (or a single object) at short, medium, or long distances. Depending on the energy resources utilized to obtain the data, remote sensing imaging equipment can be passive or active. In the last few decades, passive sensors have undergone a significant transformation. The machine learning and signal processing fields are challenged by these sensors. These sensors will be the focus of our efforts in this study. An aerial or satellite spectrometer collects a variety of wavelengths of solar radiation from passive devices that harness the sun's energy. Spectrum refers to the acquired signal or spectral signature, which is used to identify the scene's materials.

The satellite remote sensing imaging spectroscopic principle is shown in Fig. 22.1 [3]. The resulting multispectral image is made up of many spectrally contiguous bands that were acquired from a remote platform while being spatially co-registered. Machine learning and signal processing algorithms work in parallel for achieving a variety of goals and characteristics of the input. Images from remote sensing platforms have many erratic variables, including high dimensionality, several noise sources, and inherent nonlinearity. This makes it difficult to perform statistical characterizations of remote sensing images. Machine learning has shown successful in remote sensing categorization, regression, clustering, coding, and source separation. As far as we can tell, these novel learning paradigms have gotten little or no consideration. This chapter focuses on the use of machine learning in the processing of remote sensing data.

Atmospheric Remote Sensing
DOI: https://doi.org/10.1016/B978-0-323-99262-6.00005-5

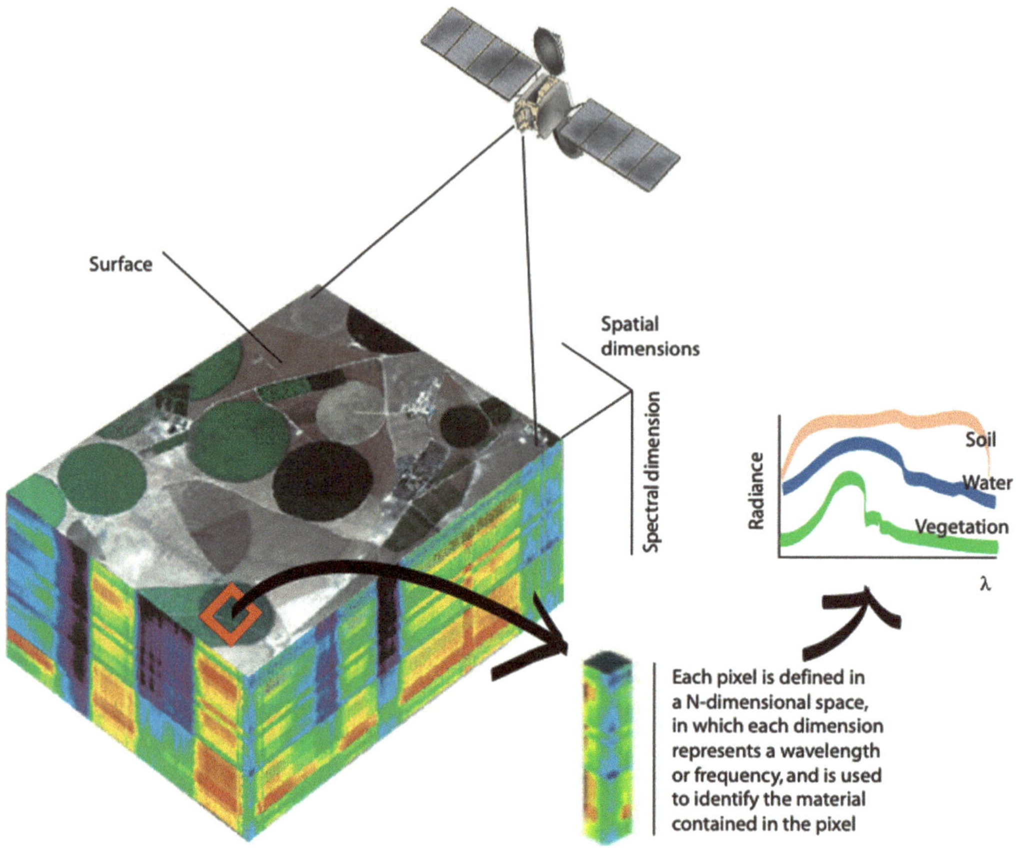

Fig. 22.1 ***Snapshot showing principle of imaging spectroscopy*** (adopted from Camps-Valls [3]).

22.2 Traditional machine learning for remote sensing

Remote sensing applications necessitate typical machine learning techniques. Regression and unmixing are only revised for the most relevant applications of classification, feature selection, and extraction.

22.2.1 Image classification

Remote sensing image processing yields the most important product, that is, "classification maps." It has become increasingly common in the remote sensing community to use data-driven methodologies. Non-parametric approaches, in particular, have shown good performance. Further, the distinction between supervised and unsupervised is clarified below.

22.2.1.1 Supervised image classification

These methods use labeled information about the class membership of individual pixels (as designated by experts) to build a model that can generalize to the full image (or set of images). Support vector machines and neural networks have become the most widely used approaches in recent years. Multispectral and hyperspectral data have been used for a variety of purposes, including object recognition, land cover and time series analysis, and urban monitoring, just to name a few. Classifier ensembles and boosting approaches are also becoming increasingly popular.

22.2.1.2 Unsupervised image classification

The unsupervised classification of remote sensing images is a critical topic in many applications, whether it is used to visualize and monitor similar areas in the scene or to prepare for supervised classifiers. There are a variety of clustering algorithms, including rule-based, neural network, and SVM-based approaches. The field is dominated by three techniques. Fuzzy clustering has been utilized for spatially membership relations optimization alone or in combination with multiobjective optimization in the first place. For context regularization, data from several sources have been combined using a variety of methods, including graph-cuts, projection chase, hierarchical clustering, and Markov random fields. Self-organizing maps (SOMs) and hybrid genetic algorithms have also been proposed for multicomponent image segmentation. Spatio-temporal reasoning and visualization can benefit from dynamic clustering strategies.

22.2.1.3 Multitemporal classification and change detection

Increased interest in multitemporal and change detection challenges has been inspired by the increased availability of complete time series pictures as well as an interest in monitoring earth changes on local and global scales. An increasing number of multitemporally guided techniques, such as evidence reasoning, extended least squares, neural networks, and support vector machines are now being used. There are also hidden Markov random fields and fuzzy-based methods to connect time-varying statistics. Images can be subtracted or ratioed, change vectors can be analyzed, or cross-correlation can be used to identify changes. Kernel methods and neural networks have recently been employed. Multisensor and multisource information can be combined with the use of composite kernels. It has been increasingly common in recent years to use semi- or un-supervised methods to reduce the amount of user intervention required.

22.2.2 Feature selection and extraction

Because of the increased processing time and poor model quality that can be caused by high collinearity and the presence of noisy bands in high-dimensional datasets like hyperspectral images, dealing with these kinds of data can be a major challenge. Even more crucial, though, may be determining which bands are most relevant for a certain activity. The features of the elements being recognized or modeled can be tied to the spectral bands, which have a physical meaning. Studies in remote sensing without the traditional discriminative criteria have looked at feature selection. Genetic algorithms and SVM-based recursive feature elimination, both of which use advanced machine learning approaches, have recently been employed. Recent years have seen an increase in interest in feature extraction techniques. Despite the widespread use of linear methods like PCA and PLS, multivariate kernel machines have recently made significant strides in the field of dealing with data that are nonlinear.

22.2.3 Signal unmixing

The pixel data is always contaminated with the signatures of the various materials identified inside the spatial limit of the ground instantaneous field view. It is a major difficulty in remote sensing to build automatic extraction algorithms for spectral pure pixels (also known as endmembers). Pure pixels can serve as the foundation for all other pixels to be combined into a linear or non-linear combination for subpixel identification and mineral mapping. An orthogonal subspace projection methodology and the N-FINDR algorithm are the most often used methods for this purpose, as is a vertex component algorithm (VCA). Most important in today's unmixing procedure are 'parameters selection' and 'spatial information incorporation'. Pure pixels can now be selected using the support vector domain description (SVDD).

22.2.4 Regression and model inversion

Remote sensing requires robust, rapid, and accurate regression tools. Estimating biophysical parameters is particularly important for better understanding of local and world-scale environmental dynamics. Analytical models inverted introduce a higher level of complexity, resulting in a significant processing overhead, and an increased sensitivity to noise. It has become increasingly popular to apply empirical models tailored to discover the correlation between the spectra and real-world observations in recent years. In general, parametric models have a number of fundamental drawbacks that often result in poor predictions on unknown data. Approaches like as neural networks, support vector regression, relevance vector machines, and gaussian processes have thus far shown great success in the real world. However, despite their superior accuracy to analytical models, they are limited in their use due to their lack of interpretability and reliance on training data from the observed scene.

22.3 New trends in machine learning for remote sensing

Researchers are always working on new machine learning methods to handle specific remote sensing challenges. An overview of some intriguing remote sensing machine learning algorithms is provided in this section.

22.3.1 Manifold learning

Manifold learning has recently emerged as an effective approach for analyzing data nonlinearities [4]. Computer science, machine learning, image processing, and cybernetics all have applications in dimensionality reduction and nonlinear feature extraction. Data with large dimensionality can be simplified via manifold learning while still preserving the most significant features. High-dimensional data can be viewed and understood using this method. Traditional linear dimensionality reduction strategies fail because of a lack of knowledge about the underlying structure of remote sensing data. Some preliminary works using nonlinear transforms, such as the Isomap, Laplacian techniques, or Local Linear Embedding have been presented. In addition, certain algorithms for analyzing hyperspectral images' intrinsic dimensionality might be mentioned.

22.3.2 Semisupervised learning

Semisupervised learning, which is concerned with building models that take advantage of (typically few) labels on data sets and a large number of unlabeled examples, is a related area for modeling datasets that have a variety of structures [5]. In remote sensing, there are a variety of strategies, both generative and discriminative, have been created. In generative models, the estimation of the conditional density has been intensively explored. Many graph-based algorithms for classification, regression, and target detection have recently been developed. Cluster and bagged kernel designs were demonstrated to be successful. Additionally, it's been put to use in image classification and change detection.

22.3.3 Transfer learning

A common difficulty in remote sensing is updating land-cover maps by classifying time series of pictures, especially when only training samples collected all at once are available.Transfer learning or domain adaptation are the two terms used to describe this process [6]) Partially unsupervised classifiers and neural networks were used to begin solving the problem. SVM domain adaption was a successful extension of the method.The problem of sample selection bias or covariance shift is also a concern when attempting to identify a picture using samples from different time periods. These problems were recently revealed by kernel machine definitions.

22.3.4 Active Learning

It is only recently that active learning approaches have been used in remote sensing to choose suitable training samples [7,8]. Object-oriented classifiers use SVM; pixel-based classifiers use maximum likelihood classifiers, which were first proposed in. Iteratively weighting the selected pixels was recently proposed as an extension of this approach. A model-independent active learning method for very high-resolution satellite pictures was developed for target detection using information-based active learning.

Table 22.1 Statistical machine learning methods for analysis of remote sensing and other data, collated according to the analytic aim (Adopted from [16]).

Sr. no.	Categories	Algorithms/existing methods
1	Classification	Support vector machines • Random forest • K-nearest neighbor (k-nn) • Intra or subpixel classification
2	Clustering	Mixture models • K-means • Agglomerative clustering
3	Regression	Linear regression and extensions • Boosted regression trees • State space models • Neural networks • Functional analysis
4	Dimension Reduction	Functional data analysis • Principal component analysis

22.4 Statistical machine learning methods

Using statistical machine learning approaches, known as empirical methods, a statistical link can be constructed between the spectrum bands or frequencies utilized and the variable observed (field-based) parametric, semiparametric, or nonparametric are all possibilities for this connection. Sampling and model error can be quantified scientifically in SML techniques, allowing for accurate estimations and predictions, as well as a better understanding of the uncertainty associated with these findings. Hypotheses can be tested using models, and some models, like decision trees, include only a few assumptions. The drawbacks are the ground truth data needed to build the model or check the output of the model (unlike the physics-based models). A limitation of this approach is that it relies heavily on empirical data that may be difficult to generalize.

To accomplish various goals, many statistical machine learning algorithms have been developed. Based on remote sensing data, algorithms for sustainable development The four primary analytical goals are as follows: Data classification, clustering, regression, and dimension reduction are all examples of these techniques (Table 22.1).

22.4.1 Informed statistical machine learning methods

Semiempirical or informed statistical machine learning methods integrate process knowledge with SML or empirical models [9]. You can learn about the process or the factors that make up the process, depending on what you need to know. Remote sensing data has been analyzed using these techniques for more than a decade. For example, chosen water quality and substrate cover type parameters were mapped using Landsat 7 Enhanced Thematic Mapper image data). An approach that relies on the spectral properties of some parameters to improve the statistical model is described by Dekker et al. [10] for determining the water quality. A small selection of variables is employed in their example, with carefully selected spectral areas and appropriate wavebands. The authors point out how popular and useful these approaches are, but they also point out that they still require ground truth data. In theory, semiempirical methods are easier to adapt to new situations than strictly empirical models, but they are still limited by the data used to develop them. Tripathy et al. [11] estimated wheat yields using semiempirical methods, including physiological measurements, spectral measurements, and spatial features. In spite of the fact that models based on vegetation indices are extensively employed, the authors point out that data-intensiveness and complexity limit the adoption of spectral (empirical) and physics-based (mechanistic) models. An intermediate method is proposed by their semiempirical approach.

22.4.2 Physics-based methods

Models based on physics have a thorough understanding of the system they are modeling. Model parameters can be calibrated using data or without data. This method can be used to automate huge areas if the model is parameterized correctly and properly. Over the past 15 years, Using remote sensing data, scientists have built and applied biophysical and geophysical models to a wide range of circumstances. For instance, a PHYSGROW plant growth model was augmented with NOAA and NASA satellite data to create fodder production maps for a large area. Data from temperature, rainfall, and NDVI were gathered by remote sensing, and the resulting gridded datasets can be used to explore spatial autocorrelation. With a geographic information system (GIS), this data could be linked to economic models, environmental research, or drought warning systems.

Landsat 7 Enhanced Thematic Mapper images were used by Phinn et al. (2005) to map specific water quality and substrate cover type criteria. This was the first time these technologies were used in a maritime setting. There are numerous examples of these models in use today. Using satellite data, five surface energy balance models' performance in estimating daily evapotranspiration in sorghum with a high biomass was evaluated by Wagle et al. [12]. A variety of physics-based methodologies and complex algorithms are frequently used to account for sensor performance as well as the effects of the atmosphere, sea surface, and other environmental factors.

22.4.3 Object-based image analysis

When analyzing remote sensing data, object-based image analysis (OBIA) includes combining pixels into homogenous segments or objects, rather than analyzing individual pixels. Mean, variance, and mean ratio values for each band are included in these segments in addition to the pixels. When working with high-resolution satellite images, aggregating pixels into segments reduces the amount of time and effort required to process it. Point-based, edge-based, region-based, and hybrid image segmentation techniques are all subdivided into four groups. There are several papers that use region-based segmentation to simulate the effects of cyclones, with the most recent one estimating cyclone risk in light of possible future climate change scenarios. OBIA is a technique that employs spatial, spectral (brightness and color), and textural variables to segment QuickBird satellite pictures to identify new buildings in Ghana. An extension of OBIA, geographic object-based image analysis involves the identification of pixel segments using both their spatial location and their spectral properties.

22.5 Categories of statistical machine learning methods

As indicated in Table 22.1, classification, clustering, regression, and dimensional reduction are among the most typical remote sensing data analysis goals. Remote sensing data can be analyzed using four different types of methods that can be described as follows.

22.5.1 Classification

In order to reliably assign items to a set of known (typically small) classes or groups, a classification approach is appropriate. Input variables are used to determine this allocation. Some other names for these variables in the scientific literature include explanatory factors and covariates. These will be referred to as input variables or covariates for the purposes of this review. To build or "train" the model. As a result, we employ a collection of data that comprises both input variables and output variables. This model can then be put to the test using test datasets that just contain the input variables. Crop types can be characterized using a training dataset that comprises variables extracted from the photos (the input variables) and ground truth crops (the output variables) at a set of locations. When no ground truth data is available, the crop classification model can be used to categorize the crop type in the rest of the image.

22.5.2 Clustering

If the goal is to group or classify items based on a collection of input variables, a clustering method is appropriate. There is no need for a training data set while using clustering, which is an unsupervised learning method. The output variable and classifications are unknown in classification. Because of this, we must devise a technique to categorize the objects based on their degree of resemblance. The number of groups (clusters) can be specified or left unspecified, and the number of groups can be estimated as part of the analysis. An object's participation in a cluster may be predicted using this analysis, which can then be utilized to make judgments. As an example, a set of input variables taken from a picture can be used to categorize pixels. The properties of these groups can then be examined, described, and compared.

22.5.3 Regression

Regression can be used to estimate or forecast a response variable based on a set of covariables. Categorization methods rely on continuous rather than binary replies. Regression models can be built or trained using a set of input variables for which the response is known. To accurately predict or forecast agricultural yields, for example, remote sensing photos can be used. In contrast to the classification of crops, the yield of a crop is a continuous variable. As seen in Table 22.1, the most frequent regression methods are linear and logistic regression as well as neural networks. Examples of neural network types include artificial, convolutional, and deep neural networks, to name a few. Imagery applications including pixel-level classification of satellite photographs to improve maps and the integration of various forms of remote sensing data to monitor land surface changes often use convolutional and deep neural networks (deep learning). Examples of neural networks used to classify land use and land cover include. Artificial neural networks were recently used to create deforestation risk maps for Madagascar and Mexico.

The predictive power of deep learning is making it a popular tool for remote sensing research, but this precision comes at a price. Examples include the long short term memory (LSTM) neural network, trained on satellite images, and then used to forecast new time series data based on these images. As shown in Table 22.1, all of these new neural network approaches may be accessible online, including LSTMs like recurrent neural networks andVGGs like autoencoders. It has been used to identify abnormalities in images, classify topography in synthetic aperture radar images and extract attributes from and classify satellite images by using deep learning neural networks.

22.5.4 Dimension reduction

For example, remote sensing data can be used to extract a large number of variables that can then be combined into new variables that incorporate the most information feasible (big set of input variables). The use of a strategy that reduces the dimensions is appropriate in this situation. Using these new variables as inputs into other research or creating their own end products is possible. Examples include looking at them as "features" or "indices" to better grasp what's going on in the world (e.g., two satellite reflectance variables are combined to give a single vegetation index variable, VI). This is an unsupervised learning process because there is no response variable to estimate.

22.6 Data and methodology

Rabi and Kharif are the two harvest seasons of India. The first of these typically begins in April and continues through October each year. From November through April, the Rabi season is in effect. Most farmers burn crop residues left over from rice harvest in order to get their fields ready for growing wheat and other Rabi crops soon.

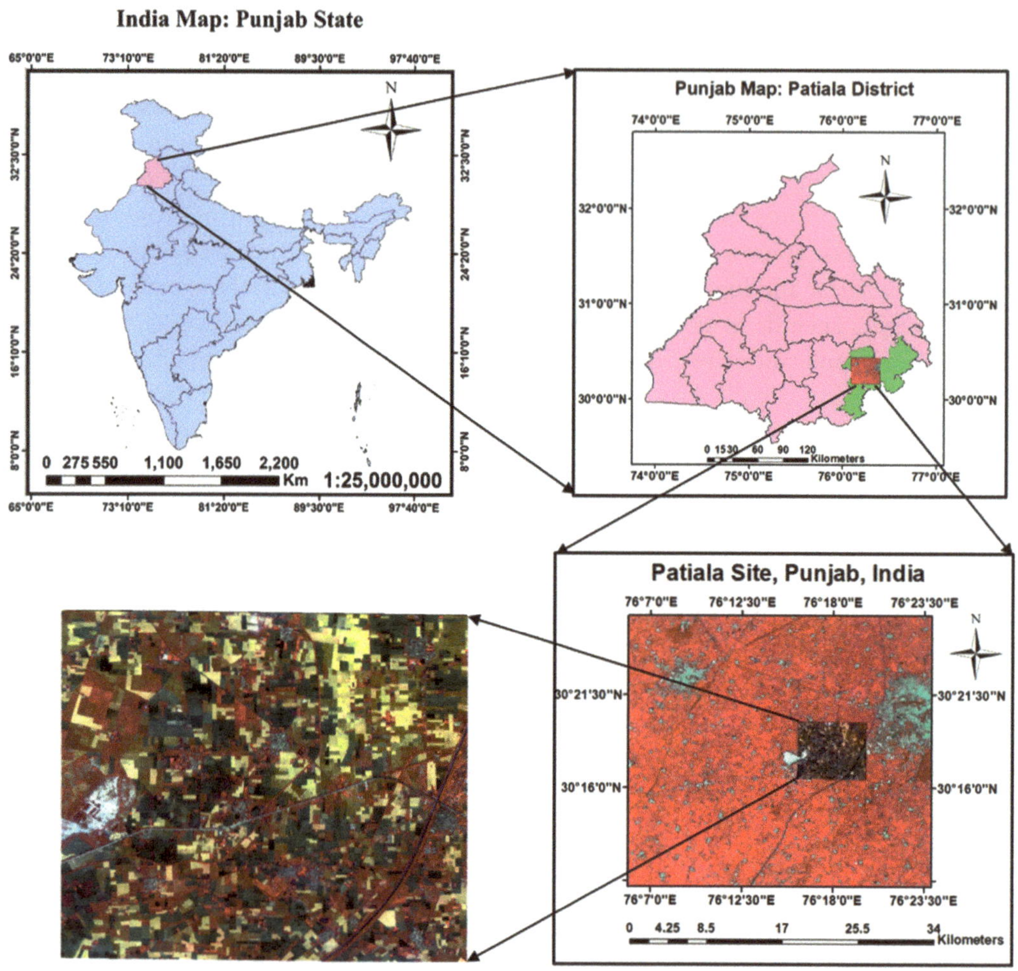

Fig. 22.2 ***The study area showing Patiala (Punjab, India) and its nearby regions*** (adopted from Singh and Kumar [13]).

Approximately 7 to 8 million metric tons of paddy residue are burned openly in Punjab each year between October and November. The present study is the identification of these burnt paddy fields in Patiala and surrounding areas as shown in Fig. 22.2; [13]. The data used are Sentinel 2A,2B. Classification of the burnt fields is done using possibilistic c-means (PCM) classifier (Details are enlisted in Table 22.2).

Using two satellites, Sentinel-2 provides high-resolution optical imagery for the SPOT and Landsat missions, as well as providing continuity. With one satellite every 10 days and two satellites every 5 days, the project covers the whole land surface of the Earth, providing invaluable data for ongoing research. High-resolution optical imaging is provided by the MSI (Multispectral Imager) instrument on board the satellites. The

Table 22.2 Details of data used for the burnt-paddy field identification.

Name of satellite	Sentinel-2
Spatial resolution	10 m (B2,B3,B4,B8), 20m (B11,B12)
Spectral bands	B2-Blue (490 nm) B3-Green (580 nm) B4-Red (665 nm) B8-NIR (842 nm) B11-SWIR-1 (1610 nm) B12-SWIR-2 (2190 nm) Temporal Multispectral data 30-09-18, 05-10-18, 12-10-18, 15-10-18, 20-10-18, 25-10-18, 30-10-18, 09-11-18 19-11-18
Swath width	290 km
Pixel quantization	12 bit

two most important bands that contribute in the identification of the burnt fields are the Short Wave InfraRed (SWIR 1 & 2). These bands have a high reflectance on the burnt pixels and thus have the capability to identify the burnt patches. The detailed methodology of the present study has been depicted in the following flowchart diagram (Fig. 22.3)

A two-dimensional set of data is all that is required for this categorization method. Nevertheless, the 3-D nature of the temporal data we've used as input is evident in the fact that it is spectral, temporal, and spatial in nature. Spectral dimensionality can be reduced by creating temporal burned indices, which is the solution to the problem. Burnt fields can be distinguished from the rest of the landscape using temporal vegetation indices. Fuzzy maps are used to keep track of the location and size of cropland that is being moved around. In 1993, Krishnarajpuram and Keller devised a possibilistic c-means classification that was utilized to differentiate the target class from the other classes of interest. The PCM classifier with a membership value of 1.5 and a threshold of 0.99 was used to classify temporal burnt photos. The test results were used to verify the accuracy of the categorized photos. By comparing satellite photos with ground-truth data, burnt fields could be detected. PCM is a classifier that has the capability to extract a single class as it works on the concept of possibility i.e. the degree of belongingness to the class, where it maximizes the membership value of a particular pixel using η. η is a regularization parameter that controls the shape and size of the class and keeps the noise and outliers away.

$$\text{Index} = (\text{Max} + \text{Min}) / (\text{Max} - \text{Min}$$

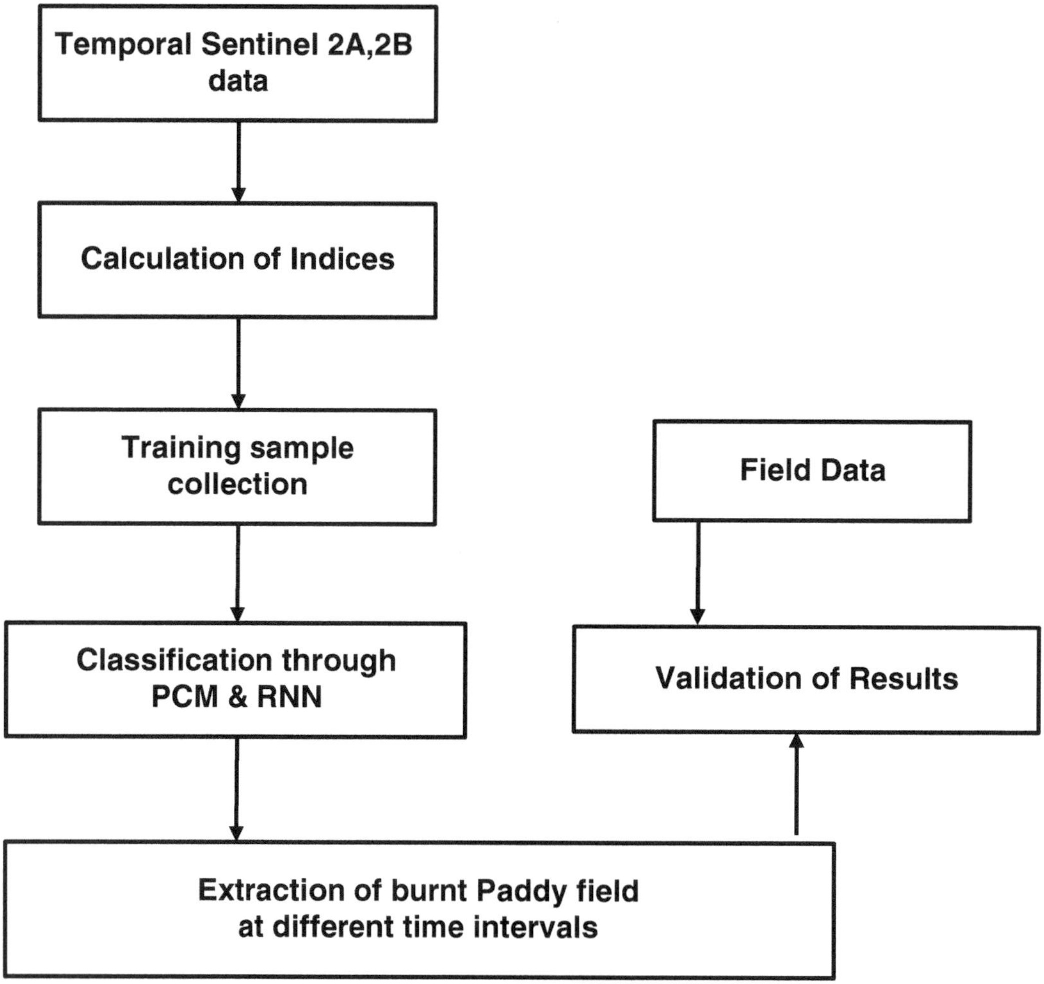

Fig. 22.3 *Detailed flow chart of the methodology used in the present study.*

where

Max = band with maximum reflectance

Min = band with minimum reflectance

PCM algorithm In the PCM algorithm, each cluster is independent of the other clusters [14].

The objective function can be written as in Eq. (22.1)

$$J_m = \sum_{i=1}^{N} \sum_{j=1}^{C} \left(u_{ji}\right)^m \left\|x_i - v_j\right\|^2 + \sum_{i=1}^{N} \eta_i \sum_{j=1}^{C} \left(1 - u_{ji}\right)^m \tag{22.1}$$

The membership update equation in the PCM is

$$u_{ji}(n+1) = \frac{1}{1+\sum_{i=1}^{N}\left[\frac{\|x_i - v_j\|^2}{\eta_i}\right]\frac{1}{m-1}} \tag{22.2}$$

In Eq. (22.2), v_j represents the prototype associated with class C, and η_i is the "bandwidth," "resolution," or "scale" parameter which controls the shape and size of the class as in Eq. (22.3);

$$\eta_i = \frac{\sum_{i=1}^{N}(u_{ji})^m \|x_i - v_j\|^2}{\sum_{i=1}^{N}(u_{ji})^m} \tag{22.3}$$

i represents a pixel and ranges from 1 to *N*, where *N* is the image's total number of pixels.

For example, the first term in Eq. (22.1) sets the minimum distance between pixels and their prototype feature vectors, and the second term sets the maximum value for *ji* by setting the commencement of trivial solution. Both terms are interdependent. The objective function, Jm in Eq. (22.1) and membership value μ_{ji} in Eq. (22.2) satisfy the following conditions given as Eqs. (22.4)–(22.6):

$$u_{ji} \in [0,1], \forall i, j \tag{22.4}$$

$$0 < \sum_{i=1}^{N} u_{ji} \leq 1, \forall i \tag{22.5}$$

$$\max_j u_{ji} > 0, \forall i \tag{22.6}$$

PCM Algorithm includes the steps as follows:

Step 1: Assign mean values to each class.

Step 2: Assign the value of degree of fuzziness $\dot{m} > 1$

Step 3: Compute the regularization parameter η_i from Eq. (22.3)

Step 4: Calculate the membership matrix μ_{ji} from Eq. (22.2)

Step 5: Assign final class to each pixel.

22.7 Results and discussion

The key to monitoring stubble burning in Punjab's rice fields in the Patiala area is remote sensing temporal data available in a 5-day interval, often known as a 5-day temporal resolution. It took 5 days to collect the burnt paddy fields from various locations., example October 20, October 25, October 30, November 9, and November 19, 2018. The cross-validation has been done with the ground truth (Singh, A., [7]) which matches the burnt pixels. PCM is a classifier that has the capability to extract a single class as it works on the concept of possibility, that is, the degree of belongingness to the class.

The output shown in Fig. 22.4 was extracted using various satellite images. The October 20, 2018 burnt patches were extracted using the 12 October and 15 October

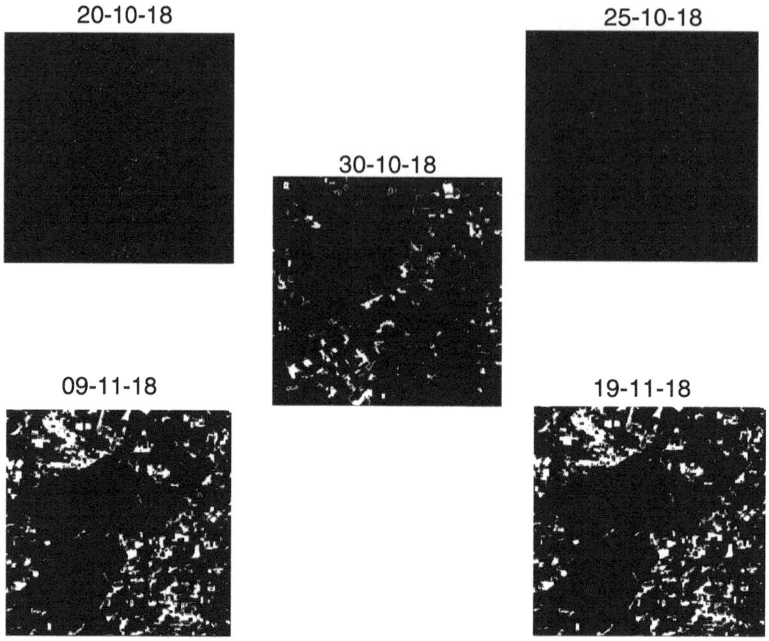

Fig. 22.4 ***The PCM classified images of crop burning on different dates of the study.***

satellite images. The number of samples used was 192. The October 25, 2018 burnt patches were extracted using the October 12, October 15, and October 20 satellite images. The number of samples used for the burnt patches were 79. The October 30, 2018 burnt patches used the base files from the October 12, October 15, October 20, October 25 satellite images. Number of samples used were 73. The November 9, 2018 burnt patches used the base files from the October 5, October 15, October 20, October 25, October 30 satellite images. The number of samples used were 70. The November 19, 2018 burnt patches were extracted using the October 12, October 15, October 20, October 25, October 30, November 9 satellite images. Number of samples used was 301. The Degree of fuzziness was m = 1.5.

Fig. 22.4 shows the classified images on the respective dates. Since it is a single class highlighted during the classification, the images show two colors: white and black for each pixel. The white color is the burnt pixel whereas the black color denotes the rest of the classes which includes everything except the burnt pixel. The PCM classified images show that the burning of the fields was low on 20-10-18 and 25-10-18. Burning has increased from 30-10-18 and the maximum is seen on 19-11-18. These results are verified with the ground truth data and have been found correct [13]. Similar work for classification using machine learning algorithms is carried out in the "Deep learning-based models for temporal satellite data processing: Classification of paddy transplanted fields" by Rawat et al. [15].

22.8 Conclusions

In this study, we described the use of fuzzy-based PCM algorithm to classify the burnt paddy fields. This is a single-class classification. The data are taken from Sentinel 2A, 2B satellites which provided a good temporal data to examine the burning of fields within the desired period. The spectral dimensionality of the data was reduced through the application of temporal burnt indices. The temporal vegetation indices made it possible to separate the burned areas from the rest of the landscape using satellite imagery. The study shows correct classification results for all the dataset ranging from the date 20-10-18 to 19-11-18. This has been verified by the ground survey. The RNN algorithm for Deep Learning is a structure-based algorithm which relies heavily on LSTM. The testing has been carried out for 19-11-18 dataset. The future scope of the work done is to check the saturation of the activation function in the algorithm.

Acknowledgments

My heartfelt thanks to Dr. Prakash Chauhan, Director Indian Institute of Remote Sensing for providing an opportunity to work on the experiment. The Institute has a blend of rich scientific experience which has been a great support in the current study. I would like to express my humble gratitude to Mr. Abhishek Singh, University ofTrento for the inputs on the study area and the technical discussions related to the work. It has been an enriching experience working and documenting the current study.

References

[1] Y.H. Kerr, P. Waldteufel, J.P. Wigneron, J.A.M.J. Martinuzzi, J. Font, M. Berger, Soil moisture retrieval from space: The Soil Moisture and Ocean Salinity (SMOS) mission, IEEE Trans. Geosci. Remote Sens. 39 (8) (2001) 1729–1735.

[2] H. Liu, D. Zhu, L. Niu, L. Wu, C. Wang, X. Chen, J. Wu, MICAP (Microwave imager combined active and passive): A new instrument for Chinese ocean salinity satellite, 2015 IEEE International Geoscience and Remote Sensing Symposium (IGARSS), IEEE, 2015, July, pp. 184–187.

[3] G. Camps-Valls, Machine learning in remote sensing data processing. In: 2009 IEEE international workshop on machine learning for signal processing, IEEE, 2009, September, pp. 1–6.

[4] J. Zhang, Y. Wang, P. Molino, L. Li, D.S. Ebert, Manifold: A model-agnostic framework for interpretation and diagnosis of machine learning models, IEEE Trans. Visual Comput. Graphics 25 (1) (2018) 364–373.

[5] X. Zhu, A.B. Goldberg, Introduction to semi-supervised learning, Synthesis lectures on artificial intelligence and machine learning 3 (1) (2009) 1–130.

[6] S.J. Pan, Q. Yang, A survey on transfer learning, IEEE Trans. Knowl. Data Eng. 22 (10) (2009) 1345–1359.

[7] Olsson, F. (2009). A literature survey of active machine learning in the context of natural language processing.

[8] Settles, B. (2009). Active learning literature survey.

[9] Von Rueden, L., Mayer, S., Beckh, K., Georgiev, B., Giesselbach, S., Heese, R., & Schuecker, J. (2019). Informed machine learning—a taxonomy and survey of integrating knowledge into learning systems. *arXiv preprint arXiv:1903.12394.*

[10] A.G. Dekker, S. Peters, R. Vos, M. Rijkeboer, Remote sensing for inland water quality detection and monitoring: state-of-the-art application in Friesland waters. In: GIS and Remote Sensing techniques in Land-and Water-management, Springer, Dordrecht, 2001, pp. 17–38.

[11] R. Tripathy, K.N. Chaudhary, R. Nigam, K.R. Manjunath, P. Chauhan, S.S. Ray, J.S. Parihar, Operational semi-physical spectral-spatial wheat yield model development, The International Archives of Photogrammetry, Remote Sensing and Spatial Information Sciences 40 (8) (2014) 977.
[12] P.Wagle, N. Bhattarai, P.H. Gowda, V.G. Kakani, Performance of five surface energy balance models for estimating daily evapotranspiration in high biomass sorghum, ISPRS J. Photogramm. Remote Sens. 128 (2017) 192–203.
[13] A. Singh, A. Kumar, Identification of paddy stubble burnt activities using temporal class-based sensor-independent indices database: modified possibilistic fuzzy classification approach, J. Indian Soc. Remote Sens. 48 (3) (2020) 423–430.
[14] R. Krishnapuram, J.M. Keller, The possibilistic c-means algorithm: insights and recommendations, IEEE Trans. Fuzzy Syst. 4 (3) (1996) 385–393.
[15] A. Rawat, A. Kumar, P. Upadhyay, S. Kumar, Deep learning-based models for temporal satellite data processing: Classification of paddy transplanted fields, Ecological Informatics 61 (2021) 101214.
[16] J. Holloway, K. Mengersen, Statistical machine learning methods and remote sensing for sustainable development goals: a review, Remote Sensing 10 (9) (2018) 1365.

CHAPTER 23

Remote sensing-based study of landslide hazard zonation in Namchi and its surrounding area of Sikkim, India

Anuska Kumari[a], Praveen Kumar Rai[b], Varun Narayan Mishra[c], Prafull Singh[d], Akansha Mehra[e]

[a]GIS Engineer, National Informatics Centre, New Delhi, India
[b]Department of Geography, K.M.C. Language University, Lucknow, India
[c]Amity Institute of Geoinformatics & Remote Sensing (AIGIRS), Amity University, Noida, Gautam Buddha Nagar, India
[d]Department of Geology, School of Earth, Biological and Environmental Science, Central University of South Bihar, Gaya, India
[e]Independent Researcher, New Delhi, India

23.1 Introduction

Landslide can be defined as movement of mass of rock, debris down the slope. Landslides are one of the most common occurring hazards in Himalayan terrain that cause significant damage to human life and property, being topographically sensitive area the Namchi district witnesses' landslide almost every year causing great damage to roads, settlements agricultural field and human lives. Geology, climatic conditions, geomorphological settings, and seismic activities builds a complex relationship that triggers the hazardous event in this area.

Anthropogenic activities such as building of houses, hotels in an environmentally sensitive area and altering the natural landscape for own benefit has also amplified the impact of the disaster. As per the records a massive landslide occurred near the village of Mantam in Sikkim on August 13, 2016, as per the news reports formation of the lake and consequent rise in water level submerged the Kanka bridge and washed away about 300 m of road. Five houses in the Mantam village have been submerged. The villages of Tingvong, Lingdem, Laven, Kayeem, Lingzya, Bay, Sakyong Pentong, and Ruklu Kayeem are cut-off due to the damage to the connecting road (Source: Bhuvan).

Another landslide occurred in upper Gurpisey, Namchi, South Sikkim on 20.09.2017 due to torrential rainfall over the past 24 hours. The landslide caused loss of three human lives, injuries to two people and damages to properties (Source: Geological Survey of India). Another devastating landslide was witnessed near Machong village, Pachey town along Pakyong-Machong road in East district, Sikkim on July 9, 2017 the landslide swept away five buildings, breached the water supply line to Nopgaon and Pakyong town and washed away a portion of the Pakyong-Machong road (Source:

Atmospheric Remote Sensing
DOI: https://doi.org/10.1016/B978-0-323-99262-6.00013-4

Geological Survey of India). Various researches have been done in mapping, analyzing, and mitigating landslide hazard.

It can be devastating with massive destruction to life and property and may also lead to large-scale landscape transformations. It is frequently responsible for considerable losses of both money and lives. The severity of the landslide problem worsens with increased urban development and change in land-use [1].

The steep slopes, deep weathering of substrata, and intense southwest Monsoon rainfall in the lesser and the Middle Himalayas in conjunction develop complex geological conditions that are favorable for the frequent occurrence of landslides [2,3]. The hydro-geological conditions and anthropogenic activities are among the key elements that too strongly influence the landslide occurrences in the Himalayan region of India [4,5]. Using different models of landslide susceptibility index is one of the methods to identify the prone areas for future landslides and preventive activities should be taken to prevent or decrease their damages in those areas [2].

These maps offer valuable realities for government specialists, organizers, decision makers, and local landowners to make crisis arrangements to diminish the negative impacts on foundation, superstructure, and human life [6]. Over a period of time satellite remote sensing and GIS have played a vital role in managing landslide hazards. Remote sensing along with GIS provides greater advantage in landslide hazard zonation studies. Remote sensing images are helpful in preparation of landslide inventory, temporal capability of satellite images provide valuable knowledge of past and present event on the other hand GIS has proved to be very effective in data handling, manipulation and statistical measures. The landslide inventory provides useful information about the location and distribution of landslides in the area but geospatial technique provides capability of reaching almost every part of the inaccessible area, identifying all types of landslides, and determining landslide-prone zones on the basis of major causative factors [7].

GIS-based multistandards investigation produce powerful procedures for landslide susceptibility mapping and estimate of landslide risks [8]. Visual interpretation of high-resolution satellite data and application of DEM are used for developing landslide inventory maps. Furthermore, geospatial tool had great importance for preparation of digital database and updating information regarding landslides [9]. Moreover, GIS-based multicriteria analysis provides powerful techniques for landslide hazard mapping and prediction of landslide hazards [10].

In order to prepare landslide susceptibility maps various methods can be used such as analytical hierarchy process (AHP) and weighted overlay analysis (WOE). In AHP method each factor that is used in landslide hazard zonation is broken into smaller classes and then these classes are weighted based on the importance. It is performed to analyze comparison between various factors the relative importance is measured between two factors on a scale between 1 and 9 where one indicates that the two factors are equally important while 9 represents that one factor is much more important than another after that the prepared layers were combined and a final map is produced in this method the weight of the layer is determined by the judgment of the researcher. On the other hand, in weighted overlay method (WOE) each cell is reclassified on a common preference scale such as 1-9 with 1 being the least important and being the most important.

The aim of the study is to carry geospatial-based landslide susceptibility zonation for Namchi district, Sikkim which is highly prone to slope instability. The study carried two different approaches, the first one is analytic hierarchy process (AHP) while another method is weighted overlay analysis, the methods were executed in the study, and the results were obtained and discussed the landslide susceptibility maps described how prone the area is for future landslides hence susceptibility maps are of great importance for management and future planning of the area.

23.2 Material and methods

23.2.1 Study Area

The study area of Namchi and surrounding area of Sikkim Himalayas are located between 27°4′42" and 27°7′27" N latitudes and 88°19′21" and 88°30′30" E longitudes at an elevation of 1675 m (5500 feet) above mean sea level, being the second most populous district of Sikkim, Namchi is the headquarter of south Sikkim, it is dominated by hot and humid climate with an average rainfall of 1422.21 mm annually [10]. The onset of the monsoon is from last week of June and continues till October where June to September is peak monsoon months, the area receives rainfall majorly from south-west monsoon. The area witness summer in the month of March till June the temperature of summer season varies between 4°C–12°C which can shoot up to 28°C, making the weather warm whereas the temperature of monsoon season ranges between 5°C–13°C, on the other hand, the temperature of winter season i.e., October to March ranges between 8 and 15°C respectively.

The south Sikkim is a mountainous area of Eastern Himalaya characterized with series of crisscross ravines and ridges. The topography is rugged in nature having steep slopes and moderately dissected hills, highly dissected valleys and denudational hills. Several lineaments occur in study area indicates the weak zone in the area weathering and erosion has also been observed in the rocks. Tista is the mainstream of Sikkim as well as South Sikkim district it originates from a glacial lake, Chho Lhamo located at the north-eastern comer of the state, Sikkim. The river Tista enters in the South Sikkim district near a place Lingi. Various tributaries join river Tista among them, river Rangit is one of the important right bank tributaries of South Sikkim the right bank tributaries of the river Tista originates from high mountain glaciers, that is why, they are perennial in nature and more voluminous. But the left bank tributaries are nonperennial in

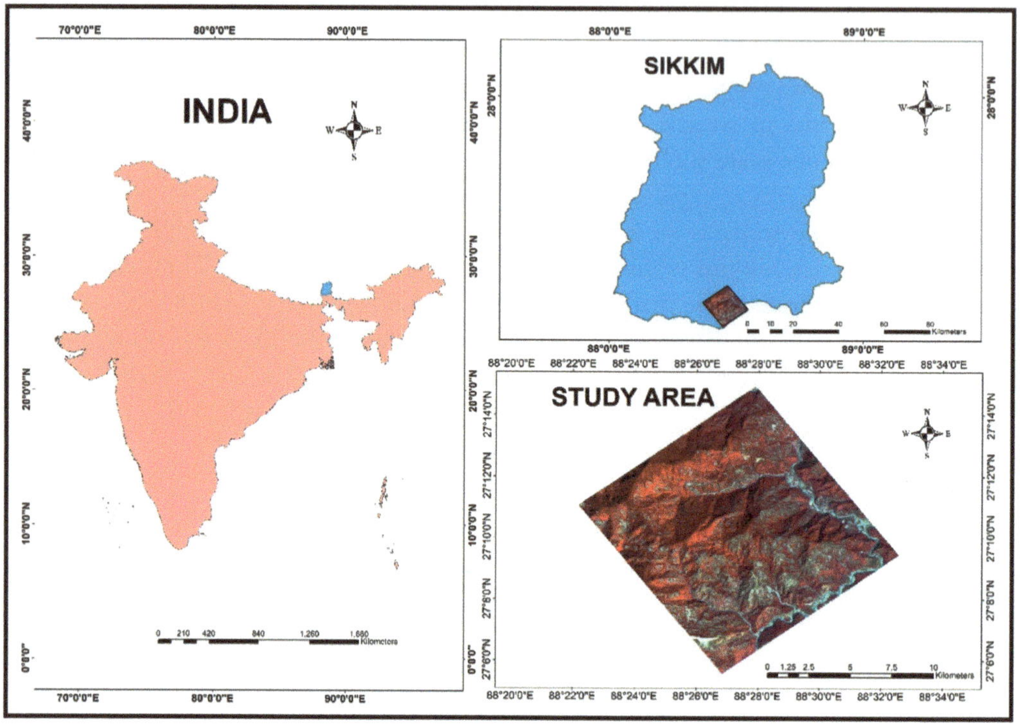

Fig. 23.1 ***Location of the study area.***

nature and less voluminous because it originates from seasonal rain and semipermanent snow field.

Fine loamy sand to silty clay loam soil is concentrated in South Sikkim where 80% of the majority contains slate and phyllite. The barren, steep slope are involving of thin and slack rough soil, at the same time as thickly vegetated areas having moderately solid soils that are rich in organic matters. The location map of the study area is shown in Fig. 23.1.

23.2.2 Data used

In the present study, VRICON Ture Ortho Image and multispectral satellite image of Landsat 8 were used to create landslide inventory for Namchi, South Sikkim. From these images, data were extracted that is useful for creation of landslide inventory. However, Bhuvan portal was also used to acquire Carto DEM data for the extraction of elevation and slopes. Soil data were also obtained from ancillary maps. Arc-GIS software was used for the visual interpretation of satellite images, digitization, digital image processing, and mapping. Results were obtained from AHP and weightage overlay analysis. The type of data used in the present study is mentioned in Table 23.1.

Table 23.1 Data use and source.

Data type	Data source	Sensor	Resolution	Data derivative
Image	DRDO	VRICON Ture Ortho Image	50 CM	LULC, Photo-Lineament
	USGS earth explorer	Landsat 8	30 M	Roads
DEM	NRSC, Bhuvan	Cartosat 1	30M	Slope Aspect Drainage Relative Relief Curvature
Ancillary	Geological survey of India	Geology Map		Geology map Lithology map Landslide Point
	NATMO			Soil map

23.3 Methodology

In the present study, AHP and weighted overlay analysis (WOA) techniques have been used and a comparative study has been done by using both methods. The first step of the study was to prepare landslide causative factors layers, In the present study, causal factors derived from remote sensing imagery are land/ use land/cover map. slope, aspect, and relative relief maps were derived from DEM, LULC, and photo-lineaments were extracted from Vircon True Ortho Image. The detailed flow chart of the methodology adopted in this study is shown in Fig. 23.2.

In this study, distance to photo-lineament and distance to drainage layer was used as causal factors. Furthermore, geology and soil maps were acquired from ancillary data, these maps were georeferenced with satellite imagery derived in GIS environment and vector layers were created. The detailed flow chart of the methodology applied in this study is mentioned below.

23.3.1 AHP method

There are various methods for landslide susceptibility mapping such as fuzzy logic, frequency ratio, AHP, weighted overlay analysis, etc. The present study applies two methods, that is, the AHP and weighted overlay analysis. The AHP method was firstly created by Saaty. AHP is a multiobjective, multicriteria semiqualitative method used for decision-making approach which enables the user to arrive at a scale of preference drawn from a set of alternatives it includes matrix-based pair-wise comparison of the effect of each factor responsible for landsliding.

In this method, each layer is broken into smaller factors, and then these factors are compared based on their importance, each factor is rated against every other factor by assigning a relative dominant value between 1 and 9 where one indicates that the two

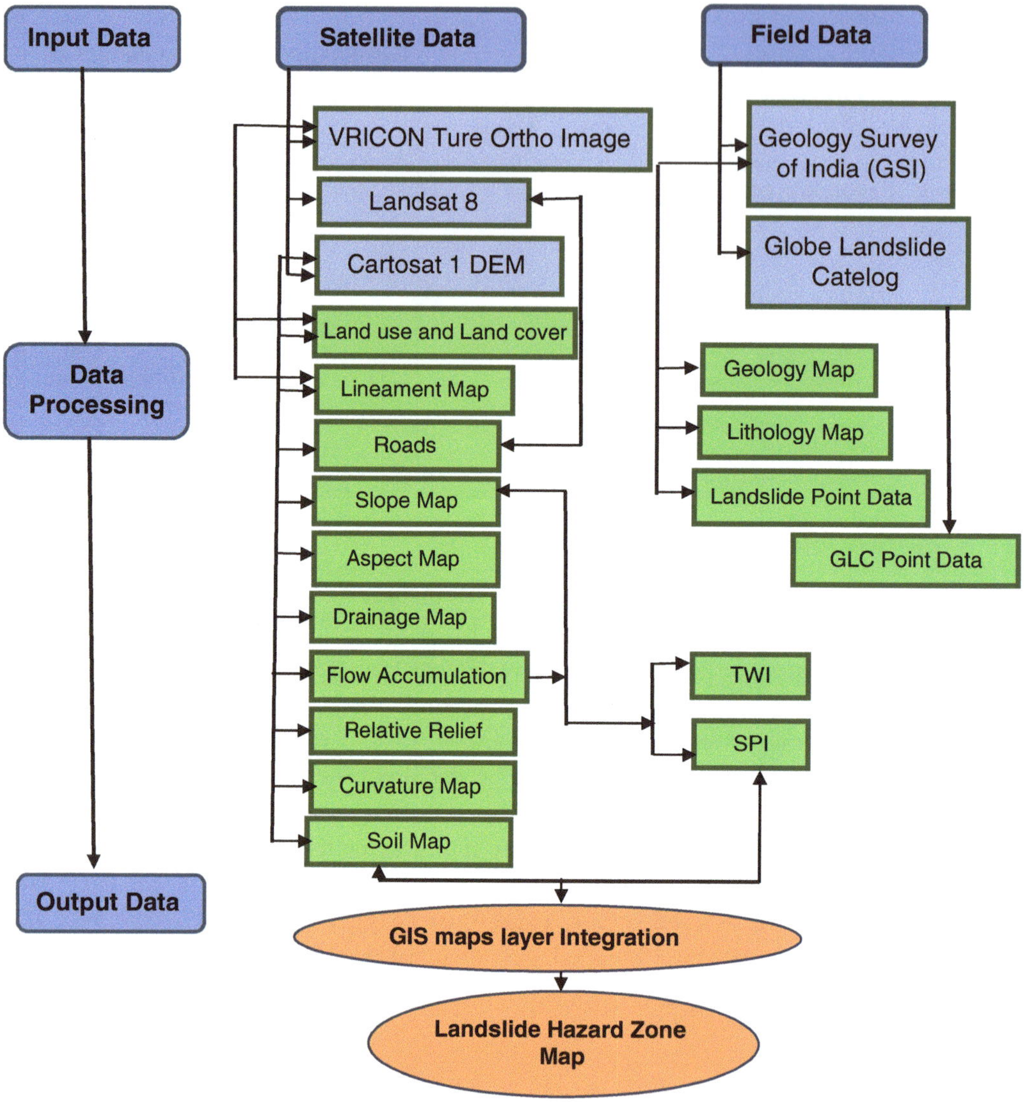

Fig. 23.2 ***Flow chart of methodology.***

factors are equally important while 9 represents that one factor is much more important than another. At the point when the factor at the vertical pivot is more basic than the factor on the horizontal axis, this worth changes among 1 and 9 Conversely, the value varies between the reciprocals ½ and 1/9 [11].

The scale of every perimeter is determined by its importance, the criteria are as followed (Table 23.2) where 1 being Equally Important, 3 being Moderately important, 5 being Strongly Important, 7 being Very Strongly Important, 9 being Extremely Important 2,4,6,8. Intermediate Important. In AHP for checking consistency in matrix

Table 23.2 Scale represents preference of judgement.

Preference/ordinal scale	Degree of preference	Remarks
1	Equally	Factors inherit equal contribution
3	Moderately	One factor moderately favored over other
5	Strongly	Judgement strongly favors one over other
7	Very strongly	One factor very strongly favored over other
9	Extremely	One factor favored over other in highest degree
2,4,6,8	Intermediate	Compensation between weights 1,3,5,7 and 9
Reciprocals	Opposite	Refers inverse comparison

Reference: [17]

Table 23.3 Random consistency index.

N	1	2	3	4	5	6	7	8	9	10	11	12	13	14	15
RI	0	0	0.58	0.9	1.12	1.24	1.32	1.41	1.45	1.49	1.51	1.53	1.56	1.57	1.59

Reference: [11].

the consistency ratio (CR) is used which is dependent on number of parameters. The CR is obtained by comparing the consistency index (CI) with the average random consistency index (RI) which was compiled by Satty [11] as shown in (Table 23.3). CR value of 0.1 represents the maximum threshold of consistency of the matrix. CR value > 0.1 is thought to be inconsistent whereas value 0 indicates perfectly consistent comparison result.

23.3.2 Weighted overlay analysis method

Landslide incidence is decided from landslide associated element and the destiny landslide can occur within the same condition with past landslide. Based on the belief using probability approach, the relationship among regions with landslide occurrences and landslide-associated features can be illustrious from the association among place short of occurrences of landslide and landslide-connected elements. To current the difference quantifiable, the weightage overlay technique turned into used for this study.

In this method the factors and their classes were assigned numerical value on a scale between1 to 9 based on the importance, this may be done by assigning a rating scheme. A rating scheme was developed based on the associated causative factors for landslides surveyed in the field and on the knowledge by expertise on landslide causes from previous work [12,19,20]. In the present study, the weights were assigned between 0 to 9, where the highest weight indicates greater influence for landslide occurrence.

The WOE is a simple, direct, and adequate methodology for evaluations of potential landslide area [21,22].

In this study various environmental factors were used for landslide susceptibility mapping, in this process, each cell from each map layer is reclassified into a common preference scale, such as 1–9, with 9 being the most favorable. All thematic maps were integrated using weighted overlay model into GIS platform and landslide susceptibility map was prepared.

23.3.3 TWI index

Topographical wetness index (TWI) has been frequently adopted in landslide susceptibility mapping. In the present study of Namchi, south Sikkim area, this method has been utilized has been utilized. Topographic wetness index (TWI) is a steady-state wetness index; that describes topography's influence. It identifies the water saturation zones in an area when there is a high catchment area with a low slope gradient [13]. It is defined as:

$$\text{TWI} = \text{TWI ln}\left(\text{Flow Accumulation/Slope}\right.$$

In which CA stands for catchment region and slp for slope gradient. TWI index has been computed in Arc GIS 10.1 software and result is generated. The TWI map is shown in Fig. 23.3.

23.3.4 SPI index

The stream power index (SPI) represents the erosive power of the streams. The slope gradient in the catchment area and the streams' erosive power show a positive relationship. According to (Moore et al.,1988 and [23]). SPI is calculated with this method given beneath

$$\text{SPI} = \ln\left(a * \tan\beta\right)$$

where ln = Natural logarithm; a = Flow accumulation/specific catchment area of each cell; β = Slope gradient in degrees. The SPI map is shown in Fig. 23.4.

23.4 Input variables

23.4.1 Geology map

The geology map is prepared with the help of geological and mineral maps of Sikkim which are georeferenced with satellite image. The observation location is split into specific geological units with different lithological formations. There are two types of formation Reyang Formation and Gorubathan Formation 85% of area cover with Reyang formation. Gorubathan Formation consists of green slate, phyllite, quartzite, pyritiferrous black slate, and basic meta-volcanic. Reyang formation consists of thick-bedded ortho-quartzite, black slate to ([10].)

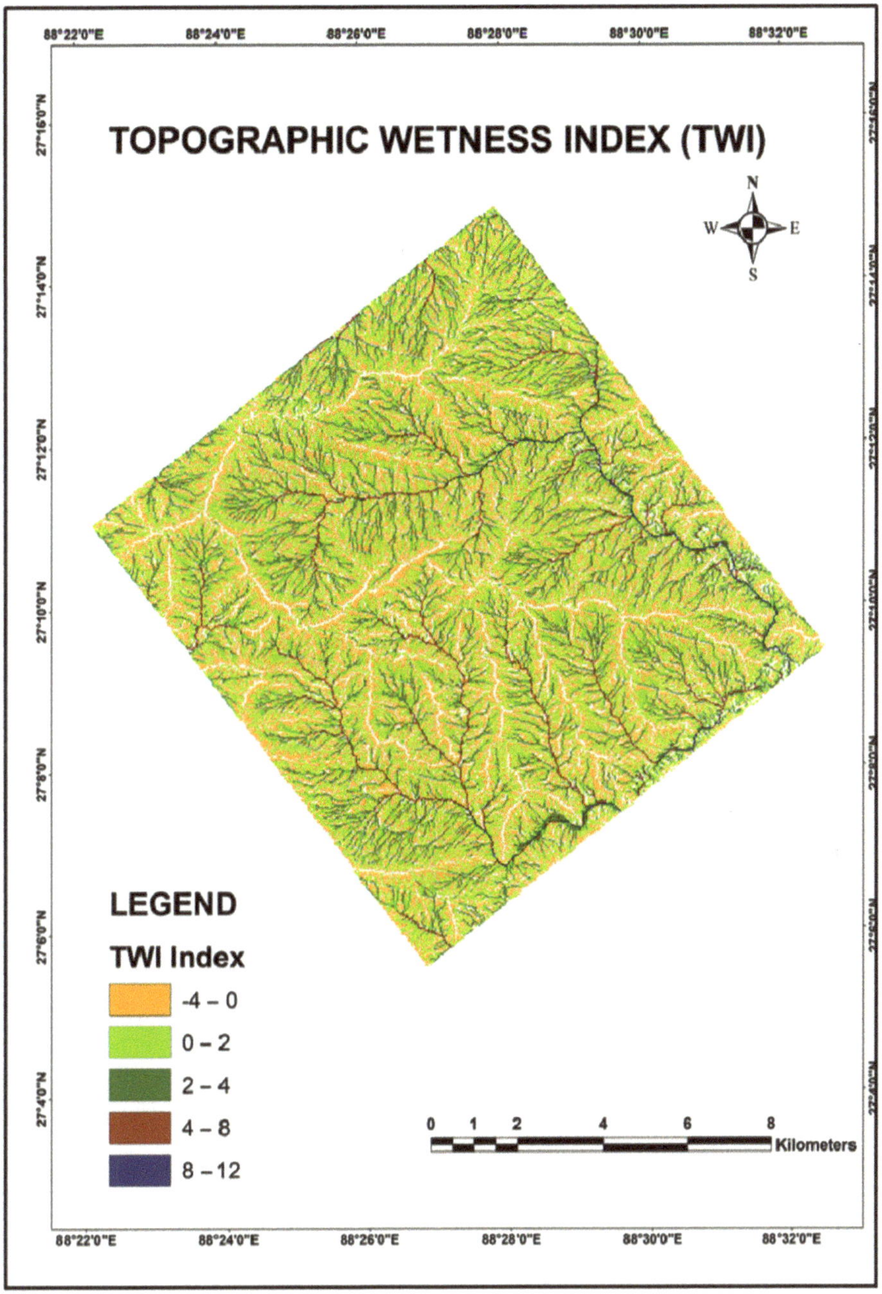

Fig. 23.3 *Topographic wetness index map.*

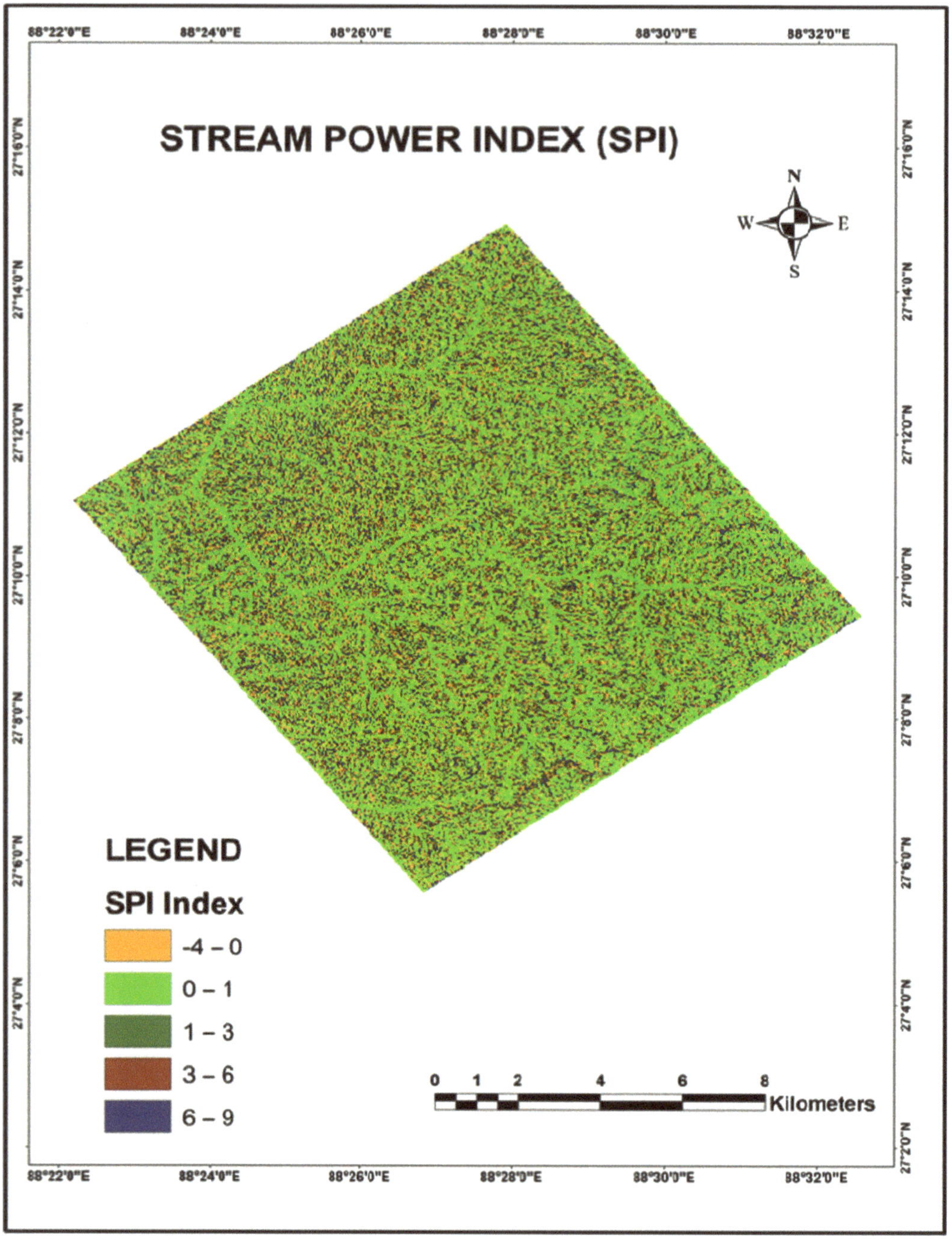

Fig. 23.4 *Stream power index map.*

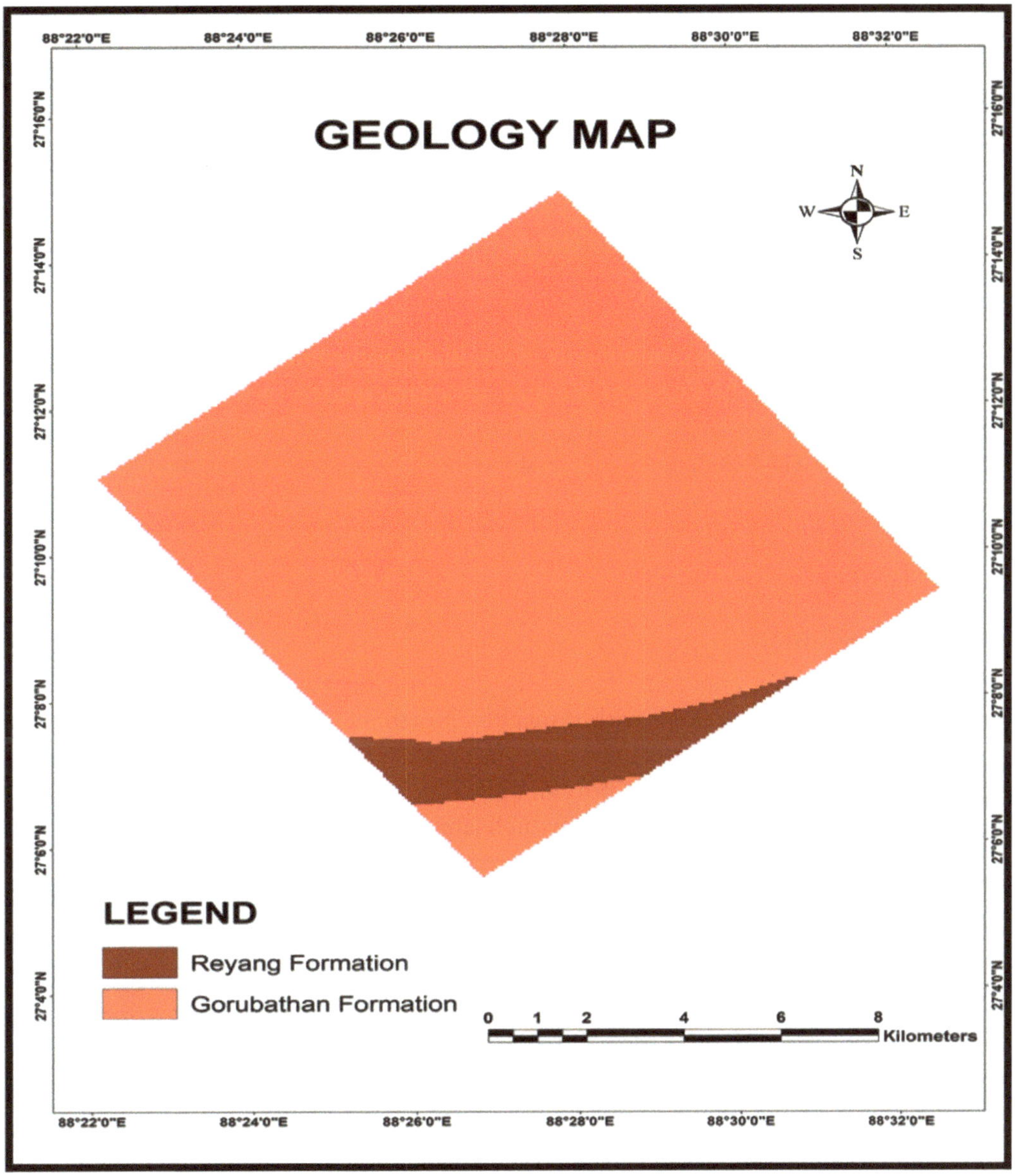

Fig. 23.5 ***Geology map of the study area.***

It has been reported that the Reyang subgroup consists of the conformable bands of mafic meta-volcanic with some relict igneous textures defined by the primary pyroxene and plagioclase grains and some foliated amphibolites in more altered regions. The geology map of the study area is shown in Fig. 23.5.

23.4.2 Soil map

The soil of Namchi main constituted of coarse loamy, fine loamy, loamy skeletal weak soil, and loamy skeletal weathered soil. These soils are prone to sheet erosion, gully erosion, and creeping. In the case of soil texture, the landslide occurrence probability value is higher in rocky sandy loam and sandy loam, and is lower in fine sandy loam, silt loam, gravelly silt loam, and loam [14]. The description about the soils of the region is mentioned in the below map. The soil map of the study area is shown in Fig. 23.6.

23.4.3 Slope map

Slope is one of the essential components of landslide occurrences as it shows the degree of stability of an area. The slopes of the area are characterized in terms of degrees, and are divided into five slope types, namely, steep (> 60°), high (30°–40°), moderately high (20°–30°), gentle (10°–20°), and rolling slopes (0–10). It is said that steep slope areas are highly prone to the downfall of debris causing great damage to life and property. As the slope angle increases, shear stress in soil or other unconsolidated material generally increases as well [14]. The slope map of the study area is shown in Fig. 23.7.

23.4.4 Aspect map

Aspect is one of the important factors in preparing a landslide susceptibility map. Aspect is the direction of slope with respect to North, it generally shows the direction of mountain slope, the direction ranges from 0° to 360°, wherein 0° is north-bound, 90° is east-bound, 180° is south-bound, and 270° is west-bound. Exposure to solar radiation, rainfall and discontinuities which are related parameters of aspect might control the occurrence of landslides [10]. Generally, west-facing slope shows the hottest time of the day in the afternoon, and in most cases, a west-facing slope will be warmer than sheltered east-facing slope [24]. The aspect map of the study area is shown in Fig. 23.8.

23.4.5 Land use and land cover map

The land use and land cover map of the study area is an important factor in controlling the landslides. The complex physical processes acting on the surface of the earth can be determined from the land-use pattern [15]. The study area is divided into six land use/land cover classes, agriculture land, barren land, build-up, fallow land, forest and water bodies. The land use/land cover with vegetation has shown that the woody type plants with large root systems have helped in the improvement in the slope stability in an area. Settlement area is seen to lie in some small pocket of the study area; it is the most fragile class as compared to other classes. Barren rocky surface, highly developed areas with network of roads, increases the probability of landslide occurrence in the study area (Source: Central Groundwater Control Board). the land use land cover map is shown below in Fig. 23.9.

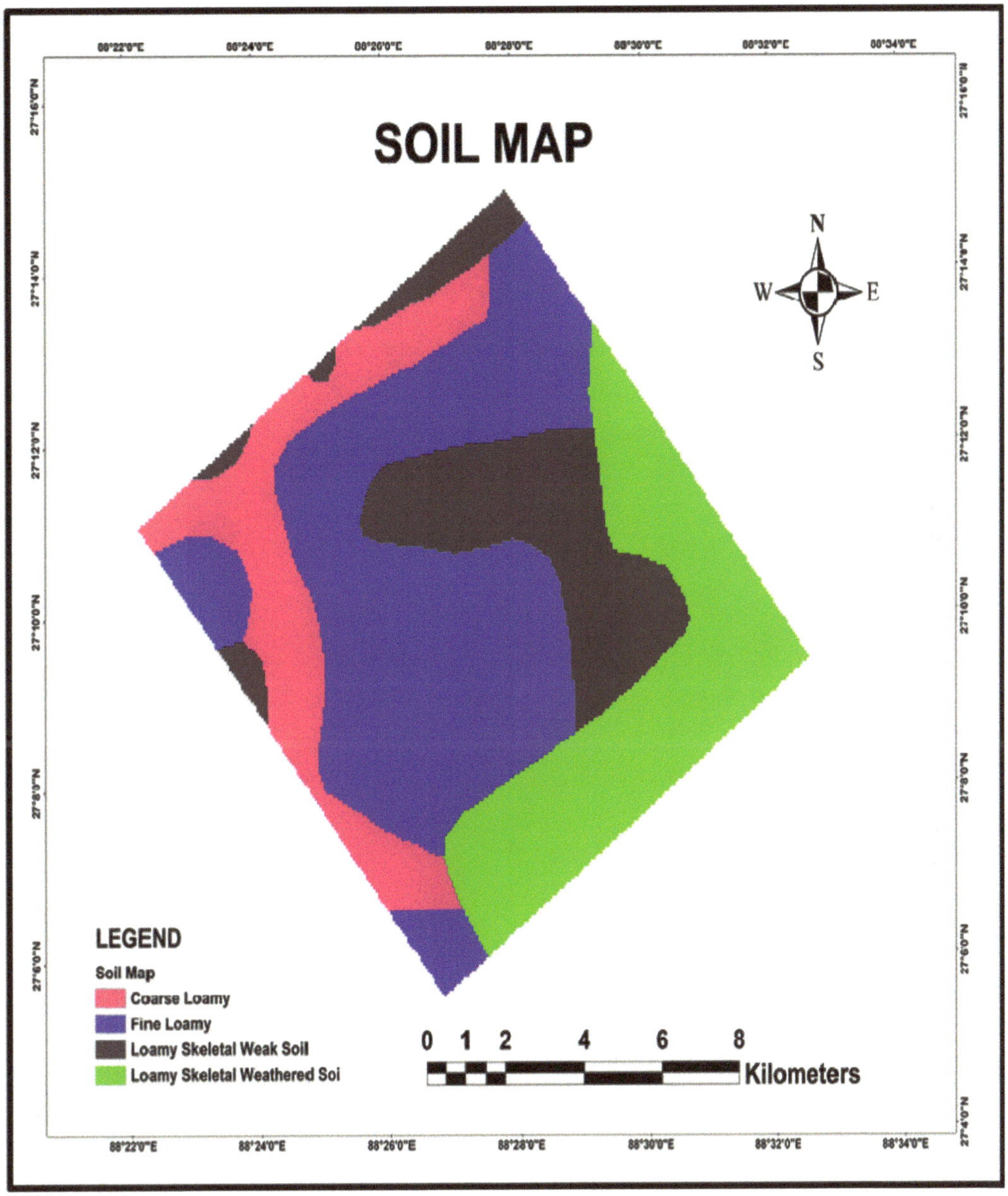

Fig. 23.6 ***Soil map of the study area.***

23.4.6 Distance from the road map

The distance from roads is one of the most important factors in determining the future occurrence of landslides roads change the nature of topography and decrease the shear strength of toe of slope and cause the tensile stress [16]. The construction of

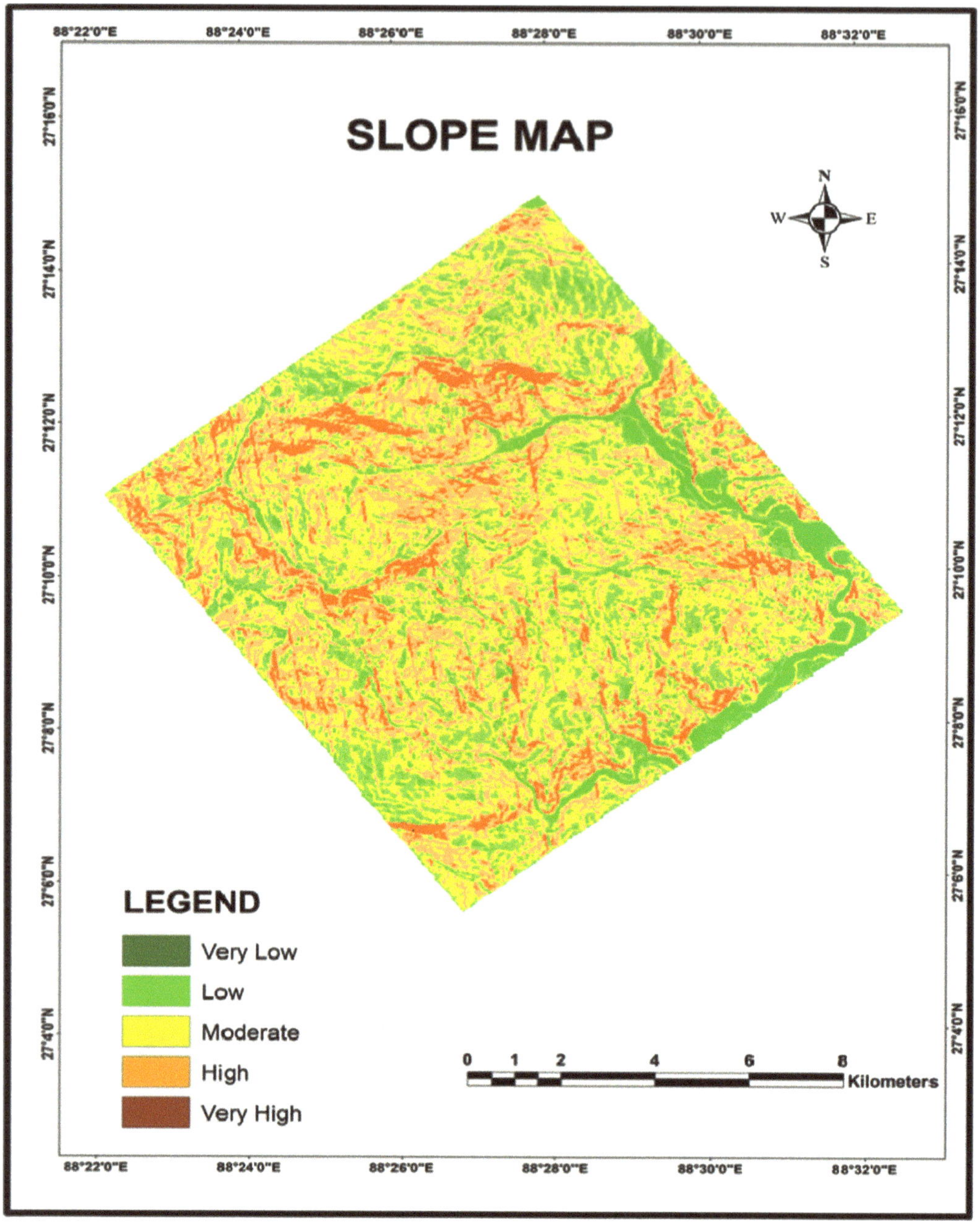

Fig. 23.7 ***Slope map of the study area.***

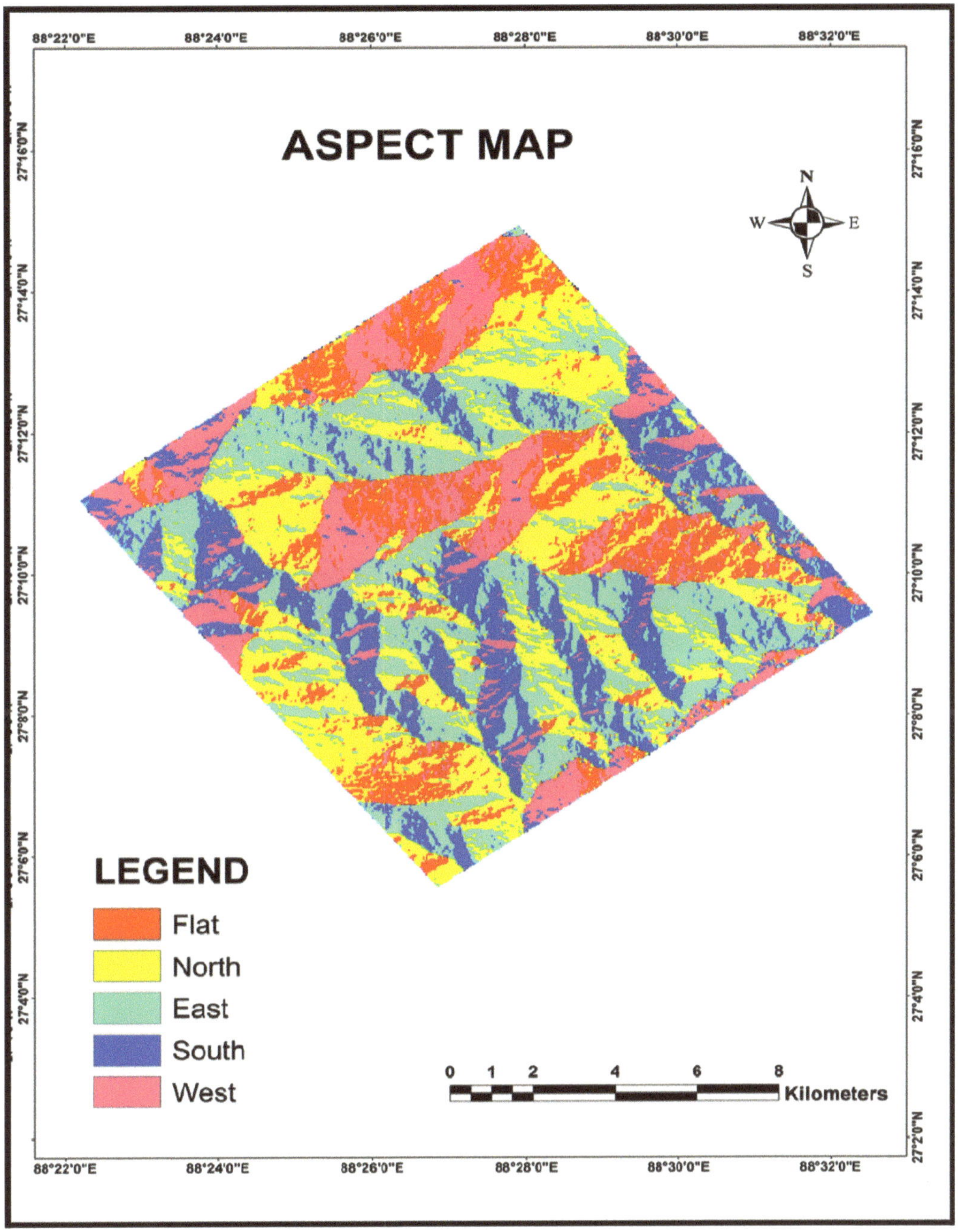

Fig. 23.8 *Aspect map of the study area.*

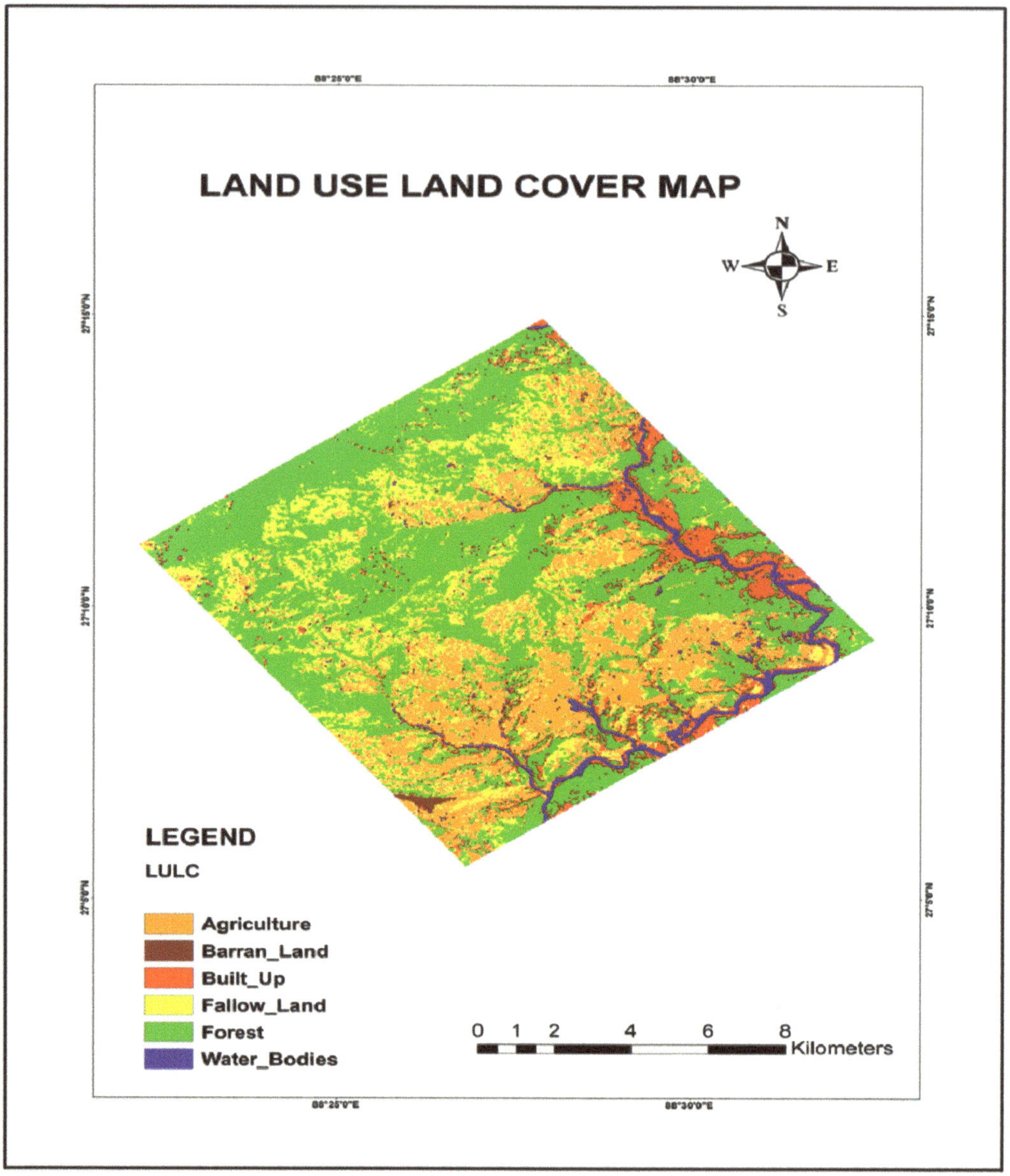

Fig. 23.9 ***Land use and land cover map.***

roads reduces the stability of slope, roads construction leads to infiltration of water in slope thus creating stress because of traffic loads. In this study main roads are observed and this factor is classified into five distances to road categories, which are shown in Fig. 23.10.

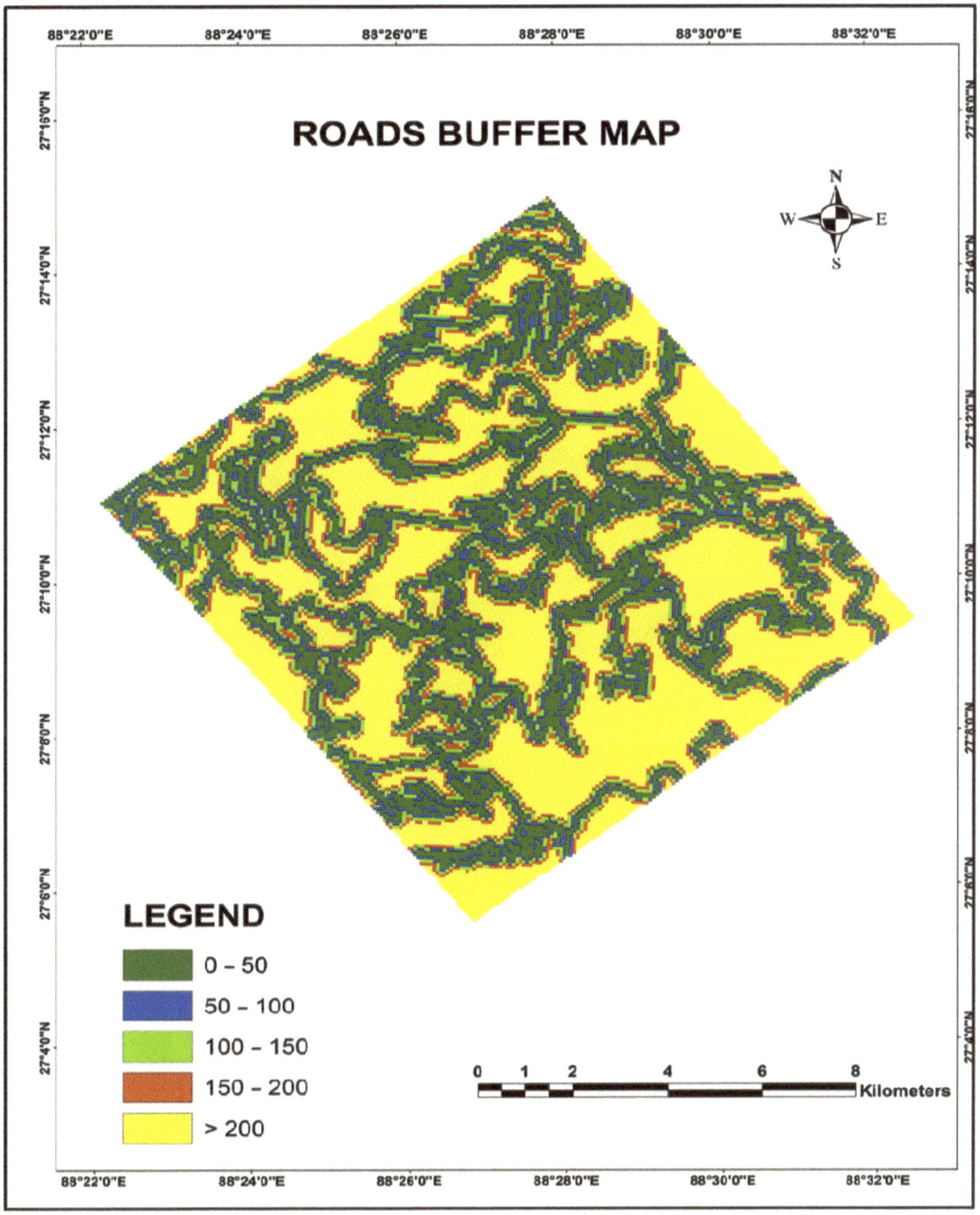

Fig. 23.10 *Road buffer map.*

23.4.7 Distance from drainage map

Drainage is one of the significant aspects that manipulate the landslide as it is closely related to the nature of the soil and its geotechnical properties. The closeness of slope to drainage may harm the stability of slope. Stream increases the potential of landslide by

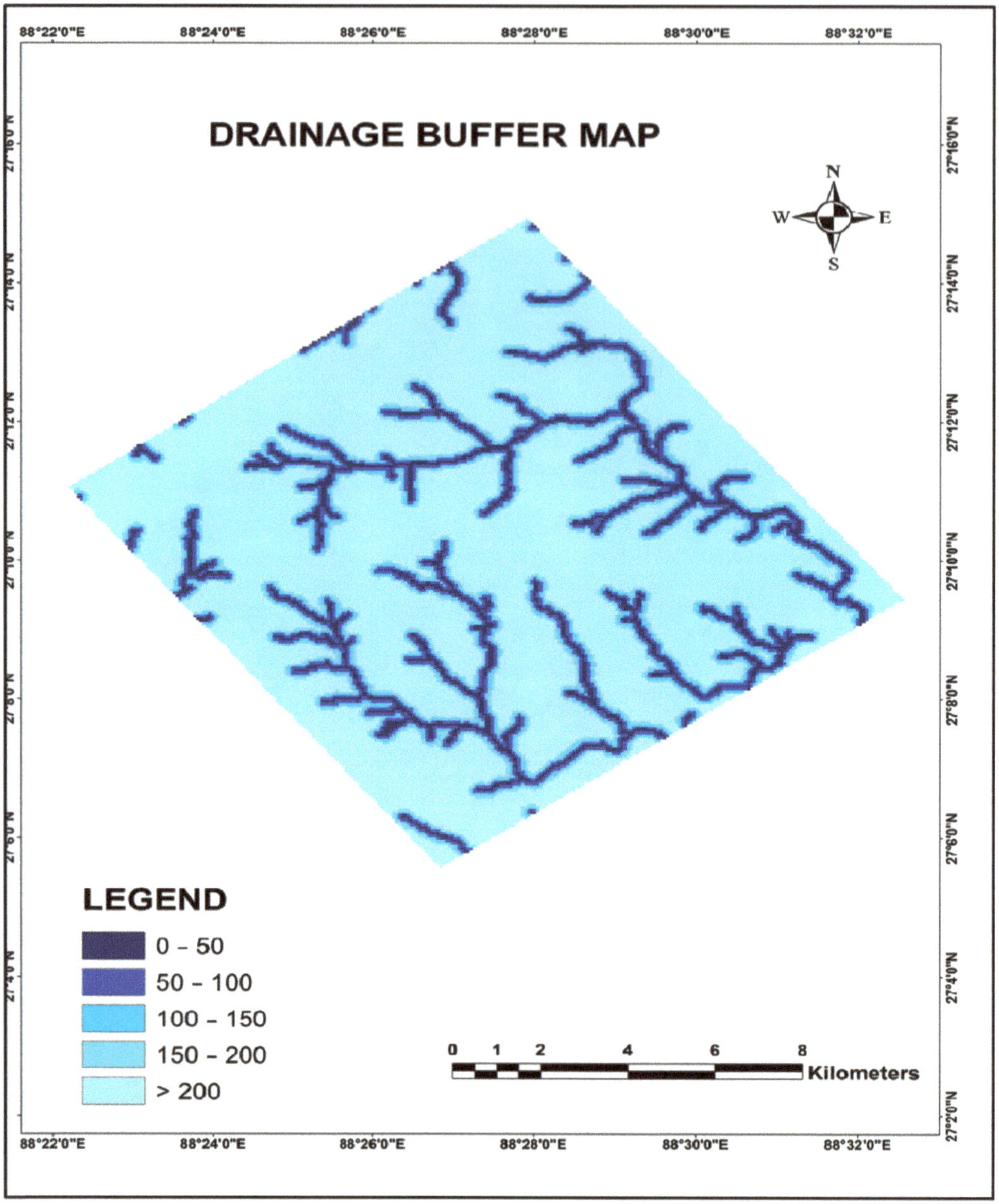

Fig. 23.11 ***Drainage buffer map.***

erosion of the slope and increasing the degree of saturation of materials of slope [16]. To analyze the relation between landslides and drainage, a buffer map with distance from drainage is created, five exclusive buffer regions had been created within the observe area which are 50 m, 100 m, 150 m, 200 m, and more than 200, the distance from drainage is shown in Fig. 23.11.

23.4.8 Relative relief map

Relative relief determines the change among the maximum and minimum altitudes within a certain cell. In this study, relative elevation is derived from the digital elevation data. It is observed that the elevation differs from 300 m in south to 2100 m in north portion of the study area. The maximum top elevation is found within the north -western part of the study area which is shown in Fig. 23.12.

23.4.9 Distance from lineament map

Lineaments refer to linear geological discontinuities, distance from lineaments is one of the main factors responsible for triggering the landslides as rocks near the lineaments are weak. It is widely observed in the Himalayas that landsliding phenomena are particularly severe close to regional geological lineaments [25,26]. As the distance from the lineament increases the chances of occurrence of landslide decreases the distance from the lineament map is produced by creating six buffer zones, that is, 50–100 m, 100–150, 150–200, and more than 200 buffer zone which is shown in Fig. 23.13.

23.4.10 Lithology map

Lithology plays an important role in landslide susceptibility of an area; landslide events are closely associated with the lithological settings of a particular area. The landslide phenomenon a part of the geologic studies and research, is related to the lithology and weathering properties of the material of the land [16]. Lithological variations often lead to a difference in the strength and permeability of rocks and soils [18]. There are eight types of rock present in the area among which schist and phyllites are the weaker rocks, the foremost lithology is schists. Most of the study is covered by Schist Type of rock formation, this rock is a sort of metamorphic rock which is chiefly produced from flaky minerals and shows a plated shape known as schistocyte. The lithology of the study area is mentioned in Fig. 23.14.

23.5 Results and discussion

Landslide hazard zonation has been delineated by opting AHP technique and weightage overlay analysis for Namchi district of Sikkim. In order to achieve these eleven landslides-inducing factors were taken into consideration, these factors are geology, slope, aspect, LULC, distance from road, distance from drainage, relative relief, curvature, distance from lineaments and lithology. These factors were evaluated by assigning weights to each factor using both techniques, all the layers were combined and landslide susceptibility maps from both the techniques were obtained. The classification of the LSZ maps is divided into Five different classes: I. Very Low Susceptibility, II. Low hazard Zone, III. Moderate hazard Zone, IV. High hazard Zone, V. Very high hazard Zone. In this study, the CR is

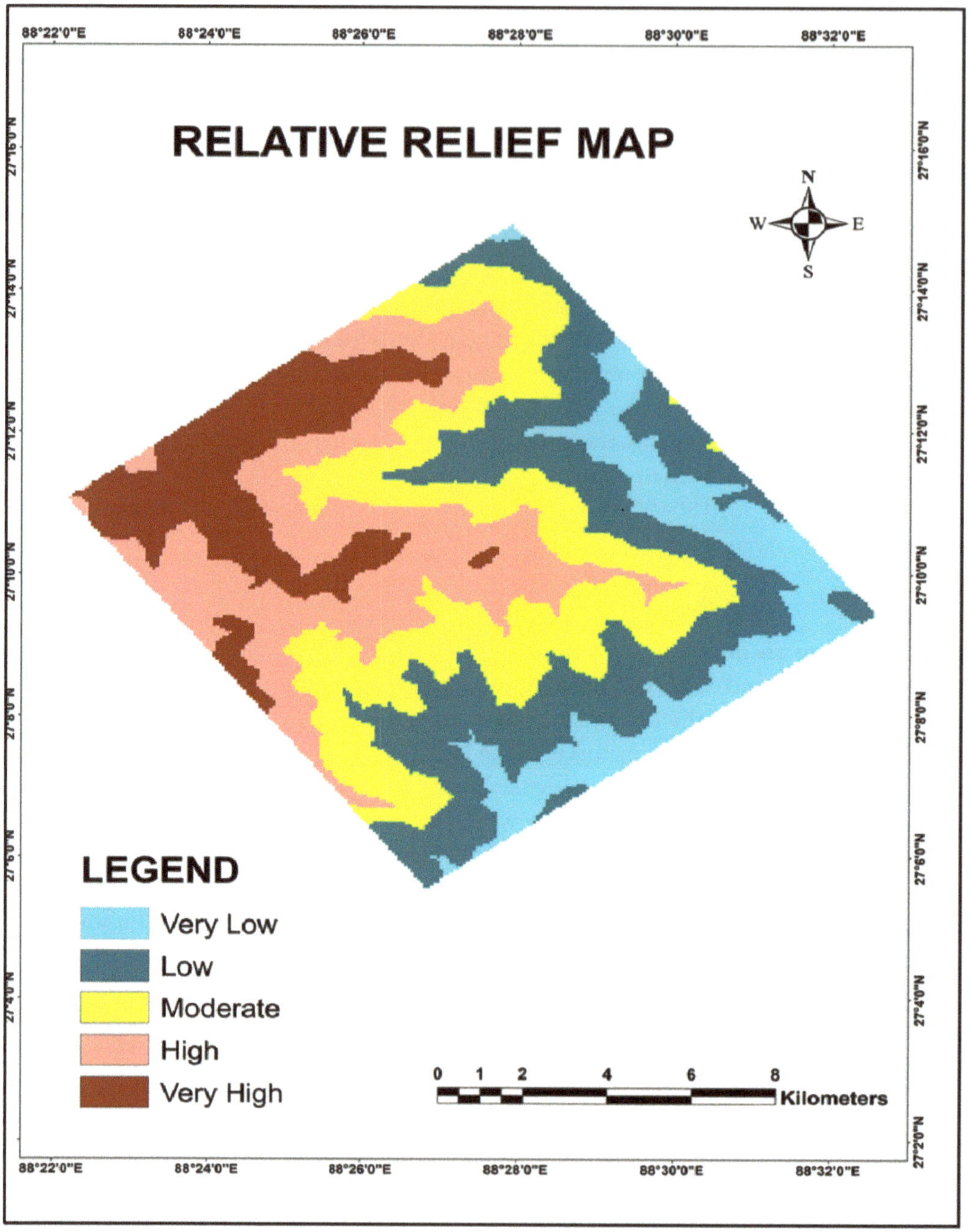

Fig. 23.12 *Relative relief map.*

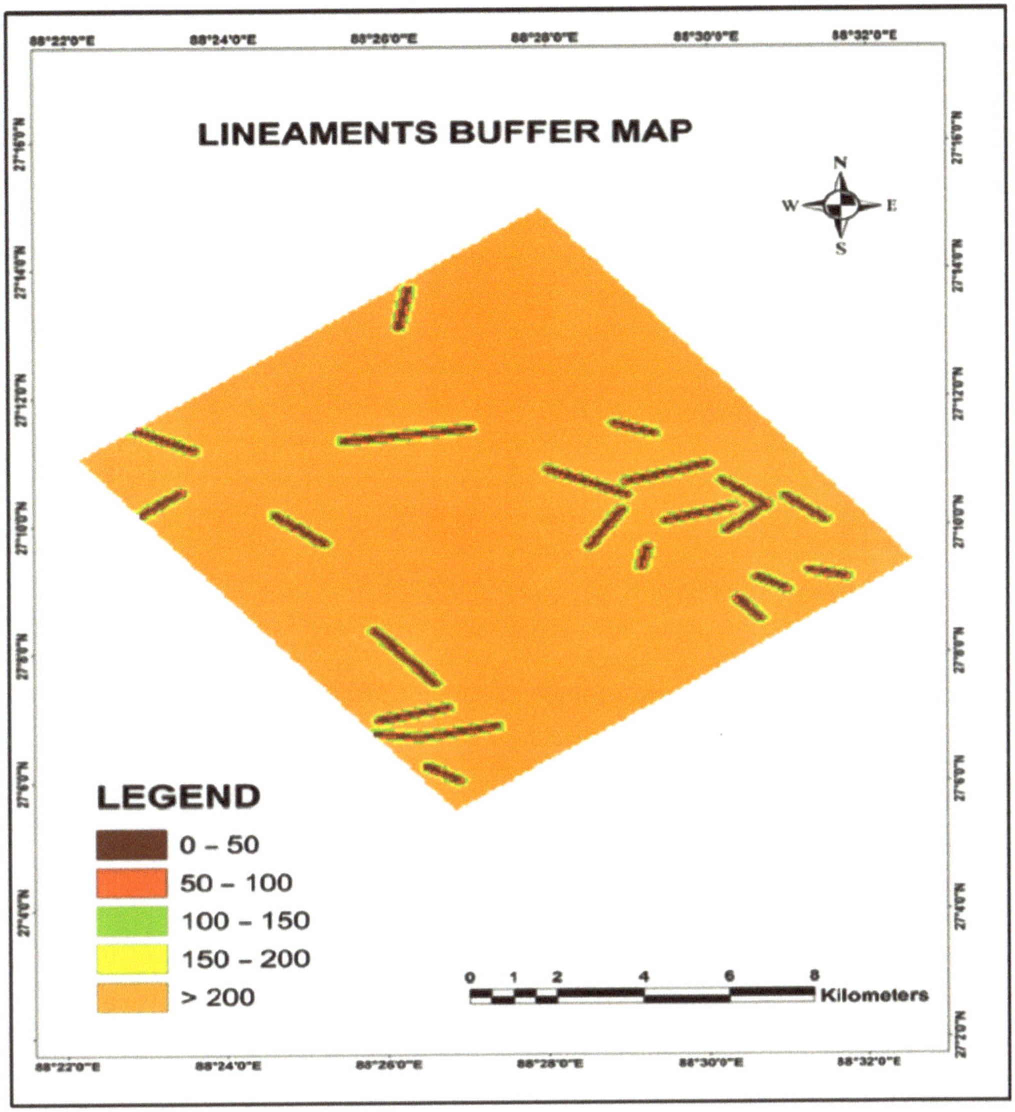

Fig. 23.13 ***Lineaments buffer map.***

0.092 the ratio specifies an appropriate degree of consistency. The AHP pairwise comparison matrixes are shown in Table 23.4. In respect to AHP, Weightage overlay analysis takes account weights assigned to every single class that are mentioned in Table 23.5. The weights indicate that the land use and landcover is an important parameter, most landslide falls in agricultural land and comes under very high and high susceptibility. Another class of land use cover comes under moderate and low. In case of lithology, mostly high susceptibility landslides are occurring in Boulder slate lithology and Low in

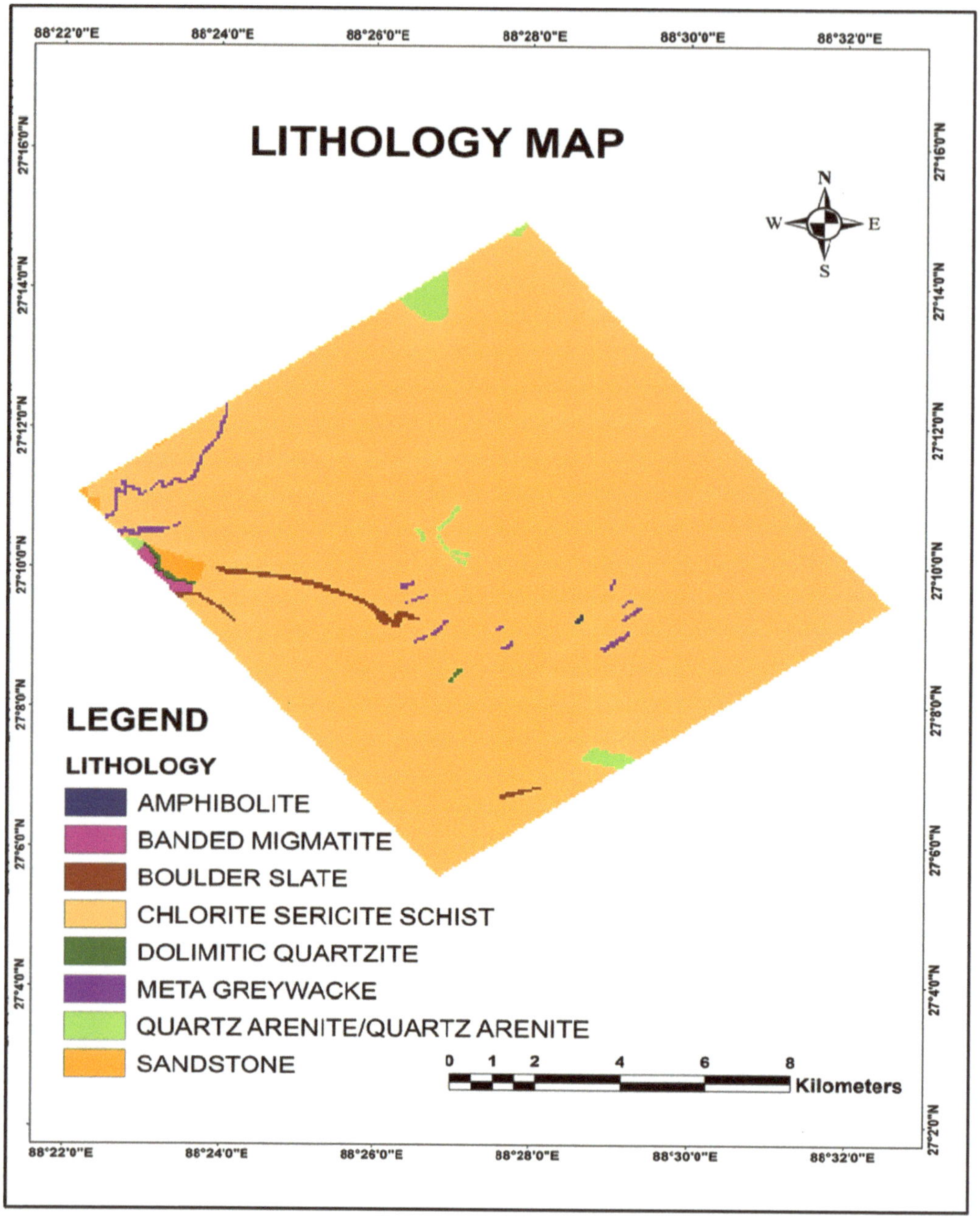

Fig. 23.14 *Lithology map.*

Table 23.4 Pairwise comparison matrix, weightage and consistency ratio of the data layer.

Factors	1	2	3	4	5	6	7	8	9	10	11	12	13	Weightages
Geology	1													0.032
soil	3	1												0.09
Lithology	3	2	1											0.133
LULC	3	3	1/2	1										0.104
Lineament	3	4	4	4	1									0.209
Drainage	4	1/4	1/3	1/2	1/3	1								0.062
Slope	4	1	1/3	1/2	1/3	1	1							0.068
Aspect	4	1/3	1/2	1/3	1/3	1	1	1						0.053
Relative Relief	1	1/2	1/4	1/4	1/6	1/3	1/3	1/4	1					0.032
TWI	2	1	1	2	1/3	4	4	2	3	1				0.114
SPI	1/3	1/2	1/2	1/6	1/4	1/5	1/5	1/2	1/4	1/3	1			0.02
Curvature	1/4	1/3	1/2	1/4	1/5	1/5	1/6	1	1/3	1/4	3	1		0.025
Roads	4	1/4	1/3	1/2	1/3	1	1	1	3	1/2	5	2	1	0.058

Principal eigen values: 14.72
Consistency ratio (CR): 0.092

Table 23.5 Weightage and rating.

Thematic Raster Layer	Field	Rates	Influence
Geology	Reyang Formation	1	7
	Gorubathan Formation	1	
soil	Loamy, Shallow Depth Soil	1	9
	Loamy, Coarse Loamy Rock Soil	2	
	Fine Loamy Mixed Soil	3	
	Coarse Loamy	4	
Lithology	Dolimitic, Quartzite	1	11
	Sandstone	2	
	Quartz Arenite, Black Slate	3	
	CHLORITE SERICITE SCHIST	4	
	BOULDER SLATE	5	
	BANDED MIGMATITE, MICA SCHIST	6	
	AMPHIBOLITE	7	
	META GREYWACKE	8	
Land Use and Land Cover	River	1	14
	Forest	2	
	Barren Land	2	
	Agricultural Land	3	
	Fallow land	6	
	Built-Up	7	
Distance From Lineament	0–50	1	13
	50–100	2	
	100–150	3	
	150–200	4	
	> 200	5	

(Continued)

Table 23.5 *Cont'd*

Thematic Raster Layer	Field	Rates	Influence
Distance From Drainage	0–50	1	
	50–100	2	
	100–150	3	
	150–200	4	
	> 200	5	
Slope	0–10	1	10
	10–20	2	
	20–30	3	
	30–40	4	
	> 60	5	
Aspect	Flat	1	5
	North	2	
	West	3	
	South	4	
	East	5	
Relative Relief	Very Low	1	4
	Low	2	
	Moderate	3	
	High	4	
	Very High	5	
TWI	-4–0	1	7
	0–2	2	
	2–4	3	
	4–8	4	
	8–12	5	
SPI	-4–0	1	2
	0–1	2	
	1–3	3	
	3–6	4	
	6–9	5	
Curvature	-21–2	1	5
	2–6	2	
	6–8	3	
	8–12	4	
	12–16	5	
Distance From Road	0–50	1	6
	50–100	2	
	100–150	3	
	150–200	4	
	> 200	5	

chlorite sericite schist lithology. In the soil map most of landslide observe in fine loamy soil. The maximum number of landslides observed in the 50 -100 m class of drainage distance because most of landslide activities are found in that area. Similar case is distance from road, mostly landslide are falls in 50–100 m buffer zone. The slope aspect is also a very important factor that is responsible for inducing landslides, typically very high zone is discovered around the steep slope, roads, agricultural and cliffs. Two different kinds of techniques are used for Landslide hazard zonation for the given study area. The obtained results indicate that the outcomes are similar, little variation is observed. Results from both the methods are shown in Fig. 23.15. The total area of landslide hazed zonation map in square kilometres with particular hazard classes are shown in Tables 23.6 and 23.7. Table 23.6 shows results obtained from AHP technique and Table 23.7 shows results obtained from the weightage overlay analysis technique.

Validation of the result is done to get the accuracy of landslide hazard zonation mapping. For the accuracy of the result reference data from deferent sites has been obtained. Accuracy of the results are totally based on input data, and work experience, size of the area. In this research, accuracy of landslide susceptibility map has been done by considering existing landslide zone, which are classified from the global landslide

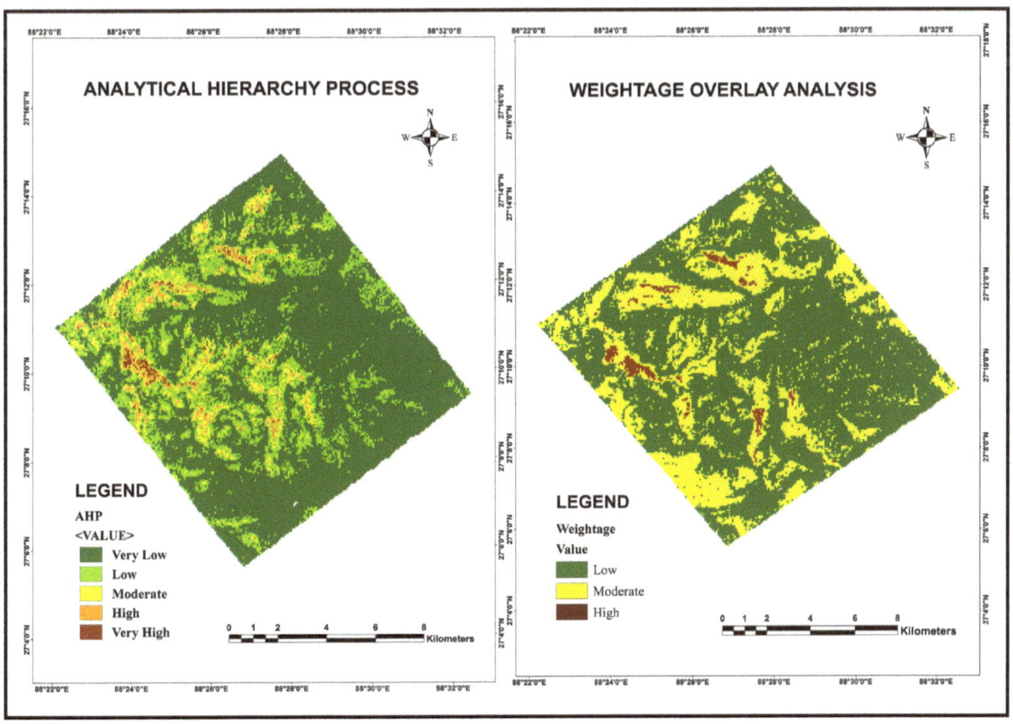

Fig. 23.15 ***Landslide hazard zonation map.***

Table 23.6 Area of landslide hazard map (analyze hierarchy modal).

Susceptibility class	Area Sq.(km)	Area percentage
Very low	0.014871	0.0098
Low	99.53824	65.95
Moderate	34.19396	22.65
High	11.25258	7.46
Very high	4.654701	3.08

Table 23.7 Area of landslide hazard map (weightage overlay analysis).

Susceptibility class	Area Sq.(km)	Area percentage
Low	101.4	67.17
Moderate	47.79	31.558
High	4.8	3.17

catalogue (GLC) and also from landslide susceptibility zone point data from Geology Survey of India. Comparison of the results with reference data 80% accuracy is achieved in AHP while 85% accuracy is achieved in weight overlay technique. The results almost match with the reference data which are shown in Fig. 23.16. The factors for the AHP are limited up to some extent because of the availability of datasets and the time frame of the study. Since the prediction of landslides is considered to be valuable for land use planning and assessing the hazard and risk. Therefore, it is essential to choose a model with high prediction capabilities.

23.6 Conclusion

The landslide susceptibility map for Namchi district, Sikkim was produced using AHP and Weighted Overlay Analysis research has been conducted to analyze and map the landslide activity of Namchi Sikkim area. Eleven important causative factors of landslide occurrence were considered for the construction of thematic data layers using GIS. The numerical weight for each factor was assigned by AHP. The landslide susceptibility indices were calculated using weighted overlay technique, and finally, landslide susceptibility map was produced using GIS. The landslide susceptibility map classified the study area five categories of landslide susceptibility zones that is very low class, low class, moderate class, high class, and very high class. The map obtained from Weightage overlay analysis classified study area into three classes low class, moderate class, and high class. Results obtained from both techniques indicates high susceptibility in western part and low in southern pert. Results indicates that AHP and weightage method are highly correlated, 80% accuracy is achieved from AHP technique whereas 85% accuracy is achieved from weighted overlay analysis technique thus indicating the effective usefulness of geospatial technique in future landslide prediction and management.

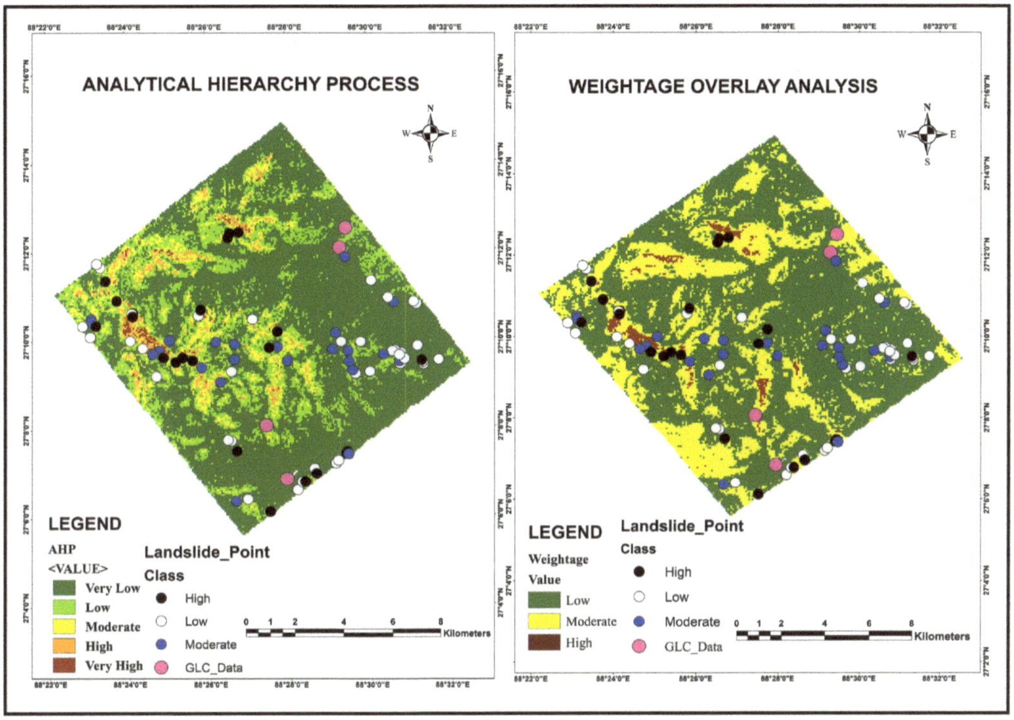

Fig. 23.16 ***Landslide hazard zonation map with reference of Global Landslide Catalog Data (GLC) and Geology survey of India (GSI, Landslide Point).***

References

[1] s.Abdul Rahaman, Aruchamy Solavagounder, Geoinformatics based landslide vulnerable zonation mapping using analytical hierarchy process (AHP), a study of Kallar river sub watershed, Kallar watershed, Bhavani basin, Tamil Nadu, Modeling Earth Systems and Environment 3 (2017), doi:10.1007/s40808-017-0298-8.

[2] M. Abedini, S. Tulabi, Assessing LNRF, FR, and AHP models in landslide susceptibility mapping index: a comparative study of Nojian watershed in Lorestan province, Iran. Environ Earth Sci 77 (2018) 405.

[3] EJ Gabet, DW Burbank, J Putkonen, B Pratt-Sitaula, T. Ojha, Rainfall thresholds for gearmotor company, IJMVSC 4 (3) (2004) 11–23.

[4] P K Rai, K Mohan, VK Kumra, Landslide hazard and its mapping using remote sensing and GIS, Journal of Scientific Research Banaras Hindu Univesity 58 (2014) 1–13.

[5] P Singh, A Sharma, U Sur, PK. Rai, Comparative landslide susceptibility assessment using statistical information value and index of entropy model in Bhanupali-Beri region, Himachal Pradesh, India, Environ Dev Sustainability (2020).

[6] T. Kavzoglu, Landslide susceptibility mapping using GIS-based multi-criteria decision analysis, support vector machines, and logistic regression, Landslides 11 (2014), 425–439.

[7] R. Anbalagan, R. Kumar, S. Parida, K. Lakshmanan, GIS based post-earthquake landslide hazard zonation mapping of Lachung Basin, Sikkim, Int J Emerg Technol Adv Eng 4 (1) (2014) 431–444.

[8] F. Guzzetti, Landslide hazard evaluation: A review of current techniques and their application in a multi-scale study, Central Italy, Geomorphology (1999).
[9] M. Basharat, H.R. Shah, N. Hameed, Landslide susceptibility mapping using GIS and weighted overlay method: a case study from NW Himalayas, Pakistan. Arab J Geosci 9 (2016) 292.
[10] A. Bera, B.P. Mukhopadhyay, D. Das, Landslide hazard zonation mapping using multi-criteria analysis with the help of GIS techniques: a case study from Eastern Himalayas, Namchi, South Sikkim, Natural Hazards 96 (2019) 935–959.
[11] T.L. Saaty, The analytic hierarchy process: planning. *Priority Setting*, Resource Allocation, MacGraw-Hill, New York International Book Company (1980).
[12] S. Sarkar, An integrated approach for landslide susceptibility mapping using remote sensing and GIS, Photogramm. Eng. Remote Sens. 70 (2004), 617–625.
[13] U. Sur, P. Singh, S.R. Meena, Landslide susceptibility assessment in a lesser Himalayan Road corridor (India) applying fuzzy AHP technique and earth-observation data, Geomatics, Natural Hazards and Risk 11 (1) (2020) 2176–2209, doi:10.1080/19475705.2020.1836038.
[14] S. Lee, J. Choi, K. Min, Probabilistic landslide hazard mapping using GIS and remote sensing data at Boun, Korea, Int. J. Remote Sens. 25 (11) (2004) 2037–2052.
[15] F.C. Dai, Landslide characteristics and slope instability modeling using GIS, Lantau Island, Hong Kong, Geomorphology 42, (3–4) (2002), 213–228.
[16] M. Moradi, M.H. Bazyar, Zargham Mohammadi, GIS-based landslide susceptibility mapping by AHP method, a case study, Dena City, Iran. J Basic Appl Sci Res. 2 (2012) 6715–6723.
[17] T.L. Saaty, A scaling method for priorities in hierarchical structures, J. Math. Psychol. 15 (1977), 234–281.
[18] Y. Achour, A. Boumezbeur, R. Hadji, et al. Landslide susceptibility mapping using analytic hierarchy process and information value methods along a highway road section in Constantine, Algeria. Arab J Geosci 10 (2017), 194.
[19] K. Pareta, J. Kumar, U. Pareta Landslide hazard zonation usingquantitative methods in GIS. Int. J. Geospatial Eng. Technol 1 (1) (2012) 1–9.
[20] D. Ghorai, S. Maiti, A.K. Paul, Landslide susceptibility mapping:a case study of cauvery basin and east flowing rivers south ofcauvery basin, India. J. Remote Sensing GIS 6 (1) (2015) 24–36.
[21] A. Erener, H.S.B.D. Uzgeun, Analysis on landslide hazardmapping methods: regression models versus weight rating. The international archives of the photogrammetry, remote sensing and spatial sciences, 37 (Part B8) (2008), Beijing.
[22] M.F. Ahmed, J.D. Rogers, E.H. Ismail, A regional levelpreliminary landslide susceptibility study of the upper Indusriver basin. Euro. J. Remote Sensing 47 (2014) 343–373.
[23] Y. Wang, X. Wu, Z. Chen, F. Ren, L. Feng, Q. Du, Optimizing the predictive ability of machine learning methods for landslide susceptibility mapping using SMOTE for Lishui City in Zhejiang Province, China. Int. J. Environ. Res. Public Health 16 (3) (2019), 368.
[24] B.R. Singh, S. Singh, P. Singh, D.K. Mishra, A study on geomorphological response for runoff prediction in small watershed. IOSR J. Comput. Eng. (IOSR-JCE), 16 (4) (2014), 86–94.
[25] A.K. Pachauri, M. Pant, Landslide Hazard Mapping Based on Geological Attributes. Engineering Geology, 32 (1992), 81–100.
[26] J. Mathew, V.K. Jha, G.S. Rawat. Weights of evidence modeling for landslide hazard zonation mapping of Bhagirathi Valley, Uttarakhand. Current Science, 92 (5) (2007), 628-638.

Index

Page numbers followed by "*f*" and "*t*" indicate, figures and tables respectively.

FSC
www.fsc.org
MIX
Papier | Fördert gute Waldnutzung
FSC® C083411

Kolibri 360 GmbH
Ferdinand-Jühlke-Straße 7
99095 Erfurt, Deutschland
produktsicherheit@kolibri360.de